Macrocyclic Receptors for Environmental and Biosensing Applications

Satish Kumar
Department of Chemistry
St. Stephen's College, University Enclave, Delhi, India

Priya Ranjan Sahoo
Institute of Multidisciplinary Research for Advanced Materials
Tohoku University, Sendai, Miyagi, Japan

Violet Rajeshwari Macwan
Department of Chemistry
St. Stephen's College, University Enclave, Delhi, India

Jaspreet Kaur
Department of Chemistry
St. Stephen's College, University Enclave, North Campus
Delhi, India

Mukesh
Department of Chemistry, Keshav Mahavidyalaya
University of Delhi, Delhi, India

Rachana Sahney
Amity Institute of Biotechnology
Amity University, AUUP, Noida, Uttar Pradesh, India

T0321380

CRC Press
Taylor & Francis Group
Boca Raton London New York

CRC Press is an imprint of the
Taylor & Francis Group, an **informa** business

A SCIENCE PUBLISHERS BOOK

Cover credit: Priya Ranjan Sahoo

First edition published 2022
by CRC Press
6000 Broken Sound Parkway NW, Suite 300, Boca Raton, FL 33487-2742

and by CRC Press
4 Park Square, Milton Park, Abingdon, Oxon, OX14 4RN

© 2022 Taylor & Francis Group, LLC

CRC Press is an imprint of Taylor & Francis Group, LLC

Library of Congress Cataloging-in-Publication Data (applied for)

ISBN: 978-0-367-85632-8 (hbk)
ISBN: 978-1-032-35860-4 (pbk)
ISBN: 978-1-003-01403-4 (ebk)

DOI: 10.1201/9781003014034

Typeset in Times New Roman
by Shubham Creation

Preface

Metals, anions and gases are ubiquitous and are involved in almost every biological and environmental process. However, any increase in their concentration beyond the permissible limits leads to detrimental effects on both human health and its environment. Human advancements and indiscriminate industrial and farming practices during the last century have largely contributed to environmental pollution and also have increased the risk of diseases in humans, plants and animals. Consequently, the design and synthesis of highly sensitive and selective receptors for the detection and estimation of different metal ions, anions and gases owing to their biological, analytical and environmental applications has become of utmost importance. Based on the inspiration and learning from the biological world, scientists and engineers have actively utilized the principles of molecular recognition from the past few decades to augment the sensitivity as well as selectivity of various available sensing techniques through the usage of macrocyclic receptors. These have vast potential to differentiate between various analytes at very low concentration due to their variable cavity dimensions. The recent developments in supramolecular chemistry, a plethora of macrocyclic receptors have been reported in the literature and the field has received considerable attention in the past few decades. The development of newer macrocyclic receptors has strengthened host-guest bonhomie in the last decade. Apart from conventional macrocycles, tailoring of new hybridized receptors such as "multifarenes" from multiple building blocks has led to its usage in nitroaromatics detection, metal ion sensing. Very recently, prismarenes were added into the macrocyclic field in the year 2020. Fortunately, new kinds of arene receptors such as saucer[n]arenes, pagoda[n]arenes, helicarenes, etc. are being discovered constantly. The discovery of new receptors is bringing new avenues to fine-tune their chemistry as well as biosensing techniques.

The book also outlines recent development in macrocyclic chemistry covering areas like self-assembly, stimuli-responsive systems, molecular motors, switches, biosensing, and medicine. Molecular switches and motors based on macrocyclic receptors are the focus of current scientific efforts to integrate them into larger assemblies of molecular machines to generate macroscopic motions. The stimuli-responsive systems, which drive the amazing concepts like molecular switches and motors based on well-known supramolecules such as calixarenes, crown ethers, cyclodextrins, pillar[n]arenes have been discussed in detail in the book. Details of the design and basic mechanism behind the development of responsive molecular switches and motors have also been discussed. Moreover, the rapidly developing field of nanomaterials further provides strong motivation to the development of stimuli-responsive materials based on macrocyclic chemistry due to their directional and preorganized nature. Stimuli-responsive systems with the macrocyclic unit at the core have also been discussed as a function of stimuli like light, heat, pH, metal ion, solvent polarity, redox, and chemical reaction. The book attempts to provide the theory behind the relevance of supramolecular chemistry in biology that provides a basis for the development of macrocyclic responsive systems with utility in areas like enzyme mimic, drug delivery, and environment.

The main purpose of this book is to give an idea about the discovery as well as biosensing perspectives of both recent and conventional macrocyclic receptors. We have introduced various kinds of macrocyclic receptors and their historical development.

Authors hope that the reader will find the book valuable and stimulating in the emerging area of macrocyclic chemistry, stimuli-responsive materials, switches molecular motors, chemical biology, and organic chemistry. We have also discussed the applications of different macrocycles in the recognition of various analytes.

Satish Kumar
Priya Ranjan Sahoo
Violet Rajeshwari Macwan
Jaspreet Kaur
Mukesh
Rachana Sahney

Contents

Chapter **1**

Macrocyclic Receptors Synthesis, History, Binding Mechanism: An Update on Current Status

INTRODUCTION

Macrocyclic receptors possess large applications in the field of biological anion transport, sensor, molecular machine, permanently porous liquids, etc. They are the primary pillars of supramolecular chemistry. Receptors like crown ethers, which are first-generation macrocycles have shown remarkable binding responses since their discovery (Liu et al. 2017). For example, s-block metal ions, doubly charged metal ions like Pb^{2+}, Sn^{2+} and organic amines. Among available macrocyclic receptors, calixarenes are the first synthetically prepared (man-made) macrocycles, which contributed largely to the development of supramolecular chemistry and are considered as the third generation of macrocycles. Macrocycles are often prepared using phenols or their derivatives as building blocks along with either formaldehyde or sulfur as another condensing unit. Extension of phenolic units to naphthols leads to the incorporation of large aromatic pi systems within the cavity of the macrocycle, which accompanies an optical property and are quite useful for a sensing purpose but lead to complexity in the system. Other supramolcules such as, Bambusurils calix[4]pyrroles, are anion receptors, owing to the presence of the functional group, which can form a H-bond easily with anions. High solvation of anions and the existence of positive charge during binding in water are some important reasons for lesser exploration of anion binding chemistry than cation binding (Lizal and Sindelar 2018). The positive charge also competes with respective anions. The fields of macrocyclic receptors are expanding every day. Recently in year 2019, the Day group at the University of New South Wales developed a dicationic tiara-shaped macrocyclic receptor (Chandrakumar et al. 2019). The tiara-shaped receptor was synthesized after a condensation reaction of two different units such as glycoluril diether and dipyrazole.

The design of the macrocyclic receptor is important for guest recognition, where binding sites can be preorganized judiciously to hold guest molecules through a variety of interactions. Receptors like crown ethers hold metal cations through ion-dipole interactions (Steed 2001). In another case, [18]crown-6 forms a stronger host-guest complex with alkyl ammonium cations with the help of hydrogen bonding. In comparison to crown ethers, cryptands are more rigid macrocyclic receptors and hence better preorganized, which exhibits greater association constant (Nielsen 2013). Similarly, calixarenes form supramolecular complexes through their oxygen atoms while binding with metal cations by electrostatic interactions or through cation–π interactions while holding to aromatic units. Unlike calixarenes, pillararenes bind with organic guest species with greater selectivity due to a more rigid core structure. Pillararenes also hold on paraquat or viologen in an organic medium. Intramolecular H-bonding, guest molecules, solvents, temperature, etc., are associated with rotational properties present in the pillar[5]arene receptors. In the case of pillar[5]arenes, inclusion complexes are mainly associated with C–H$\cdots\pi$ interactions (Cragg and Sharma 2012). Receptors such as cyclodextrins and curcurburils form a stronger complex with non-polar organic guests in an aqueous medium owing to their inherent hydrophobicity. Additionally, curcurburils form a host-guest complex with ions and molecules by interactions like charge-dipole and hydrogen bonding (Kim 2002; Lee et al. 2003). The interior of cryptophanes can hold analytes and small molecules through cation–π or van der Waals interactions. Association of quaternary ammonium ion with oxatub[4]arene are dominated by C–H$\cdots\pi$ and cation–π interactions. Receptors like tetrathiafulvalene make them chargeable by donating or loosing electron owing to their amphoteric nature (Nielsen 2013) and found them to be promising in supramolecular chemistry.

MACROCYCLIC RECEPTORS AND TYPES

Macrocyclic receptors are the most widely applicable supramolecular systems that can hold guest species in their cavity. Incorporation of heteroatoms in the receptor widens structural conformations, self-organization tendency, chemistry and binding properties. Depending on conformations and cavity sizes, the different types of macrocyclic receptors listed are listed below.

Crown Ethers

Crown-shaped macrocyclic polyethers were first accidentally discovered in 1967 by the American chemist Charles J. Pederson after long years of continuing research at DuPont chemical industry in Delaware (Pedersen 1967). Initially, Dibenzo[18]crown-6 was isolated along with other derivatives. Crown ethers are the most common form of macrocyclic receptors. These doughnut-shaped molecules are cyclic oligomers consisting of dioxane units (Pedersen 1988). In particular, crown ethers contain ethylethoxy ($-CH_2CH_2O-$) units as repeating units (Fig. 1.1) in their core structures. These are electron-rich circular rings with a polar core and hydrophobic exterior parts, make them suitable to solubilize in organic solvents. Due to the presence of electron-rich heteroatoms, these receptors can accommodate cationic guests within their core (Chen and Liu 2020).

Figure 1.1 Examples of cyclic oligomers of ethylene oxide are shown with different oxygen units (Gokel et al. 2004).

Crown ethers are used extensively for separation, binding (Li et al. 2017), as liquid crystals (Kaller et al. 2011), etc. Similarly, aza-crown ethers are used as intermediates during the preparation of cryptand receptors. Further, a spurge in applications has made it possible for aza-crown ethers towards binding to ammonium salts and transition metal ions (Krakowiak et al. 1989).

Synthesis of crown ether

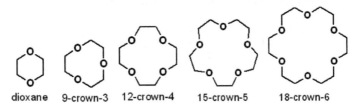

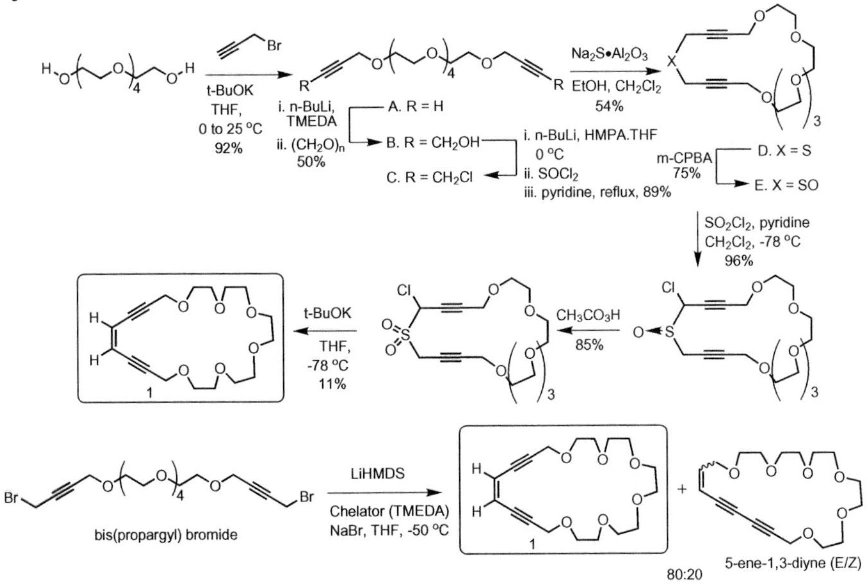

Scheme 1. Synthesis of 24-membered crown ether receptor **1** by two different routes.

McPhee et al. synthesized enediyne-based 24-membered crown ether receptor **1** in a multistep synthetic step in 11% yield (Scheme 1) (McPhee and Kerwin 1996). The same receptor **1** was also synthesized in another shorter and convenient route through macrocyclization of bis(propargyl) bromide. The successful one-step macrocyclization reaction was possible due to the existing polyether nature of starting dibromide precursor. The 3-ene-1,5-diyne crown ether receptor **1** was able to bind potassium ions more effectively than its counterpart metal ions such as sodium or lithium ions. Innate existence of enediyne moiety in the core structure of receptor **1** might be able to activate a more prominent rearrangement reaction such as Bergmann cyclization following alkali metal ion binding.

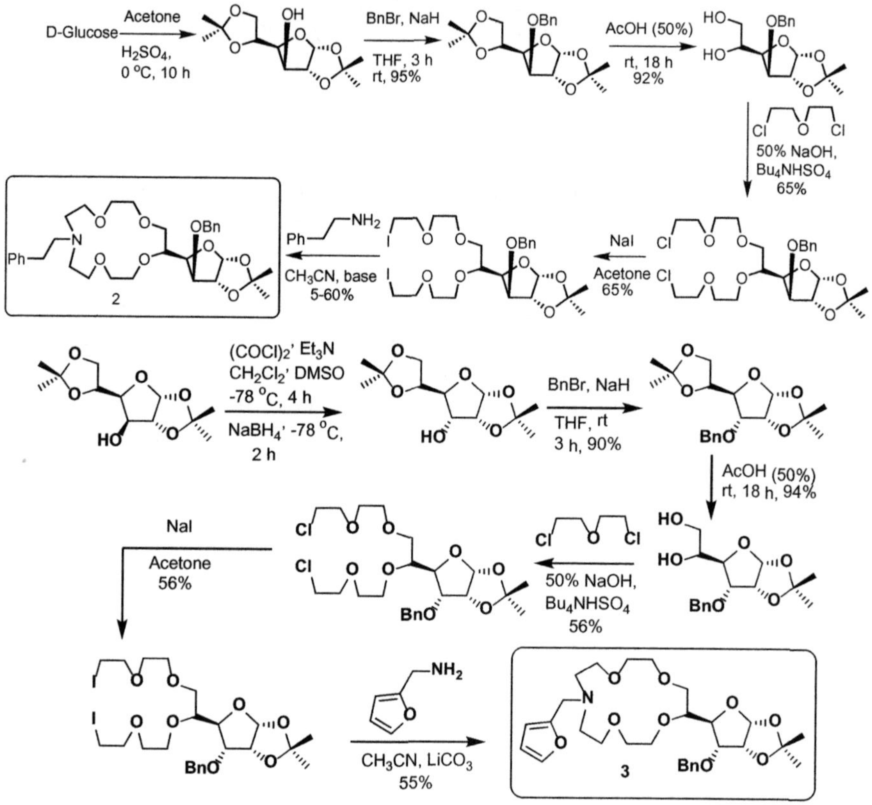

Scheme 2   Synthesis of D-glucose-based crown ether receptor **2** and D-allose-based crown ether receptor **3**.

Mishra et al. have reported chiral crown ether receptors **2** and **3** (Scheme 2) in a convenient synthetic process without byproduct formation (Mishra et al. 2016). Receptor **2** was synthesized in 60% yield from easily available D-glucose after six synthetic steps. Similarly, receptor **2** was isolated in 55% yield. The synthesized chiral receptors might be used for the introduction of stereo-selectivity during the preparation of stereoisomers. Further, generations of chiral cavities are advantageous for the chiral recognition of guest species.

Boubekeur-Lecaque et al. synthesized carboxylic acid-based macrocyclic crown ether receptor **4** in four synthetic steps (Scheme 3) from commercially available precursors such as diaza-18-crown-6-ether and 1-fluoro-2-nitrobenzene

(Boubekeur-Lecaque et al. 2014). The receptor was able to bind with barium metal ion in 1:1 stoichiometry using UV and NMR spectroscopic techniques.

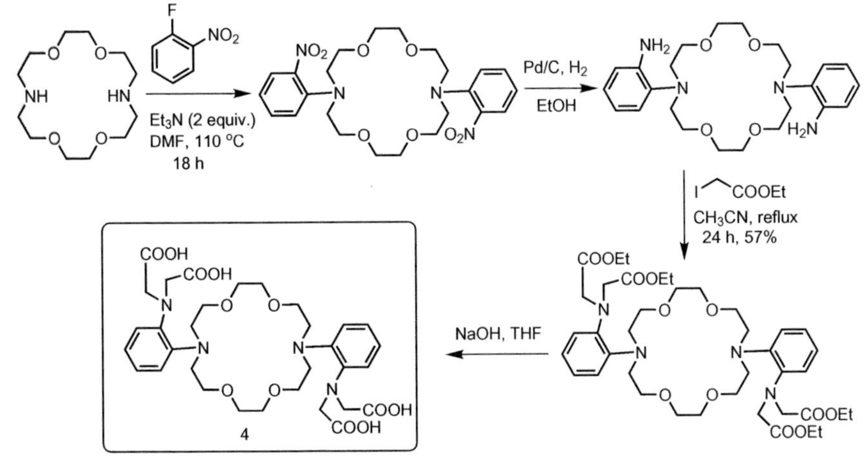

Scheme 3 Synthesis of water-soluble diaza crown ether receptor **4**.

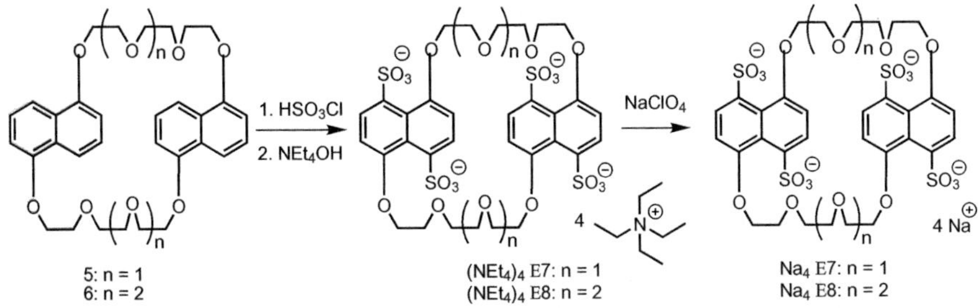

Scheme 4 Synthesis of macrocyclic receptors $Na_4 7$ and Na_4 **8**.

Chen et al. modified conventional 1,5-dinaphthocrown ethers (**5-6**) to sulfonated crown ether receptors in order to make them water-soluble and to exhibit a greater binding response (Chen et al. 2012). The authors synthesized receptors $Na_4 7$ and Na_4 **8** after a two-step synthetic process (Scheme 4), where chlorosulfonation occurred in the presence of chlorosulfonic acid followed by neutralization using tetraethylammonium hydroxide and finally counterion exchange took place in the presence of sodium perchlorate to give the target receptors in 97.1 and 98% respectively. Both the receptors exhibited excellent guest affinity in water, where bipyridinium guests were diffused into the receptor cavities through electrostatic and π-stacking interaction.

Sung et al. developed a method for the synthesis of macrocyclic bis-crown ether receptor **9** in 63% synthetic yield (Scheme 5) (Sung et al. 2007). The method involved light-induced macro-cyclization of polyether chains attached to the bis-phthalimide skeleton, where the ring formation was promoted by Single Electron Transfer (SET). Both symmetrical and unsymmetrical bis-crown ethers with different ring sizes were synthesized by following the developed method.

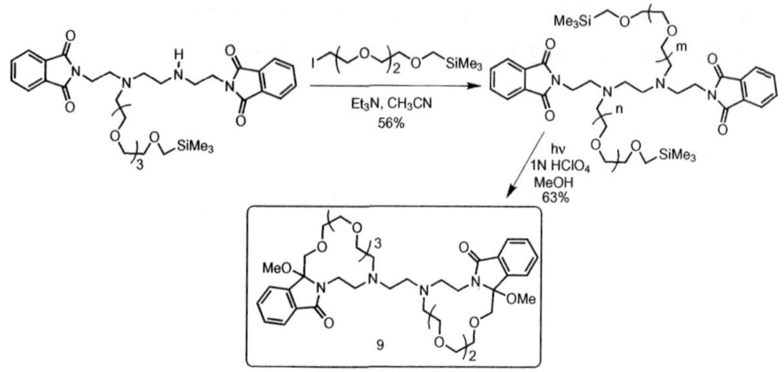

Scheme 5 Synthesis of macrocyclic bis-crown ether receptor **9**.

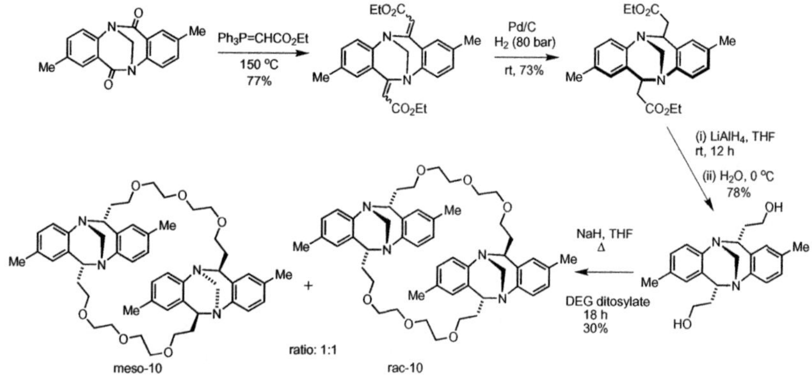

Scheme 6 Synthesis of macrocyclic receptor **10**.

Artacho et al. synthesized bis-Troger's base crown ether (*meso*-**10**) in 15% yield along with its *rac*-isomer in 15% yield (Scheme 6) (Artacho et al. 2012). The structure of *meso*-**10** was confirmed by single-crystal X-ray crystallography (Fig. 1.2). The crown ether possesses an unusual and inverted methylene bridge-to-bridge structure with bis-Troger's base skeleton. Owing to its inverted methylene bridge-to-bridge existence, it was unsuitable for binding metal ions in a solution state.

Figure 1.2 Single crystal X-ray structure of inverted crown ether *meso*-**10**. Reproduced here with the permission of the American Chemical Society (Artacho et al. 2012).

Binding Mode of Crown Ether

In terms of host-guest recognition and binding, the solvent environment plays an important role in macrocyclic chemistry. The binding strength of crown ethers towards cationic guests varies by differently associated solvent molecules. For example niple, 18-crown-6 exhibits a low binding response towards K^+ in water (Takeda and Arima 1985) (binding constant; 1.1×10^2 M^{-1}), whereas the binding affinity increases in methanol (binding constant; 1.1×10^6 M^{-1}) (Lamb et al. 1980). The crystal structure of the receptor **4-Ba** complex reminds one that macrocyclic host systems can hold variable guest species in order to attain the best-fit arrangement. Additionally, NMR titration of receptor **4** and barium (Ba^{2+}) metal ion revealed 1:1 binding stoichiometry (Fig. 1.3).

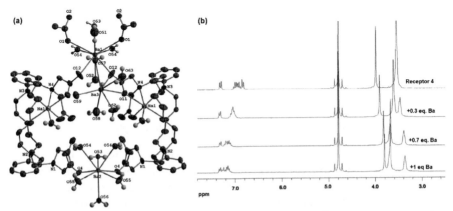

Figure 1.3 **(a)** Crystal structure of receptor **4-Ba** complex; **(b)** ^{1}H-NMR titration studies of receptor **4** in the presence of different equivalents of Ba^{2+} in D_2O. Reprinted with permission from Elsevier (Boubekeur-Lecaque et al. 2014)

A paper was introduced by Ravi et al. based on the cation exchanging ability of crown ether receptor (Ravi et al. 2018). The receptor contains a crown ether receptor and two bipyridine units. The crown ether moiety binds to the potassium ion. Interestingly, potassium ion is removed from the crown cavity during the interaction of zinc ions with the bipyridine units owing to conformational switching.

Cryptands

Cryptands were first introduced to the macrocyclic scientific community in the year 1969 by (Blanco-Gómez et al. 2020). These are cyclic (bicyclic and polycyclic) multidentate ligands used extensively for trapping metal ions (Bharadwaj 2017). Alternatively, cryptands are three-dimensional cyclic structures of crown ether derivatives. These three-dimensional structures bind guest species in a crypt (a stone chamber that holds a coffin) and hence the name is known as 'cryptand'.

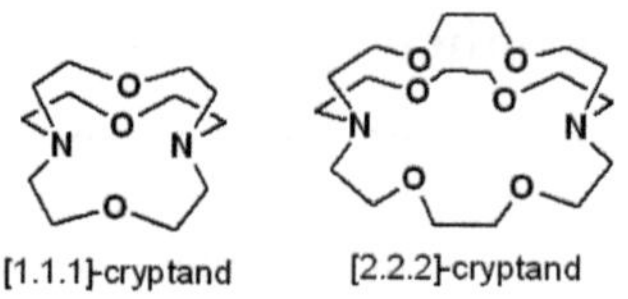

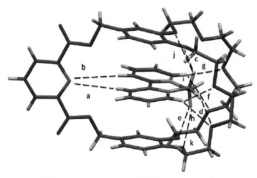

[1.1.1]-cryptand   [2.2.2]-cryptand

Figure 1.4   Structures of most common [1.1.1]-cryptand and [2.2.2]-cryptand.

Synthesis of Cryptands

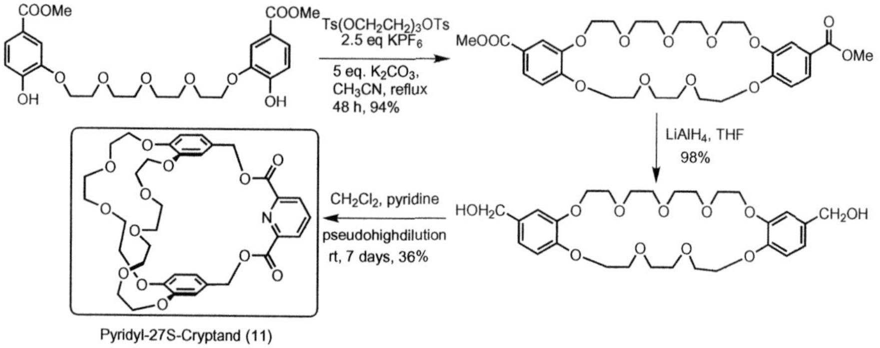

Scheme 7   Synthesis of macrocyclic cryptand receptor **11**.

Pederson et al. synthesized crown ether diester cyclic intermediate from diphenol derivative and tri(ethylene glycol) ditosylate and subsequent reduction of ester groups (Scheme 7) with LiAlH$_4$ yielded diol intermediate (Pederson et al. 2017). The diol intermediate on reaction with pyridine-2,6-dicarboxylic acid dichloride produced pyridyl-27S Cryptand receptor **11** in 36% yield. Receptor 11 formed complex with diquat (bipyridinium), paraquat and paraquat diol.

Figure 1.5   Single crystal X-ray structure of **11**·DQ(PF$_6$)$_2$ without PF$_6^-$ counter anions. DQ (diquat) is 1,1′-ethylene-2,2′-bipyridinium di-bromide. Reprinted with permission from the American Chemical Society (Pederson et al. 2017).

The authors obtained the crystals of **11**·DQ(PF$_6$)$_2$ using the vapor diffusion method, where they used pentane for dissolving the sample and finally kept it in acetone solution. The diquat guest species resides on the arms of crown ether (Fig. 1.5).

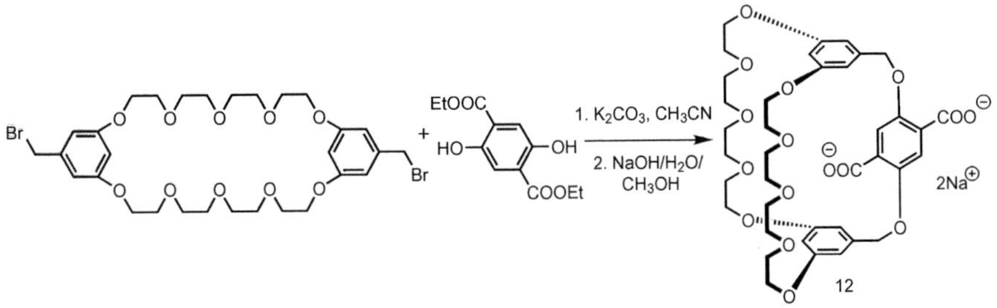

Scheme 8 Synthesis of water-soluble macrocyclic cryptand receptor **12**.

Ji et al. synthesized a water-soluble dicarboxylate-based cryptand **12** (Scheme 8) in 60% yield utilizing bis(5-bromomethyl-m-phenylene)-32-crown-10 and diethyl 2,5-dihydroxyterephthalate as synthetic precursors (Ji et al. 2013). Tethering another polyether binding site to the macrocyclic receptor resulted in better binding of paraquat (bipyridinium species) guest molecule. The self-assembly system (both assembly and disassembly) involving both receptor and guest species can be controlled reversibly by pH variation.

Cyclodextrins

The 'bucket' shaped macrocyclic receptor cyclodextrins were first discovered by the French chemist Antoine Villiers in 1891, while isolating crystalline material from starch after bacterial digestion. The only macrocyclic receptor, which contains multiple D-glucopyranose units (six or more) at the core structure, is commonly known as cyclodextrin (Harada et al. 2009). The D-glucopyranose units are connected by α-1, 4-linkages in a cyclic framework (Hapiot et al. 2006). Very often common types of cyclodextrins are available as α, β, γ cyclodextrins with 6, 7 and 8 glucose units positioned in the cycle respectively (Figs. 1.6 and 1.7).

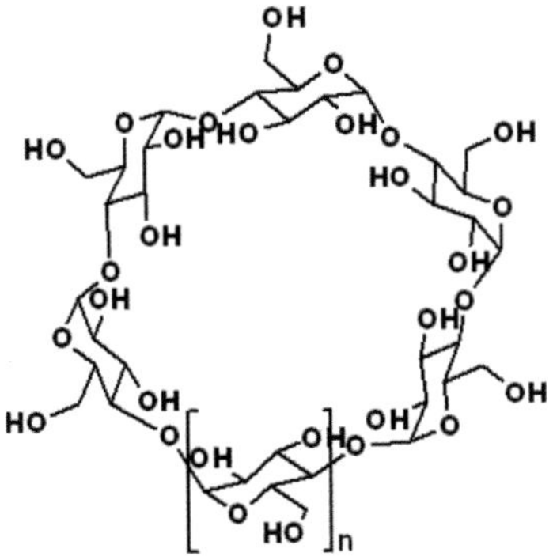

n = 1: α-cyclodextrin
n = 2: β-cyclodextrin
n = 3, γ-cyclodextrin

Figure 1.6 Representative structures of α, β, γ cyclodextrins with exceeding D-glucopyranose units.

Cyclodextrin possesses a hydrophobic core cavity for holding guest species, leading to a myriad of applications. The interesting host-guest phenomenon has led to a booming market growth in pharmaceutics, foods, cosmetics, textile industries (Cravotto et al. 2006; Bilensoy 2011; Zhang and Ma 2013; Fenyvesi et al. 2016). Owing to its ability to form inclusion complexes, cyclodextrins are extensively used for molecular recognition (Liu and Chen 2006). The condensation of glucose monomers produces a rigid cyclic structure with specific cavity diameter and volume. Interestingly, the cavity height of α, β, γ cyclodextrins are fixed at 0.78 nm.

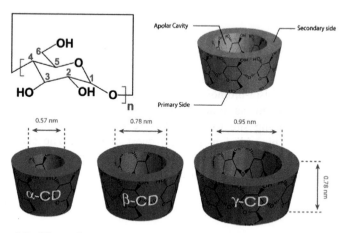

Figure 1.7 Three-dimensional representation of α, β, γ cyclodextrin receptors. Reprinted with permission from the American Chemical Society (Crini 2014).

Table 1.1 Differences among α, β, γ cyclodextrins.

	α-CD	β-CD	γ-CD
Molecular mass	973	1135	1297
No. of glucose units	6	7	8
Cavity diameter (nm)	0.57	0.78	0.95
Volume (mL/mol)	174	262	472

Synthesis of Cyclodextrins

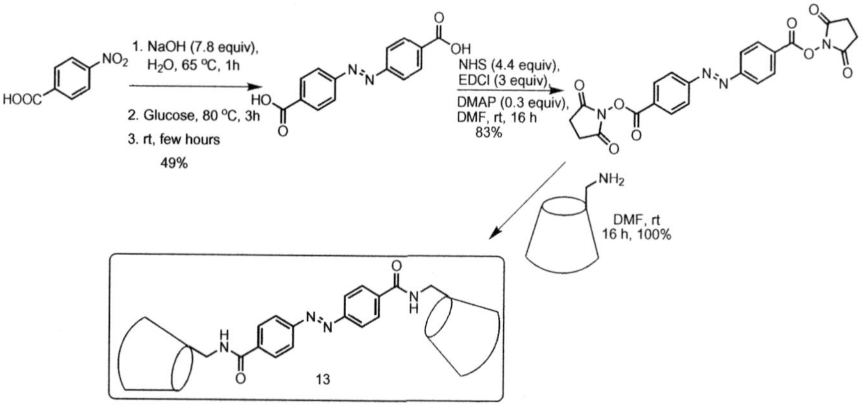

Scheme 9 Synthesis of β-cyclodextrin based dimeric receptor **13**.

Hamon et al. synthesized azo benzene-based β-cyclodextrin dimer (receptor **13**) from easily available 4-nitrobenzoic acid precursor after three steps in 62% yield (Scheme 9) (Hamon et al. 2014). The receptor exhibited a different inclusion complex in the presence of adamantly dimer depending on the existence of different isomers induced by light. The *cis-trans* isomerization process can be controlled through light irradiation (365 and 254 nm) and is reversible in nature. *Cis*-isomer of the receptor exhibited 1:1 complex formation whereas *trans*-isomer of the receptor produced supramolecular polymers with adamantyl guests.

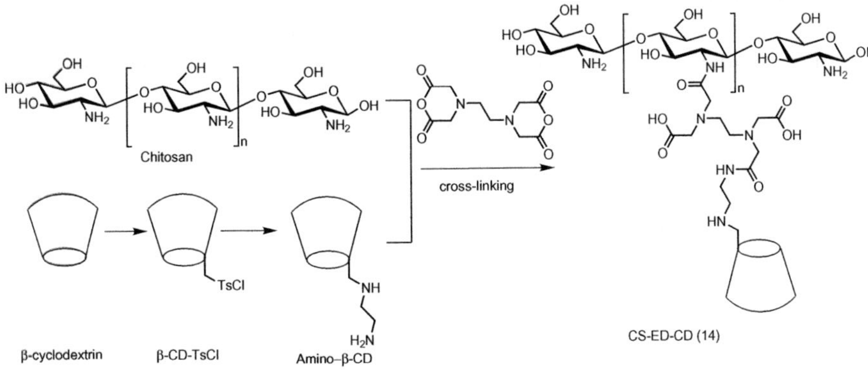

Scheme 10 Synthesis of trifunctional macrocyclic receptor **14**.

Zhao et al. developed chitosan-EDTA-β-cyclodextrin receptor **14** in a one-pot synthetic process mediated by a cross-linking agent such as EDTA (Scheme 10) (Zhao et al. 2017a). This bio-based receptor **14** acted as an adsorbent for the detection and removal of toxic pollutants and metal ions from waste water. The cyclodextrin receptor holds organic pollutants in its cavity through the inclusion complex formation, whereas EDTA acts as a chelator for the metal ion.

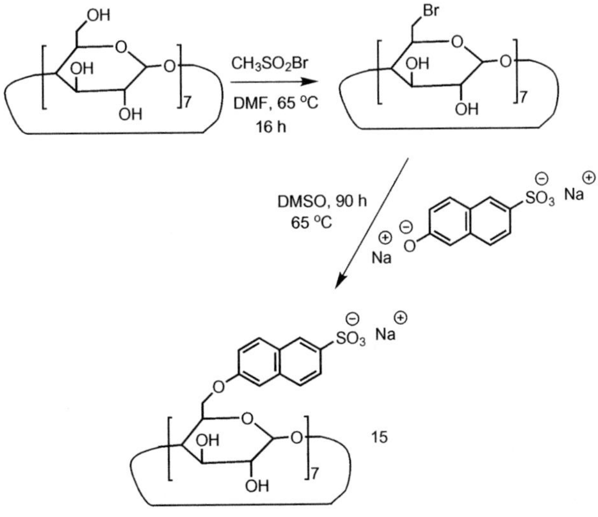

Scheme 11 Synthesis of water-soluble β-cyclodextrin receptor **15**.

Gravett and Guillet 1993 prepared naphthalene-based water-soluble β-cyclodextrin receptor **15** utilizing heptakis(6-bromo-6-deoxy)-β-cyclodextrin and disodium salt of 6-hydroxy-2-naphthalene sulfonic acid by nucleophilic substitution reaction (Scheme 11). The introduction of sulfonate (anionic) moieties facilitated monomer emission by keeping naphthalene groups far from each other.

Calixarenes

The condensation reaction of *p-tert*-butylphenol and formaldehyde in the presence of a base yields *p-tert*-butylcalix[4]arene (Fig. 1.8). Both lower rims and upper rims of calix[4]arene can be functionalized with most of the organic groups (Gutsche 2008). Calix[4]arenes possess excellent binding pockets to trap guest molecules due to their inherent hydrophobic and hydrophilic cavities. The calix[4] arene skeleton can take a different conformation depending on the orientation of the calix[4]arene ring (Fig. 1.9). These macrocyclic receptors can also trap metal ions through cation–π interactions (Zadmard et al. 2020).

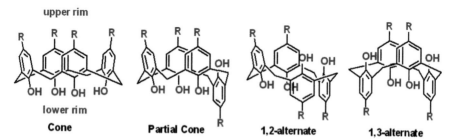

Figure 1.8   Structures of calixarene receptors with different numbers of aromatic units.

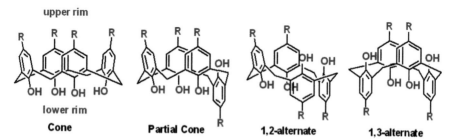

Figure 1.9   Various conformations of calix[4]arenes (Español and Villamil 2019).

Synthesis of calixarenes

Chetcuti et al. reported the synthesis of alkyne substituted calixarene receptors **16-19** (Chetcuti et al. 2009). Refluxing *p-tert*-butylcalix[4]arene with propargyl bromide for 15 hours in the presence of a base such as K_2CO_3 in acetone (Scheme 12) yielded receptor **16** in 22% yield. Refluxing the same synthetic precursors but with three equivalents of propargyl bromide for an extended period resulted in receptor **17** (1,3-*cone* conformer) in 80% yield. The reaction with six equivalents of propargyl bromide in a reflux condition even for more time resulted in *cone* conformer of receptor **18** in 45% yield. Interestingly, repeated synthesis of *cone* conformers altered its conformational design when K_2CO_3 was replaced with Cs_2CO_3. Thus, reaction of **17** with propargyl bromide in Cs_2CO_3 produced *1,3-alternate* conformer **19** in 60% yield. The alkyne derivatives might be suitable for click reaction and developing new calixarene-based materials.

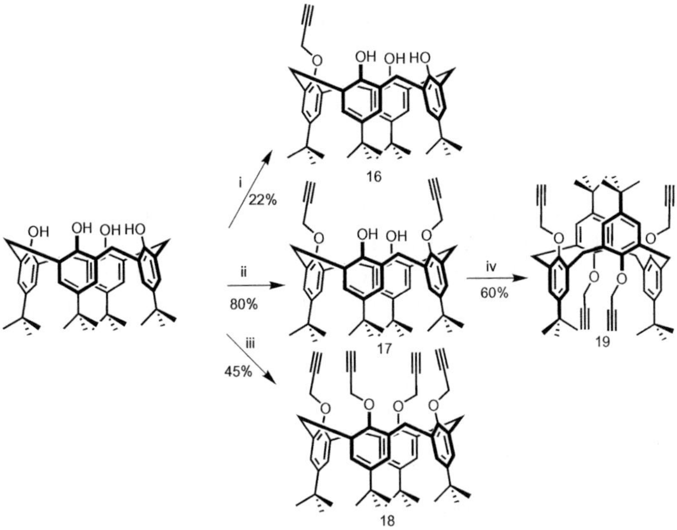

Scheme 12 Synthesis of macrocyclic calixarene receptors **16–19**. (i) $BrCH_2C\equiv CH$, K_2CO_3, acetone, reflux (ii) $BrCH_2C\equiv CH$, K_2CO_3, acetone, reflux (iii) $BrCH_2C\equiv CH$, K_2CO_3, acetone, reflux (iv) $BrCH_2C\equiv CH$, Cs_2CO_3 acetone, reflux.

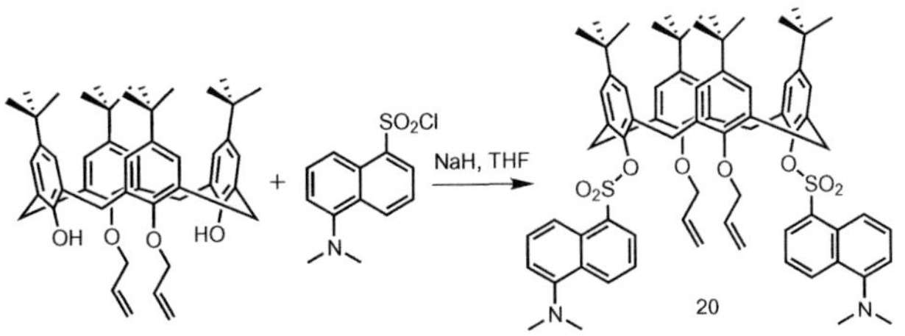

Scheme 13 Synthesis of calix arene receptor **20**, where the lower rim is functionalized with dansyl moieties.

The introduction of dansyl moieties in the parent molecular structure enhances fluorescence behavior in the visible region of the electromagnetic spectrum. Therefore, Gruber et al. synthesized receptor **20** in 74% yield, where calix[4]arene was used for direct functionalization of the lower rim with dansyl groups with the help of sodium hydride in THF (Scheme 13) (Gruber et al. 2009).

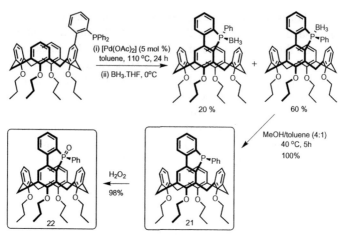

Scheme 14   Synthesis of macrocyclic receptor **21** and its derivative phosphole oxide receptor **22**.

Elaieb et al. synthesized phosphole attached calixarene receptor **21** from phosphole borane (Elaieb et al. 2017). The phosphole receptor **21** was synthesized in 100% synthetic yield. Receptor **21** formed a white solid known as phosphole oxide **22** after reaction with hydrogen peroxide (H$_2$O$_2$) in 98% yield (Scheme 14).

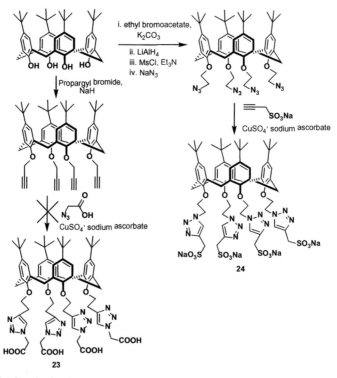

Scheme 15   Synthesis of water-soluble calixarenes receptors **23–24** based on click reaction.

To introduce polar functional groups, Ryu and Zhao 2005 utilized the click chemistry approach and synthesized water-soluble receptor **24**. Isolation of receptor **23** was not possible at room temperature and further complications occurred at higher temperature (60°C) due to the formation of a number of impurities. However, by alternating the precursors, where the reaction of azido calixarene occurred smoothly with the alkyne derivative giving rise to receptor **24** in 79% yield (Scheme 15). The method introducing water-soluble alkynes can generate a wide range of water-soluble macrocyclic systems.

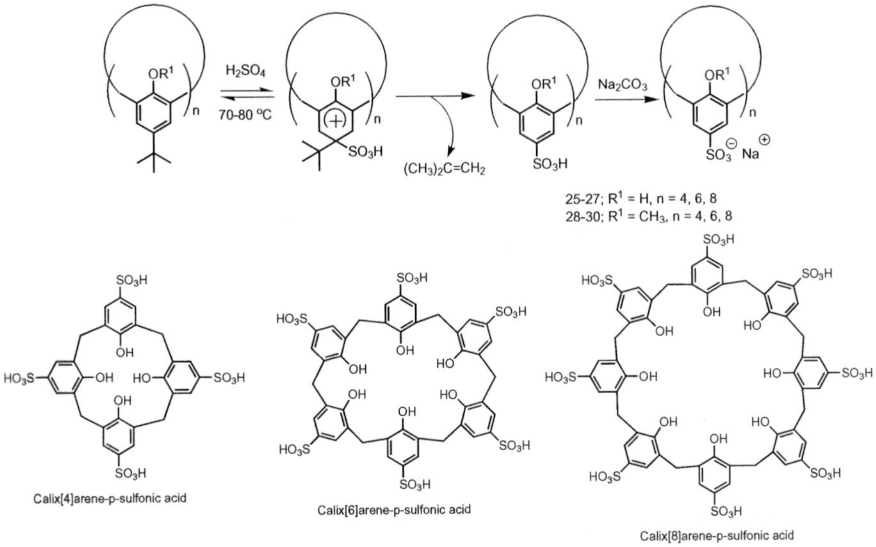

25-27; R^1 = H, n = 4, 6, 8
28-30; R^1 = CH_3, n = 4, 6, 8

Calix[4]arene-p-sulfonic acid

Calix[6]arene-p-sulfonic acid

Calix[8]arene-p-sulfonic acid

Scheme 16 Synthesis of sulfonic acid-based water-soluble calixarenes receptors **25-30**.

Cheaper reagents often enhance value addition starting from bench-top synthesis to large-scale commercialization. Kumar et al. developed *p*-sulfonated calixarene **25-30** in one synthetic step (Kumar et al. 2003). The authors introduced the *ipso*-substitution technique for direct construction (Scheme 16) of sulfonated calixarene from *p-tert*-butylcalix[n]arene precursors. During the *ipso*-substitution, the *tert*-butyl group was replaced with sulfonic acid. The receptors were isolated in 60-70% yield involving the most common and cheaper reagents such as sulfuric acid and sodium carbonate.

With the aim towards using calixarene in catalyzing Mannich reactions involving aldehydes, aromatic ketones, and amines, Akceylan and Yilmaz 2016 synthesized receptors **31-32**. Both receptors **31-32** were isolated as light brown solid in 60 and 75% yield respectively (Scheme 17).

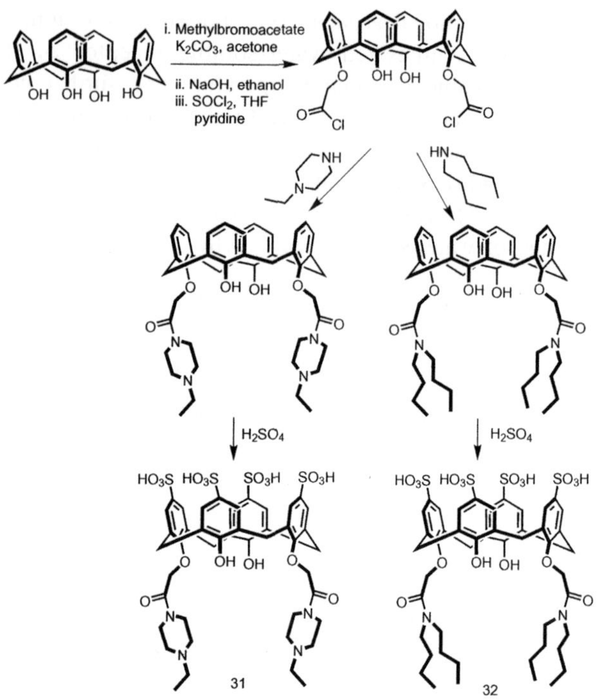

Scheme 17   Synthesis of sulfonic acid substituted calix[4]arene receptors **31-32**.

Resorcinarenes

Condensation product of resorcinol with an aldehyde in an acidic environment usually generates cyclic structures which are otherwise known as resorcinarenes. Calix[4]resorcinarenes were first developed by Professor Baeyer in the late 19th century (Wieser et al. 1997). Since then, resorcinarenes have been used in many practical fields such as chemical separation (Sliwa and Kozlowski 2009; Li et al. 2014) and optical applications (Liu et al. 2016b).

Synthesis of Resorcinarenes

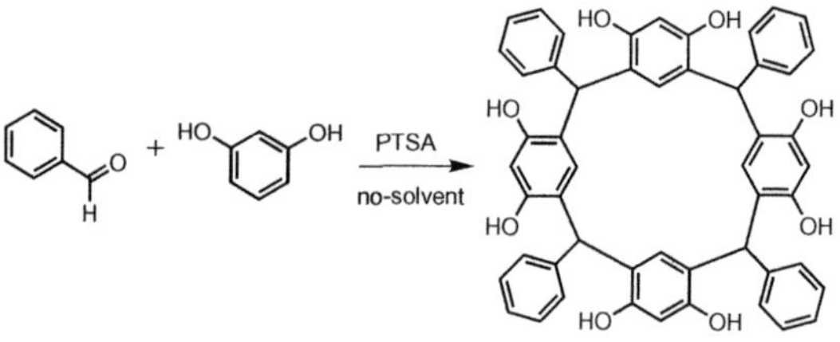

Scheme 18   Synthesis of calix[4]resorcinarene **33** receptor.

In 2001, Roberts et al. developed a solvent-free one-pot synthetic route for the preparation of calix[4]resorcinarene **33** with the mere crushing of starting materials (Roberts et al. 2001). The synthetic route involves the equimolar amount of benzaldehyde and resorcinol as starting materials and a solid catalyst such as PTSA (Scheme 18). The synthesized product was obtained in 90% yield after recrystallization from hot methanol following a wash with water. Since the synthetic route requires no solvent or heating, it is considered a viable as well as greener method for synthesis of a bulk amount of resorcinarene.

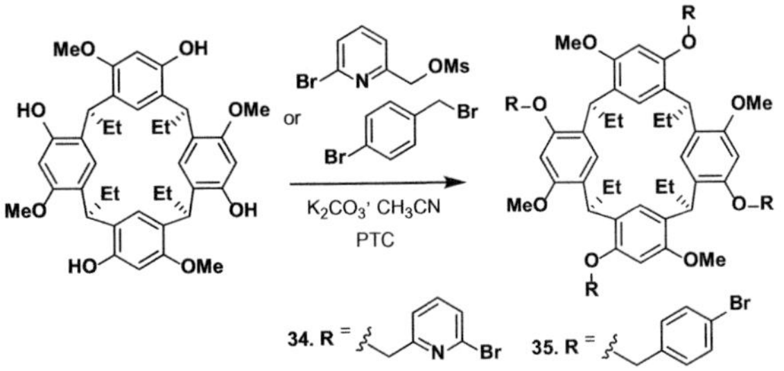

Scheme 19   Synthesis of pyridine (**34**) and benzyl (**35**) based resorcinarene receptors.

Tero et al. synthesized bromo resorcinarene receptors **34-35** from tetramethoxy resorcinarene derivative by a nucleophilic substitution reaction in the presence of K_2CO_3 and a phase transfer catalyst (Tero et al. 2012). Receptors **34** and **35** were isolated as a white solid with 43 and 86% yield respectively. Receptor **34** was crystallized in three different ways with chloroform and dichloromethane solvate in crystal structures while grown from a methanol-chloroform and methanol-dichloromethane solution. In all the three crystal structures, the core resorcinarene attained a twisted boat conformation similar to a crown arrangement. Interestingly, bromine-bromine interaction was found between adjacent resorcinarene molecules in one of the crystal structures. However, receptor **35** was crystallized in the p1 space group in acetonitrile-chloroform solution.

Figure 1.10   Structure of receptors **36-37**.

To investigate solvent-induced conformational alteration, Salorinne and Nissinen 2009 had grown single crystals of receptors **36-37** using a variety of solvents. The tetramethoxy receptor **36** was crystallized in a crown pinched , and twisted pinched crown conformation (Fig. 1.11) when a solution of receptor **36** was grown from acetone solution with the addition of lesser amount of ethanol, potassium hexafluorophosphate (KPF$_6$) and acetonitrile respectively. Due to the presence of hydroxyl groups in the receptor **36** and the existence of intermolecular hydrogen bonds facilitated twisting of resorcinarene core structure. However, bis-crown-5 tethered tetramethoxy receptor **37**, which is devoid of intramolecular hydrogen bonding was crystallized in boat conformation in ethanol-water mixed solvent and DMSO solutions.

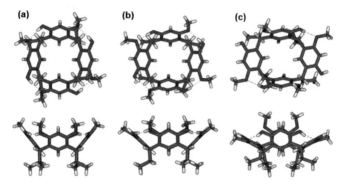

Figure 1.11   Various conformations of receptor **37** were obtained from crystal structures. (a) Crown (b) pinched crown (c) twisted pinched crown. Hydrogen bonding is represented by dotted lines. Reproduced here with the permission of the RSC (Salorinne and Nissinen 2009).

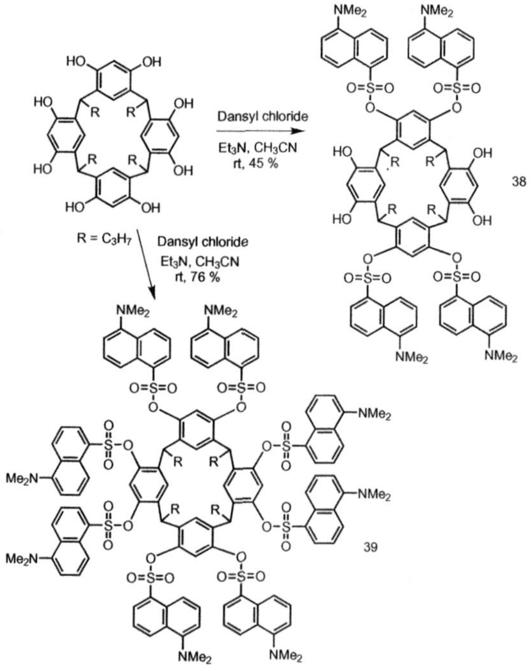

Scheme 20   Partial and complete dansylation (receptors **38-39**) using dansyl chloride.

Beyeh et al. synthesized multiple dansyl substituted resorcinarene receptors **38-39** in a convenient step using dansyl chloride and resorcinarene (Beyeh et al. 2007). The tetra substituted dansyl resorcinarene was obtained in 45%

yield, whereas octa substituted dansyl resorcinarene was synthesized in a higher yield (76%) (Scheme 20). Receptors **38** and **39** exhibited C_{2v} and C_{4v} symmetry respectively in the solution state whereas both of these receptors exhibit C_{2v} symmetry in solid-state (X-ray structure, Fig. 1.11).

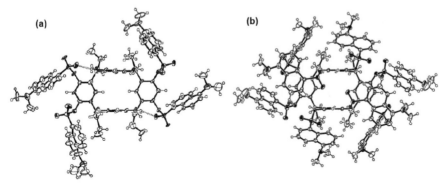

Figure 1.12   Single crystal X-ray structure of (a) tetra-substituted resorcinarene receptor **38** (b) octa-substituted resorcinarene receptor **39**. Reproduced with permission from The Royal Society of Chemistry (Beyeh et al. 2007).

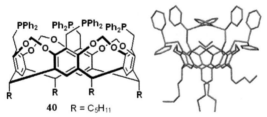

Figure 1.13   Structure of resorcinarene receptor **40**. Single crystal X-ray structure of tetra-phosphinated calix[4]resorcinarene receptor **40**. Reproduced here with the permission of RSC (El Moll et al. 2009).

El Moll et al. synthesized tetraphosphine based resorcinarene receptor **40** in a multistep synthetic process in 98% yield (El Moll et al. 2009). Receptor **40** (ligand) acted as an effective Heck catalyst in the presence of [Pd (OAc)$_2$] and Cs$_2$CO$_3$ and successfully performed the coupling reaction with 4-bromoanisole using Pd/ligand as 1:1 ratio. The crystal structure of receptor **40** attained a bowl-shaped arrangement (Fig. 1.12) similar to well-known resorcinarene receptors.

Jordan et al. synthesized water-soluble resorcinarene receptor **41** in two synthetic steps (Scheme 21) from commercially available resorcinol (Jordan et al. 2019). The chloro-substituted resorcinarene precursor was synthesized using 4-chlorobutanal dimethyl acetal and aqueous hydrochloric acid in 95% yield (Scheme 22). Later receptor **41** was synthesized by reacting with an excess amount of pyridine in 92% yield. Receptor **41** acted as a ditopic binding host with the ability to hold both guest cations and anions at the upper rim and the lower crown of pyridinium groups respectively. Similar binding preference of the host receptor towards both pyridinium cation and iodide ions resulted in self-assembly formation in water in a head-to-tail fashion

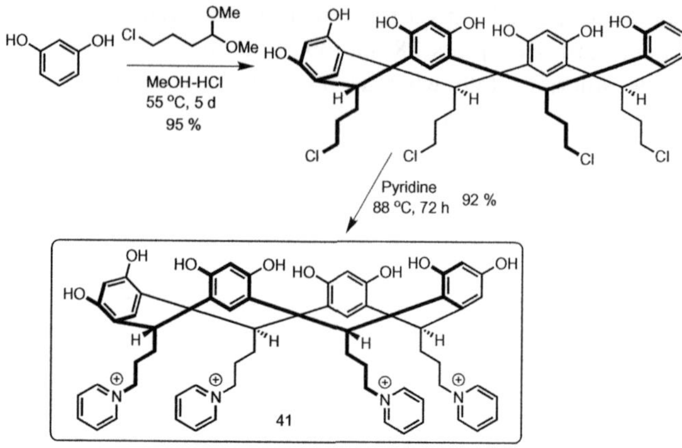

Scheme 21 Synthesis of receptor **41**.

Multifarenes

Reacting to the right pair of precursors often results in a fascinating molecules. By tailoring alternate synthons, Keinan, Poppo and their team developed a new class of macrocyclic receptors with varying cavity sizes in the year 2013 (Scheme 22) and named those molecules as 'multifarenes' (Parvari et al. 2014). Such types of receptors comprise of multiple building blocks and hence the name multifarenes. Owing to the convergence of multiple building units, the resulting properties such as cavity size, conformational shape, molecular rigidity, analyte binding response and their solubility also become diverse in nature (Swamy et al. 2018). These days multifarenes are used for the detection of metal ions, (Huang et al. 2018, Huang et al. 2019) neutral guest molecules (Luo et al. 2020), explosives (Qiu et al. 2020), and others (Zhao et al. 2020).

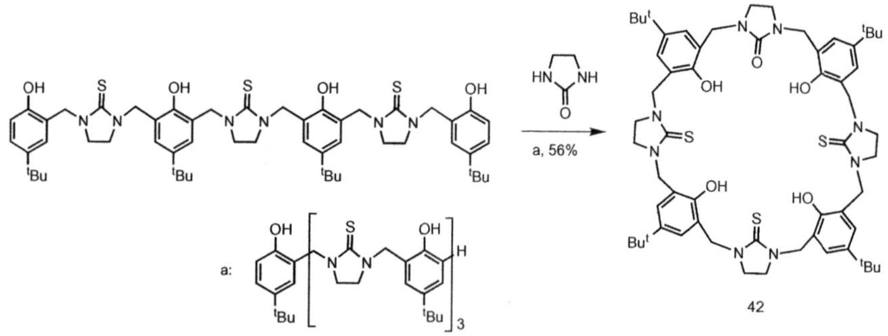

Scheme 22 Synthesis of multifarenes receptor **42**.

Calixpyrroles

Calix[4]pyrrole was first developed by Baeyer in the year 1886 (Gale et al. 1998). In the beginning, m*eso*-octamethyl calix[4]pyrrole was the initial calix[4]pyrrole obtained from the condensation reaction of pyrrole and acetone in the presence of an acid such as HCl (hydrochloric acid). Calix[4]pyrrole has a big role in molecular recognition, ion transport (Kim and Sessler 2015) and anticancer areas (Kohnke 2020).

Synthesis of Calixpyrroles

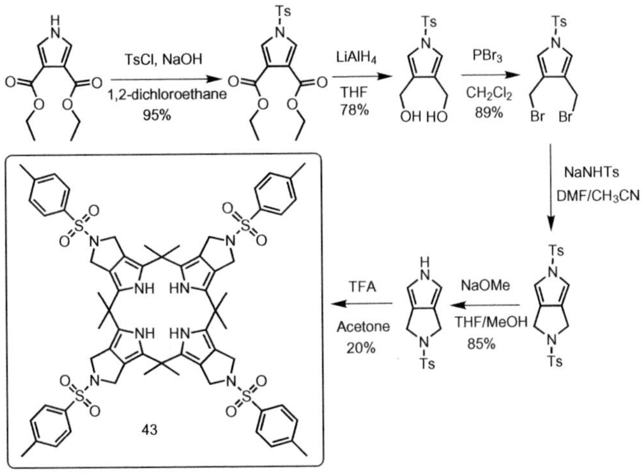

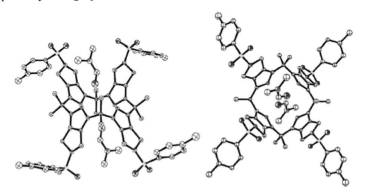

Scheme 23. Synthesis of tetra-tosyl substituted calix [4] pyrrole receptor **43**.

Kim et al. synthesized N-tosylpyrrolidine substituted calix[4]pyrrole receptor **43** through a condensation reaction of alkylated pyrrole derivative with trifluoroacetic acid (TFA) in acetone (Scheme 23) (Kim et al. 2011). Receptor **43** exhibited excellent recognition property towards halide ions (Fig. 1.14) as confirmed with the help of crystallographic studies.

Figure 1.14 (Two different views) Single crystal X-ray structure of N-tosylpyrrolidine calix[4]pyrrole **43** with two molecules of DMF. Reprinted with permission from the American Chemical Society (Kim et al. 2011).

In 2004, Nielsen et al. developed a macrocyclic system for capturing electron-deficient guest species. The macrocyclic system encompasses four tetrathiafulvalene derivatives tethered to the calix[4]pyrrole core structure (Nielsen et al. 2004). The tetra-TTF-calix[4]pyrrole receptor **44** was synthesized from pyrrolo-tetrathiafulvalene and TFA in an acetone-dichloromethane solvent system in 18% yield (Scheme 24). The receptor in its 1,3-alternate conformation forms host-guest complex through charge-transfer interactions.

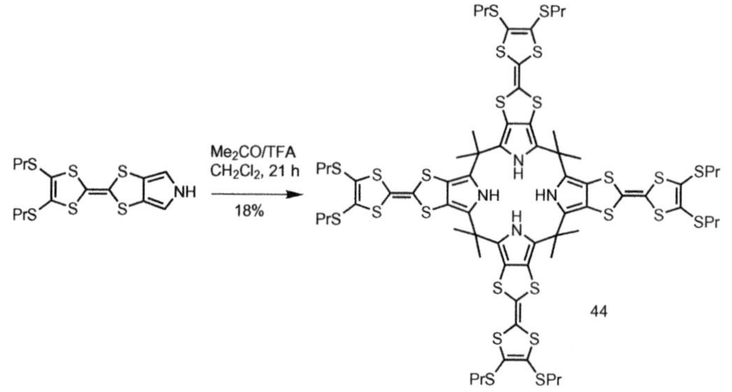

Scheme 24   Synthesis of tetrathiafulvalene based calix[4]pyrrole receptor **44**.

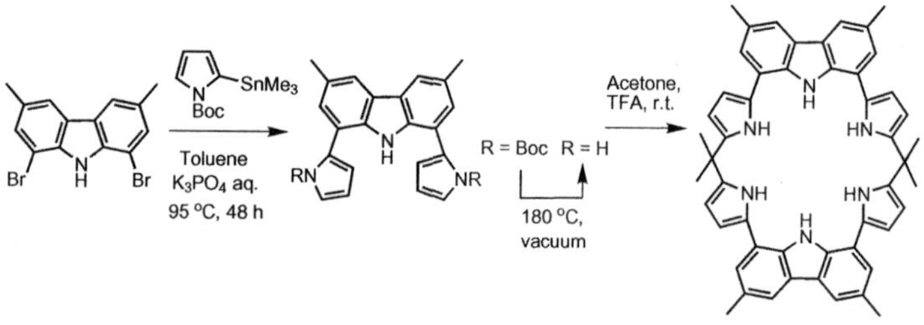

Scheme 25   Synthesis of carbazole-based calix[4] pyrrole[2]carbazole receptor **45**.

Piątek et al. tailored receptor **45** from dibromo carbazole derivative using Tetrakis(triphenylphosphine)palladium(0) (Scheme 25) followed by deprotection and subsequent condensation with acetone (Piątek et al. 2004). The receptor was isolated in 40% yield. The authors also noticed a complex formation with acetate anion in organic solvents such as CH_3CN and CH_2Cl_2 through fluorescence quenching response highlighting anion recognition tendency of the receptor **45**.

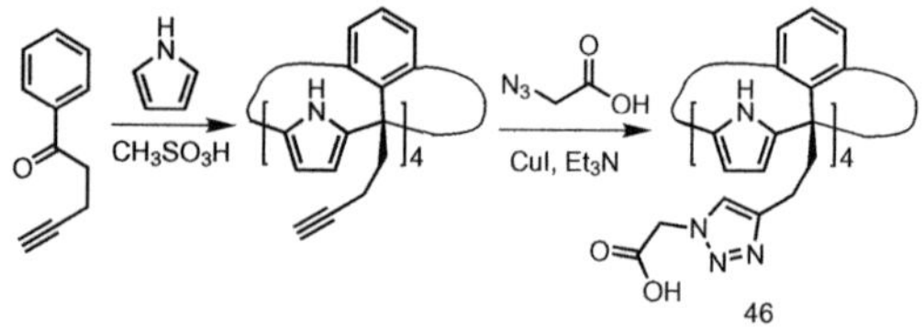

Scheme 26   Synthesis of calix[4]pyrrole receptor **46**.

In 2015, Hernandez-Alonso et al. published an improved protocol for the synthesis of carboxylic acid-based water-soluble receptor **46** (Scheme 26) (Hernandez-Alonso et al. 2015). Condensation of butynyl phenyl ketone with pyrrole in acidic medium (methane sulfonic acid) produced cyclic product. Subsequent reaction with azide in the presence of copper iodide and triethyl amine in DMSO resulted in the desired tetracarboxylic acid appended calix[4] pyrrole receptor **46** in 66% yield. Judiciously, the upper rim of receptor **46** is unblocked, making it suitable for further functionalization. Surprisingly, receptor **46** formed complexes with a variety of pyridyl *N*-oxide guest molecules in an aqueous medium through H-bonding.

Synthesis of N-confused calix [4]pyrrole

At first, N-confused calix[4]pyrrole was synthesized in the year 1999 (Depraetere et al. 1999). N-confused calix[4]pyrrole is the isomer with one inverted pyrrole moiety. Therefore, in the case of N-confused calix[4]pyrrole, one of the nitrogen atoms are placed at the outer periphery whereas the respective carbon atom is located at the inner periphery of the ring (Anzenbacher Jr et al. 2006).

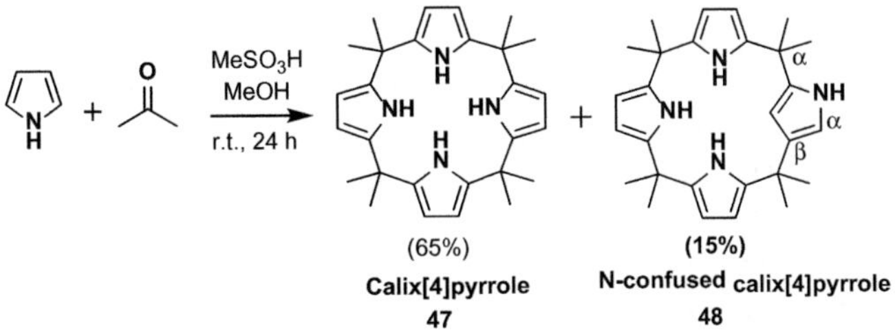

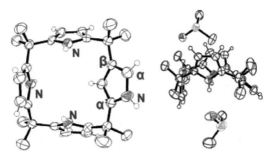

Scheme 27   Synthesis of N-confused calix[4]pyrrole **48**.

Nishiyabu et al. developed a route for improving the synthetic yield of N-confused calix[4]pyrrole **48** (Nishiyabu et al. 2006). N-confused calix[4]pyrrole was synthesized in 15% yield from a reaction of pyrrole and acetone in the presence of MeSO$_3$H in methanol (Scheme 27). The inverted pyrrole moiety was confirmed from single-crystal x-ray crystallographic analysis (Fig. 1.15).

Figure 1.15   (Left) Single crystal X-ray structure of N-confused calix[4]pyrrole macrocyclic receptor. (Right) Crystal structure of inclusion complex obtained from the receptor with two DMSO molecules. Reprinted with permission from the American Chemical Society (Nishiyabu et al. 2006).

Calix[4]pyrroles possess anion-binding motifs in their cavities as reported by Kim et al. using crystallographic data (Fig. 1.16) (Kim et al. 2011). Crystal data validated the cone conformation of calix[4]pyrrole **43** indicating NH interactions (hydrogen bonding) bound to chloride ion.

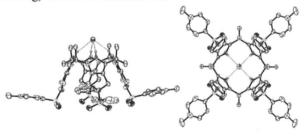

Figure 1.16   (Two different views) Single crystal X-ray structure of N-tosylpyrrolidine calix[4]pyrrole **43** with chloride ion. Reprinted with permission from the American Chemical Society (Kim et al. 2011).

Helic[6]arenes

Insertion of chiral auxiliary into the macrocyclic scaffold results in chiral macrocyclic receptors. Helicarenes are one kind of chiral macrocyclic receptors, where 2,6-dihydroxytriptycene moieties are assembled in a cyclic manner through a methylene linker. The chiral 2,6-dihydroxytriptycene moieties help in providing the receptor into a helical chiral cavity as a result of cyclic tethering. Such helical arenes (called 'helicarenes') were first developed by Chen and Han in the year 2016 (Chen and Han 2018).

Synthesis of Helic[6]arenes

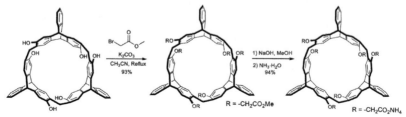

Scheme 28   Synthesis of water-soluble 2,6-helic[6]arene **49**. Reproduced here with the permission of RSC (Zhang et al. 2017).

Zhang et al. synthesized water-soluble 2,6-helic[6]arene receptor **49** in 94% yield by modifying hydroxy group of helic[6]arene to carboxylate groups (Zhang et al. 2017). The receptor possesses *C*3 symmetry in its skeleton. Further, the authors used the chiral receptor for encapsulation of quaternary phosphonium salts in 1:1 binding stoichiometry.

Direct and large-scale synthesis of enantiopure receptors is advantageous for commercial purposes. A synthetic route for the preparation of (O-methyl)6-2,6-helic[6]arenes (+)-**51** and (−)-**52** was provided by Chen et al. 2021 (Scheme 29) (Wang et al. 2018). Receptors (+)-**51** and (−)-**52** were obtained in 51 and 53% respectively after a reaction of enantiopure precursor **50** with *o*-dichlorobenzene and a Lewis acid catalyst such as FeCl₃ under reflux condition for 12 hours.

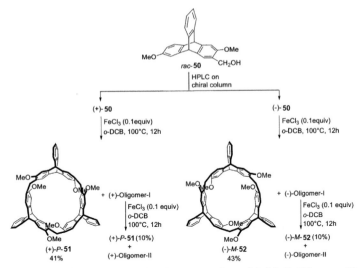

Scheme 29   Synthetic scheme for the preparation of 2,6-helic[6]arenes **50-52**. Reproduced with the permission of ACS (Wang et al. 2018).

Cyanostars

The 'star' shaped macrocyclic receptor 'cyanostar' was first synthesized in 2013 by Flood et al. Structurally, these star-shaped receptors possess pseudo-C_{5h}-symmetry, where weak H-bond donors surround the central core (Fujimoto 2019). The electropositive central core with its 20 CH units facilitates hydrogen bonding with the guest anion (Zahran et al. 2018). Cyanostar receptors form a strong host-guest 2:1 sandwich complex with larger anionic guest molecules such as ClO_4^-, PF_6^- and BF_4^-, etc. (Fig. 1.17) Lee et al. 2013).

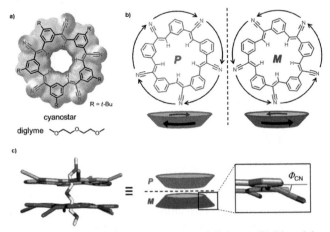

Figure 1.17   (a) Chemical structure of cyanostar and diglyme. (b) Pictorial representation of cyanostars with bowl chirality (P and M isomer) (c) Single-crystal X-ray structure of 2:1 cyanostar-diglyme complex. Reprinted with permission from the American Chemical Society (Liu et al. 2016b).

Synthesis of cyanostars

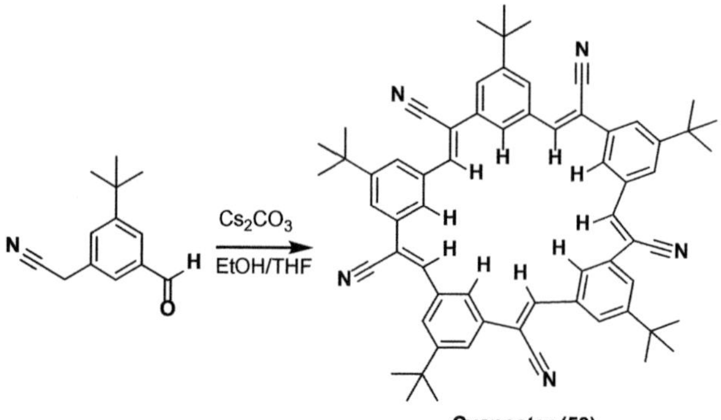

Cyanostar (53)

Scheme 30   One-pot synthesis of macrocyclic receptor **53** 'cyanostar'.

Lee et al. reported multi-gram scale synthesis of cyanostar **53** using a single synthetic step through Knoevenagel condensation (Lee et al. 2013). The condensation involves *meta*-cyanomethyl appended benzaldehyde and a base such as Cs$_2$CO$_3$. The receptor was isolated with 81% synthetic yield (Scheme 31). The receptor cyanostar possesses C5-symmetry in its molecular structure.

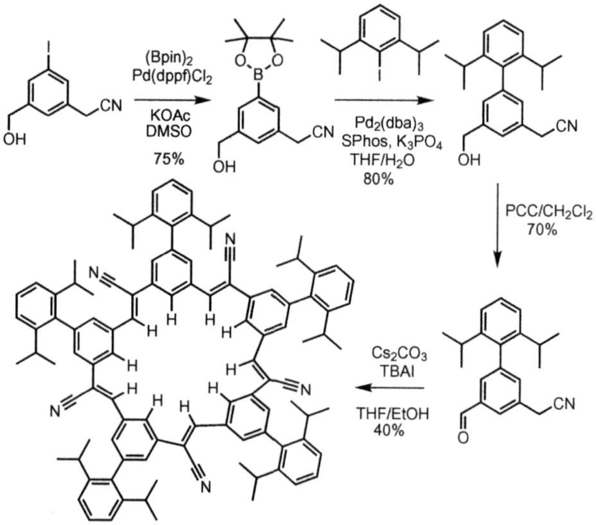

Scheme 31   Synthesis of macrocyclic receptor **54**.

Qiao et al. developed isopropylphenyl based bulky cyanostar receptor (**54**) in four synthetic steps using Miyaura borylation and Suzuki-Miyaura cross-coupling, oxidation reaction using pyridinium chlorochromate (Scheme 31). The final step includes the reaction of aldehyde precursor with Cs$_2$CO$_3$, where the reaction was facilitated further with a template such as a tetrabutylammonium iodide (TBAI) and produced the receptor **54** in 40% yield (Qiao et al. 2016).

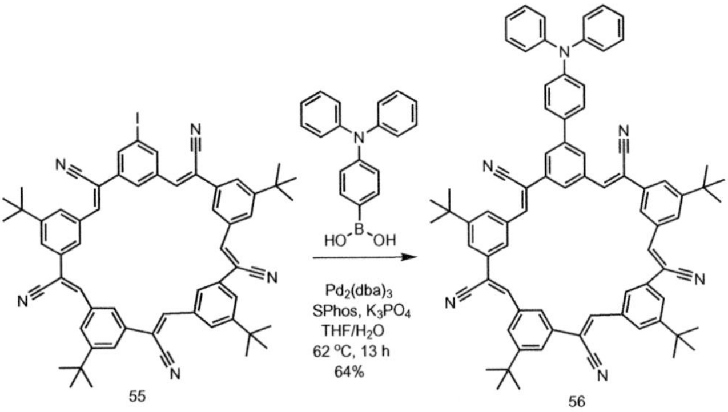

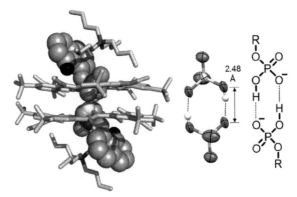

Scheme 32   Synthesis of macrocyclic receptor **56**.

Recently, Dhara et al. developed a triphenylamine-based zero-overlap cyanostar receptor (**56**) with 308 nm of Stokes shift through Suzuki coupling (Scheme 32) (Dhara et al. 2020). By using this cyanostar receptor, it is possible to investigate host-guest complexation at higher (millimolar) concentrations due to the absence of spectral overlapping. The change in concentration of the macrocyclic receptor does not alter its emission wavelength (633 nm) indicating concentration-independent characteristic. The authors demonstrated a novel strategy for the development of zero-overlap receptors by the removal of local emissions. The formation of inclusion complex with dihydrogenphosphe is consistent with 2:2 stoichiometry was also demonstrated (Fig. 1.18)

Figure 1.18   Single crystal X-ray structure of the 2:2:2 complex formed between cyanostar and mononaphthylphosphate $NpHPO_4^-$ (with tetrabutylammonium). Reproduced here with permission of John Wiley and Sons (Zhao et al. 2017b).

Oxatub[4]arene

The tub-shaped macrocyclic receptor with flexible CH_2–O–CH_2 linker was discovered by Jiang et al. in 2015 (Jia et al. 2015). The same group introduced oxatub[4]arenes for encapsulation of several cations (Jia et al. 2016). Recently, Chen et al. utilized oxatub[4]arene for neurotransmitter determination (Chen et al. 2021).

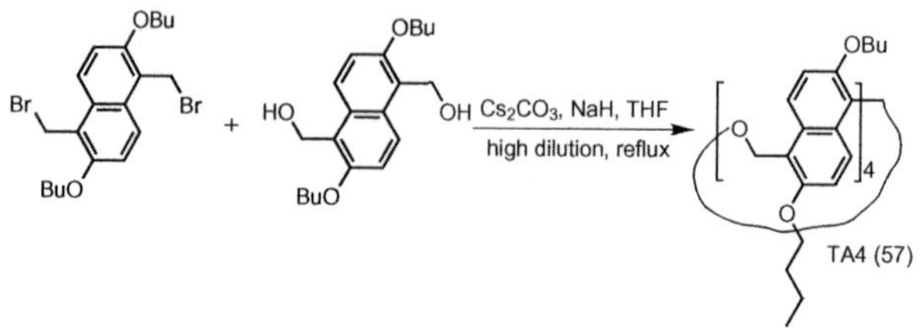

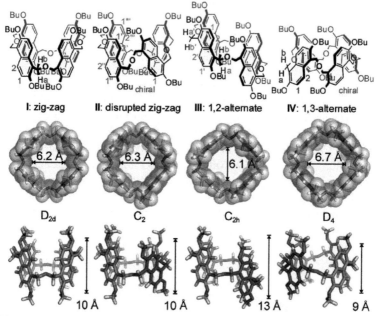

Scheme 33   Synthesis of macrocyclic receptor **57**.

Figure 1.19   Four different conformers of naphthalene-based oxatub[4]arene. Reproduced here with the permission of RSC (Jia et al. 2015).

Jia et al. reported naphthalene-based smart macrocyclic receptors with four different conformations through one-pot reaction and named it 'oxatub[4]arene' (TA4) (Jia et al. 2015). These conformers are also regarded a as molecular 'transformer' due to their potential in cavity transformation (Fig. 1.19). The receptor exhibits multiple interconvertible cavities owing to the flipping of naphthalene moieties. Unlike the well-known Calixarenes, this naphthalene-based oxatub[4]arene exhibited different cavity sizes. Due to the existence of different and well-defined cavities, these four different conformers display different binding preferences towards guest species.

The ether (CH_2–O–CH_2) linker present in the oxatub[4]arene facilitated in achieving a deeper cavity for guest trapping in comparison to calixarene, where mostly steric hindrance results due to the presence of methylene linker. Also these oxatub[4]arenes avoid self-inclusion and diminished the number of linkage regioisomers due to the blockage of 1,5-position of naphthalene moieties (Yang et al. 2016a). The interesting nature of these conformers is such that interconversion among conformers is quite fast at room temperature, although the interconversion can be controlled by the addition of different guest species. However, the interconversion rate among different conformers reduces at a lower temperature such as $-80°C$.

Pillararenes

Pillararenes, one of the new kinds of macrocyclic receptors with pillar-shaped structures, was first unveiled in 2008 by Ogoshi et al. (Ogoshi et al. 2008). In pillararenes, methylene (CH_2) bridges hold the dialkoxybenzene (*O*-alkylated hydroquinone) units at the *para*-position in a cylindrical or pillar-like shape (Ogoshi et al. 2016).

Due to the electron-rich (-rich) skeleton of pillararenes composed of hydroquinones, this type of macrocyclic receptors forms an inclusion complex with electron-poor guest species. In particular, inner surfaces of pillar[n]arene cavities are composed of negatively charged species, whereas outer surfaces and rims are neutral (Yu et al. 2015). Electron-rich decorated cavities make them a potential contender for binding with viologen cations, pyridiniums, imidazoliums, ammonium salts and other neutral guest molecules.

The uniqueness of pillar[n]arenes makes them different from other macrocyclic receptors such as crown ethers, calixarenes, cyclodextrins and cucurbiturils. Some of its unique features are:

(a) Possesses symmetrical and rigid architecture, hence forming selective binding with guest species
(b) Hassle-free functionalization by suitable substituents makes them easy to modify and tune host-guest chemistry
(c) High solubility in organic solvents

Therefore, pillar[n]arenes are used a great deal in the field of chemosensors, cell imaging, supramolecular gels, supramolecular polymers, smart materials and artificial transmembrane proton channels.

Pillar[5]arene, one of the conformationally stable macrocyclic receptors, possessing five hydroquinone units. Pillar[8]arenes possess two pentagon-like structures, whereas pillar[9]arenes possess one pentagon and one hexagon. Similarly, pillar[10]arenes possess two hexagons like structures

Synthesis of Pillararenes

Pillararenes are symmetrical macrocyclic receptors, where synthesis of these types of receptors requires a single synthetic step and hence can be easily accessible with higher synthetic yield. The condensation product of 1,4-dimethoxybenzene and paraformaldehyde in the presence of a Lewis acid catalyst produces the first pillararene as 1,4-dimethoxypillar[5]arene. The subsequent removal of methoxy groups using BBr_3 (boron tribromide) yielded pillar[5]arene. Commonly used Lewis acid catalysts for pillararenes synthesis include ferric chloride ($FeCl_3$), boron trifluoride (BF_3), sulfuric acid (H_2SO_4), trifluoromethanesulfonic acid (CF_3SO_3H), trifluoroacetic acid (CF_3COOH) and *p*-toluenesulfonic acid (PTSA).

Pillararenes are easy to be derivatized and the hydroxyl position of the pillararenes and can be easily functionalized selectively at a single position or multiple positions over benzene moiety.

Homo-cyclooligomerization

Pillararenes can be synthesized using a single monomer unit known as homo-cyclooligomerization (Yang et al. 2016a). In 2008, homo-cyclooligomerization was used for the synthesis of the first symmetrical 1,4-dimethoxypillar[5]arene from the easily available precursors such as 1,4-dimethoxybenzene, paraformaldehyde, and boron trifluoride etherate (Scheme 35).

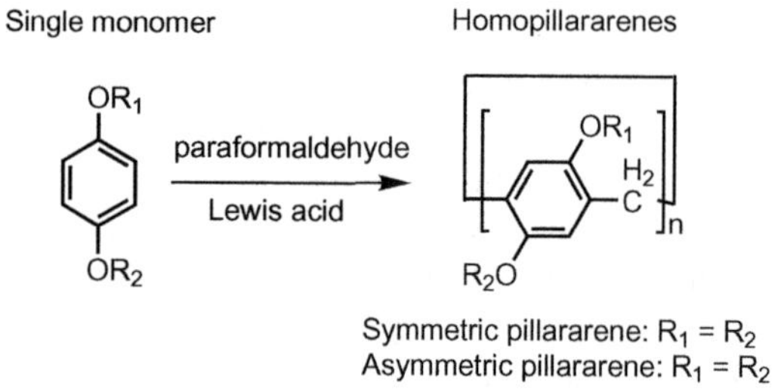

Symmetric pillararene: $R_1 = R_2$
Asymmetric pillararene: $R_1 = R_2$

Scheme 34   Synthesis of Pillar[n]arenes.

Co-cyclooligomerization

The two different monomer units can be combined in the presence of paraformaldehyde and a Lewis acid to yield pillararenes through a process known as co-cyclooligomerization. Huang et al. used the co-cyclooligomerization method for introducing different repeating units in the pillar[5]arenes (Scheme 36) (Zhang et al. 2010).

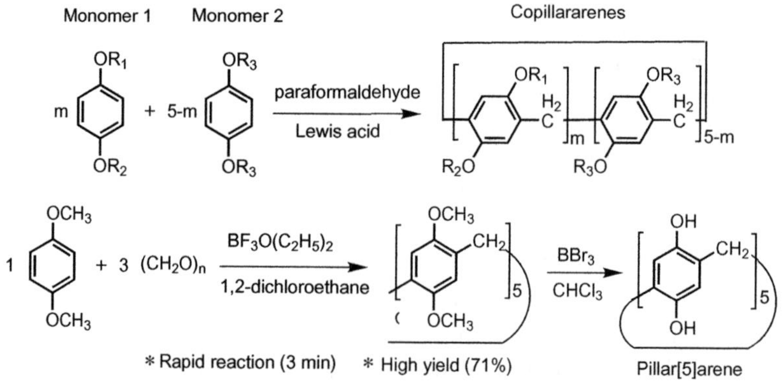

* Rapid reaction (3 min)   * High yield (71%)   Pillar[5]arene

Scheme 35   Synthesis of macrocyclic receptor pillar[5]arene (**58**).

Further, Ogoshi et al. synthesized dimethoxypillar[5]arene **58** in a higher yield (71%) through a rapid process, where completion of reaction took just 3 minutes. (Ogoshi et al. 2011) Subsequent deprotection of dimethoxypillar[5]arene (DMpillar[5]arene) yielded a quantitative amount of pillar[5]arene. The cyclization reaction involved commercially available precursors such as 1,4-dimethoxybenzene and three equivalents of paraformaldehyde. The involvement of three equivalents of paraformaldehyde instead of one resulted in a reduced amount of polymer as a byproduct, thereby increasing overall yield to 71% (Scheme 35). The authors used size-exclusion chromatography for monitoring the progress of the cyclization reaction at different intervals of time. The structure of the pillar[5]arene was confirmed from single-crystal X-ray crystallography (Fig. 1.20).

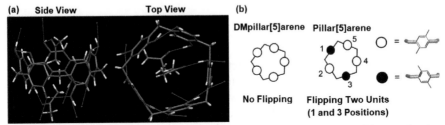

Figure 1.20 (a) Single crystal X-ray structure of pillar[5]arene. Intra and intermolecular hydrogen bonds are shown by blue and red lines respectively. (b) Pictorial representation of conformations of DMpillar[5]arene and pillar[5]arene in the crystalline state. Reprinted with permission of American Chemical Society (Ogoshi et al. 2011).

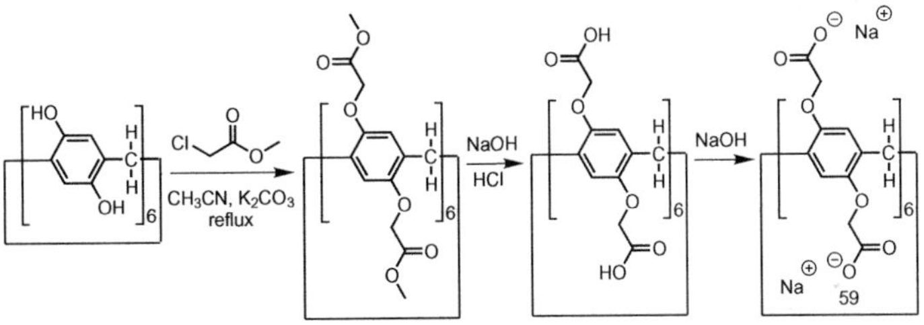

Scheme 36 Synthesis of macrocyclic receptor **59**.

Yu et al. synthesized carboxylate-based water-soluble pillar[6]arene receptor **59** for the first time in 2012 (Yu et al. 2012). The synthesis includes etherification followed by basic hydrolysis to give carboxylic acid-based pillar[6]arene. Water-soluble pillar[6]arene was obtained after reaction with NaOH (1 equivalent) in water. The pillar[6]arene receptor has a strong affinity towards pyridiniums salt. The receptor formed a host-guest vesicular complex in the presence of pyridinium guest molecules. Moreover, the authors demonstrated the pH-controlled solubility of the receptor in water (Scheme 36).

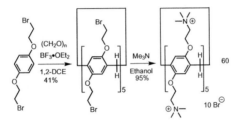

Scheme 37 Synthesis of macrocyclic receptor **60**.

Ma et al. synthesized a cationic water-soluble pillar[5]arene receptor **60** in two synthetic steps (Scheme 37) (Ma et al. 2011). The reaction with paraformaldehyde in the presence of boron trifluoride etherate yielded an intermediate in 41% yield. The receptor contains trimethylammonium moiety at both ends of pillar[5] arene and was isolated as a colorless solid with 95% yield. It also formed a 1:1 pillar[5]arene-sodium 1-octanesulfonate supramolecular complex in aqueous media. Both hydrophobic and electrostatic interactions played a major role during supramolecular complexation.

Prismarenes

The 'prism' shaped macrocyclic receptors 'prismarenes' was first discovered by Professor Carmine Gaeta and his team while investigating naphthol-based derivatives at *Universita di Salerno* (Italy) by (Della Sala et al. 2020). Prismarenes possess excellent π-electron rich cavities, which are deeper and larger in size in comparison to pillar[n]arenes. Interestingly, prismarenes are in-built chiral structures, show strong binding affinity towards positively charged quaternary ammonium ions through cation and $^+$NC–H$\cdots$π interactions (Yang et al. 2016b).

Synthesis of Prismarenes

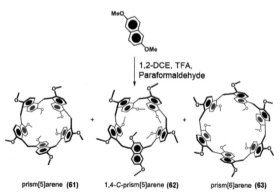

Scheme 38   Synthesis of macrocyclic receptors **61-63**. Reproduced here with the permission of the American Chemical Society (Della Sala et al. 2020).

Sala et al. synthesized prism[5]arene by reacting 2,6-dimethoxynaphthalene (2.5 mM) in the presence of paraformaldehyde (1.2 equivalents) and an acid catalyst such as CF_3COOH in 1,2-dichloroethane as a solvent in an optimized condition (Della Sala et al. 2020). Although, the reaction yield of prism[5]arene is lower (0.3%), another isomer (1,4-confused ring) 1,4-*C*-prism[5]arene was obtained in 40% yield from the same reaction condition (Scheme 38). Harsh conditions such as reduced concentration elevated temperature, and a longer time was the key to the development of prism-like cyclic receptors unlike well-known oligomers, which are obtained at a much lower temperature. Further, the authors improved reaction yield by using the template method during synthesis. In the presence of ammonium templating agents such as 1,4-diazabicyclo[2.2.2] octane-based cationic template (iodide salt), prism[5]arene was synthesized in 47% yield by following the optimized condition.

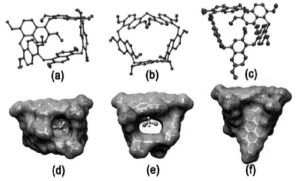

Figure 1.21 Single crystal X-ray structure of (a) and (b) forms of prism[5]arene and (c) form of 1,4-C- prism[5]arene. Reprinted with permission from the American Chemical Society (Della Sala et al. 2020).

Cucurbiturils

Cucurbiturils are a family of macrocyclic supramolecular receptors obtained by condensation of glycoluril and formaldehyde. They can be easily modified to obtain amazing guest-binding behavior towards ionic or molecular guests (Dsouza et al. 2011). They can be produced in different cavity sizes depending on the reaction conditions. For example, Shen et al. reported the synthesis of two pumpkin-shaped macrocyclic receptors using $BaCl_2$ and $CaCl_2$ as a template (Fig. 1.22) (Shen et al. 2017).

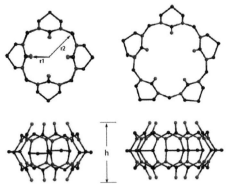

Fig. 1.22 Single crystal X-ray structure of top and side views of macrocyclic receptor $TD[4].2CaCl_2$, $TD5.2BaCl_2$. (Shen et al. 2017)

Bambus[n]urils

The macrocyclic receptor bambus[6]uril was first discovered by Prof. Vladimir Sindelar and his team from Masaryk University (Czech Republic) in the 2010

(Lizal and Sindelar 2018). The X-ray crystal structure of the receptor was similar (looks wise) to the natural bamboo plant stem (Fig. 1.23). Bambus[n]urils were first synthesized by the condensation reaction of disubstituted glycoluril and formaldehyde in the presence of an acid catalyst. Bambus[n]urils exhibits excellent structural flexibility over cucurbiturils.

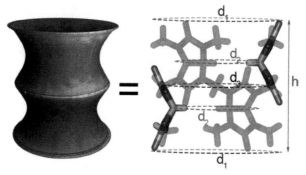

Figure 1.23   Cartoon representation of bambus[6]uril along with a cross-section of x-ray crystal structure (Lizal and Sindelar 2018).

Bambusurils are often used as anion transporters. These rigid receptors form a stronger complex with anions through Van der Waals interactions (Mondal et al. 2019). In a study published in 2016, Solel et al. found that replacing heteroatom in the core skeleton could significantly intensify the anion binding ability of bambusuril receptors (Solel et al. 2016). Heteroatom replacement could easily induce polarization and hence can bind more strongly with anions (Fig. 1.24). The strength of anion binding depends on the nature of heteroatom (sulfur, oxygen or nitrogen) that is installed over the receptor.

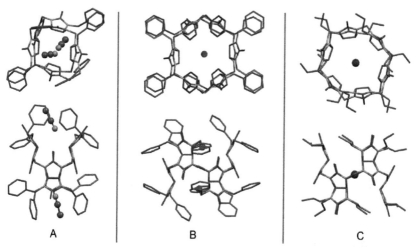

A   B   C

Figure 1.24   Structures of (A) (A) $Bn_8BU[4]:2CH_3CN$ complex, (B) $Bn_{12}BU[6]:2Cl^-$ complex, and (C) $Pr12BU[6]: I^-$ complex viewed from the top side obtained by using crystallography. Reproduced here with permission from ACS (Havel et al. 2011).

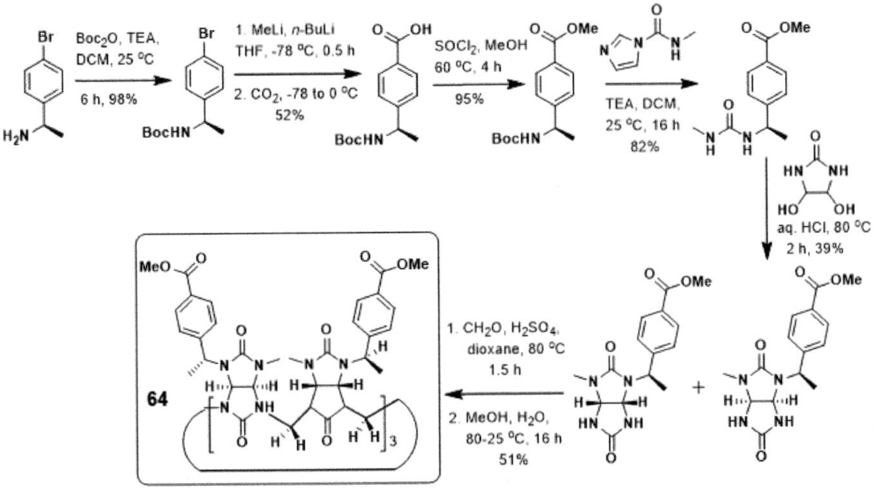

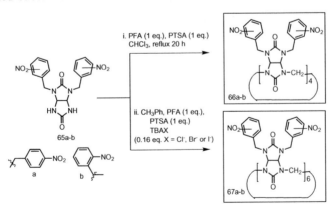

Scheme 39 Synthesis of chiral macrocyclic bambusuril receptor **64**.

Enantioselective detection of chiral compounds is important in the field of asymmetric catalysis. Sokolov et al. synthesized ester functionalized chiral bambusuril receptor **64** with 51% yield in a multistep synthetic process (Scheme 39) (Sokolov et al. 2020). The receptor was able to encapsulate chiral carboxylates within its cavity with an enantioselectivity factor of 3.1. Detection of selective carboxylate guest molecule mainly depends on the steric hindrance resulting from the attached substituents over the bambusuril core.

Scheme 40 Synthesis of nitro group-based bambusuril receptors **66-67**.

Yawer et al. synthesized multiple nitro group functionalized bambusuril receptors (Scheme 40) (Yawer et al. 2018). Glycouril precursors **65a-b** reacted in the presence of PTSA and paraformaldehyde to produce bambusuril receptors. The authors also obtained single crystals of **66a** from DMSO solution. The 4-membered bambusuril receptors **66a-b** could not bind with anions. Fortunately, 6-membered bambusuril receptors **67a-b** largely interacted with anions. It is interesting to note that templates such as tetrabutylammonium iodide salt (TBAI) can be used to control the bambusuril cavity, hence, the supramolecular complexation can be fine-tuned accordingly.

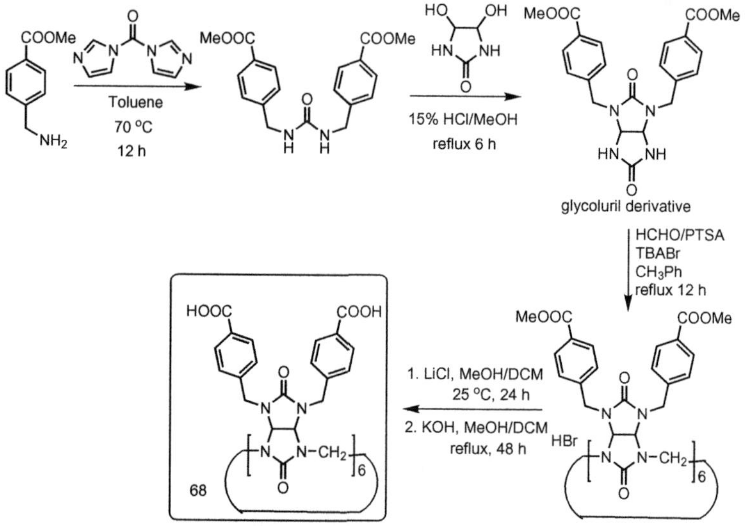

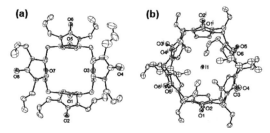

Scheme 41   Synthesis of carboxylic acid-based water-soluble bambusuril receptor **68**. PTSA = *p*-toluenesolfonic acid.

Yawer et al. synthesized benzoate-based bambusuril receptor (Scheme 41) from 4-methoxycarbonylbenzylamine through four synthetic steps. The first step involves the synthesis of dialkylurea followed by a glycoluril derivative (Yawer et al. 2015). Bambusuril methyl ester was formed after treatment of glycoluril intermediate with paraformaldehyde and PTSA in toluene. The addition of a template drives the reaction in such a way that only a 6-membered macrocyclic ring is formed as a major compound instead of an undesired 4-membered ring. Therefore, tetrabutylammonium bromide was mixed in the reaction mixture to get the desired 6-membered ring. Further reaction with LiCl and KOH yielded the carboxylic acid-based water-soluble receptor **68** in 15% yield.

Figure 1.25   Single crystal X-ray structure of unsaturated bambusuril receptors **69a-b** (a) $Allyl_8BU[4]$ and (b) $Allyl_{12}BU[6]$. Reprinted with permission from the American Chemical Society (Rivollier et al. 2013).

Rivollier et al. prepared unsaturated bambusurils $Allyl_8BU[4]$ and $Allyl_{12}BU[6]$ through microwave synthesis in higher yield (85%) (Fig. 1.25) (Rivollier et al. 2013).

The receptor **69a**, i.e., Allyl₈BU[4] was synthesized (85%) from a mixture of reagents such as diallyglycoluril, formaldehyde, and PTSA in diluted chloroform (0.04 M) solvent in a microwave synthesizer at 75°C for 2 hours. Additionally, receptor **69b** Allyl₁₂BU[6] was obtained in 60% yield, when the microwave reaction was performed in the presence of a template such as tetrabutylammonium iodide (TBAI).

Cyclotriveratrylenes

The British organic chemist Robinson synthesized 2,3,6,7-tetramethoxy-9,10-dihydroanthracene through a condensation reaction of veratryl alcohol in the presence of an acid catalyst in the year 1915 (Collet et al. 1993). For many years, the structural conformation of cyclotriveratrylene had its own unique story to be told. The exact structure of the molecule was in ambiguity for some time. After almost 50 years, this previously synthesized molecule was later established as a cyclic trimer and called cyclotriveratrylene by another chemist Lindsey in the year 1965 (Lindsey 1965). Further, NMR studies confirmed its structure as locked crown conformation. Zhang and Atwood obtained the crystal structure of guest-free cyclotriveratrylene after slow evaporation of toluene solution in the year 1990 (Fig. 1.26) (Zhang and Atwood 1990).

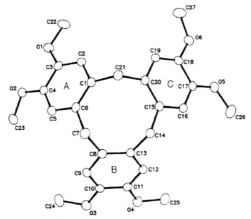

Figure 1.26   Single crystal X-ray structure of guest-free cyclotriveratrylene receptor. Reproduced here with the permission of Springer Nature (Zhang and Atwood 1990).

Cyclotriveratrylenes (CTVs) are macrocyclic receptors with a whitish appearance and possess an open pyramidal arrangement. In particular, these receptors are bowl-shaped rigid structures with a three-dimensional shape, which possess a C3 symmetry. These receptors contain 9-membered core-ring as part of the unique structural feature. These macrocyclic receptors possess a hydrophobic binding pocket and are suitable for holding bulkier species such as *ortho*-carborane and fullerene (Hardie 2016). Recently, there is a surge in the applications of cyclotriveratrylene in separation, sensor, gels, liquid crystals, etc. (Hardie 2010; Cai et al. 2012).

Synthesis of Cyclotriveratrylene

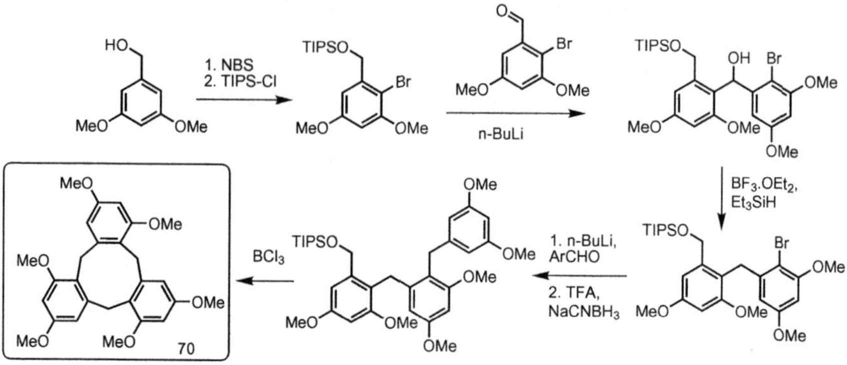

Scheme 42   Synthesis of macrocyclic receptor **70**.

Veratrole alcohol, when undergoing an acid catalysis produces cyclotriveratrylene. With a goal of extended derivatization, Wright et al. synthesized cyclotriveratrylene receptor **70** with 60% yield in a step-wise synthetic route (Scheme 42) (Wright et al. 2014). The overall synthetic steps include some key synthetic procedures such as bromination, protection of hydroxyl group and lithiation followed by cyclization in the presence of Lewis acid catalyst BCl_3. Single-crystal X-ray crystallography confirmed crown-like conformation similar to cyclotriveratrylene.

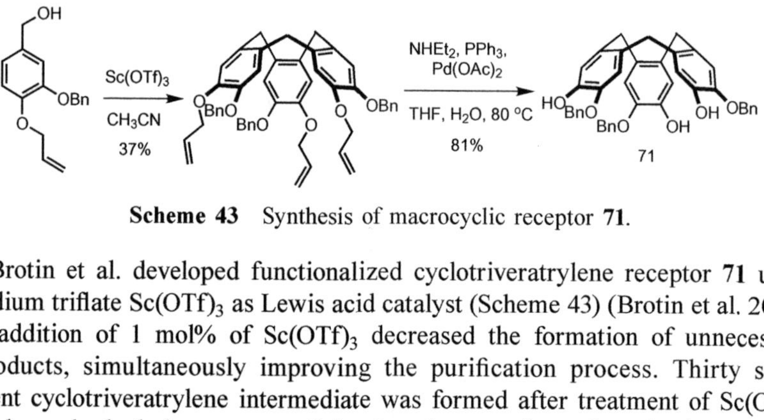

Scheme 43   Synthesis of macrocyclic receptor **71**.

Brotin et al. developed functionalized cyclotriveratrylene receptor **71** using scandium triflate $Sc(OTf)_3$ as Lewis acid catalyst (Scheme 43) (Brotin et al. 2005). The addition of 1 mol% of $Sc(OTf)_3$ decreased the formation of unnecessary byproducts, simultaneously improving the purification process. Thirty seven percent cyclotriveratrylene intermediate was formed after treatment of $Sc(OTf)_3$ with benzyl alcohol precursor. Selective deprotection of cyclotriveratrylene derivative in the presence of $Pd(OAc)_2$ produced trihydroxy cyclotriveratrylene in 81% synthetic yield.

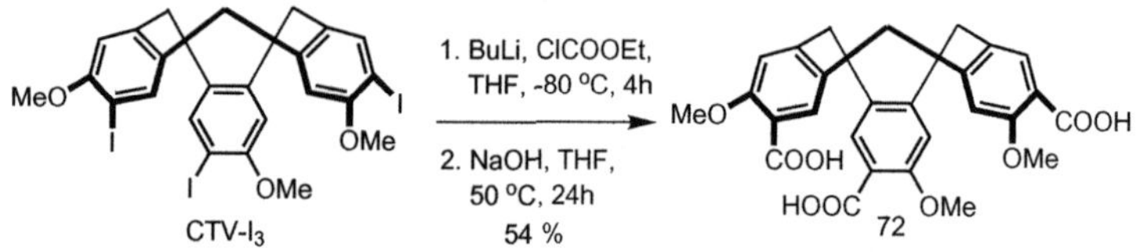

Scheme 44   Synthesis of macrocyclic receptor **72**.

Peyrard et al. introduced -extended electron-withdrawing moieties such as COOH moiety at the ortho position of methoxy group attached directly to the cyclotriveratrylene core structure (Peyrard et al. 2012). The reaction proceeded through a lithiation reaction (Scheme 44), with ethyl chloroformate and finally, hydrolysis of ester group to provide carboxylic acid-based cyclotriveratrylene receptor **72** in 54% yield.

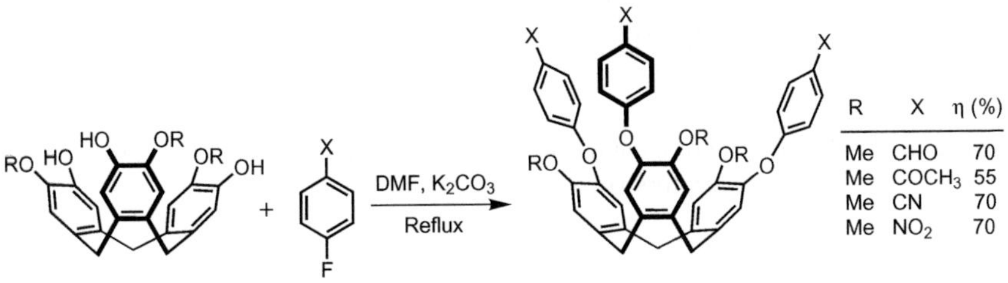

R	X	η (%)
Me	CHO	70
Me	COCH$_3$	55
Me	CN	70
Me	NO$_2$	70

Scheme 45 Synthetic scheme for the preparation of cyclotriveratrylene receptors **73-76** via S$_N$Ar reaction (Arduini et al. 2004).

Arduini et al. synthesized a series of cavity extended cyclotriveratrylene receptors **73-76** bearing electron-withdrawing moieties through functionalization of cyclotriguaiacyclene by aromatic nucleophilic substitution reaction (S$_N$Ar) (Scheme 45) (Arduini et al. 2004). The reaction with substituted *p*-fluorobenzene groups provided good synthetic yields with multiple electronic characters, where electron-withdrawing moieties such as CHO, COCH$_3$, CN and NO$_2$ served as electrophiles. The synthesized receptors can be further modified easily to NH$_2$ or OH precursors for developing more useful macrocyclic cavities.

Cryptophanes

Cryptophanes were first synthesized and discovered in the early 1980s (Gabard and Collet 1981). These are also known as macropolycyclic cyclophanes. Cryptophanes are cage-like structures, where two cup-shaped [1.1.1]-orthocyclophane units connect each other through three aliphatic linker units. In other words, cryptophanes consist of two cyclotriveratrylene units facing straight towards each other. Based on the length of linkers, these macrocyclic receptors can exhibit ellipsoid or spherical cavities for trapping suitable guest species (Schaly et al. 2016). Two different types of cryptophanes are normally available as *syn* and *anti* diastereomers (Brotin and Dutasta 2009). Based on symmetry, *syn* cryptophane exhibit the C$_{3h}$ point group, whereas *anti* cryptophane displays the D3 point group (Fig. 1.27).

One of the emerging roles of cryptophanes includes gas encapsulation, in particular, the storage of hydrogen gas for fuel cell applications. Additionally, cryptophanes are used as containers for carrying out organic reactions. These organic cage receptors are excellent binding agents for xenon gas and hydrocarbons such as methane (Fig. 1.28) (Gao et al. 2015; Demissie et al. 2017).

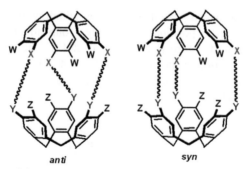

anti   *syn*

Figure 1.27   Pictorial representation of *anti* and *syn* conformation of cryptophane receptors (Brotin and Dutasta 2009).

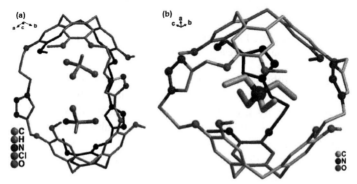

Figure 1.28   Single crystal X-ray structure of (a) *anti* isomer and (b) *syn* isomer of triazole-bridged cryptophane receptor. Reproduced here with permission from John Wiley and Sons (Satha et al. 2017).

Synthesis of Cryptophanes

Satha et al. reported *syn* and *anti*-isomer of triazole bridged cryptophane through template method (Satha et al. 2017). The authors isolated and characterized both the isomers using NMR and X-ray crystallography. The crystal structure of the *anti*-isomer (Fig. 1.27) was obtained from chloroform solution of cryptophane after heating at 60°C for 30 minutes followed by slow evaporation. The structure of *syn*-isomer was obtained from slow evaporation of a solution of cryptophane and tetraethylammonium bromide (1:1) in CH_3CN: $CHCl_3$ mixture.

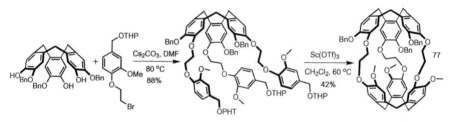

Scheme 46   Synthesis of cryptophane receptor **77**.

Cyclotriveratrylene derivatives can be used as a synthetic precursor for the development of cryptophanes receptors. Brotin et al. utilized cyclotriveratrylene intermediates and synthesized cryptophane receptor **77** after two synthetic steps (Scheme 46) (Brotin et al. 2005). The authors used Sc(OTf)$_3$ as a Lewis acid catalyst, which is environment friendly and minimizes solvent requirement due to efficient recovery of both catalyst and solvent during the reaction.

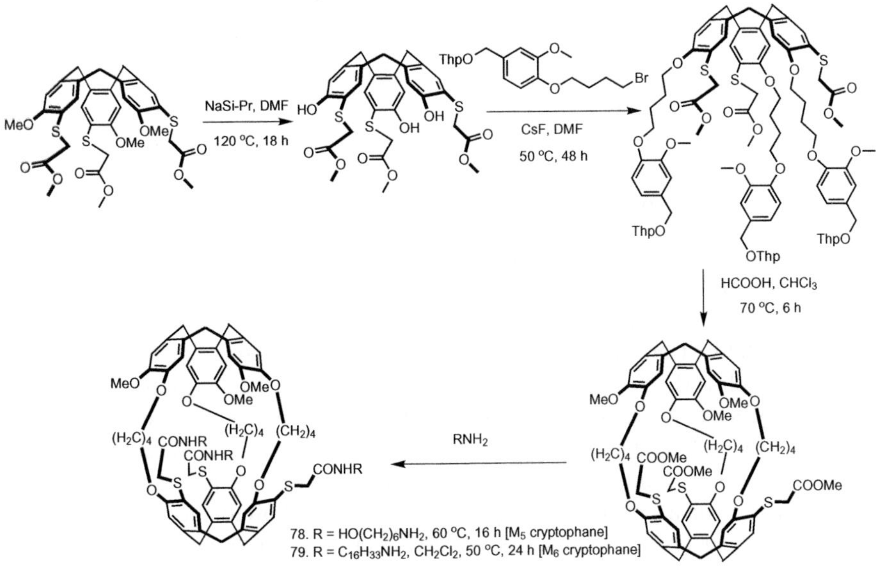

Scheme 47   Synthesis of macrocyclic receptor **78-79**.

78. R = HO(CH$_2$)$_6$NH$_2$, 60 °C, 16 h [M$_5$ cryptophane]
79. R = C$_{16}$H$_{33}$NH$_2$, CH$_2$Cl$_2$, 50 °C, 24 h [M$_6$ cryptophane]

Gosse et al. developed amphipathic cryptophane receptors **78-79** from thia-cyclotriveratrylene intermediate (Gosse et al. 2005). The reaction of triester-cryptophane with 6-aminohexan-1-ol at 60°C yielded tris-hexanol substituted cryptophane with 75% synthetic yield (Scheme 47). Further, the reaction of trimester-cryptophane with hexadecyl amine at 50°C produced tris-hexadecyl-based cryptophane with 55% isolated yield. The tris-hexadecyl-based cryptophane receptor displayed reversible Langmuir monolayers when the surface pressure was high. Incorporation of a longer alkyl chain in the cryptophane receptor facilitated intermolecular H-bonds and well-organization molecules, which resulted in stable and reversible Langmuir monolayer formation.

Rotaxanes and Catenanes

Rotaxanes and catenanes are mechanically interlocked macrocyclic receptors. Such supramolecular architectures attracted considerable attention owing to their potential application in emerging areas such as switches and molecular machines owing to the free movement of interlocked components. Several triggers like metal ion, hydrogen bonding and π–π stacking are continuously being utilized for the development of rotaxanes and catenanes (Evans 2019).

Synthesis of Rotaxanes and Catenanes

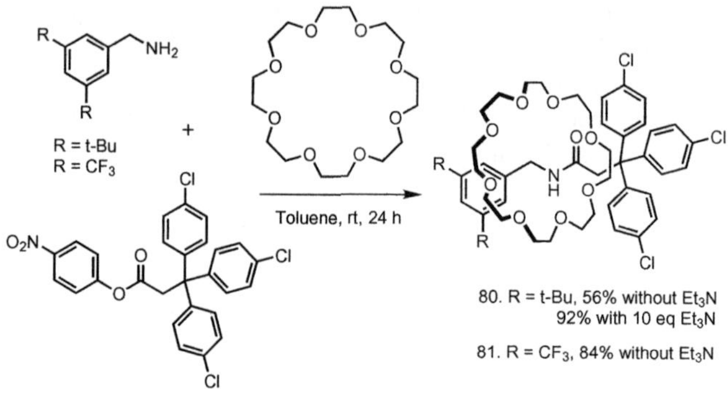

Scheme 48   Synthesis of rotaxane receptors **80-81**.

A variety of syntheses of rotaxanes are available in literature (Sauvage 1998; Altieri et al. 2011; Beves et al. 2011; Evans 2019). Tian et al. also synthesized rotaxane such as receptor **80-81** (Scheme 48) up to 95% synthetic yield via template method (Tian et al. 2020). The synthetic precursors include primary amine, electrophile (C=O, C=S, etc.,) and crown ether in a 1:1:1 stoichiometric ratio.

Similarly, several literature studies are available which elucidate the synthesis and applications of catenanes (Coronado et al. 2009; Crowley et al. 2009; Evans and Beer 2014; Gil-Ramírez et al. 2015). The development of polyrotaxanes and polycatenanes, which consist of multiple macrocycles allowed the incorporation of unusual properties within such novel architectures. One interesting example of polycatenanes **82** based on cyclodextrin was reported recently by Higashi et al. (Higashi et al. 2019) through a one-pot synthetic method, which had more than 10 cyclodextrins attached to a poly(ethylene glycol)-poly(propylene glycol) copolymer core (Fig 1.29). The polycatenanes have a potential application in molecular machines, molecular actuators, switches materials of biological importance and as a drug delivery vehicle.

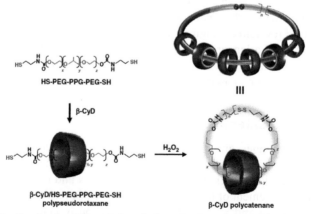

Figure 1.29   Synthesis of β-cyclodextrin based polycatenanes **82**. Reproduced here with the permission of Nature (Higashi et al. 2019).

Functionalized Tetrathiafulvalenes

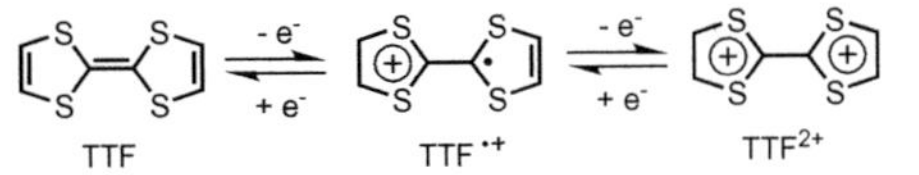

Scheme 49　Tetrathiafulvalene with different states showing reversible redox process.

Tetrathiafulvalene (TTF) is a strong electron donor, which exhibits remarkable optoelectronic properties on subsequent functionalization (Segura and Martín 2001). These simple electroactive compounds were developed in 1970 (Wudl et al. 1970). In general, functionalization of TTF moiety is carried out to yield a number of D–A (donor-acceptor) or D–π–A supramolecular structures (Jana et al. 2018). Owing to its sensitive nature, it shows different redox potentials depending on the environment (Scheme 49). These redox-active molecules can undergo oxidation to produce TTF^{+} and TTF^{2+} reversibly. Due to the electron-donating tendency, TTF scaffolds are substantially used as optical sensors, optoelectronic materials, electrochemical switches, etc. (Canevet et al. 2009; Bergkamp et al. 2015). TTF scaffolds have also been attached to macrocyclic molecules such as calix[4] pyrrole, porphyrin, phthalocyanine, etc. (Fig. 1.30) to develop hybrid macrocyclic systems.

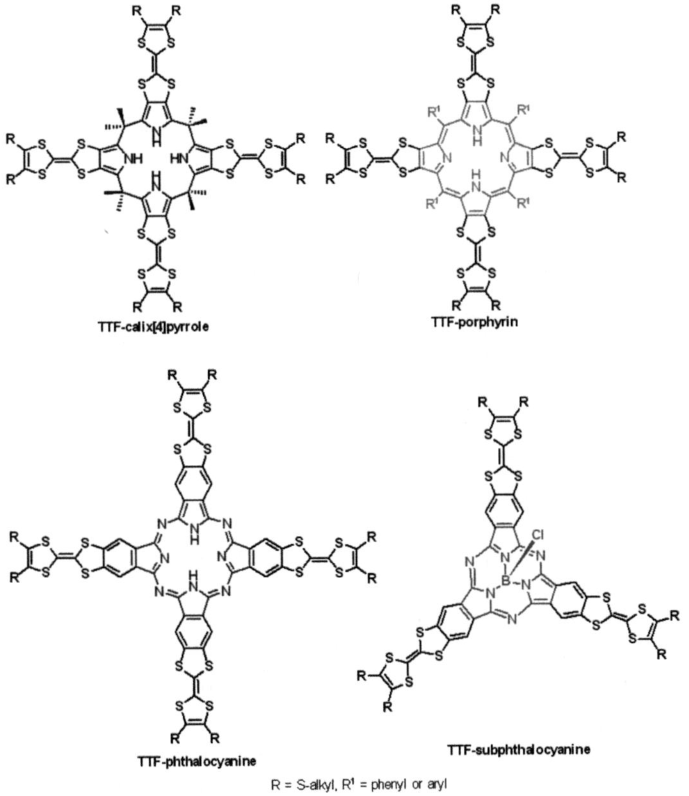

TTF-calix[4]pyrrole

TTF-porphyrin

TTF-phthalocyanine

TTF-subphthalocyanine

R = S-alkyl, R^1 = phenyl or aryl

Figure 1.30　Commonly available TTF-annulated macrocyclic structures (Jana et al. 2017).

Synthesis of Tetrathiafulvalenes

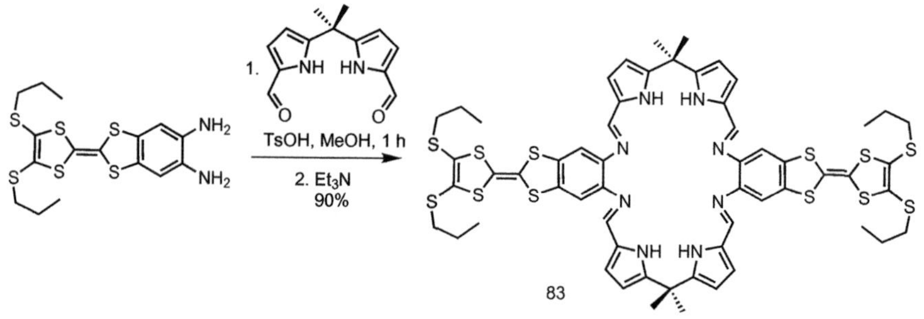

Scheme 50   Synthesis of macrocyclic receptor **83**.

Bejgar et al. synthesized tetrathiafulvalene tethered calixpyrrole receptor **83** in the presence of *para*-toluenesolfonic acid in 90% yield (Bejger et al. 2011). The authors used tetrathiafulvalene phenylenediamine and diformyldipyrromethane as synthetic precursors for the preparation of the orange-colored Schiff's base product (Scheme 50). The synthesized product with an appropriate orientation of two TTF parts can stabilize the dimeric form of mixed-valence radical, i.e., $[TTF_2]^+$ upon oxidation.

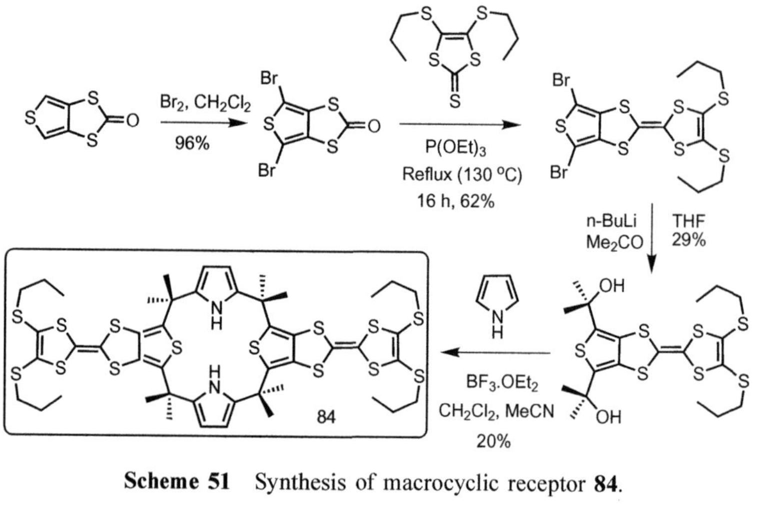

Scheme 51   Synthesis of macrocyclic receptor **84**.

Kim et al. prepared calix[2]thiophene[2]pyrrole receptor **84** flanked by bis-tetrathiafulvalene units, where cyclization was achieved (Scheme 51) using a Lewis acid catalyst such as $BF_3.OEt_2$ (Kim et al. 2009). The yellow-colored solid product was isolated with a 20% yield after column purification. The receptor produced a change color in the presence of explosive materials such as trinitrobenzene and picric acid in chloroform. The binding occurred with electron-deficient guest molecules by a sandwich-type coordination complex formation (Fig. 1.31). Unlike other tetrakisTTF-calix[4]pyrrole receptors, **68** is insensitive to common interfering guests such as chloride ions.

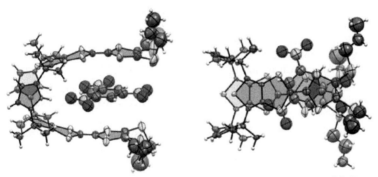

Figure 1.31   Single crystal X-ray structure of tetrathiafulvalene receptor **84**. Reproduced here with the permission of the American Chemical Society (Jana et al. 2017).

CONCLUSIONS

Synthetic and historical aspects of macrocyclic receptors, including newly made structural motifs were discussed here. Functionalization of newly discovered macrocyclic receptors such as multifarenes, helicarenes, cyanostars, bambusurils and prismarenes, etc., need further attention for engineering water-soluble scaffolds by introducing carboxyl or sulfonic acid groups into the core structure. Exploration of lipophilic host receptors can enhance guest binding in a water medium and can further act as biological precursors for anion transport. Furthermore, such cyclic receptors can also be modified with selective organelle targeting groups for biological events monitoring. Additionally, new macrocyclic systems can be tailored through a straightforward approach using widely available organic synthons.

REFERENCES

Akceylan, E. and M. Yilmaz. 2016. Synthesis of water-soluble calixarenes catalyzed one-pot Mannich-Type reaction in aqueous media. Polycycl. Aromat. Compd. 36(5): 801–816.

Altieri, A., V. Aucagne, R. Carrillo, G.J. Clarkson, D.M. D'Souza, J.A. Dunnett, D.A. Leigh and K.M. Mullen. 2011. Sulfur-containing amide-based[2]rotaxanes and molecular shuttles. Chem. Sci. 2(10): 1922–1928.

Anzenbacher Jr, P., R. Nishiyabu and M.A. Palacios. 2006. N-confused calix[4]pyrroles. Coord. Chem. Rev. 250(23-24): 2929–2938.

Arduini, A., F. Calzavacca, D. Demuru, A. Pochini and A. Secchi. 2004. Synthesis of cavity extended cyclotriveratrylenes. J. Org. Chem. 69(4): 1386–1388.

Artacho, J., E. Ascic, T. Rantanen, C.-J. Wallentin, S. Dawaigher, K.-E. Bergquist, M. Harmata, V. Snieckus and K. Wärnmark. 2012. Tröger's base twisted amides: endo functionalization and synthesis of an inverted crown ether. Org. Lett. 14(18): 4706–4709.

Bejger, C., C.M. Davis, J.S. Park, V.M. Lynch, J.B. Love and J.L. Sessler. 2011. Palladium induced macrocyclic preorganization for stabilization of a tetrathiafulvalene mixed-valence dimer. Org. Lett. 13(18): 4902–4905.

Bergkamp, J.J., S. Decurtins and S.-X. Liu. 2015. Current advances in fused tetrathiafulvalene donor–acceptor systems. Chem. Soc. Rev. 44(4): 863–874.

Beves, J.E., B.A. Blight, C.J. Campbell, D.A. Leigh and R.T. McBurney. 2011. Strategies and tactics for the metal-directed synthesis of rotaxanes, knots, catenanes, and higher order links. Angew. Chem. Int. Ed. Engl. 50(40): 9260–9327.

Beyeh, N.K., J. Aumanen, A. Åhman, M. Luostarinen, H. Mansikkamäki, M. Nissinen, J. Korppi-Tommola and K. Rissanen. 2007. Dansylated resorcinarenes. New. J. Chem. 31(3): 370–376.

Bharadwaj, P.K. 2017. Metal ion binding by laterally non-symmetric macrobicyclic oxa–aza cryptands. Dalton Trans. 46(18): 5742–5775.

Bilensoy, E. 2011. Cyclodextrins in pharmaceutics, cosmetics, and biomedicine: current and future industrial applications. John Wiley & Sons.

Blanco-Gómez, A., P. Cortón, L. Barravecchia, I. Neira, E. Pazos, C. Peinador and M.D. García. 2020. Controlled binding of organic guests by stimuli-responsive macrocycles. Chem. Soc. Rev. 49: 3834–3862

Boubekeur-Lecaque, L., C. Souffrin, G. Gontard, K. Boubekeur and C. Amatore. 2014. Water soluble diaza crown ether derivative: synthesis and barium complexation studies. Polyhedron 68: 191–198.

Brotin, T., V. Roy and J.-P. Dutasta. 2005. Improved synthesis of functional CTVs and cryptophanes using Sc (OTf) 3 as catalyst. J. Org. Chem. 70(16): 6187–6195.

Brotin, T. and J.-P. Dutasta. 2009. Cryptophanes and their complexes--present and future. Chem. Rev. 109(1): 88–130.

Cai, F., J.-S. Shen, J.-H. Wang, H. Zhang, J.-S. Zhao, E.-M. Zeng and Y.-B. Jiang. 2012. Hydrogelators of cyclotriveratrylene derivatives. Org. Biomol. Chem. 10(7): 1418–1423.

Canevet, D., M. Sallé, G. Zhang, D. Zhang and D. Zhu. 2009. Tetrathiafulvalene (TTF) derivatives: key building-blocks for switchable processes. Chem. Commun. (17): 2245–2269.

Chandrakumar, P.K., R. Dhiman, C.E. Woodward, H. Iranmanesh, J.E. Beves and A.I. Day. 2019. Tiara [n] uril: a glycoluril-based macrocyclic host with cationic walls. J. Org. Chem. 84(7): 3826–3831.

Chen, L., H.-Y. Zhang and Y. Liu. 2012. High affinity crown ether complexes in water: thermodynamic analysis, evidence of crystallography and binding of NAD+. J. Org. Chem. 77(21): 9766–9773.

Chen, C.-F. and Y. Han. 2018. Triptycene-derived macrocyclic arenes: from calixarenes to helicarenes. Acc. Chem. Res. 51(9): 2093–2106.

Chen, L. and Y. Liu. 2020. Water-soluble aromatic crown ethers: from molecular recognition to molecular assembly. pp. 3–26. *In*: L. Yu, C. Yong and Z. Heng-Yi [eds]. Handbook of Macrocyclic Supramolecular Assembly. Singapore, Springer.

Chen, Q., L.-P. Yang, D.-H. Li, J. Zhai, W. Jiang and X. Xie. 2021. Potentiometric determination of the neurotransmitter acetylcholine with ion-selective electrodes containing oxatub[4]arenes as the ionophore. Sens. Actuat. B-Chem. 326: 128836.

Chetcuti, M.J., A.M. Devoille, A.B. Othman, R. Souane, P. Thuéry and J. Vicens. 2009. Synthesis of mono-, di-and tetra-alkyne functionalized calix [4] arenes: Reactions of these multipodal ligands with dicobalt octacarbonyl to give complexes which contain up to eight cobalt atoms. Dalton Trans. (16): 2999–3008.

Collet, A., J.-P. Dutasta, B. Lozach and J. Canceill. 1993. Cyclotriveratrylenes and cryptophanes: their synthesis and applications to host-guest chemistry and to the design of new materials. Supramolecular Chemistry I—Directed Synthesis and Molecular Recognition. Springer: 103–129.

Coronado, E., P. Gaviña and S. Tatay. 2009. Catenanes and threaded systems: from solution to surfaces. Chem. Soc. Rev. 38(6): 1674–1689.

Cragg, P.J. and K. Sharma. 2012. Pillar[5]arenes: fascinating cyclophanes with a bright future. Chem. Soc. Rev. 41(2): 597–607.

Cravotto, G., A. Binello, E. Baranelli, P. Carraro and F. Trotta. 2006. Cyclodextrins as food additives and in food processing. Curr. Nutr. Food Sci. 2(4): 343–350.

Crini, G. 2014. A history of cyclodextrins. Chem. Rev. 114(21): 10940–10975.

Crowley, J.D., S.M. Goldup, A.-L. Lee, D.A. Leigh and R.T. McBurney. 2009. Active metal template synthesis of rotaxanes, catenanes and molecular shuttles. Chem. Soc. Rev. 38(6): 1530–1541.

Della Sala, P., R. Del Regno, C. Talotta, A. Capobianco, N. Hickey, S. Geremia, M. De Rosa, A. Spinella, A. Soriente and P. Neri. 2020. Prismarenes: a new class of macrocyclic hosts obtained by templation in a thermodynamically controlled synthesis. J. Am. Chem. Soc. 142(4): 1752–1756.

Demissie, T.B., K. Ruud and J.H. Hansen. 2017. Cryptophanes for methane and xenon encapsulation: a comparative density functional theory study of binding properties and NMR chemical shifts. J. Phys. Chem. A. 121(50): 9669–9677.

Depraetere, S., M. Smet and W. Dehaen. 1999. N-Confused calix[4]pyrroles. Angew. Chem. Int. Ed. 38(22): 3359–3361.

Dhara, A., T. Sadhukhan, E.G. Sheetz, A.H. Olsson, K. Raghavachari and A.H. Flood. 2020. Zero-overlap fluorophores for fluorescent studies at any concentration. J. Am. Chem. Soc. 142(28): 12167–12180.

Dsouza, R.N., U. Pischel and W.M. Nau. 2011. Fluorescent dyes and their supramolecular host/guest complexes with macrocycles in aqueous solution. Chem. Rev. 111(12): 7941–7980.

El Moll, H., D. Sémeril, D. Matt, M.-T. Youinou and L. Toupet. 2009. Synthesis of a resorcinarene-based tetraphosphine-cavitand and its use in Heck reactions. Org. Biomol. Chem. 7(3): 495–501.

Elaieb, F., D. Sémeril, D. Matt, M. Pfeffer, P.-A. Bouit, M. Hissler, C. Gourlaouen and J. Harrowfield. 2017. Calix [4] arene-fused phospholes. Dalton Trans. 46(30): 9833–9845.

Español, E.S. and M.M. Villamil. 2019. Calixarenes: generalities and their role in improving the solubility, biocompatibility, stability, bioavailability, detection, and transport of biomolecules. Biomolecules 9(3): 90.

Evans, N.H. and P.D. Beer. 2014. Progress in the synthesis and exploitation of catenanes since the millennium. Chem. Soc. Rev. 43(13): 4658–4683.

Evans, N.H. 2019. Recent advances in the synthesis and application of hydrogen bond templated rotaxanes and catenanes. Eur. J. Org. Chem. 2019(21): 3320–3343.

Fenyvesi, E., M. Vikmon and L. Szente. 2016. Cyclodextrins in food technology and human nutrition: benefits and limitations. Crit. Rev. Food Sci. Nutr. 56(12): 1981–2004.

Fujimoto, K. 2019. Cyanostar: anion recognition property and modulation of lithium ion conductivity in battery electrolyte. J. Syn. Org. Chem. Jpn. 77(10): 1023–1025.

Gabard, J. and A. Collet. 1981. Synthesis of a (D 3)-bis (cyclotriveratrylenyl) macrocage by stereospecific replication of a (C 3)-subunit. J. Chem. Soc. Chem. Commun. (21): 1137–1139.

Gale, P.A., J.L. Sessler and V. Král. 1998. Calixpyrroles. Chem. Commun.(1): 1–8.

Gao, L., W. Liu, O.-S. Lee, I.J. Dmochowski and J.G. Saven. 2015. Xe affinities of water-soluble cryptophanes and the role of confined water. Chem. Sci. 6(12): 7238–7248.

Gil-Ramírez, G., D.A. Leigh and A.J. Stephens. 2015. Catenanes: fifty years of molecular links. Angew. Chem. Int. Ed. 54(21): 6110–6150.

Gokel, G.W., W.M. Leevy and M.E. Weber. 2004. Crown ethers: sensors for ions and molecular scaffolds for materials and biological models. Chem. Rev. 104(5): 2723–2750.

Gosse, I., J.-P. Chauvet and J.-P. Dutasta. 2005. Synthesis and interfacial properties of amphiphatic cryptophanes. New J. Chem. 29(12): 1549–1554.

Gravett, D.M. and J.E. Guillet. 1993. Synthesis and photophysics of a water-soluble, naphthalene-containing. beta-cyclodextrin. J. Am. Chem. Soc. 115(14): 5970–5974.

Gruber, T., C. Fischer, M. Felsmann, W. Seichter and E. Weber. 2009. Calix [4] arenes featuring a direct lower rim attachment of dansyl groups. Synthesis, fluorescence properties and first report on crystal structures. Org. Biomol. Chem. 7(23): 4904–4917.

Gutsche, C.D. 2008. Calixarenes: An Introduction. Royal Society of Chemistry.

Hamon, F., C. Blaszkiewicz, M. Buchotte, E. Banaszak-Léonard, H. Bricout, S. Tilloy, E. Monflier, C. Cézard, L. Bouteiller and C. Len. 2014. Synthesis and characterization of a new photoinduced switchable β-cyclodextrin dimer. Beilstein J. Org. Chem. 10(1): 2874–2885.

Hapiot, F., S. Tilloy and E. Monflier. 2006. Cyclodextrins as supramolecular hosts for organometallic complexes. Chem. Rev. 106(3): 767–781.

Harada, A., Y. Takashima and H. Yamaguchi. 2009. Cyclodextrin-based supramolecular polymers. Chem. Soc. Rev. 38(4): 875–882.

Hardie, M.J. 2010. Recent advances in the chemistry of cyclotriveratrylene. Chem. Soc. Rev. 39(2): 516–527.

Hardie, M.J. 2016. Self-assembled cages and capsules using cyclotriveratrylene-type scaffolds. Chem. Lett. 45(12): 1336–1346.

Havel, V., J. Svec, M. Wimmerova, M. Dusek, M. Pojarova and V. Sindelar. 2011. Bambus [n] urils: a new family of macrocyclic anion receptors. Org. Lett. 13(15): 4000–4003.

Hernandez-Alonso, D., S. Zankowski, L. Adriaenssens and P. Ballester. 2015. Water-soluble aryl-extended calix [4] pyrroles with unperturbed aromatic cavities: synthesis and binding studies. Org. Biomol. Chem. 13(4): 1022–1029.

Higashi, T., K. Morita, X. Song, J. Zhu, A. Tamura, N. Yui, K. Motoyama, H. Arima and J. Li. 2019. One-pot synthesis of cyclodextrin-based radial poly [n] catenanes. Commun. Chem. 2(1): 1–8.

Huang, Y.-H., X.-Y. Jin, Y.-Y. Zhao, H. Cong and Z. Tao. 2018. A fluorescence-enhanced chemosensor based on multifarene [2, 2] and its recognition of metal cations. Org. Biomol. Chem. 16(29): 5343–5349.

Huang, Y.-H., Q.-M. Ge, Y.-Y. Zhao, H. Cong, J.-L. Zhao, Z. Tao and Q.-Y. Luo. 2019. Recognition of silver cations by multifarene [2, 2] chemosensors with unexpected fluorescence response. Spectrochim. Acta A Mol. Biomol. Spectrosc. 218: 213–220.

Jana, A., M. Ishida, J.S. Park, S. Bähring, J.O. Jeppesen and J.L. Sessler. 2017. Tetrathiafulvalene-(TTF-) derived oligopyrrolic macrocycles. Chem. Rev. 117(4): 2641–2710.

Jana, A., S. Bähring, M. Ishida, S. Goeb, D. Canevet, M. Sallé, J.O. Jeppesen and J.L. Sessler. 2018. Functionalised tetrathiafulvalene-(TTF-) macrocycles: recent trends in applied supramolecular chemistry. Chem. Soc. Rev. 47(15): 5614–5645.

Ji, X., M. Zhang, X. Yan, J. Li and F. Huang. 2013. Synthesis of a water-soluble bis (m-phenylene)-32-crown-10-based cryptand and its pH-responsive binding to a paraquat derivative. Chem. Commun. 49(12): 1178–1180.

Jia, F., Z. He, L.-P. Yang, Z.-S. Pan, M. Yi, R.-W. Jiang and W. Jiang. 2015. Oxatub[4] arene: a smart macrocyclic receptor with multiple interconvertible cavities. Chem. Sci. 6(12): 6731–6738.

Jia, F., H.-Y. Wang, D.-H. Li, L.-P. Yang and W. Jiang. 2016. Oxatub[4]arene: a molecular "transformer" capable of hosting a wide range of organic cations. Chem. Commun. 52(33): 5666–5669.

Jordan, J.H., A. Wishard, J.T. Mague and B.C. Gibb. 2019. Binding properties and supramolecular polymerization of a water-soluble resorcin[4]arene. Org. Chem. Front. 6(8): 1236–1243.

Kaller, M., P. Staffeld, R. Haug, W. Frey, F. Giesselmann and S. Laschat. 2011. Substituted crown ethers as central units in discotic liquid crystals: effects of crown size and cation uptake. Liq. Cryst. 38(5): 531–553.

Kim, K. 2002. Mechanically interlocked molecules incorporating cucurbituril and their supramolecular assemblies. Chem. Soc. Rev. 31(2): 96–107.

Kim, D.-S., V.M. Lynch, K.A. Nielsen, C. Johnsen, J.O. Jeppesen and J.L. Sessler. 2009. A chloride-anion insensitive colorimetric chemosensor for trinitrobenzene and picric acid. Anal. Bioanal. Chem. 395(2): 393–400.

Kim, S.K., D.E. Gross, D.-G. Cho, V.M. Lynch and J.L. Sessler. 2011. N-Tosylpyrrolidine calix[4]pyrrole: synthesis and ion binding studies. J. Org. Chem. 76(4): 1005–1012.

Kim, D.S. and J.L. Sessler. 2015. Calix[4]pyrroles: versatile molecular containers with ion transport, recognition, and molecular switching functions. Chem. Soc. Rev. 44(2): 532–546.

Kohnke, F.H. 2020. Calixpyrroles: from anion ligands to potential anticancer drugs. Eur. J. Org. Chem. (28): 4261–4272.

Krakowiak, K.E., J.S. Bradshaw and D.J. Zamecka-Krakowiak. 1989. Synthesis of aza-crown ethers. Chem. Rev. 89(4): 929–972.

Kumar, S., H. Chawla and R. Varadarajan. 2003. A single step preparation of p-sulphonated calixarenes. Indian J. Chem. 42B: 2863–2865.

Lamb, J., R. Izatt, C. Swain and J. Christensen. 1980. A systematic study of the effect of macrocycle ring size and donor atom type on the log K,. DELTA. H, and T. DELTA. S of reactions at 25. degree. C in methanol of mono-and divalent cations with crown ethers. J. Am. Chem. Soc. 102(2): 475–479.

Lee, J.W., S. Samal, N. Selvapalam, H.-J. Kim and K. Kim. 2003. Cucurbituril homologues and derivatives: new opportunities in supramolecular chemistry. Acc. Chem. Res. 36(8): 621–630.

Lee, S., C.-H. Chen and A.H. Flood. 2013. A pentagonal cyanostar macrocycle with cyanostilbene CH donors binds anions and forms dialkylphosphate[3]rotaxanes. Nat. Chem. 5(8): 704–710.

Li, N., R.G. Harrison and J.D. Lamb. 2014. Application of resorcinarene derivatives in chemical separations. J. Incl. Phenom. Macrocycl. Chem. 78(1-4): 39–60.

Li, J., D. Yim, W.-D. Jang and J. Yoon. 2017. Recent progress in the design and applications of fluorescence probes containing crown ethers. Chem. Soc. Rev. 46(9): 2437–2458.

Lindsey, A. 1965. 316. The structure of cyclotriveratrylene (10, 15-dihydro-2, 3, 7, 8, 12, 13-hexamethoxy-5 H-tribenzo [a, d, g] cyclononene) and related compounds. J. Chem. Soc. (Resumed): 1685–1692.

Liu, Y. and Y. Chen. 2006. Cooperative binding and multiple recognition by bridged bis (β-cyclodextrin) s with functional linkers. Acc. Chem. Res. 39(10): 681–691.

Liu, W., H. Yang, W. Wu, H. Gao, S. Xu, Q. Guo, Y. Liu, S. Xu and S. Cao. 2016a. Calix[4]resorcinarene-based branched macromolecules for all-optical photorefractive applications. J. Mater. Chem. C. 4(45): 10684–10690.

Liu, Y., A. Singharoy, C.G. Mayne, A. Sengupta, K. Raghavachari, K. Schulten and A.H. Flood. 2016b. Flexibility coexists with shape-persistence in cyanostar macrocycles. J. Am. Chem. Soc. 138(14): 4843–4851.

Liu, Z., S.K.M. Nalluri and J.F. Stoddart. 2017. Surveying macrocyclic chemistry: from flexible crown ethers to rigid cyclophanes. Chem. Soc. Rev. 46(9): 2459–2478.

Lizal, T. and V. Sindelar. 2018. Bambusuril anion receptors. Isr. J. Chem. 58(3-4): 326–333.

Luo, H., Y.Y. Zhao, X.Y. Jin, J.M. Yang, H. Cong, Q.M. Ge, L. Sun, M. Liu and Z. Tao. 2020. Voltammetric detection of catechol and dopamine based on a supramolecular composite prepared from multifarene [3, 3] and reduced graphene oxide. Electroanalysis 32(7): 1449–1458.

Ma, Y., X. Ji, F. Xiang, X. Chi, C. Han, J. He, Z. Abliz, W. Chen and F. Huang. 2011. A cationic water-soluble pillar[5]arene: synthesis and host–guest complexation with sodium 1-octanesulfonate. Chem. Commun. 47(45): 12340–12342.

McPhee, M.M. and S.M. Kerwin. 1996. Synthesis and metal ion binding studies of enediyne-containing crown ethers. J. Org. Chem. 61(26): 9385–9393.

Mishra, A., N. Mishra and V.K. Tiwari. 2016. Synthesis of novel macrocyclic crown ethers from D-glucose. J. Carbohydr. Chem. 35(4): 238–248.

Mondal, P., E. Solel, N. Fridman, E. Keinan and O. Reany. 2019. Intramolecular van der waals interactions challenge anion binding in perthio-bambusurils. Chem. Eur. J. 25(58): 13336–13343.

Nielsen, K.A., W.-S. Cho, J.O. Jeppesen, V.M. Lynch, J. Becher and J.L. Sessler. 2004. Tetra-TTF calix[4]pyrrole: a rationally designed receptor for electron-deficient neutral guests. J. Am. Chem. Soc. 126(50): 16296–16297.

Nielsen, M.B. 2013. Organic Synthesis and Molecular Engineering. Wiley Online Library.

Nishiyabu, R., M.A. Palacios, W. Dehaen and P. Anzenbacher. 2006. Synthesis, structure, anion binding, and sensing by calix[4]pyrrole isomers. J. Am. Chem. Soc. 128(35): 11496–11504.

Ogoshi, T., S. Kanai, S. Fujinami, T.-a. Yamagishi and Y. Nakamoto. 2008. para-Bridged symmetrical pillar[5]arenes: their Lewis acid catalyzed synthesis and host–guest property. J. Am. Chem. Soc. 130(15): 5022–5023.

Ogoshi, T., T. Aoki, K. Kitajima, S. Fujinami, T.-a. Yamagishi and Y. Nakamoto. 2011. Facile, rapid, and high-yield synthesis of pillar[5]arene from commercially available reagents and its X-ray crystal structure. J. Org. Chem. 76(1): 328–331.

Ogoshi, T., T.-a. Yamagishi and Y. Nakamoto. 2016. Pillar-shaped macrocyclic hosts pillar[n] arenes: new key players for supramolecular chemistry. Chem. Rev. 116(14): 7937–8002.

Parvari, G., S. Annamalai, I. Borovoi, H. Chechik, M. Botoshansky, D. Pappo and E. Keinan. 2014. Multifarenes: new modular cavitands. Chem. Commun. 50(19): 2494–2497.

Pedersen, C.J. 1967. Cyclic polyethers and their complexes with metal salts. J. Am. Chem. Soc. 89(26): 7017–7036.

Pedersen, C.J. 1988. The discovery of crown ethers (Noble Lecture). Angew. Chem. Int. Ed. 27(8): 1021–1027.

Pederson, A.M.-P., T.L. Price Jr, C. Slebodnick, D.V. Schoonover and H.W. Gibson. 2017. The long and the short of it: regiospecific syntheses of isomers of dicarbomethoxydibenzo-27-crown-9 and binding abilities of their pyridyl cryptands. J. Org. Chem. 82(16): 8489–8496.

Peyrard, L., M.-L. Dumartin, S. Chierici, S. Pinet, G. Jonusauskas, P. Meyrand and I. Gosse. 2012. Development of functionalized cyclotriveratrylene analogues: introduction of withdrawing and π-conjugated groups. J. Org. Chem. 77(16): 7023–7027.

Piątek, P., V.M. Lynch and J.L. Sessler. 2004. Calix[4]pyrrole [2] carbazole: a new kind of expanded calixpyrrole. J. Am. Chem. Soc. 126(49): 16073–16076.

Qiao, B., J.R. Anderson, M. Pink and A.H. Flood. 2016. Size-matched recognition of large anions by cyanostar macrocycles is saved when solvent-bias is avoided. Chem. Commun. 52(56): 8683–8686.

Qiu, F., Y.-H. Huang, Q. Ge, M. Liu, H. Cong and Z. Tao. 2020. The high selective chemo-sensors for TNP based on the mono-and di-substituted multifarene [2, 2] with different fluorescence quenching mechanism. Spectrochim. Acta A Mol. Biomol. Spectrosc. 226: 117583.

Ravi, A., P.S. Krishnarao, T.A. Shumilova, V.N. Khrustalev, T. Rüffer, H. Lang and E.A. Kataev. 2018. Cation molecular exchanger based on a conformational hinge. Org. Lett. 20(19): 6211–6214.

Rivollier, J., P. Thuéry and M.-P. Heck. 2013. Extension of the bambus[n]uril family: microwave synthesis and reactivity of allylbambus[n]urils. Org. Lett. 15(3): 480–483.

Roberts, B.A., G.W. Cave, C.L. Raston and J.L. Scott. 2001. Solvent-free synthesis of calix[4]resorcinarenes. Green Chem. 3(6): 280–284.

Ryu, E.-H. and Y. Zhao. 2005. Efficient synthesis of water-soluble calixarenes using click chemistry. Org. Lett. 7(6): 1035–1037.

Salorinne, K. and M. Nissinen. 2009. Twisting of the resorcinarene core due to solvent effects upon crystallization. Cryst. Eng. Comm. 11(8): 1572–1578.

Satha, P., G.T. Illa, S. Hazra and C.S. Purohit. 2017. Syn/anti pair of triazole-bridged cryptophanes: synthesis, characterization with crystal structures. Chem. Select 2(33): 10699–10703.

Sauvage, J.-P. 1998. Transition metal-containing rotaxanes and catenanes in motion: toward molecular machines and motors. Acc. Chem. Res. 31(10): 611–619.

Schaly, A., Y. Rousselin, J.C. Chambron, E. Aubert and E. Espinosa. 2016. The stereoselective self-assembly of chiral metallo-organic cryptophanes. Eur. J. Inorg. Chem. 2016(6): 832–843.

Segura, J.L. and N. Martín. 2001. New concepts in tetrathiafulvalene chemistry. Angew. Chem. Int. Ed. 40(8): 1372–1409.

Shen, Y., L. Zou and Q. Wang. 2017. Template-directed synthesis of cucurbituril analogues using propanediurea as a building block. New J. Chem. 41(16): 7857–7860.

Sliwa, W. and C. Kozlowski. 2009. Calixarenes and resorcinarenes: synthesis, properties and applications. John Wiley & Sons.

Sokolov, J., A. Štefek and V. Šindelář. 2020. Functionalized chiral bambusurils: synthesis and host-guest interactions with chiral carboxylates. ChemPlusChem. 85(6): 1307–1314.

Solel, E., M. Singh, O. Reany and E. Keinan. 2016. Enhanced anion binding by heteroatom replacement in bambusurils. Phys. Chem. Chem. Phys. 18(19): 13180–13185.

Steed, J.W. 2001. First-and second-sphere coordination chemistry of alkali metal crown ether complexes. Coord. Chem. Rev. 215(1): 171–221.

Sung, N.K., D.W. Cho, J.H. Choi, K.W. Choi, U.C. Yoon, H. Maeda and P.S. Mariano. 2007. A facile approach to the preparation of bis-crown ethers based on SET-Promoted photomacrocyclization reactions. J. Org. Chem. 72(23): 8831–8837.

Swamy, P.C.A., E. Solel, O. Reany and E. Keinan. 2018. Synthetic evolution of the multifarene cavity from planar predecessors. Chem. Eur. J. 24(57): 15319–15328.

Takeda, Y. and O. Arima. 1985. Temperature dependence of walden product of 18-crown-6–k+ complex in water. Bull. Chem. Soc. Jpn. 58(11): 3403–3404.

Tero, T.-R., K. Salorinne and M. Nissinen. 2012. The effect of halogen bonding on the packing of bromine-substituted pyridine and benzyl functionalized resorcinarene tetrapodands in the solid state. Cryst. Eng. Comm 14(21): 7360–7367.

Tian, C., S.D. Fielden, G.F. Whitehead, I.J. Vitorica-Yrezabal and D.A. Leigh. 2020. Weak functional group interactions revealed through metal-free active template rotaxane synthesis. Nature Commun. 11(1): 1–10.

Wang, J.-Q., J. Li, G.-W. Zhang and C.-F. Chen. 2018. A route to enantiopure (O-methyl) 6-2, 6-helic [6] arenes: synthesis of hexabromo-substituted 2, 6-helic [6] arene derivatives and their Suzuki–Miyaura coupling reactions. J. Org. Chem. 83(19): 11532–11540.

Wieser, C., C.B. Dieleman and D. Matt. 1997. Calixarene and resorcinarene ligands in transition metal chemistry. Coord. Chem. Rev. 165: 93–161.

Wright, N.E., A.M. El-Sohly and S.A. Snyder. 2014. Syntheses of cyclotriveratrylene analogues and their long elusive triketone congeners. Org. Lett. 16(14): 3644–3647.

Wudl, F., G. Smith and E. Hufnagel. 1970. Bis-1,3-dithiolium chloride: an unusually stable organic radical cation J. Chem. Soc. D: 1453–1454

Yang, K., Y. Pei, J. Wen and Z. Pei. 2016a. Recent advances in pillar [n] arenes: synthesis and applications based on host-guest interactions. Chem. Commun. 52(60): 9316–9326.

Yang, L.-P., W.-E. Liu and W. Jiang. 2016b. Naphthol-based macrocyclic receptors. Tetrahedron Lett. 57(36): 3978–3985.

Yang, L.P. and W. Jiang. 2020. Prismarene: an emerging naphthol-based macrocyclic arene. Angew. Chem. Int. Ed. 59(37): 15794–15796.

Yawer, M.A., V. Havel and V. Sindelar. 2015. A bambusuril macrocycle that binds anions in water with high affinity and selectivity. Angew. Chem. 127(1): 278–281.

Yawer, M.A., K. Sleziakova, L. Pavlovec and V. Sindelar. 2018. Bambusurils bearing nitro groups and their further modifications. Eur. J. Org. Chem. 2018(1): 41–47.

Yu, G., M. Xue, Z. Zhang, J. Li, C. Han and F. Huang. 2012. A water-soluble pillar [6] arene: synthesis, host–guest chemistry, and its application in dispersion of multiwalled carbon nanotubes in water. J. Am. Chem. Soc. 134(32): 13248–13251.

Yu, G., K. Jie and F. Huang. 2015. Supramolecular amphiphiles based on host–guest molecular recognition motifs. Chem. Rev. 115(15): 7240–7303.

Zadmard, R., F. Hokmabadi, M.R. Jalali and A. Akbarzadeh. 2020. Recent progress to construct calixarene-based polymers using covalent bonds: synthesis and applications. RSC Adv. 10(54): 32690–32722.

Zahran, E.M., E.M. Fatila, C.-H. Chen, A.H. Flood and L.G. Bachas. 2018. Cyanostar: C–H Hydrogen Bonding Neutral Carrier Scaffold for Anion-Selective Sensors. Anal. Chem. 90(3): 1925–1933.

Zhang, H. and J.L. Atwood. 1990. Crystal and molecular structure of cyclotriveratrylene. J. Chem. Crystallogr. 20(5): 465–470.

Zhang, Z., B. Xia, C. Han, Y. Yu and F. Huang. 2010. Syntheses of copillar [5] arenes by co-oligomerization of different monomers. Org. Lett. 12(15): 3285–3287.

Zhang, J. and P.X. Ma. 2013. Cyclodextrin-based supramolecular systems for drug delivery: recent progress and future perspective. Adv. Drug Deliv. Rev. 65(9): 1215–1233.

Zhang, G.-W., Y. Han, Y. Han, Y. Wang and C.-F. Chen. 2017. Synthesis of a water-soluble 2, 6-helic[6]arene derivative and its strong binding abilities towards quaternary phosphonium salts: an acid/base controlled switchable complexation process. Chem. Commun. 53(75): 10433–10436.

Zhao, F., E. Repo, D. Yin, L. Chen, S. Kalliola, J. Tang, E. Iakovleva, K.C. Tam and M. Sillanpää. 2017a. One-pot synthesis of trifunctional chitosan-EDTA-β-cyclodextrin polymer for simultaneous removal of metals and organic micropollutants. Sci. Rep. 7(1): 1–14.

Zhao, W., B. Qiao, C.H. Chen and A.H. Flood. 2017b. High-fidelity multistate switching with anion–anion and acid–anion dimers of organophosphates in cyanostar complexes. Angew. Chem. 129(42): 13263–13267.

Zhao, Y.-Y., H. Li, Q.-M. Ge, H. Cong, M. Liu, Z. Tao and J.-L. Zhao. 2020. A chemo-sensor constructed by nanohybrid of multifarene [3, 3] and rGO for serotonin hydrochloride with dual response in both fluorescence and voltammetry. Microchem. J. 158: 105145.

start chapt. 2

Chapter 2

Macrocyclic Receptors for Environmentally Sensitive Metal Ions

INTRODUCTION

Metals exist almost everywhere in nature and play significant roles in many industrial, biological and environmental processes. For example, metals find widespread industrial applications due to their unique properties such as high tensile strength, lightness, good conduction of heat and electricity etc. Metals such as magnesium, iron, cobalt, copper, manganese and zinc are essential for numerous enzymatic reactions and biological processes in living organisms. Though these metals are involved in important cellular functions, yet their concentration has a significant impact on the health of an organism. If the concentration range of metals is within the permissible limits then they are considered safe. However, if they are present in high concentrations beyond the toxicity range, then this leads to toxic effects. On the contrary, some highly toxic heavy metals like mercury, lead, chromium, arsenic and cadmium are capable of causing serious harmful effects on humans and the environment even at lower concentrations (Gumpu et al. 2015).

While industrial advancements have led to the manufacture of useful products and technologies, but at the cost of more use of chemicals. The ecological contamination as a result of the release of cationic and anionic pollutants due to industrial as well as farming practices is responsible for environmental pollution and increases the risk of diseases in humans, plants and animals. Consequently, there is a growing need for specific and sensitive identification of ionic species in health and environmental applications.

Electrochemical methods like voltammetry, conductometry and potentiometry have acted as useful tools for metal ion recognition. Moreover, various sophisticated conventional analytical methods, such as UV-Vis spectrometry, atomic absorption spectrometry, inductively coupled plasma spectrometry, X-ray fluorescence spectrometry and mass spectrometry have been successfully used to detect metal ions. These methods offer high selectivity, precision, sensitivity and fast recognition of metal ions (Eddaif et al. 2019). However, they suffer from significant disadvantages, for instance, expensive instruments, challenging and time-consuming sample pre-treatment and a complex analytical operation (Dai et al. 2018). Nevertheless, the development of rapid real-time monitoring methods for selective sensing of specific metal ions in biological and environmental contexts poses a daunting challenge.

To overcome the limitations mentioned above, recent years have witnessed unprecedented activity in the design and synthesis of chemosensors for the selective detection of biologically and environmentally active ionic species including waste management. Versatile and efficient chemosensors based on carbon nanotubes, polymers, nanoparticles and quantum dots have been developed. Chemosensors offer a simple, low-cost practical and convenient recognition of cationic and anionic analytes with high selectivity and sensitivity through unique host-guest interactions such as van der Waals forces, hydrogen bonding, metal-ligand binding and electrostatic attraction. Metal ion chemosensors are composed of a receptor, metal binding unit and a signaling unit. They exhibit a high binding affinity selectively towards specific cations accompanied by a simultaneous change in one or more characteristics of the system, such as color, redox potentials, enhancement or quenching of absorbance or fluorescence. The stability of the metal complex largely depends on the cavity size of a macrocyclic host molecule. Among the various types of chemosensors, chromogenic chemosensors are especially interesting because they allow naked-eye detection of metal ion without the use of expensive equipment (Kaur and Kumar 2011, Zor et al. 2013)

In the few past decades, extensive research efforts have focused on the design and synthesis of powerful and highly selective supramolecular assemblies for metal ions in a variety of applications, for example, chemosensing, ion transport, extraction, imaging, catalysis, metalloenzyme mimics and radioactive waste management. Over the last 30-40 years, a wealth of synthetic macrocyclic receptor molecules based on several classes of macrocyclic ligands such as crown ethers, lariat crown ethers, cyclam, cyclen, cryptands, cyclophanes, podands, cyclodextrins, calix[n]arenes, cucurbit[n]urils, pillar[n]arenes and their functionalized derivatives have been strategically developed and extensively investigated for complexation with various metal ions.

Innumerable macrocyclic receptor complexes with metal ions have been reported in research literature. In this chapter, however, the application of macrocyclic chemosensors for the detection of metal ions is summarized and prospected at length. Various macrocyclic host molecules and their functionalized derivatives are described primarily with respect to their metal ion complexation ability, binding affinity, selectivity, sensitivity, spectroscopic properties, including others such as extraction ability, ion transport and redox nature. Consequently, only selected examples from a broad range of macrocyclic receptors and their complexes with alkali metal ions, alkaline earth metal ions, transition and heavy metal ions and lanthanide and actinide ions are presented.

MACROCYCLIC RECEPTORS FOR ALKALI METAL IONS

Among the alkali metal ions, Na^+, K^+ and Cs^+ are of particular interest for the design and synthesis of macrocyclic receptors. Both sodium and potassium ions are ubiquitous in nature and play a vital role in many biological processes (Crichton 2008; Kaim et al. 2014) and are important in clinical analysis. For instance, sodium controls the blood pressure and regulates the pH in blood plasma and its imbalance in the body leads to hypertension. The Na^+/K^+-ATPase pump maintains the concentration of these ions across the cell membrane. Macrocyclic chemosensors for K^+ have gained much attention in analytical biochemistry for their rapid and reasonably-priced detection in blood plasma and urine (Leray et al. 2002). However, the ability of a macrocycle to selectively detect K^+ without the interference from the large excess of Na^+ ($Na^+/K^+ \approx 30$) under typical physiological conditions remains a challenge (Xia et al. 2002). Likewise, selective recognition of the smallest and lightest alkali metal ion, lithium is quite a task in analytical chemistry (Kothur et al. 2016) On the other hand, Cs^+ is involved with radioactive waste materials and its probable leakage from nuclear waste storage facilities, where it must be selectively detected in the usual presence of large excess of Na^+ and K^+ ions (Izatt et al. 1995). The toxicity of cesium arises because of its capacity to substitute potassium ions in muscles and red blood cells (Souchon et al. 2006).

Macrocyclic Receptors for Li$^+$ Ions

A novel chiral calix[4](azoxa)crown-7 **1** was synthesized and characterized by Sirit and coworkers. The metal recognition properties of the compound were examined by solvent extraction experiments. The picrate extraction studies showed that the ligand was highly selective and effective in transferring Li^+ ions to the organic phase over other alkali and transition metal ions (Sirit et al. 2005).

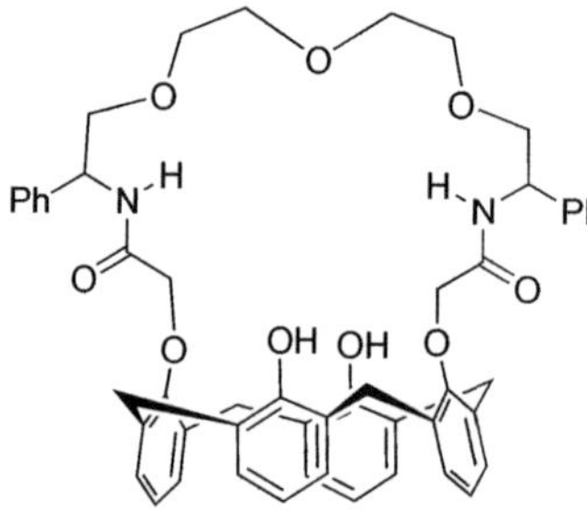

1 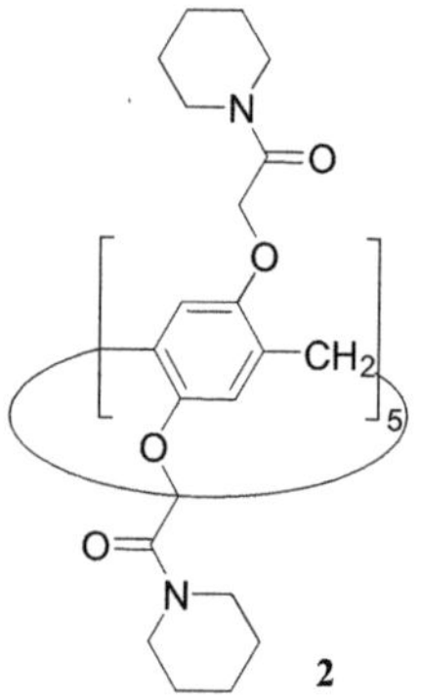   **2**

Kothur et al. synthesized and characterized a novel copillar[5]arene incorporating alkylthiol substituents. The macrocycle was attached to gold electrodes. The authors used cyclic voltammetry to characterize the capacitive signal on alkali metal ions solution (10 mM) prepared from their PF_6^- salts. The electrochemical response of the macrocycle showed high selectivity for Li^+ over other alkali metal ions (Kothur et al. 2016).

Similarly, alkali metal ions complexation behavior with pillararene receptors in methanol was examined using the UV-Vis technique by Yakimova and coworkers. Three pillar[5]arenes **2-4** containing piperidine, glycine, glycylglycine units, respectively were prepared. The three receptors were found to bind with alkali metal ions with 1:1 stoichiometry with the most effective binding efficiency for Li^+ ions over other alkali and transition metal ions. The complex results revealed that the receptor 4 bearing long glycylglycine fragments was the most efficient receptor towards all the alkali metal ions tested (Yakimova et al. 2017).

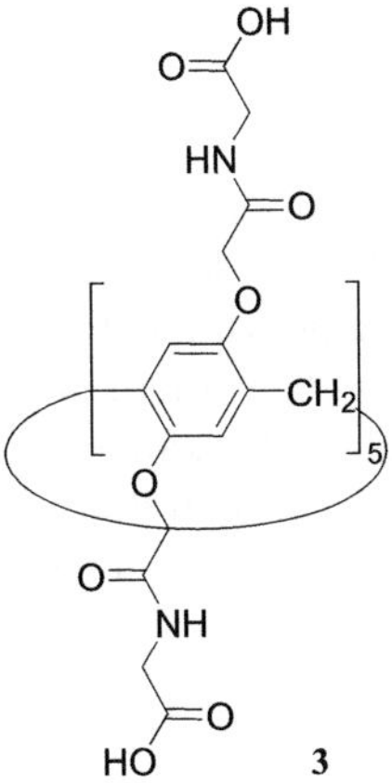

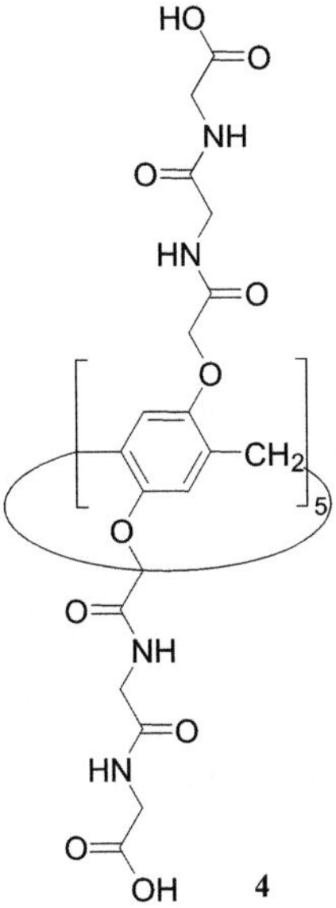

Macrocyclic Receptors for Na$^+$ Ions

Adam and coworkers synthesized a series of cryptands incorporating a common N_2O_6 donor set and explored their ability to complex sodium ions. They used computational and NMR studies to investigate the conformational aspects that influenced sodium ion complexes within the ligand series. A comparison of the solvent extraction experiments of sodium picrate from an aqueous to a chloroform phase by incorporating the ligands indicated that the relatively flexible ligand **5** was the most effective extractant for Na$^+$ ions (Adam et al. 2001).

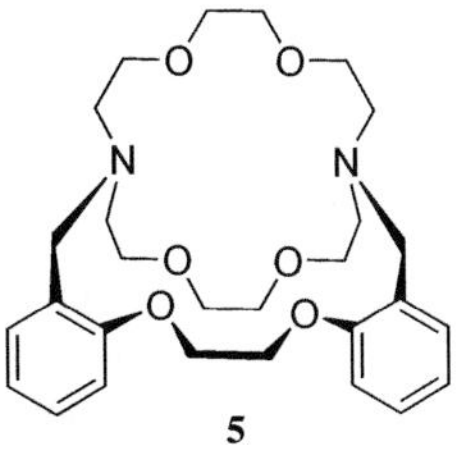

Ueno et al. prepared a fluorescent chemosensor N^α-Dansyl-L-lysine-β-cyclodextrin **6** incorporated with a monesin unit. The antibacterial hydrophobic substance, monensin forms a ring-like conformation when it binds with sodium ion at its center. As compared to other alkali metal ions, the dansylated CD with a hydrophobic cap was highly responsive to sodium ions (1.0 M) while exhibiting a strong fluorescence (Ueno et al. 1999).

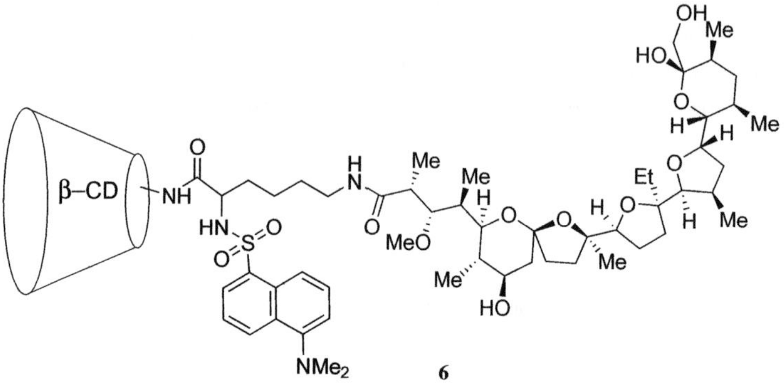

6

A novel highly selective fluorescent chemosensor **7** for Na$^+$ ions was synthesized and characterized by Wang and coworkers. The macrocycle was based on pyrene-modified calix[4]arene at the upper rim and tetraester ionophores present at the lower rim. Fluorescence spectroscopy was used to investigate the chemosensing behavior of the ligand for alkali and alkaline earth metal ions in DMSO/acetonitrile (1/9) at 2.0 x10^{-6} mol L^{-1} concentration. According to the authors, the complexation-induced conformational change in the calixarene moiety was responsible for the fluorescent chemosensing process (Wang et al. 2009).

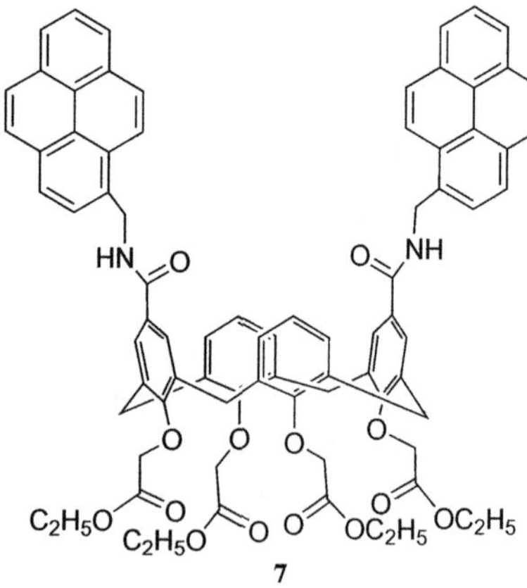

7

Macrocyclic Receptors for K$^+$ Ions

Ehala and coworkers applied Affinity Capillary Electrophoresis (ACE) to quantitatively evaluate and characterize the noncovalent binding interactions between benzo-18-crown-6-ether **8** and alkali metal ions in a 50% methanol-water binary solvent system. The receptor was highly selective to K$^+$ ion over other alkali metal ions and bound most strongly to it (Ehala et al. 2010).

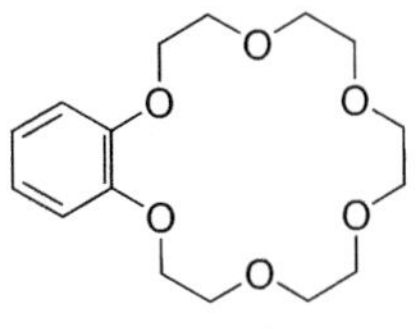

8

A series of novel 1,3-alternate calix[4]arene azacrown ethers were synthesized and explored by Kim et al. using solvent extraction, bulk liquid membrane and ^{1}H NMR studies. These compounds were found to be highly selective extractants for potassium ions over other alkali metal ions (Kim et al. 2000).

Malval and coworkers reported the synthesis of a fluorescent sensor Calix-Bodipy **9** by associating 1,3 alternate calix[4]bisazacrown-5 ionophore with substituted boron-dipyrromethene dyes as fluorophores. The metal binding ability of the fluoroionophore was examined for various alkali and alkaline earth metal ions in acetonitrile and ethanol. The compound was highly sensitive and selective for K$^+$ ions over other metal ions that were tested. It was suggested that the size of the two binding sites on the ionophore fits perfectly the K$^+$ ions (Malval et al. 2005).

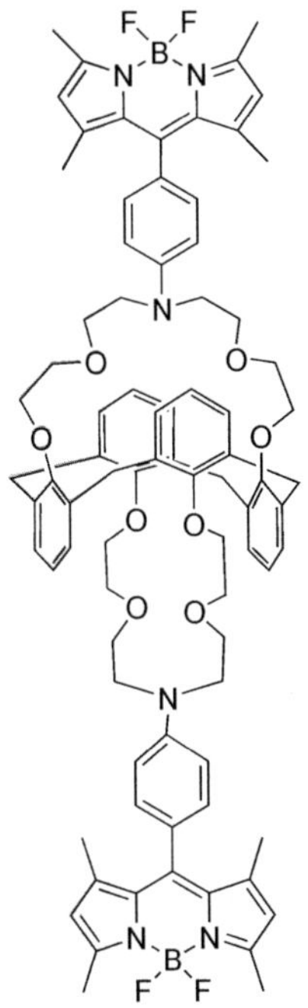

9 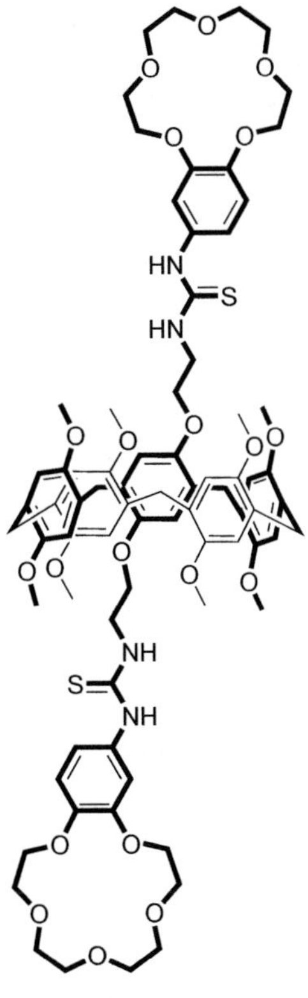   **10** 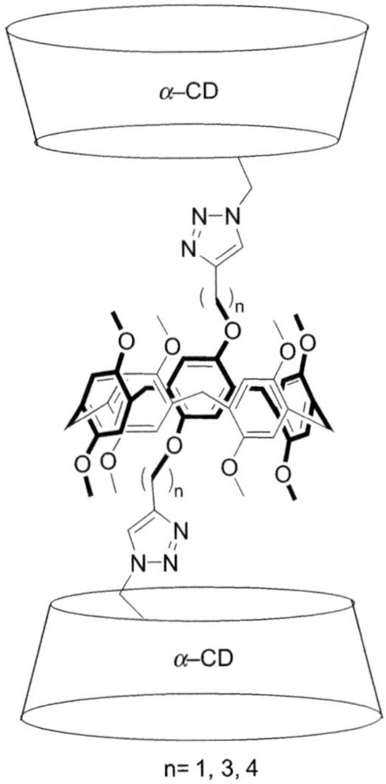   **11**

Feng et al. reported the synthesis of bis(benzo-15-crown-5-ether-ureido)-pillar[5]arene **10** with two benzo-15-crown-5 cation binding sites. On spontaneous insertion of the compound into lipid bilayers, a novel artificial selective K^+ channel was formed. These channels were able to transport the medium-sized K^+ and Rb^+ ions with high transport rates while the Li^+ (small), Na^+ (fit), and Cs^+ (large) ions were transported rather slowly. The results showed significant selectivity towards K^+ ions over Na^+ ions (Feng et al. 2017).

A new class of artificial channels for K^+ ions based on α-cyclodextrin-pillar[5] arene hybrid molecules **11** were designed and synthesized by Xin and coworkers. The channels were efficiently inserted into the lipid bilayers and examined for ion selectivity by fluorescence and electrophysiological experiments. Interestingly, the channels' cation transport selectivity could be altered by varying the length of the linkers. Though the relative selectivity sequences of the channels towards alkali metal ions were different, they exhibited high selectivity for K^+ over Na^+ ions (Xin et al. 2019).

Macrocyclic Receptors for Cs^+ Ions

Azadbakht and Khanabadi synthesized and characterized a novel nano-fluorescent sensor for cesium ions based on naphthalene macrocyclic derivative **12**. The fluorescent behavior of the ligand in an aqueous buffer solution was investigated by emission spectrophotometry. The chemosensor exhibited high sensitivity and selectivity for cesium ions in the presence of Na^+, K^+, Mg^{2+}, Ca^{2+}, Al^{3+}, Pb^{2+}, Cr^{3+}, Mn^{2+}, Cu^{2+}, Zn^{2+} and Hg^{2+} with a concentration up to 100 mM. The Job's plot method to determine the stoichiometry of the complex indicated that only a 1:1 complex was formed (Azadbakht and Khanabadi 2015).

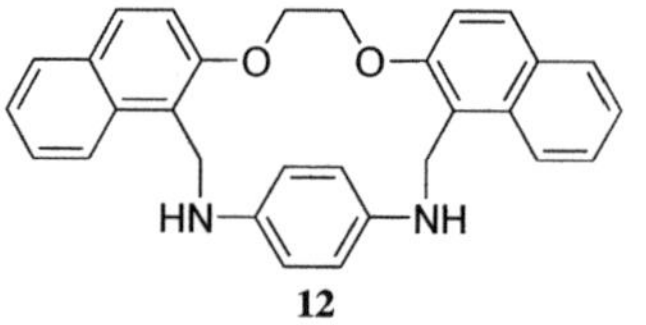

12 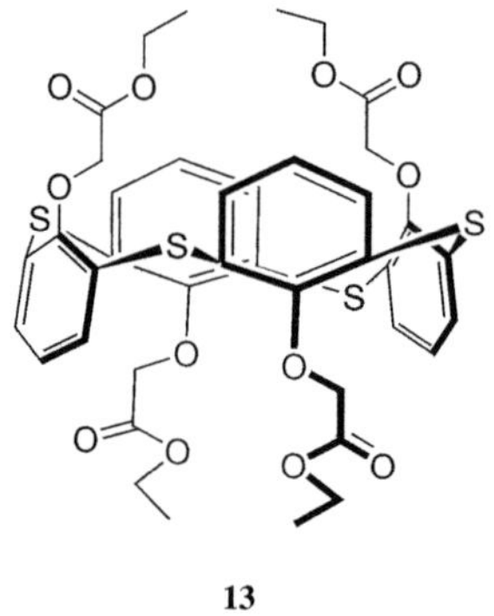   13

Akgemci and coworkers used cyclic voltammetry to examine the transfer of alkali metal ions facilitated by 25,26,27,28-tetraethoxycarbonylmethoxyth iacalix[4]arene **13** as a carrier across the water/1,2-dichloroethane interface. The ionophore was found to be highly selective towards Cs^+ ions. The authors concluded that complexes with metal ions is influenced not only by the size/shape-fit but also the calixarene moiety's conformation (Guler Akgemci et al. 2008).

Calix[4]crown-6 ethers with 1,3-alternate conformation is known to bind Cs^+ ions with notable strength and selectivity (Ji et al. 2000). Ji et al. synthesized two new highly selective fluorescent probes 1,3-alternate di-deoxygenated calix[4](9-cyano-10-anthrylmethyl)benzocrown-6, **14** and 1,3-alternate calix[4]

(9-cyano-10-anthrylmethyl)benzocrown-6, **15** for cesium ions. The emission behavior of both the macrocycles was studied in CH_2Cl_2–MeOH (1:1) in the presence of other alkali metal ions. The fluorescence sensor **14** was found to exhibit a significantly greater fluorescence turn-on response to Cs^+ as compared to sensor **15** (Ji et al. 2000).

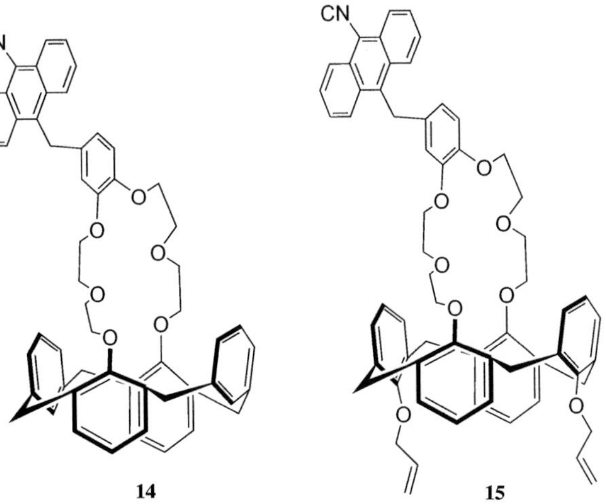

14 **15**

Leray and coworkers synthesized two new fluorescent sensors based on calix[4]biscrowns by incorporating a dioxycoumarin fluorophore into one crown or both crowns, leading to **16**, **17**. The binding properties of the ligands were investigated by carrying out fluorimetric titrations with Cs^+, K^+ and Na^+ ions in ethanol and acetonitrile. Both the sensors exhibited excellent selectivity towards Cs^+ ions. Due to the poorer fit of Na^+ into the crown cavity of the ligands, the selectivity towards K^+ is better than Na^+ ions and quite promising for practical applications (Leray et al. 2002).

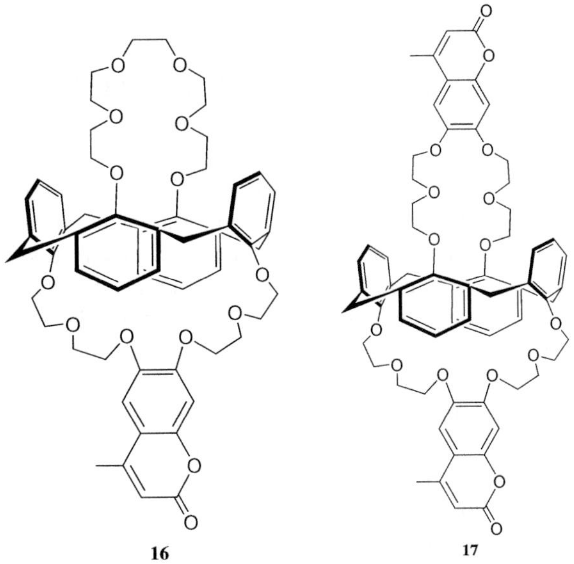

16 **17**

  Souchon et al. prepared a highly water-soluble tetrasulphonated 1,3-alternate calix[4]arene-bis(crown-6-ether) with dioxycoumarin fluorophore inserted in both crowns **18** for practical applications. The compound was found to be highly selective for Cs ions and showed enhancement of fluorescence intensity in the presence of alkali and alkaline earth metal ions (Souchon et al. 2006).

  Li et al. reported the synthesis of triazole-modified calix[4]diethylester **19** which exhibited marked selectivity for Cs^+ ions over other alkali metal ions. Picrate extraction experiments were conducted and the extraction ratio Cs^+/Na^+ of the ligand was found to be 10.52. Results indicated that the triazole groups influenced the extraction ability and selectivity of the ligand (Li et al. 2010).

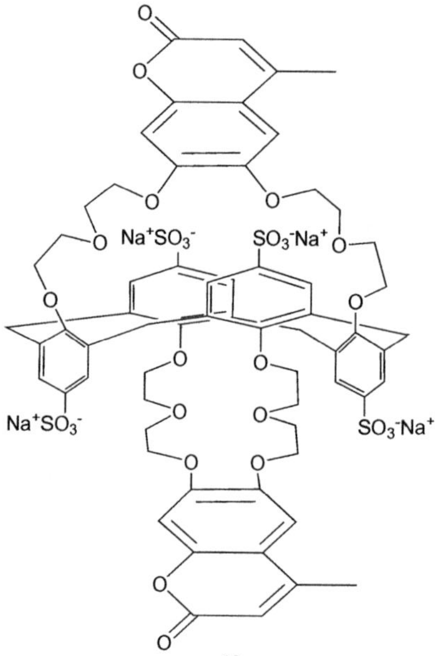

18 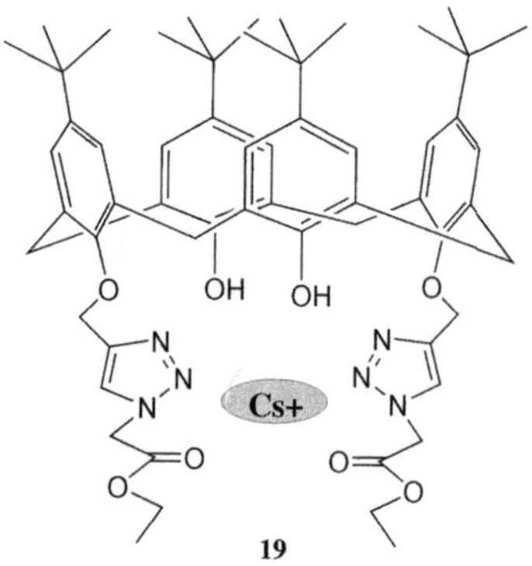   19

MACROCYCLIC RECEPTORS FOR ALKALINE EARTH METAL IONS

The development of macrocyclic host molecules for alkaline earth metals has primarily focused on Mg^{2+}, Ca^{2+}, $Sr^{2+,}$ and Ba^{2+} ions. These metal ions, magnesium and calcium, in particular, are associated with very essential chemical life processes. The divalent magnesium is the most abundant cation in living cells and plays vital roles in many cellular processes, neuromuscular transmission and functioning of ATPase enzyme and nucleotides (Crichton 2008, Kaim et al. 2014). It is also widely used in various industrial processes and aerospace applications. Hence the fabrication of macrocyclic chemosensors for efficient recognition of magnesium in clinical, biological, industrial and environmental applications has gained significant attention. However, designing chemical sensors to distinguish Mg^{2+} from other physiologically abundant sodium, potassium and calcium ions is a major challenge (Song et al. 2007).

Table 2.1 Macrocyclic receptors for alkali metal ions

Macrocyclic Receptor based on	Selectivity for	Application	Selectivity over	Reference
calix[4](azoxa)crown-7	Li^+	Extraction	Na^+, K^+, Cs^+, Hg^{2+}, Pb^{2+}, Cu^{2+}, Co^{2+}, Cd^{2+}, Ni^{2+}	(Sirit et al. 2005)
copillar[5]arene dithiols	Li^+	Voltametric sensor	Na^+, K^+, Rb^+, Cs^+	(Kothur et al. 2016)
pillar[5]arene	Li^+	Receptor	Na^+, K^+, Cs^+, Ag^+, Cu^{2+}, Fe^{3+}	(Yakimova et al. 2017)
cryptand with an N_2O_6 donor set	Na^+	Extractant	----	(Adam et al. 2001)
dansylated cyclodextrin	Na^+	Fluorescent sensor	Li^+, K^+, Cs^+	(Ueno et al. 1999)
pyrene modified calix[4] arene	Na^+	Fluorescent sensor	Li^+, K^+, Cs^+, Mg^{2+}, Ca^{2+}, Sr^{2+}, Ba^{2+}	(Wang et al. 2009)
15-crown-5-ether	Na^+	Luminescence sensor	-----	(Li and Wong 2002)
lariat ether	Na^+, K^+	Receptor	-----	(Gokel et al. 2002)
benzo-18-crown-6-ether	K^+	Receptor	Li^+, Na^+, Rb^+, Cs^+	(Ehala et al. 2010)
1,3-alternate calix[4] arene azacrown	K^+	Extractant	Li^+, Na^+, Rb^+, Cs^+	(Kim et al. 2000)
1,3 alternate calix[4] azacrown-5	K^+	Fluoroiono-phore	Na^+, K^+, Cs^+, Ca^{2+}, Ba^{2+}	(Malval et al. 2005)
Bis-15-crown-5-ether-pillar[5]arene	K^+	Ion transport	Na^+	(Feng et al. 2017)
α-CD-pillar[5]arene	K^+	Ion transport	Na^+	(Xin et al. 2019)
calix[4]tubes	K^+	Ionophore	Li^+, Na^+, Cs^+, Rb^+, Ba^{2+}	(Matthews et al. 2002)
pillar[5]arene	K^+	Ion transport	----	(Xin et al. 2017)
naphthalene based derivative	Cs^+	Fluorescent sensor	Na^+, K^+, Mg^{2+}, Ca^{2+}, Al^{3+}, Pb^{2+}, Cr^{3+}, Mn^{2+}, Cu^{2+}, Zn^{2+}, Hg^{2+}	(Azadbakht and Khanabadi 2015)
thiacalix[4]arene	Cs^+	Ionophore	Na^+, K^+, Rb	(Guler Akgemci et al. 2008)
1,3-alternate calix[4] benzocrown-6	Cs^+	Fluorescent probe	K^+, Rb^+	(Ji et al. 2000, Ji et al. 2001)
calix[4]biscrown	Cs^+	Fluorescent sensor	Na^+, K^+	(Leray et al. 2002)
calix[4] arene-bis(crown-6-ether)	Cs^+	Fluorescent sensor	Li^+, K^+, Mg^{2+}, Ca^{2+}, Sr^{2+}	(Souchon et al. 2006)
1,2,3-triazolium calix[4] arene diester	Cs^+	Extractant	Li^+, Na^+, K^+	(Li et al. 2010)

Like magnesium, calcium also plays a vital role in several biological processes and is very essential for the growth and maintenance of the human body. It is the chief constituent of teeth and bones in humans and animals and is present in various food resources such as milk, meat, vegetables. Calcium deficiency can lead to insufficient blood clotting, hypocalcemia, osteoporosis and osteopenia. Hence its effective determination in biological and environmental samples is important. Similarly, efficient chemosensors are required for Ba^{2+} and Sr^{2+} owing to their importance in various physiological and industrial processes (Gupta et al. 1999). For instance, a high dose of barium in the body can progressively block the potassium channels and impair the central nervous system (Kaur et al. 2018). ^{90}Sr is associated with radioactive waste material. Macrocyclic ligands with high selectivity for Sr^{2+} over sodium ions presented in high concentration are required to efficiently extract strontium during radioactive waste treatment.

Macrocyclic Receptors for Mg^{2+} Ions

Farruggia et al. reported the spectroscopic and photochemical characterization of two hydroxyquinoline derivatives of diaza-18-crown-6 (DCHQ) **20a, b** in living cell imaging. They used UV-Visible and fluorescence spectroscopy to study the sensitivity, selectivity and affinity of intracellular DCHQ fluorescent sensors to Mg^{2+} ions and Ca^{2+} ions. The results showed that the two probes in DMSO solution (1mg/mL) were highly selective and sensitive towards Mg^{2+} ion, with negligible inference from Ca^{2+} ions or pH changes within the physiological range (Farruggia et al. 2006).

a R = H
b R = Cl

20

Song and coworkers reported the synthesis of a new fluoroionophore with a phenanthroimidazole subunit based on calix[4]-arene-diamide **21**. The chemosensing behavior of the compound was examined by UV-Vis and fluorescence measurements. The chemosensor was found to be highly selective for Mg^{2+} ions in the presence of alkali, alkaline earth and transition metal ions in 95% aqueous DMSO solution (Song et al. 2007).

Zor et al. synthesized a novel chemosensor azocalix[4]arene derivative, 5,11,17,23- tetrakis[(acetophenone)azo]-25,26,27,28-tetrahydroxycalix[4]arene (APC4) **22**. UV-Visible titrations and voltammetric measurements were carried out to examine the optical and electrochemical behavior of the macrocycle, respectively, in the presence of alkali and alkaline earth metals. The results showed that the APC4 chemosensor displayed excellent selectivity towards magnesium ions over other metal ions in CH_3CN, especially the interfering calcium ion. Based on the voltammetric methods, the authors suggested that Mg^{2+} interacted with the calixarene derivative through the phenolic groups present at its lower rim (Zor et al. 2013).

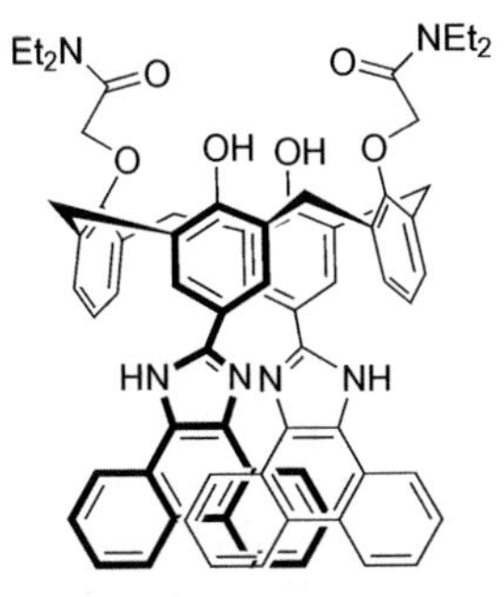

21 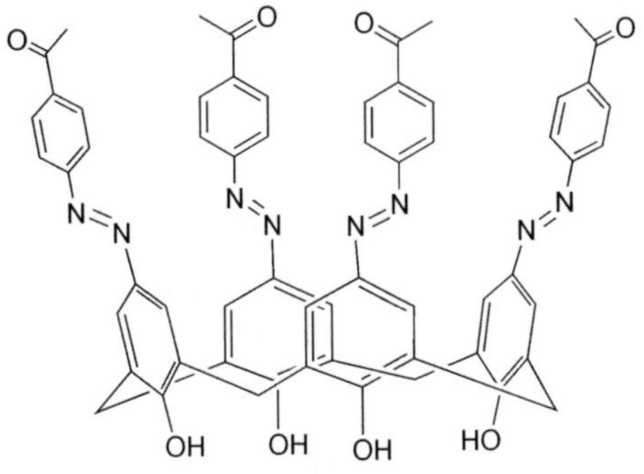   22

Macrocyclic Receptors for Ca^{2+} Ions

Liu and coworkers synthesized a water-soluble nano-sized metallo-capped polyrotaxane by the inclusion complexation of azo-calixarenes with metallo-bridged β-cyclodextrin dimers [bis(β-CD)s-Cu(II) complex]. Spectral titrations were used to study the complexes between polyrotaxane and the alkali and alkaline earth metals. The results indicated that polyrotaxane exhibits high selectivity and the binding ability for Ca^{2+} ions. The effective binding constants of polyrotaxane for cations decrease in the order: Ca^{2+}>Mg^{2+}>Sr^{2+}≈Ba^{2+}>Na$^+$>K$^+$>Li$^+$≈Rb$^+$>Cs$^+$ (Liu et al. 2004).

Kim et al. synthesized a new 1,3-alternate calix[4]arene based on/off fluorescence switch **23** by metal ion exchange. The cation binding ability of the compound with respect to fluorescence response was examined and the ligand displayed Pb^{2+} (quenching) and Ca^{2+} (enhancing) selectivity over other metal ions. The results showed that when the ligand complexes with Pb^{2+} ion through the two amide oxygen atoms are linked to the fluorophores, there is a geometrical change which was indicated by a marked quenching of excimer emission. However, when Ca^{2+} is added to this ligand-Pb^{2+} complex, the excimer emission band is revived (Kim et al. 2006a).

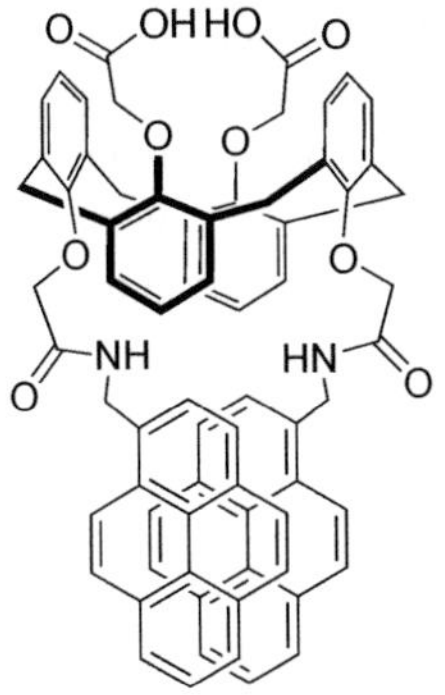

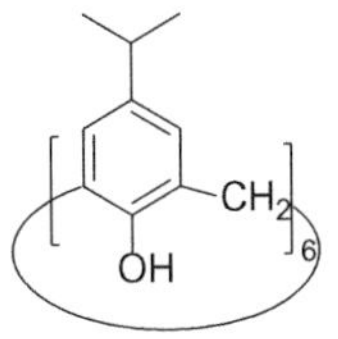

23   24

A new PVC membrane incorporating p-isopropylcalix[6]arene **24** as ionophore was developed by Jain and coworkers. It was examined as a potentiometric sensor material for Ca^{2+} ion. The electrode was found to exhibit a good potentiometric response over a working concentration of 3.9×10^{-6} to 1.0×10^{-1} M with a near

Nernstian slope of 30 mV per decade of concentration and a fast response time of 15 seconds. The selectivity data values indicated that the electrode was highly selective for Ca^{2+} ions over a wide range of metal ions (Jain et al. 2008).

Macrocyclic Receptors for Sr^{2+} Ions

Casnati et al. developed a series of novel amide derivatives of p-hydroxycalix[6] arene and p-hydroxycalix[8]arene. The extraction potential of these calixarene amide ionophores was evaluated in the nuclear waste treatment. These compounds exhibited high efficiency and selectivity towards ^{90}Sr over alkali metal ions (Casnati et al. 2001). Gupta and coworkers developed PVC membranes by incorporating dibenzo-24-crown-8 **25** and 4-tert-butylcalix[8]arene **26** and evaluated as a potentiometric sensor material for a wide range of metal ions. The membrane of dibenzo-24-crown-8 exhibited good selectivity for Ba^{2+} with a Nernstian response over a working concentration range $1.4 \times 10^{-5} - 1.0 \times 10^{-1}$ M and pH range 3.5-8.9. Similarly, the membrane having 4-tert-butylcalix[8]arene showed high selectivity and the best performance for Sr^{2+} ions. Both sensors displayed a fast response time of 15 seconds (Gupta et al. 1999).

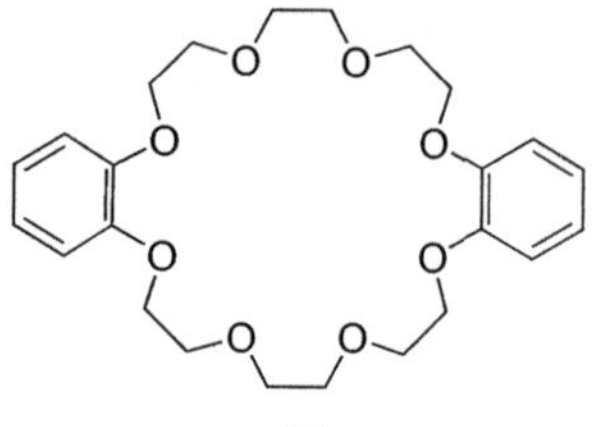

25

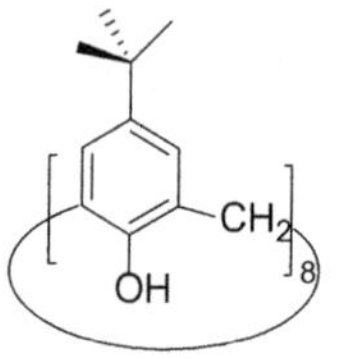

26

Macrocyclic Receptors for Ba^{2+} Ions

A voltammetric sensor based on a self-assembled redox-active monolayer of a novel Calix[4]arene-disulfide-diquinone **27** was reported by Chung and coworkers. The ionophore was found to be remarkably selective for Ba^{2+} ions in aqueous media (activity range 1.0×10^{-6} M to 1.0×10^{-4} M) with no significant interference from alkali and alkaline earth metal ions except for strontium and calcium ions. Only when the concentration of Sr^{2+} and Ca^{2+} ions was 100-fold and 500-fold respectively, the voltammetric responses could become comparable to that of Ba^{2+} (Chung et al. 2001).

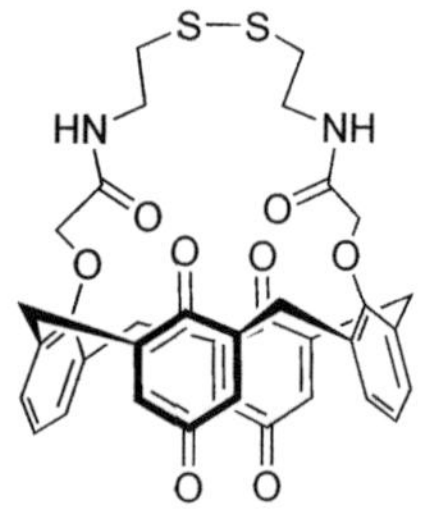

27

Table 2.2 Macrocyclic receptors for alkaline earth metal ions

Macrocyclic Receptor based on	Selectivity for	Application	Selectivity over	Reference
diaza-18-crown-6 hydroxyquinoline derivative	Mg^{2+}	Fluorescent sensor	Ca^{2+}	(Farruggia et al. 2006)
calix[4]-arene-diamide	Mg^{2+}	Fluoroionophore	Li^+, Na^+, K^+, Ca^{2+}, Ba^{2+}, Ni^{2+}, Cu^{2+}, Zn^{2+}, Cd^{2+}, Hg^{2+}, Pb^{2+}	(Song et al. 2007)
azocalix[4]arene derivative	Mg^{2+}	Electroactive chemosensor	Li^+, Na^+, K^+, Rb^+, Cs^+, Ca^{2+}, Sr^{2+}, Ba^{2+}	(Zor et al. 2013)
p-phenylenediamine-substituted azacrown ethers	Mg^{2+}, Ca^{2+}	Fluorescent sensor	Na^+, K^+	(Pearson and Xiao 2003)
polyrotaxane	Ca^{2+}	Receptor	Mg^{2+}, Sr^{2+}, Ba^{2+}, Na^+, K^+, Li^+, Rb^+, Cs^+	(Liu et al. 2004)
1,3-alternate calix[4] arene	Ca^{2+}	Fluorescent sensor	Li^+, Na^+, K^+, Rb^+, Cs^+, Ag^+, Pb^{2+}, Zn^{2+}	(Kim et al. 2006a)
p-isopropylcalix[6]arene	Ca^{2+}	Potentiometric sensor	Mg^{2+}, Sr^{2+}, NH^{4+}, Al^{3+}, Pb^{2+} alkali & various transition metal ions	(Jain et al. 2008)
calixarene amide derivatives	Sr^{2+}	Extractant	Na^+	(Casnati et al. 2001)
4-tert-butylcalix[8]arene	Sr^{2+}	ISE	Mg^{2+}, Ca^{2+}, Ba^{2+}, alkali & transition metal ions	(Gupta et al. 1999)
dibenzo-24-crown-8	Ba^{2+}	ISE	Mg^{2+}, Ca^{2+}, Sr^{2+}, alkali & transition metals	(Gupta et al. 1999)
Calix[4] arene-disulfide-diquinone	Ba^{2+}	Voltammetric sensor	Group I & II cations	(Chung et al. 2001)

MACROCYCLIC RECEPTORS FOR TRANSITION AND HEAVY METAL IONS

Transition metal ions such as Fe^{3+}, Cu^{2+}, Zn^{2+}, Mn^{2+}, Co^{2+} though present in trace amounts in living organisms, are essential nutrients for driving numerous biological processes and sustaining a healthy life (Bertini et al. 2007; Crichton 2008; Kaim et al. 2014). For instance, iron is an important constituent of hemoglobin in the blood for the transport of oxygen. It acts as a cofactor in several enzymatic reactions and is also involved in the synthesis of RNA and DNA. Abnormal levels of iron could lead to neurodegenerative diseases, like Parkinson's, Alzheimer's and Huntington's diseases (Wallace et al. 2018). Copper plays a vital role in the formation of red blood cells and the growth and maintenance of kidney, heart and brain tissues (Kaur et al. 2018). In humans, deficiency of copper can cause heart disease and an over abundance can lead to Menkes, Wilson's, Alzheimer's and Parkinson's diseases. Similarly, zinc is the second most abundant essential element in the human body. It plays a crucial role in many enzymes and regulates many biochemical processes, such as gene transcription, nerve signal transmission and immune function (Geng et al. 2016). Increased levels of zinc in the body can cause health problems, like skin irritations, nausea, stomach cramps, vomiting, Alzheimer's, Menkes and Parkinson's diseases.

However, certain heavy metal ions such as Hg(II), Cd(II), Pb(II), As(III), Cr(VI), etc. have been found to be highly toxic to humans and the environment even at lower concentrations. Heavy metals have atomic weights ranging between 63.5 and 200.6 and a density greater than 5 gcm^{-3} (Gupta et al. 2011, Gumpu et al. 2015). These metal ions enter the environment through various pathways such as, mining activities, metallurgical processes, fertilizer and paper industries, batteries, industrial effluents, waste disposal and cause widespread pollution. Since they are non-biodegradable, they tend to accumulate in plants and animals and enter the food chain. Human beings are at the top of the food chain are the worst affected due to *biomagnification*. Prolonged exposure to these metals can cause several health risks, including cancer. The hazardous effect of heavy metal ions is due to their ability to coordinate with thiol groups in proteins and alter the biochemical processes after entering the cell (Gumpu et al. 2015).

The most notorious and prevalent toxic metal in the environment is mercury, present in elemental, inorganic and organic forms. It is a neurotoxin that can cause serious damage to the central nervous system, impairment of pulmonary and kidney function, heart diseases and brain dysfunction. Cadmium is used in pigments, fertilizers, Ni-Cd batteries and electroplating industries. Accumulation of cadmium in the kidney, lungs, and liver can lead to chronic effects on human health such as kidney dysfunction, reduced lung capacity, weak brittle bones and calcium metabolism disorders (Berhanu et al. 2019). Exposure to lead can cause damage to the central nervous system, kidney, liver, reproductive system and brain functions. It can lead to memory loss, dullness, irritability, insomnia, headaches, dizziness and mental retardation in children. Likewise, thallium is widely used in technological and industrial applications. It is extremely toxic because it acts like alkali metal ions in biological systems (Roper et al. 2008). Thallium can also be easily absorbed by the skin. It can cause adverse effects on the lungs, liver, heart, kidney and nervous system.

Hence the use of macrocyclic chemosensors in the development of rapid and accurate detection methods to selectively and efficiently recognize transition and heavy metal ions in biological and environmental samples is a critical issue and has gained ever-growing research interest. Many reliable and cost-effective recognition methods for both qualitative and quantitative detection of these metals ions are being explored. Moreover, real-time monitoring methods to determine trace amounts of lead, cadmium, mercury, chromium and arsenic in drinking water and industrial wastewaters are of primary research importance to avoid pollution and maintain a safe water supply.

Macrocyclic receptors for Fe^{3+} ions

Dai et al. reported the synthesis and characterization of three novel macrocyclic fluorescent sensors **28a, b, c** containing optically active dansyl units. The metal recognition ability of the sensors towards various alkali, alkaline earth, transition metal, Al^{3+}, and Pb^{2+} ions was examined by fluorescent titrations, ESI-MS and DFT calculations. The ligands were found to be highly selective for Cu^{2+} and Fe^{3+} ions and formed stoichiometry 1:1 complex at physiological levels in DMSO/H_2O solution (v/v = 1:1, HEPES buffer, pH 7.4) through PET mechanism (Dai et al. 2017).

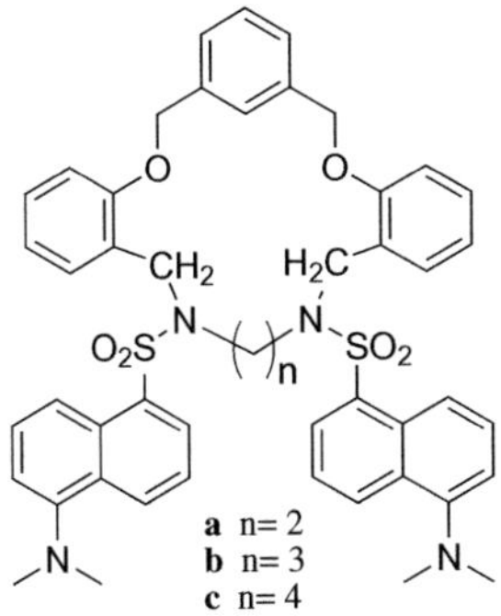

Suresh and coworkers developed a novel colorimetric and ratiometric sensor **29** by using per-6-amino–cyclodextrin as a supramolecular host and p-nitrophenol as a spectroscopic probe. The sensor effectively binds to Fe^{3+} and Ru^{3+} ions over a wide range of cations in water, with an appreciable visible color change from intense yellow to colorless (Suresh et al. 2010)

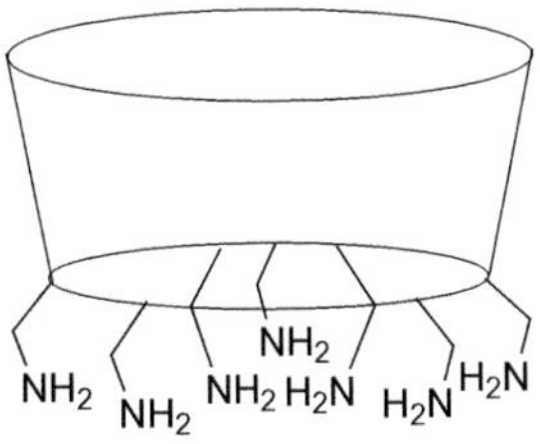

Wei et al. reported the design and synthesis of a novel pillar[5]arene functionalized with 2–aminobenzothiazole **30**. The monomeric units of the compound further self-organized into annularity supramolecular polymers through pillar[5]arene-based self-assembly interactions at a solution state. The binding ability of the pillar[5]arene towards 14 metal ions was investigated by carrying out UV–Vis and fluorescence experiments in DMSO/H_2O (8:2, v/v). The sensor was highly selective and sensitive to Fe^{3+} ions. With the addition of Fe^{3+} ions the blue fluorescence emission of the ligand was quenched, however, it was revived with the subsequent addition of F^- ions. Thus, the compound could be used as a sequential sensor for Fe^{3+} ions followed by F^- ions in an aqueous solution. The detection limit for Fe^{3+} ions was found out to be 9.0×10^{-7} mol/L (Wei et al. 2016b).

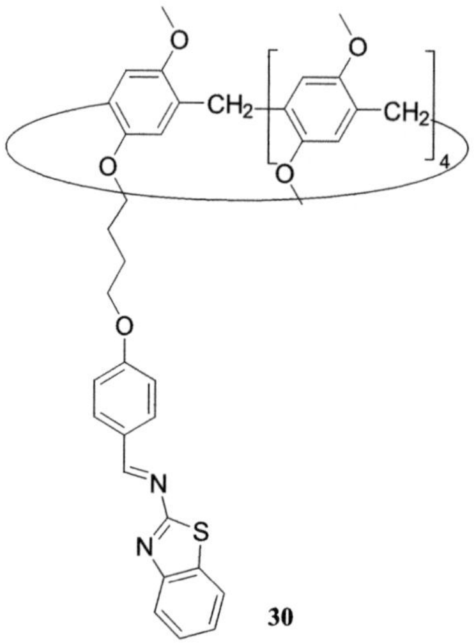

Likewise, Fe^{3+} selective fluorescent sensors based on quinoline functionalized pillar[5]arene **31** (Joseph et al. 2020) and triazole-linked decaamine derivative of pillar[5]arene **32** (Joseph 2020) have been successfully synthesized and examined for their metal-recognition abilities towards several biologically important metal ions.

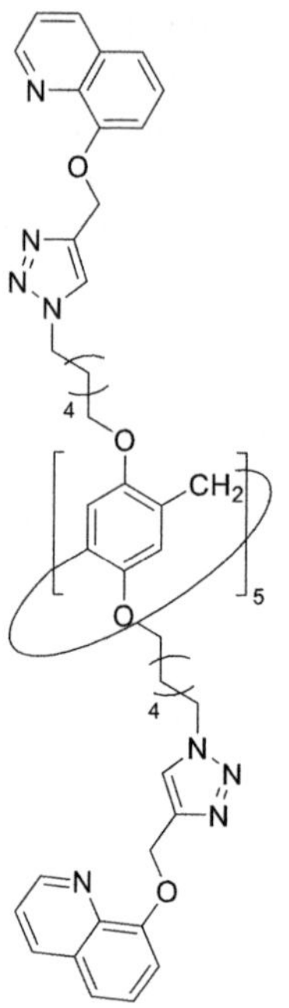

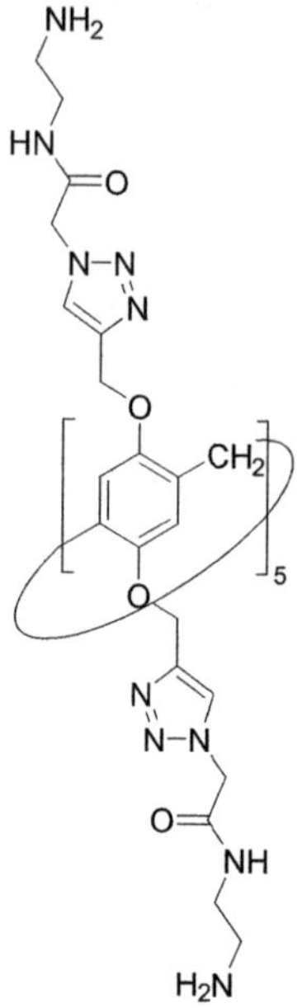

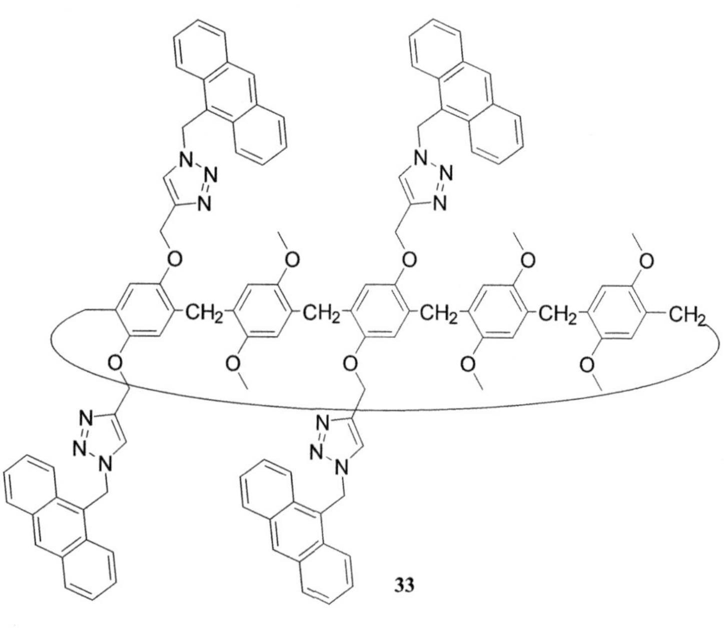

A single-ion recognition unit may lead to insufficient selectivity of many chemosensors, however, this shortcoming could be overcome by introducing multi-ion recognition units in a chemosensing molecule. With this point, Wei and coworkers synthesized and characterized a novel anthracene-appended 2:3 copillar[5]arene **33**. In this receptor, the three N atoms of 1,2,3-triazole were expected to provide more binding sites which could be used as recognition units, while the anthracene moieties served as fluorophores. The host-guest recognition ability of the chemosensor displayed specific high selectivity towards Fe^{3+} over a wide range of metal ions in DMSO. On the addition of Fe^{3+} ions to the ligand in DMSO solution, the original fluorescence emission band in the 380-550 nm range was significantly quenched and this fluorescence change could be seen by the naked eye. Further, the multi-ion recognition units of the sensor were confirmed by DFT and molecular dynamics (MD) computational calculations (Wei et al. 2015).

Macrocyclic Receptors for Cu^{2+} ions

Corradini et al. reported the synthesis of a fluorescent chemosensor for Cu^{2+} ions based on dansyldiethylenetriamine-modified β-cyclodextrin (CD-dien-DNS). The molecule conformation was studied by 2D NMR (ROESY spectra) in D_2O, circular dichroism and fluorescence measurements. The sensor displayed good selectivity for Cu^{2+} ions over Fe^{2+}, Co^{2+}, Ni^{2+} and Zn^{2+} ions. In the presence of Cu^{2+} ions, the fluorescence of CD-dien-DNS was completely quenched with the formation of a complex of 1:1 stoichiometry (Corradini et al. 1997).

Kozlowski and coworkers synthesized two β-cyclodextrin polymers by cross-linking of β-CD with nonenyl and dodecenyl derivatives of succinic anhydride. The hydrophobic macrocyclic polymers were used as ion carriers for the separation of heavy metal ions from aqueous solutions through polymer inclusion membranes into distilled water. The ion carrier containing nonenyl derivative-β-CD polymer showed high preferential selectivity for Cu^{2+} ions (Kozlowski et al. 2005).

Bhatt et al. synthesized two functionalized calix[4]arenes **34, 35** by incorporating pyridine and amine-containing amide moieties at the two opposite OH groups located at the lower rims. The interaction of these compounds with various transition metal ions was examined by UV-Vis spectroscopy in acetonitrile as well as acetonitrile-water (1:1) system. Amongst all the metal ions tested, only Cu^{2+} ions displayed a significant spectral change and a new high-intensity peak around 435 nm along with a distinct visual color change from colorless to yellow. Computational studies revealed that both the compounds formed 1:2 complexes with Cu^{2+} ions. One of the Cu^{2+} ions largely interacted with the phenolic OH causing strong absorption (ICT) in the visible region with a distinct colour change (Bhatt et al. 2017).

Similarly, Shahid and coworkers developed an efficient fluorescent sensor calix[4]arene-naphthalimide for Cu^{2+} ions **36**. The optical behavior and binding affinity of the ligand towards various metal ions were investigated using UV-Vis and fluorescence spectroscopy in an aqueous-acetonitrile solution (20%, v/v). The sensor exhibited excellent selectivity towards Cu^{2+} over other metal ions through fluorescence quenching with LOD of 1.61 μM. On addition of Cu^{2+} ions, the original strong absorption band at 340nm is replaced by a new absorption band at 290 nm. On subsequent addition of CN^- ions to this solution a significant fluorescence enhancement at 385nm was observed. Thus, it acts as a turn OFF/ turn ON fluorescent sensor for Cu^{2+} and CN^- ions (Shahid et al. 2016).

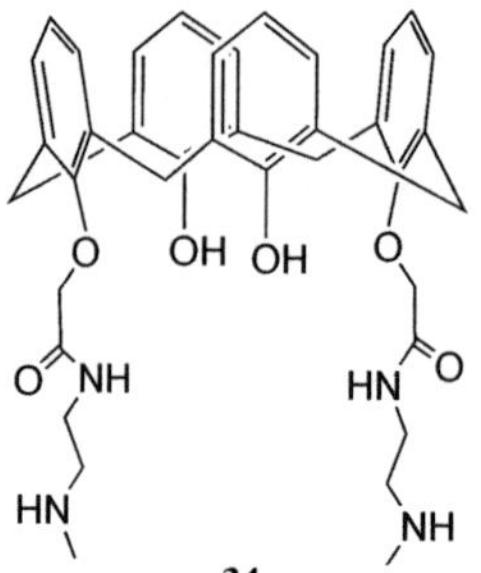

Macrocyclic receptors for Zn^{2+} ions

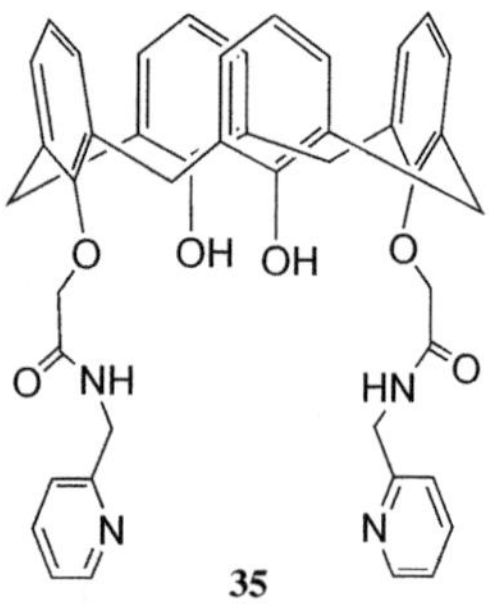

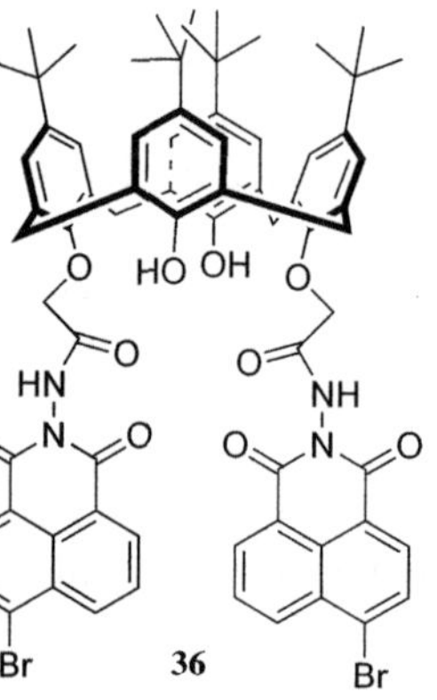

Tamanini et al. developed cyclam-based triazole derivatives as switch-on PET fluorescent sensors **37, 38** which were highly selective for Zn^{2+} ions over other biologically relevant metal ions under physiological conditions. As compared to ligand 37 (Tamanini et al. 2009), structure 38 (Tamanini et al. 2010) showed a two-fold increase in its fluorescence emission in the presence of zinc ions. It was suggested that the photoinduced electron transfer (PET) from the cyclam/ triazole unit to naphthalimide fluorophore quenched the fluorescence in the free ligand. However, on binding of zinc ion, retarded PET and fluorescence output enhanced. Similarly, Jobe et al. reported the synthesis and *in vitro* evaluation of novel cyclen-based triazole derivative which exhibited excellent selectivity for Zn^{2+} ions amongst other metal ions tested (Jobe et al. 2011).

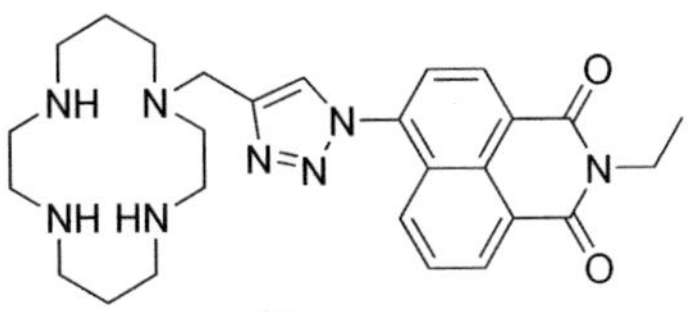

37

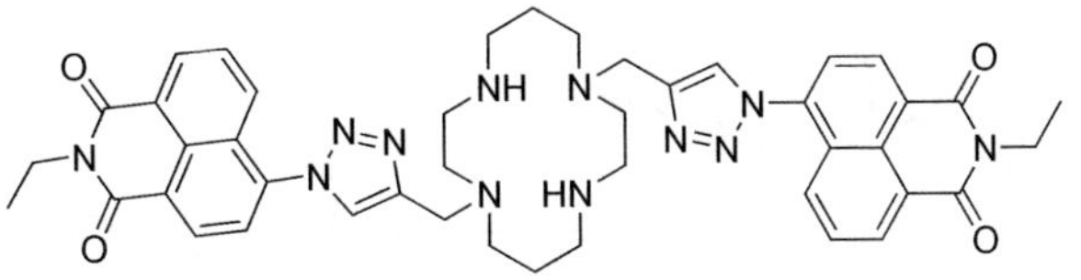

38

A new fluorescent sensor based on β-cyclodextrin substituted by 2-pyridyl triazole **39** was synthesized by David and coworkers. Spectroscopic studies revealed that the ligand displayed interesting solvent-polarity dependent fluorescence properties. The sensor was highly sensitive and selective for Zn^{2+} ions over other interfering metal ions and showed a large red-shift (320-366 nm) and significant fluorescence intensity enhancement (David et al. 2007).

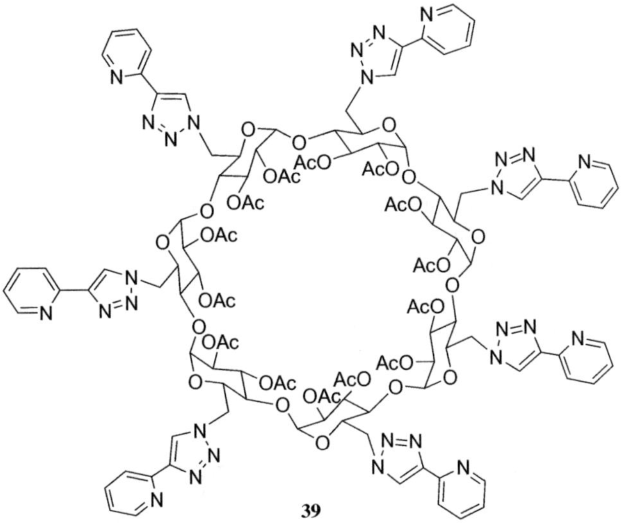

39

Two novel chromogenic 1,3-alternate calix[4]-azacrown and calix[4]-bis-crown ethers **40, 41** appended with a double bond conjugated indoaniline chromophore at the nitrogen atom of the azacrown unit with one methylene spacer were synthesized by Kim and coworkers. As compared to other alkali and alkaline earth cations tested, the metal ion complexation by both the compounds was greatest for Zn^{2+} resulting in the largest UV-Vis band shifts. It was concluded that zinc ions were selectively encapsulated into the calixazacrown cavity with the help of a conjugated indoaniline sidearm (Kim et al. 2002).

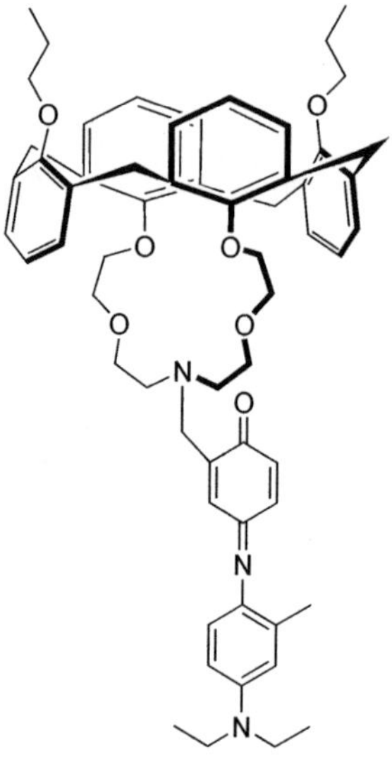

40 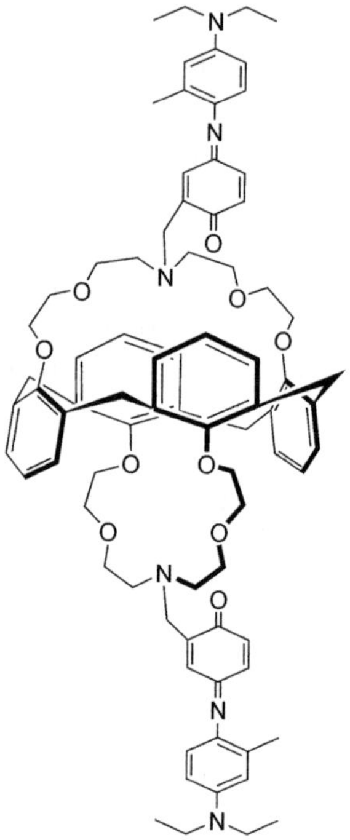   **41**

With the aim to design heteroditopic ligand with two or more binding subunits within the same macrocyclic framework, Bonaccorso et al. synthesized two novel calix[4]arenes derivatives **42, 43** having two and four 3-pyridylmethyl pendant groups at the lower rim, blocked in the cone conformation. The complexing properties of both the structures towards Zn^{2+} and Cu^{2+} in acetonitrile were examined using UV-Vis absorption measurements. ^1H NMR spectroscopic studies revealed that both the ligands formed multiple complex species with both Zn^{2+} and Cu^{2+} ions with large stability (Bonaccorso et al. 2013).

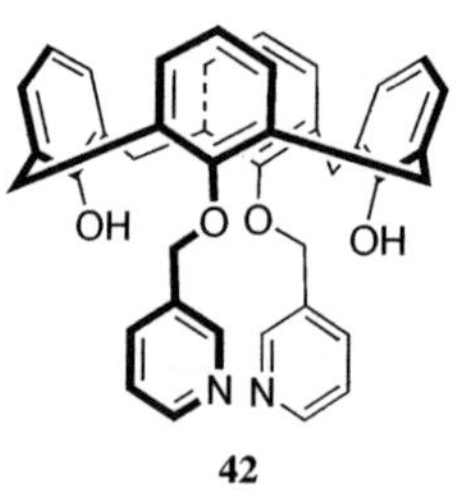

42 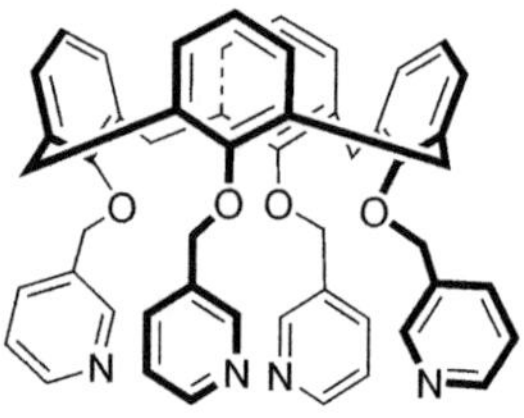   **43**

Macrocyclic Receptors for Cd^{2+} ions

A new cucurbit[7]uril based fluorescent sensor bearing a 2-(bromomethyl) benzimidazole fluorophore and a bis(pyridylmethyl)amine chelating group was synthesized by Geng and coworkers. The compound was examined by fluorescence and ^{1}H NMR spectroscopy for its metal complexes ability towards Cu^{2+}, Cd^{2+}, Zn^{2+}, Mn^{2+}, Ni^{2+}, Co^{2+}, Fe^{3+}, Cr^{3+}, Ho^{3+} or Lu^{3+} ions. The results showed that the fluorescence intensity of the sensor was significantly enhanced in the presence of Cd^{2+} and Zn^{2+} ions as they were effectively encapsulated within the curcubit[7] uril cavity (Geng et al. 2016).

Georghiou reported the synthesis and characterization of a new bis(pyrene)-appended 'capped' triazole-bridged calix[4]arene derivative **44** as an effective fluorescent sensor for Cd^{2+} and Zn^{2+} ions. The complexation properties of the host molecule towards different heavy metal ions in the $CH_3CN:CHCl_3$ (9:1) mixed solvent system were investigated using fluorescence and ^{1}H-NMR spectroscopic techniques and DFT studies. The sensor exhibited a high binding affinity towards Cd^{2+} and Zn^{2+} ions. Both the metal ions demonstrated enhanced monomer and decreased excimer emissions whereas most of the other metal ions quenched both the monomer and excimer emissions (Georghiou et al. 2020).

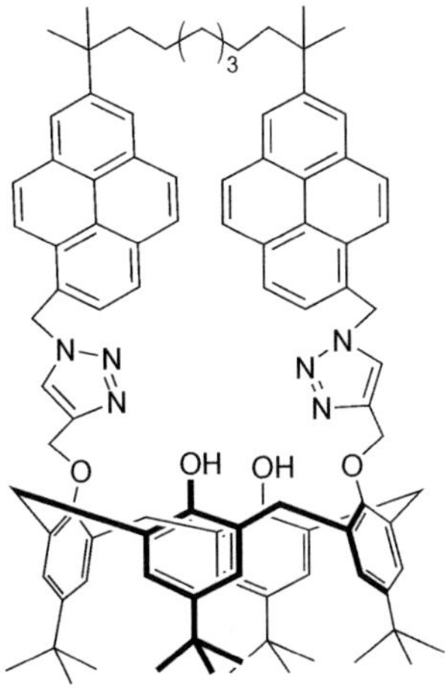

44

Macrocyclic Receptors for Hg^{2+} ions

A novel new cyclam derivative functionalized by two different fluorophores of pyrene and NBD subunits **45** was synthesized by Kim et al. The compound exhibited excellent Hg^{2+}-selective OFF-ON type and Cu^{2+}-selective ON-OFF type signalling behavior at 7.9×10^{-6} and 2.6×10^{-7} M in aqueous acetonitrile solution ($H_2O-CH_3CN= 10:90$, v/v), respectively over various other metal ions. The fluorescence intensity increased 10-fold on the addition of Hg^{2+} at the characteristic 538nm NBD peak while with Cu^{2+} ions marked quenching of the NBD region was observed. The fluoroionophoric molecule displayed a selective Hg^{2+}/Cu^{2+}-induced OFF-ON-OFF signaling pattern that can be used to construct a more elaborate supramolecular switching system (Kim et al. 2006b).

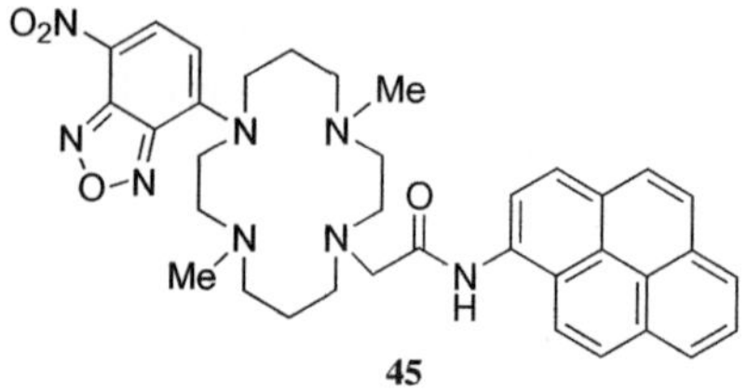

45

A water-soluble optical chemical sensor **46** based on reduced texaphyrin was examined for a wide range of metal ions by Root and coworkers. On complexing with a metal ion, the reduced form of the ligand underwent oxidation accompanied by a color change from red to green. The ligand was highly selective for Hg^{2+} and In^{3+} ions in aqueous media. Moreover, In^{3+} gave a ratiometric fluorescence response with an overall increase in intensity (Root et al. 2020).

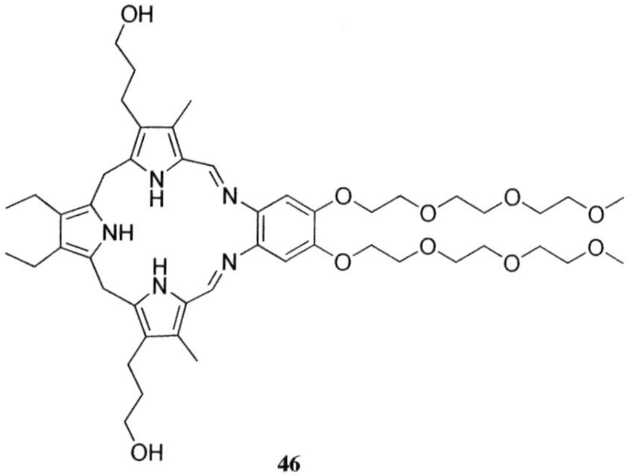

46

Deng et al. investigated the interaction of a wide range of metal ions with a known inclusion complex of H33258 and cucurbit[8]uril, as a fluorescent probe at different pH (1.5, 4.5 and 7.2). The fluorescent intensity of H33258 enhanced rapidly with an increasing amount of cucurbit[8]uril up to a ratio of 2:1 (cucurbit[8] uril: H33258). The probe displayed selective recognition of metal ions at different pH values. However, at pH 4.5, the fluorescent sensor responded to only Hg^{2+} and Fe^{3+} ions. The probe lost all its recognition properties at pH 1.5 (Deng et al. 2019).

Maniyazagan and coworkers explored the fluorogenic chelating ability of 3,3′–dihydroxybenzidine:α–cyclodextrin (DHB:α–CD) solid inclusion complex in acetonitrile-water system. Amongst the various metal ions tested, the sensor exhibited high characteristic fluorescent behavior towards Hg^{2+} and Fe^{3+} ions. The host-guest coordination system showed high binding constant values, $k=3.1626\times10^4$ M^{-1} and 1.8225×10^4 M^{-1} for Hg^{2+} and Fe^{3+} ions respectively. The ligand fluorescence property and non-toxicity made it a useful imaging tool for Hg^{2+} and Fe^{3+} ions in living cells (Maniyazagan et al. 2017).

For the first time, Jaiswal and coworkers reported the synthesis of a new hybrid macrocyclic receptor **47** based on the coupling of bile acid and calix[4] arene via the formation of 1,2,3-triazole ring. The metal recognition properties of the receptor in acetonitrile were investigated for Pb^{2+}, Cd^{2+}, Cu^{2+}, Zn^{2+}, Mn^{2+},

Hg^{2+} and Li^+ by UV-Vis spectroscopy. The receptor displayed the strongest affinity for Hg^{2+} ions, with 1:1 complex formation stoichiometry and binding constant $K_a = 1.2 \times 10^4 \ M^{-1}$ (Jaiswal et al. 2017).

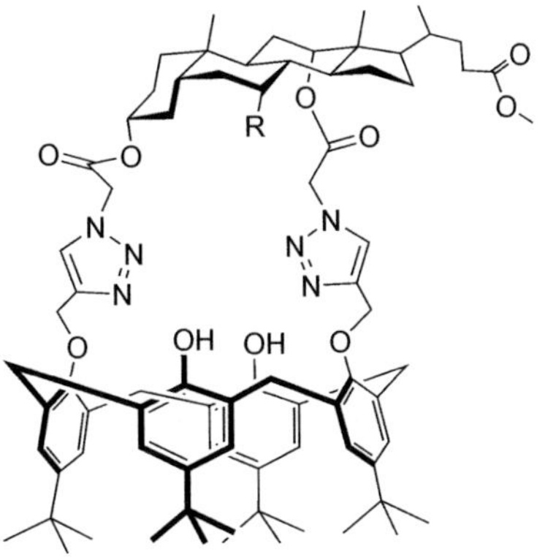

47

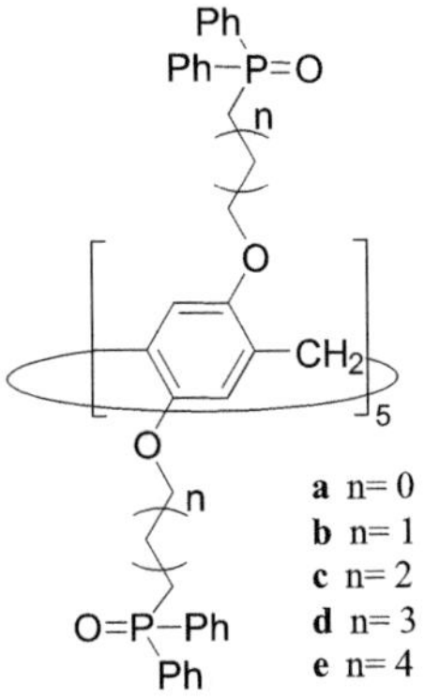

48

A new class of pillar[5]arene based receptors **48a-e** bearing phosphine groups were synthesized by Jia and coworkers. The metal recognition abilities of these compounds in the selective separation of heavy metal ions were examined by the liquid-liquid solvent extraction method. The compounds demonstrated efficient and excellent selectivity in the extraction of Hg^{2+} from aqueous solution into dichloromethane over other metal ions tested. The extractability towards Hg^{2+} ions ranged from 75.2 to 90.9% at a ligand concentration of 2×10^{-4} M. Hence, these pillar[5]arene receptors have the potential to be successfully used for the separation and detection of Hg^{2+} ions from environmental samples (Jia et al. 2014).

Similarly, Ma et al. reported the synthesis and characterization of a novel fluorescent probe **49** based on 8-hydroxyquinoline functionalized pillar[5]arene, which displayed strong AIE (aggregation-induced emission) fluorescence in an aqueous solution. The receptor was found to be highly selective and showed sensitive fluorescence 'on-off' behavior towards Hg^{2+} ions over 14 other metal ions tested in DMF/H_2O (8:2, v/v) binary solution. On addition of Hg^{2+} ions, significant quenching fluorescence occurred immediately, with a detection limit of 2.53×10^{-8} M for Hg^{2+} (Ma et al. 2019).

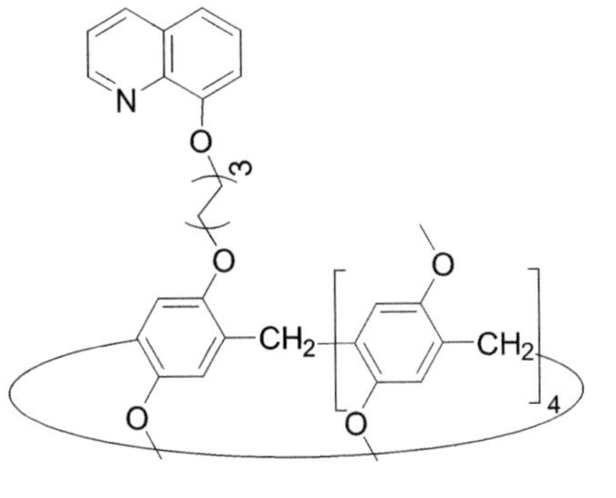

49

Macrocyclic Receptors for Pb^{2+} ions

Zhang and coworkers investigated the complexation ability of cage-like compound decamethylcucurbit-[5]uril **50** towards alkali, alkaline earth, Pb^{2+} and Cd^{2+} cations using calorimetric and potentiometric titrations. The compound was highly selective for Pb^{2+} ions and formed both 1:1 and 2:1 complexes with the ligand in 50% formic acid/water and aqueous solutions (Zhang et al. 2000).

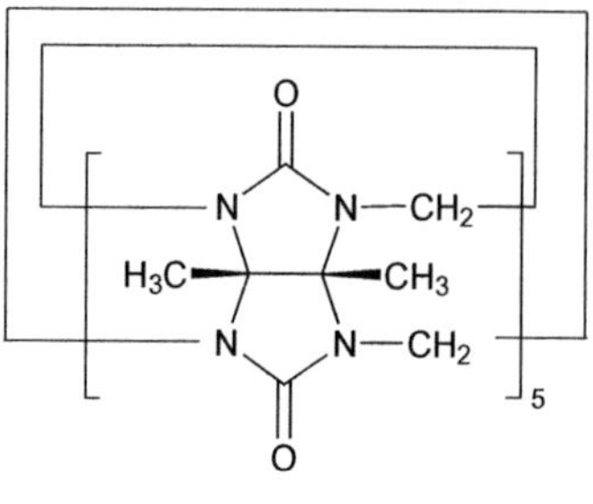

50 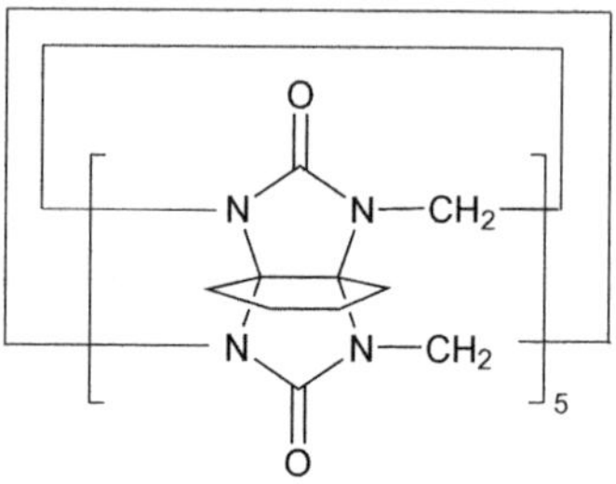   **51**

Similarly, Zhao et al. reported the synthesis and characterization of a new curcubit[5]uril derivative **51** containing five cyclohexanoglycoluril units. Despite the presence of five equator-bound fused cyclohexane rings, the compound was highly soluble in water as compared to common organic solvents. The membrane electrode prepared with this derivative behaved as an ion-selective electrode for Pb^{2+} ions with remarkable sensitivity and selectivity (Zhao et al. 2001).

Using a microwave-assisted heating method in alkaline media, Aswathy and coworkers reported the synthesis of β-cyclodextrin functionalized gold nanoparticles. The metal recognition ability of β-CD-AuNPs towards Cu^{2+}, Cd^{2+}, Ca^{2+}, Zn^{2+}, Mg^{2+}, Hg^{2+}, Pb^{2+}, and Ba^{2+} was examined by the colorimetric response and UV-Vis spectroscopy. The sensor was found to be highly selective towards Pb^{2+} ions in the presence of interfering metal ions, resulting in a visual color range from red to blue. TEM, DLS and FTIR analyses confirmed that Pb^{2+} ions induced aggregation of β-CD-AuNPs (Aswathy et al. 2012).

Métivier et al. reported a detailed study of cation-induced photophysical changes and complexing properties of two fluoroionophores, calix[4]arene dansyl derivatives **52, 53** towards Na$^+$, K$^+$, Ca^{2+}, Cu^{2+}, Zn^{2+}, Cd^{2+}, Hg^{2+} and Pb^{2+} ions in CH$_3$CN/H$_2$O mixture (60:40 v/v). The receptor 53 exhibited extremely high affinity and selectivity for Pb^{2+} ion with an unprecedented detection limit of 4 μgL^{-1}. On the other hand, receptor 52 demonstrated high selectivity towards Hg^{2+} ions through fluorescent quenching (Métivier et al. 2004).

Yusof and coworkers synthesized novel ion imprinting polymers through copolymerization of diallylaminomethyl-calix[4]resorcinarene **54** and divinylbenzene. The polymers were examined for the adsorption experiments of Pb^{2+}, Ni^{2+} and Cu^{2+} ions, effects of pH and metal ion concentration. The imprinted polymer exhibited high selectivity and adsorption capacity towards Pb^{2+} ions for a single ion system as well as competitive extraction experiments in the presence of Ni^{2+} and Cu^{2+} ions (Yusof et al. 2016)

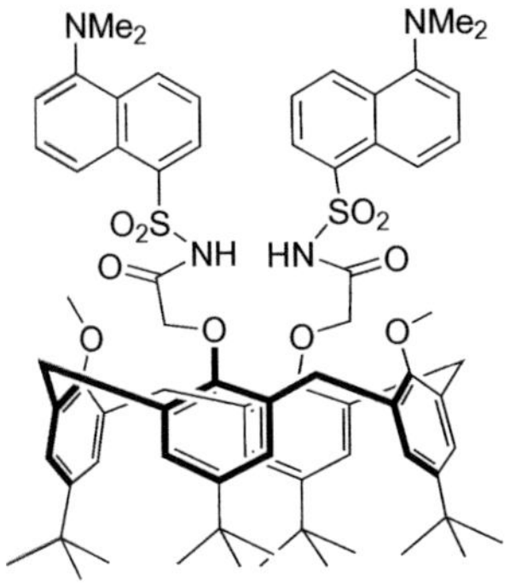

52

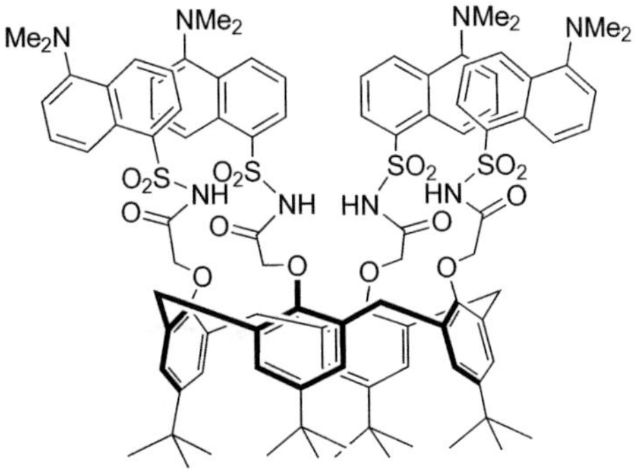

53

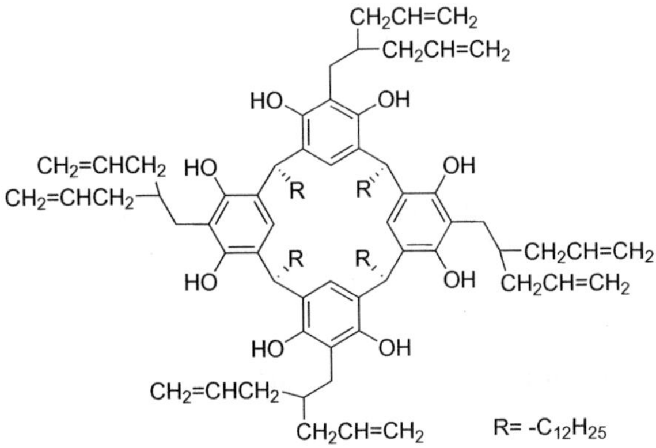

54

Macrocyclic receptors for Tl⁺ ions

A novel fluorogenic, dansyl derivative of 1,3-alternate calix[4]arene-bis(crown-6-ether) **55** was synthesized and found to demonstrate a highly selective fluorescent recognition ability for Tl^+ ions over many other metal ions tested (Talanova et al. 2005). Later, Roper et al. reported that this fluorescent sensor was very efficient and selective in detecting Tl^+ and Cs^+ ions at a low concentration level in MeCN-H_2O (1:1) solvent system and formed a 1:1 complex with the respective metal ions (Roper et al. 2007). Furthermore, the receptor was also explored for its metal ion extraction ability from an aqueous solution into chloroform (Roper et al. 2008). Yet again, 1,3-alternate calix[4]arene-bis(crown-6-ether) exhibited a very high selectively in extracting Tl^+ over other univalent cations, including Cs^+ and Rb^+. This phenomenon of enhanced complexing ability was attributed to the stronger tendency of π-coordination of soft thallium ion with the aromatic rings of calixarene framework.

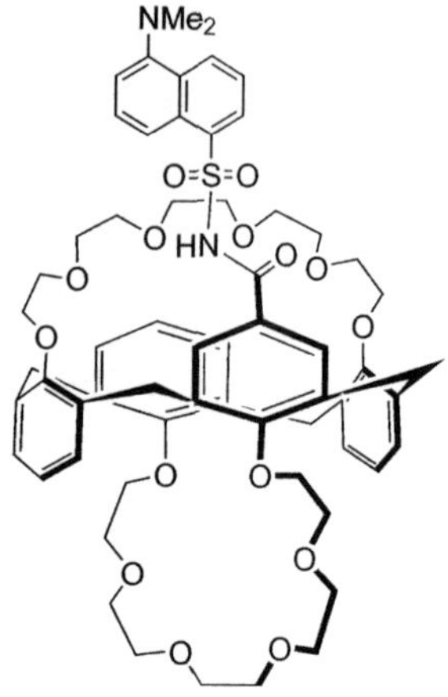

55

Makrlík and coworkers carried out extraction experiments to determine the effectiveness of calix[4]arene-bis(coumarin-crown-6) **17** as a receptor for various metal ions. The exchange extraction constants corresponding to the water-nitrobenzene system were evaluated and the stability constants of the metal-ligand complexes were also calculated. The results indicated that the macrocycle was very effective in binding with Tl^+ ions over other ions (Makrlík et al. 2015).

MACROCYCLIC RECEPTORS FOR LANTHANIDE AND ACTINIDE IONS

The f-block elements comprising of lanthanides (Ln) and actinides (Ac) altogether consist of 30 elements and are also known as the *inner transition metals*. The lanthanide elements along with scandium and yttrium are often called *rare earth* elements. Lanthanide and actinide cations usually display a common +III oxidation state. They act as hard acids, showing a strong affinity for hard bases like oxygen and primarily form electrostatic bonds (Bonal et al. 2001). The f-block elements play a significant role in electronic industries, manufacturing, agriculture, nuclear medicine, military and energy. Many of these elements enter the environment and eventually accumulate *in vivo* and have the potential to damage to the liver, cause a metabolic disturbance, tumors and genetic mutation. Additionally, the development of nuclear weapons and the use of nuclear-powered electricity has increased the risk of radioactive elements like uranium and plutonium, being released into the environment (Sessler et al. 2004).

Due to close similarity between ions, the efficient segregation of lanthanides and actinides remains a daunting problem in hydrometallurgical and nuclear fuel management processes (Fang et al. 2013). For example, efficient segregation of Am^{3+} from Eu^{3+} in strongly acidic media in reprocessing of nuclear fuels remains an elusive goal for researchers (Lambert et al. 2000, Wang et al. 2004). Moreover, the radiotoxicity of actinides makes their investigation difficult (Sliwa and Girek 2010). To overcome this problem, analogous cations such as Eu(III) for Pu(III),

Table 2.3 Macrocyclic receptors for transition and heavy metal ions

Macrocyclic Receptor	Selectivity for	Application	Selectivity over	Reference
dansylated hetero-atom-bridged macrocycle	Fe^{3+} and Cu^{2+}	Fluorescent sensor	Li^+, Na^+, K^+, Ag^+, Mg^{2+}, Ca^{2+}, Ba^{2+}, Pb^{2+}, Zn^{2+}, Co^{2+}, Cd^{2+}, Hg^{2+}, Ni^{2+}, Mn^{2+}, Cr^{3+} or Al^{3+}	(Dai et al 2017)
per-6-amino- β-cyclodextrin/p-nitrophenol	Fe^{3+}, Ru^{3+}	colorimetric and ratiometric sensor	Mn^{2+}, Fe^{2+}, Cu^{2+}, Zn^{2+}, Cd^{2+}, Hg^{2+}, Pb^{2+}, Cr^{3+}, Hg^{2+}, La^{3+}, Eu^{3+}, Ag^+	(Suresh et al. 2010)
pillar[5]arene- 2–aminobenzothiazole derivative	Fe^{3+}	Fluorescent sensor	Hg^{2+}, Ag^+, Ca^{2+}, Cu^{2+}, Zn^{2+}, Cd^{2+}, Ni^{2+}, Pb^{2+}, Co^{2+}, Cr^{3+}, Mg^{2+}, Fe^{2+} and Al^{3+}	(Wei et al. 2016b)
pillar[5]arene-2-mercaptobenzothiazole derivative	Fe^{3+}	Fluorescent sensor	Hg^{2+}, Co^{2+}, Ca^{2+}, Ni^{2+}, Pb^{2+}, Cd^{2+}, Zn^{2+}, Cr^{3+}, Cu^{2+}, Mg^{2+}, Ag^+	(Wei et al. 2016a)
anthracene-appended 2:3 copillar[5]arene	Fe^{3+}	Fluorescent sensor	Cr^{3+}, Ca^{2+}, Cu^{2+}, Co^{2+}, Ni^{2+}, Cd^{2+}, Pd^{2+}, Zn^{2+}, Mg^{2+}, Hg^{2+}, Ag^+	(Wei et al. 2015)
2-(hydroxy)-naphthyl imino functionalized pillar[5]arene	Fe^{3+}	Fluorescent sensor	Hg^{2+}, Ag^+, Ca^{2+}, Cu^{2+}, Co^{2+}, Ni^{2+}, Cd^{2+}, Pb^{2+}, Zn^{2+}, Cr^{3+}, Mg^{2+}, Eu^{3+}, Tb^{3+}	(Zhu et al. 2018)
quinoline appended pillar[5]arene	Fe^{3+}	Fluorescent sensor	Na^+, K^+, Ca^{2+}, Mg^{2+}, Mn^{2+}, Fe^{2+}, Co^{2+}, Ni^{2+}, Cu^{2+}, Zn^+	(Joseph et al. 2020)
triazole-linked decamine-pillar[5]arene	Fe^{3+}	Fluorescent sensor	Na^+, K^+, Ca^{2+}, Mg^{2+}, Mn^{2+}, Fe^{2+}, Co^{2+}, Ni^{2+}, Cu^{2+}, Zn^{2+}	(Joseph 2020)
β-cyclodextrin- urea-bipyridine podand	Ru^{3+}, Os^{3+}	Luminescent sensor	-----	(Pereira Silva et al. 2003)
β-cyclodextrin-dansyldiethylenetriamine derivative	Cu^{2+}	Fluorescent sensor	Fe^{2+}, Ni^{2+}, Co^{2+}, Zn^{2+}	(Corradini et al. 1997)
β-cyclodextrin polymer	Cu^{2+}	Ion transport	Co^{2+}, Ni^{2+}, Zn^{2+}	(Kozlowski et al. 2005)
calix[4]arene-pyridine-amide	Cu^{2+}	Colorimetric sensor	Ni^{2+}, Zn^{2+}, Co^{2+}, Hg^{2+}, Ag^+, Fe^{2+}, Cr^{3+}, Pb^{2+}, Cd^{2+}	(Bhatt et al. 2017)

Table 2.3 (Contd.)

Macrocyclic Receptors for Environmental and Biosensing Applications

Table 2.3 (Contd.) Macrocyclic receptors for transition and heavy metal ions

Macrocyclic Receptor	Selectivity for	Application	Selectivity over	Reference
calix[4]arene-naphthalimide	Cu^{2+}	Fluorescent sensor	Zn^{2+}, Pb^{2+}, Co^{2+}, Ni^{2+}, Na^+, Ag^+, Hg^{2+}, Cd^{2+}, K^+, Mn^{2+}, Fe^{3+}, Mg^{2+}	(Shahid et al. 2016)
pyrene-γ-cyclodextrin complex	Cu^{2+}	Fluorescent sensor	K^+, Mg^{2+}, Al^{3+}, Fe^{3+}, Zn^{2+}, Fe^{2+}, Cd^{2+}, Ag^+, Pb^{2+}, Hg^{2+}, Mn^{2+}	(Yang et al. 2004)
N/O/S-donor macrocycles with 1,10-phenanthroline sub-unit	Cu^{2+}, Ag^+	Ion-selective electrode	Ni^{2+}, Pd^{2+}, Pt^{2+}, Rh^{3+}, and Ru^{2+}, Pb^{2+}, Cd^{2+}, Hg^{2+}	(Lippolis and Shamsipur 2006)
p-tert-butylcalix[4]arene	Cu^{2+}	Extractant, membrane transport	Zn^{2+}, Ni^{2+}, Hg^{2+}, Fe^{3+}, Cu^{2+}	(Singh and Jang 2009)
calix[4]resorcinarenes	Cu^{2+}	Monolayers	-----	(Turshatov et al. 2004)
1,3-alternate calix[4]arene	Cu^{2+}, Hg^{2+}	Extractant	Group I, Ca^{2+}, Co^{2+}, Ni^{2+}, Zn^{2+}, Cd^{2+}, Pb^{2+}, La^{3+}, Gd^{3+}, Lu^{3+}	(Podyachev et al. 2008)
cyclam-triazole-naphthalimide derivative	Zn^{2+}	Fluorescent sensor	Na^+, K^+, Mg^{2+}, Ca^{2+}, Mn^{2+}, Ni^{2+}, Co^{2+}, Fe^{3+}, Cd^{2+}, Hg^{2+}, Cu^{2+}	(Tamanini et al. 2009); (Tamanini et al. 2010)
cyclen-based triazole derivative	Zn^{2+}	Fluorescent sensor	Na^+, K^+, Mg^{2+}, Ca^{2+}, V^{3+}, Mn^{2+}, Ni^{2+}, Co^{2+}, Fe^{3+}, Cd^{2+}, Hg^{2+}, Cu^{2+}	(Jobe et al. 2011)
β-cyclodextrin-2- pyridyl triazole derivative	Zn^{2+}	Fluorescent sensor	Mg^{2+}, Mn^{2+}, Fe^{2+}, Co^{2+}, Ni^{2+}, Cu^{2+}, Hg^{2+}, Pb^{2+}, Ag^+	(David et al. 2007)
1,3-alternate calix[4]azacrown	Zn^{2+}	Chromoionophore	Na^+, K^+, Rb^+, Cs^+, Mg^{2+}, Ca^{2+}, Sr^{2+}	(Kim et al. 2002)
calix[4]arenes/3-pyridylmethyl	Zn^{2+}, Cu^{2+}	Ditopic receptor	-----	(Bonaccorso et al. 2013)
1,8-disubstituted-cyclam/ naphthalimide derivative	Zn^{2+}	Fluorescent probe	Group I, II, transition metal ions, Pb^{2+}	(Wong et al. 2016)
cucurbit[7]uril- N,N-bis(2-pyridylmethyl)amine derivative	Cd^{2+}, Zn^{2+}	Fluorescent sensor	Mn^{2+}, Fe^{3+}, Ni^{2+}, Co^{2+}, Cu^{2+}, Cr^{3+}, Ho^{3+}, Lu^{3+}	(Geng et al. 2016)
triazole-pyrene-calix[4]arene	Cd^{2+}, Zn^{2+}	Fluorescent sensor	Ag^+, Co^{2+}, Fe^{2+}, Hg^{2+}, Pb^{2+}, Ni^{2+}, Mn^{2+}	(Georghiou et al. 2020)

Macrocyclic Receptor	Selectivity for	Application	Selectivity over	Reference
NBD-cyclam-pyrene derivative	Hg^{2+}, Cu^{2+}	Fluoroionophore	Na^+, K^+, Ca^{2+}, Mg^{2+}, Ni^{2+}, Zn^{2+}, Cd^{2+}, Pb^{2+}	(Kim et al. 2006b)
cyclam-triazole derivative	Hg^{2+}, Cu^{2+}	Fluorescent sensor	Li^+, Na^+, K^+, Ca^{2+}, Ba^{2+} Mn^{2+}, Ni^{2+}, Co^{2+}, Zn^{2+}, Cd^{2+}, Pb^{2+}, Ag^+	(Lau et al. 2011)
texaphyrin	Hg^{2+}, In^{3+}	Redox active receptor	Mg^{2+}, Bi^{3+}, Al^{3+}, Ga^{3+}, In^{3+}, Mn^{2+}, Co^{2+}, Fe^{3+}, Ni^{2+}, Cd^{2+}, Hg^{2+}, Lu^{3+}, Eu^{3+}, La^{3+}, Y^{3+}, Dy^{3+}, Nd^{3+}, Ho^{3+}, Gd^{3+}, Er^{3+}	(Root et al. 2020)
cucurbit[8]uril	Hg^{2+}, Fe^{3+}	Fluorescent sensor	Group I, Group II, Hg^{2+}, Fe^{2+}, Cr^{3+}, Eu^{3+}, Tm^{3+}, Yb^{3+}	(Deng et al. 2019)
3,3'–dihydroxybenzidine:α–cyclodextrin	Hg^{2+}, Fe^{3+}	Fluorescent sensor	K^+, Zn^{2+}, Mg^{2+}, Cu^{2+}, Cd^{2+}, Mn^{2+}, Cr^{3+}, Ga^{3+}, Al^{3+}, Sn^{4+}	(Maniyazagan et al. 2017)
calix[4]arene-bile acid-based triazole	Hg^{2+}	Receptor	Pb^{2+}, Cd^{2+}, Zn^{2+}, Mn^{2+}, Li^+, Cu^{2+}	(Jaiswal et al. 2017)
pillar[5]arene-based phosphine oxides	Hg^{2+}	Extractant	Na^+, K^+, Ca^{2+}, Mg^{2+}, Co^{2+}, Cu^{2+}, Ni^{2+}, Zn^{2+}, Cd^{2+}, Pb^{2+}, Fe^{3+}, Ag^+	(Jia et al. 2014)
quinoline functionalized pillar[5]arene	Hg^{2+}	Fluorescent sensor	Zn^{2+}, Pb^{2+}, Cd^{2+}, Ni^{2+}, Co^{2+}, Ag^+, Ca^{2+}, Mg^{2+}, Cr^{3+}, Ba^{2+}, Tb^{3+}, Eu^{3+}, La^{3+}, Al^{3+}	(Ma et al. 2019)
decamethylcucurbit[5]uril	Pb^{2+}	Receptor	Group I, II ions, Cd^{2+}	(Zhang et al. 2000)
cucurbit[5]uril derivative	Pb^{2+}	Ion selective electrode	Na^+, K^+, Cu^{2+}	(Zhao et al. 2001)
β-cyclodextrin capped AuNPs	Pb^{2+}	Chromoionophore	Cu^{2+}, Cd^{2+}, Ca^{2+}, Zn^{2+}, Mg^{2+}, Hg^{2+}, Ba^{2+}	(Aswathy et al. 2012)
calix[4]arene dansyl derivatives	Pb^{2+}, Hg^{2+}	Fluoroionophore	Na^+, K^+, Ca^{2+}, Cu^{2+}, Zn^{2+}, Cd^{2+}	(Métivier et al. 2004)
diallylaminomethyl-calix[4]resorcinarene	Pb^{2+}	Ion imprinting polymer	Cu^{2+}, Ni^{2+}	(Yusof et al. 2016)
dansylated 1,3-alternate calix[4]arene-bis(crown-6-ether)	Tl^+	Fluorescent sensor	Na^+, K^+, Ca^{2+}, Ag^+, Hg^{2+}, Pb^{2+}	(Talanova et al. 2005)
dansylated 1,3-alternate calix[4]arene-bis(crown-6-ether)	Tl^+	Extractant	Li^+, Na^+, K^+, Rb^+, Cs^+, Ag^+	(Roper et al. 2008)
Calix[4]arene-bis(coumarin-crown-6)	Tl^+	Extraction	Li^+, K^+, Na^+, Rb^+, Cs^+, Ag^+, NH_4^+	(Makrlík et al. 2015)

Am(III) and Cm(III); Th(IV) for U(IV) and Pu(IV); and UO_2^{2+} for NpO_2^{2+} and PuO_2^{2+} are often studied (Ramírez et al. 2008).

When it comes to nuclear waste management, several families of metals ions such as alkali, alkaline earth, transition, lanthanide and actinide cations are present in the spent nuclear fuel and fission products (Ramírez et al. 2008). Declassification of nuclear waste before disposal is very essential and relies mostly on solvent extraction procedures. Furthermore, effective use of nuclear fuel poses a challenging task of uranium recovery from spent nuclear fuel in acidic feeds (Chen et al. 2018). A great deal of research efforts have been dedicated to developing efficient extraction and separation methods incorporating macrocyclic ligands for reprocessing radioactive waste. Liquid-liquid extraction is the most commonly used technique where extractants play a crucial role in the segregation process (Chen et al. 2018).

Several calixarene complexes with lanthanides and actinides in radioactive waste management have been reported and examined in literature. The selectivity and efficiency of the calixarene moiety can be achieved by the incorporation of several binding groups at its lower and/ or upper rim. In this category, calix[4] arene bearing organophosphorus ligands like CMPO [(N, N-di-isobutylcarbamoyl) octylphenylphosphine oxide] have proved to be remarkably efficient extracting agents for lanthanides and actinides (Lambert et al. 2000, Wang et al. 2004, Mikulášek et al. 2007, Christelle et al. 2008).

Sansone and coworkers studied the impact of enlargement of the macrocyclic scaffold and the increase in the number of chelating functions on the complexes and selectivity of trivalent lanthanides and actinides. They prepared three calix[6] arene derivatives **56a-c** and two calix[8]arene derivatives **57a,b** by introducing six and eight CMPO units, respectively at the lower/narrow rim through ether links (Sansone et al. 2006). The UV spectrophotometry results showed that the increase in CMPO binding sites had a remarkable influence on the influence on ligands' extraction efficiency towards Eu^{3+} and Am^{3+} from aqueous nitric acid to o-nitrophenylhexyl ether. The authors concluded that such ligand properties were more superior to dendritic octa-CMPO-derivatives (Wang et al. 2004) of calix[4]arenes.

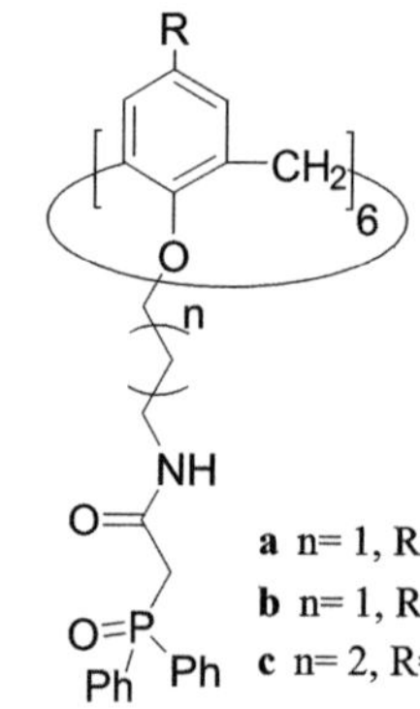

a n= 1, R= Bu^t
b n= 1, R= H
c n= 2, R= H

56

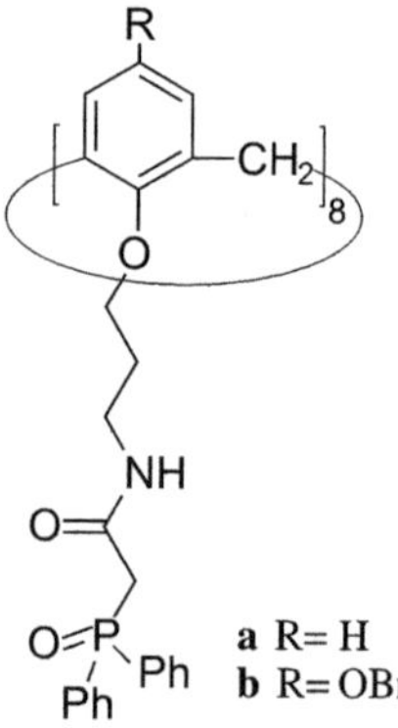

a R= H
b R= OBn

57

Carbamoylmethylphosphine oxides (CMPOs) are also being used as chelating agents to discriminate between lanthanides and actinides. Fang et al. designed and synthesized a new series of homoditopic tethered with dense bidentate CMPO chelating groups on both sides of pillar **58a-c**. Results from MS, NMR and fluorescence spectroscopy titrations and FT-IR spectroscopy, confirmed that the host-guest complexing process proceeded in a stepwise manner to form a bimetallic complex. The CMPO-pillar[5]arenes displayed excellent selectivity towards Am^{3+} over Eu^{3+} under acidic conditions. The efficacy of the pillararenes also rose with the increase in spacer length and was consistent with the corresponding stability constants (Fang et al. 2015).

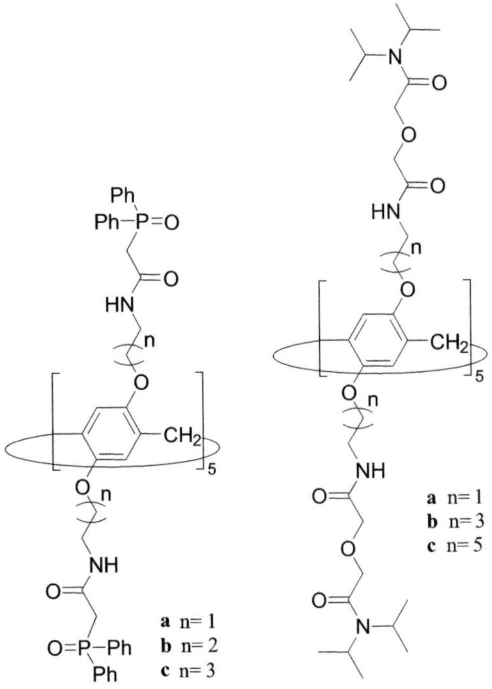

58   **59**

Wu and coworkers developed three new pillar[5]arenes **59a-c** functionalized with 10 diglycolamide (DGA) arms on both sides (rims) of the pillar. Even at highly acidic conditions (3 M HNO_3), the novel extractants displayed remarkable separation and extraction ability towards partitioning Eu^{3+} and Am^{3+}, suggesting its potential for nuclear water remediation (Wu et al. 2014).

Rouis et al. investigated the complexing properties of chromogenic azo-calix[4]arene **60a,b** towards Cd^{2+}, Pb^{2+}, Mg^{2+}, and Eu^{3+} in acetonitrile solution using UV-Vis spectroscopy. By adding the metal ions to 60a solution, the original band at 393 nm decreased and a new red-shifted absorption band appeared at 500 nm. Compound 60b also gave similar observations. Amongst the tested metal ions, the complexation of Eu^{3+} ions with the ligands exhibited strong spectral changes in the absorption spectrum (Rouis et al. 2006).

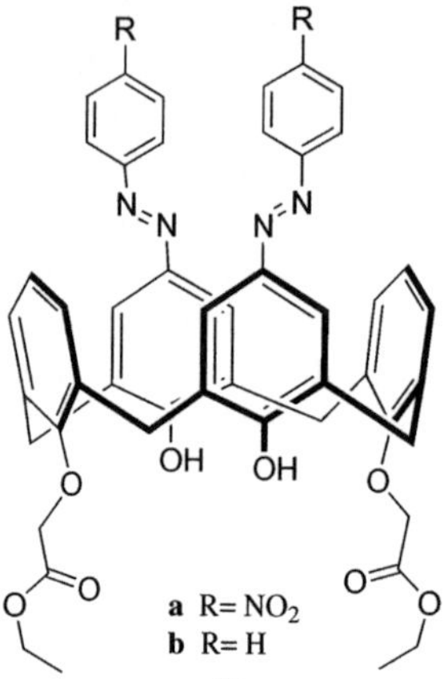

a R= NO₂
b R= H
60
61 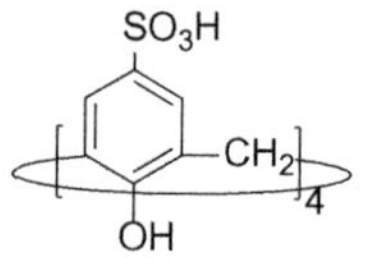
62 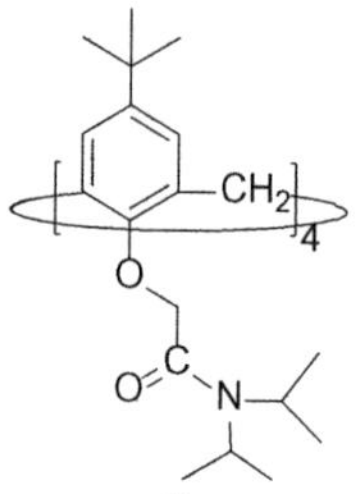

Due to their solubility in water, p-sulfonatocalix[n]arenes have generated considerable interest and are being explored for their potential application in the separation science of rare-earth metal ions. For instance, complexes involving azacrown ethers, p-sulfonatocalix[5]arene and europium(III) ions were explored for the separation of lanthanide ions by Dalgarno and coworkers (Dalgarno et al. 2004). Bonal et al. carried out a microcalorimetry study of the complexation of rare earth cations with p-sulfonatocalix[n]arenes in aqueous media (Bonal et al. 2001, Bonal et al. 2006). Israëli and coworkers thermodynamically characterised the complexation of La (III) with p-sulfonatocalix[4]arene **61** using ^{139}La NMR. The complexation was entropy-driven and results were consistent with 1:1 complex formation due to the electrostatic interactions between La^{3+} and the sulfonato groups (Israëli et al. 2002).

Tieke and coworkers investigated the transport of lanthanide ions across self-assembled multi-layered films prepared by alternate electrostatic layer-by-layer assembly of anionic p-sulfonatocalix[n]arenes with n=4, 6 and 8 and cationic polyvinylamine (PVA). The effect of pH and ring size was also examined. The results showed that the permeability of lanthanide ions tested is low. Since sodium ions could easily pass through the membrane, the separation factors α (Na^+/Ln^{3+}) were high. Consequently, p-sulfonatocalix[8] arene-based membranes were found to be well suited for the enrichment of lanthanide ions through complexation (Tieke et al. 2008).

Danil de Namor and Jafou reported the complexation properties of two calix[4]arene derivatives towards lanthanide (III) ions, Sc^{3+} and Y^{3+} ions in nonaqueous media (acetonitrile and N, N-dimethylformamide) by using ^{1}H NMR complexation experiments, conductance measurements and the titration microcalorimetry technique. The calix[4]arene derivative, *p-tert*-butylcalix(4) arene tetradiisopropylethanoamide **62** in particular, was found to exhibit selective behavior by forming most stable complexes with Gd^{3+} and Eu^{3+} ions over other cations in acetonitrile (Danil de Namor and Jafou 2001).

A novel naked eye chemosensor **63**, Schiff-based derivative of calix[4]arene containing two photochromic imine groups at the upper rims of the calixarene was synthesized by Liang and coworkers. UV-Visible and fluorescence spectroscopy were used to carry out the complexation studies. By adding Dy^{3+} and Er^{3+} to the ligand in CH_2Cl_2, the original peak at 375 nm decreased and a new broad absorption band appeared at 450-550 nm. No notable variation was observed for other lanthanide ions tested. Interestingly the color of the chemosensor turned pink immediately when Dy^{3+} was added while for Er^{3+} it turned yellow after standing in the dark for 24 hours (Liang et al. 2007).

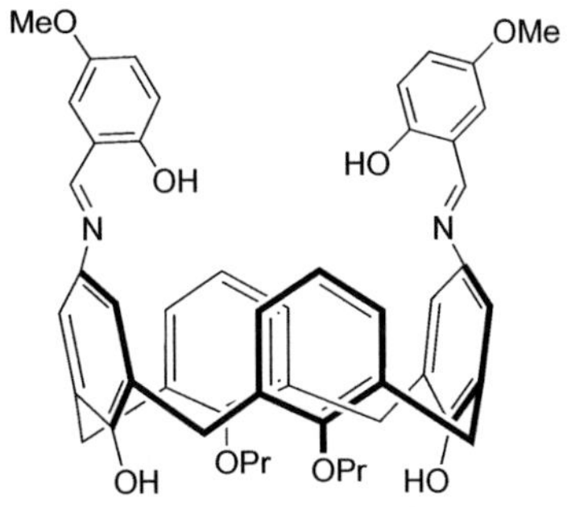

63

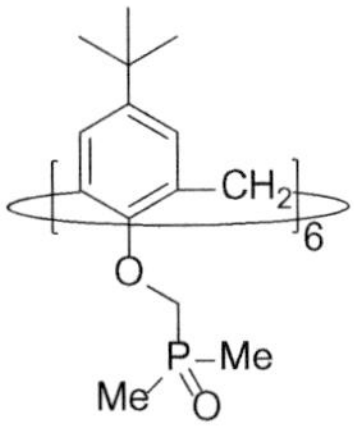

64

Han et al. developed novel β-cyclodextrin–4,4′-dipyridine inclusion complex modified silver nanoparticles. The ion recognition process was highly efficient and sensitive in the presence of Yb^{3+} ions in an aqueous solution via Yb^{3+}-induced aggregation to form chain-like supramolecular aggregates. The addition of Yb^{3+} ions induced a distinct color change from yellow to red corresponding to an increase in the absorbance intensity at ~ 610 nm. The limit for visual detection for Yb^{3+} was 2×10^{-7} M (Han et al. 2009).

Ramírez and coworkers investigated the extraction efficiency of hexa-phosphinoylated p-tert-butylcalix[6]arene **64** for towards UO_2^{2+} and Th(IV) over La^{3+}, Eu^{3+}, and Y^{3+}. Spectrometric titrations were performed and the actinide complexes were isolated and characterized by elemental analysis, UV-Vis and IR. The liquid-liquid extraction results demonstrated good extraction capability of the macrocycle with respect to the actinides (UO_2^{2+} and Th^{4+}) but not for the lanthanides (Ramírez et al. 2008).

New calix[4]arene based sequestering agents bearing sulfocatecholamide **65** and hydroxypyridinone **66a, b** moieties for uranyl ion (UO_2^{2+}) were prepared by Leydier and coworkers. The chelating properties were determined in aqueous media by UV spectrophotometry. The calixarene 66a, b exhibited a high affinity for uranyl ion in acidic and neutral pH whereas 65 was found to be more efficient at basic pH (Leydier et al. 2008).

Sessler et al. investigated a new colorimetric actinide sensor, Hexaphyrin (1.0.1.0.0.0) (isoamethyrin) **67** using UV–Vis spectroscopic and ESI-mass spectrometric analysis. In the presence of UO_2^{2+}, PuO_2^{2+}, NpO_2^{2+} the chemosensor solution in MeOH-CH_2Cl_2 exhibited significant color change as compared to other actinide and transition metal ions tested. On complexation with uranyl ion, the macrocycle undergoes spontaneous oxidation, consequently the overall electronic structure changed from antiaromatic to aromatic. While the addition of Pu (IV) and Np(IV) to the isoamethyrin solution induced instant color change, the uranyl complex requires 24 hours to show a significant color change (Sessler et al. 2004).

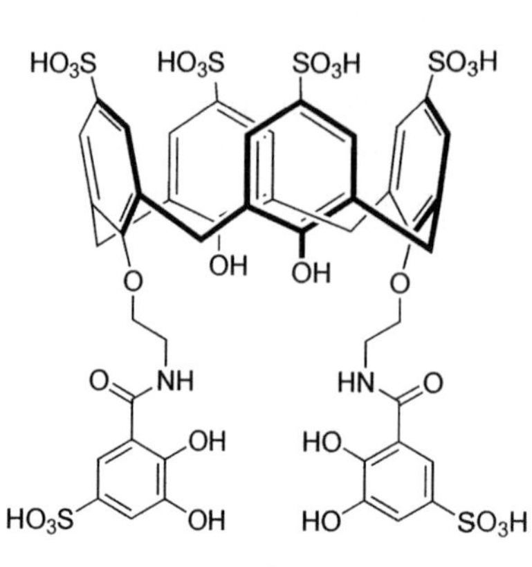

Organophosphorus species have been established as well-known ligands for chelating f-block elements effectively. Fang and coworkers reported the synthesis of a new class of pillar[5]arene-based phosphine oxides **68a-c** anchored with 10 chelating groups on both the rims of the pillar. These preorganized multidentate extractants displayed notable selectivity and efficiency towards Th^{4+} and UO_2^{2+} ions over other lanthanide ions. Log-log plot analysis indicated 1:1 stoichiometry (ligand/metal) for the extracted Th^{4+} or UO_2^{2+} complex. The extraction efficiency was further enhanced on increasing acidity in a wider range of 0.1–1.5 M HNO_3. In the presence of a synergist (Br_6-COSAN), the ligands showed more affinity for Eu^{3+} than Am^{3+} in 1 M HNO_3 (Fang et al. 2013). Very recently, Yuan and coworkers explored the thermodynamic behavior and the effect of organic diluents with different polarity in the extraction of UO_2^{2+} from 1 M nitric acid solutions by phosphine oxides functionalized pillar[5]arenes. As compared to its analogue with a shorter spacer ($-C_nH_{2n}-$, n=2), the extractant with a longer spacer (n=4, 6) was a better extractant for all diluents. Also, the uranyl extraction was found to be more favorable at a lower temperature (Yuan et al. 2020).

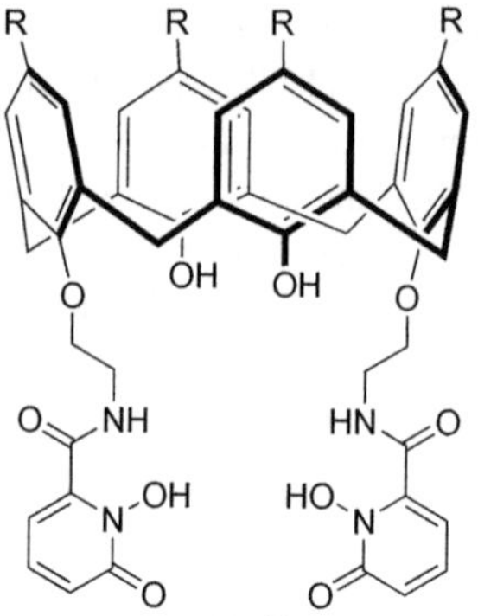

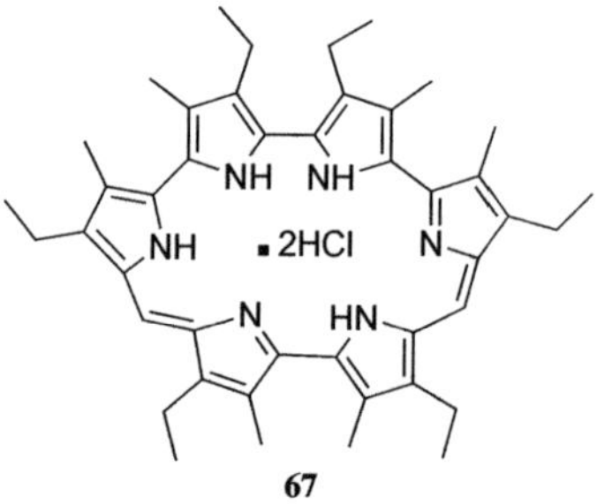

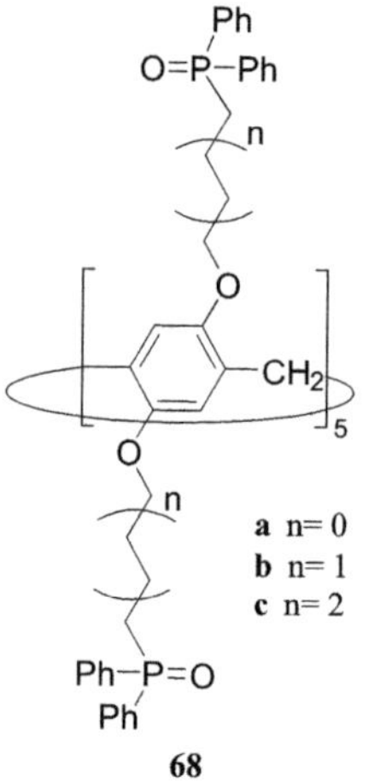

Table 2.4 Macrocyclic receptors for lanthanides and actinides

Macrocyclic Receptor	Selectivity for	Application	Selectivity over	Reference
CMPO-derivatives of calix[n]arenes	Eu^{3+}, Am^{3+}	Extractant	-----	(Sansone et al. 2006); (Wang et al. 2004)
CMPO-pillar[5]arenes	Am^{3+}	Extractant	Eu^{3+}	(Fang et al. 2015a)
pillar[5]arene-based diglycolamides	Eu^{3+}	Extractant	Am^{3+}	(Wu et al. 2014)
azo- calix[4]arene	Eu^{3+}	Chromoionophore	Cd^{2+}, Pb^{2+}, Mg^{2+}	(Rouis et al. 2006)
p-sulfonatocalix[4]arene	La^{3+}	Receptor	-----	(Israëli et al. 2002)
p-sulfonatocalix[n]arene, n=4,6,8	Y^{3+}, La^{3+}, Pr^{3+}, Sm^{3+}, Ce^{3+},	Membrane transport	Na^{+}	(Tieke et al. 2008)
calix[4]arene	Gd^{3+}, Eu^{3+}	Receptor	Ln^{3+}, Sc^{3+}, Y^{3+}	(Danil de Namor and Jafou 2001)
calix[4]arene Schiff base derivative	Dy^{3+}, Er^{3+}	Chromoionophore	La^{3+}, Pr^{3+}, Eu^{3+}, Gd^{3+}, and Yb^{3+}	(Liang et al. 2007)
β-cyclodextrin–4,4'-dipyridine modified Ag NPs	Yb^{3+}	chromoionophore	Pr^{3+}, Eu^{3+}, Sm^{3+}, La^{3+}, Nd^{3+}, Ce^{3+}	(Han et al. 2009)
phosphinoylated p-tert-butylcalix[6]arene	UO_2^{2+}, Th(IV)	Extractant	La^{3+}, Eu^{3+}, Y^{3+}	(Ramírez et al. 2008)
calix[4]arene	UO_2^{2+}	Receptor	-----	(Leydier et al. 2008)
hexaphyrin(1.0.1.0.0.0)	UO_2^{2+}, PuO_2^{2+}, NpO_2^{2+}	Redox active chromoionophore	Ce^{4+}, Sm^{3+}, La^{3+}, Gd^{3+}, Cu^{2+}, Cd^{2+}, Ni^{2+}, Pb^{4+}, Zn^{2+}	(Sessler et al. 2004)
pillar[5]arene-based phosphine oxides	Th^{4+}, UO_2^{2+}	Extractant	La^{3+}, Ce^{3+}, Pr^{3+}, Nd^{3+}, Sm^{3+}, Eu^{3+}, Gd^{3+}, Yb^{3+}, Lu^{3+}	(Fang et al. 2013)
pillar[5]arene based on triazolelinked 8-oxyquinolines	Th^{4+}	Fluorescent sensor	La^{3+}, Ce^{3+}, Pr^{3+}, Nd^{3+}, Sm^{3+}, Eu^{3+}, Gd^{3+}, Er^{3+}, Yb^{3+}, Lu^{3+}, UO_2^{2+}, Cu^{2+}, Zn^{2+}, Pb^{2+}, Cd^{2+}, Na^{+}, Ca^{2+}	(Fang et al. 2015b)
pillar[5]arene-based phosphine oxides	UO_2^{2+}	Extractant	------	(Yuan et al. 2020)
pillar[5]arene-based diglycolamides	Pu^{4+}, PuO_2^{2+}	Extractant (in RTIL)	Lu^{3+}, Th^{4+}, Eu^{3+}, Yb^{3+}, Gd^{3+}, Pr^{3+},	(Sengupta et al. 2017)
pillar[5]arene-based phosphine oxides	UO_2^{2+}	Extractant (in RTIL)	La^{3+}, Ce^{3+}, Sm^{3+}, Nd^{3+}, Er^{3+}	(Chen et al. 2018)

A novel non-symmetric pillar[5]arene bearing five triazole-linked 8-oxyquinolines at one rim and five alkyl chains at the other end was designed and synthesized by Fang and coworkers. The recognition behavior of the receptor was investigated towards various lanthanide, actinide, transition metal, Na^+ and Ca^{2+} ions in $CH_3CN–H_2O$ (9:1) using UV-Vis, fluorescence, NMR and mass spectroscopy techniques. The receptor exhibited high selectivity and sensitivity towards Th^{4+} over other metal ions tested. The mechanism of recognition probably involved MLCT (Metal-Ligand Charge Transfer) by the electron transfer from quinoline N to Th^{4+}, resulting in the formation of a weak fluorescent receptor–Th^{4+} complex of 1: 1 stoichiometry (Fang et al. 2015b).

As an alternative to conventional organic solvents, the room temperature ionic liquids (RTILs) also known as the 'designer solvents' hold tremendous potential and promise for RTILs-based extraction of lanthanide and actinide metal ions. Sengupta and coworkers carried out thermodynamics and radiolytic stability study of DGA functionalized pillar[5]arenes 59a-c in RTIL, 1-n-Octyl-3-methylimidazolium bis(trifluoromethane)-sulfonamide ($C_8mimNTf_2$) as a diluent. The results demonstrated that the system was highly efficient in extracting Pu^{4+} and PuO_2^{2+} species from acidic solutions. The extraction followed the 'cation-exchange mechanism (Sengupta et al. 2017). Similarly, Chen et al. demonstrated the highly selective and efficient extraction of UO_2^{2+} from the nitric acid medium by pillar[5]arene-based phosphine oxides 68a-c in RTIL, $C_8mimNTf_2$ over other lanthanides and Th^{4+} (Chen et al. 2018).

CONCLUSION

In this chapter, various macrocyclic receptor molecules were considered based on several classes of macrocyclic ligands like crown ethers, lariat crown ethers, cyclam, cyclen, cryptands, cyclophanes, podands, cyclodextrins, calix[n]arenes, cucurbit[n]urils and pillar[n]arenes. The macrocyclic host molecules and their functionalized derivatives were described and summarized with regard to their metal ion complexation ability, binding affinity, selectivity, sensitivity, spectroscopic properties, including others such as extraction ability, ion transport and redox nature. Many of these macrocycles presented highly selective recognition of targeted metal ion species. Some of them were even capable of detecting the analytes at sub-micro levels. The binding and recognition efficacy of a receptor for different metal ions depends on factors such as the ring size of the macrocycle, incorporation of substituent units/ functional groups, spacer length, presence of heteroatoms (N, O, S) and the solvent system. Preorganization of multiple chelating units within the ligand framework with limited conformational liberty enables the construction of more rigid systems with cavities of precisely tailored shapes and sizes.

Such tailor-made synthetic macrocycles offer effective opportunities and solutions to study and handle metal ions in biological, analytical and environmental applications. Moreover, they act as excellent model systems to study the nature of host-guest non-covalent interactions, which is very crucial for the design and creation of sophisticated functional assemblies. In terms of application, the macrocyclic receptors for metal ions have been primarily utilized as colorimetric and fluorimetric sensors, extractants, ion-selective electrodes and membrane transport. Among the variety of chemosensors reported in literature, naked eye sensors have gained significant importance for metal ion recognition due to their simplicity, speed and cost-effectiveness. Though sensors based on organic macrocyclic molecules have been found to be effective in the recognition of metal ions in common organic solvents, their poor solubility in aqueous media still remains a practical problem. With the advancement of new and efficient methods of chemical synthesis, it is hoped that in future more sophisticated, efficient and eco-friendly chemosensors will be developed for selective detection of metal ions for a variety of *in vivo* and *in vitro* applications.

REFERENCES

Adam, K.R., I.M. Atkinson, J.Kim, L.F. Lindoy, O.A. Matthews, G.V. Meehan, F. Raciti, B.W. Skelton, N. Svenstrup and A.H. White. 2001. Macrocyclic ligand design. A synthetic, solvent extraction, computational and NMR study of the effect of cryptand flexibility on sodium ion affinity. J. Chem. Soc., Dalton Trans. (16): 2388–2397.

Aswathy, B., G.S. Avadhani, S. Suji and G. Sony. 2012. Synthesis of β-cyclodextrin functionalized gold nanoparticles for the selective detection of Pb^{2+} ions from aqueous solution. Front. Mater. Sci. 6(2): 168–175.

Azadbakht, R. and J. Khanabadi. 2015. A novel fluorescent nano-chemosensor for cesium ions based on naphthalene macrocyclic derivative. Spectrochim. Acta A Mol. Biomol. Spectrosc. 139: 279–285.

Berhanu, A.L., Gaurav, I. Mohiuddin, A.K. Malik, J.S. Aulakh, V. Kumar and K.-H. Kim. 2019. A review of the applications of Schiff bases as optical chemical sensors. Trends Analyt. Chem. 116: 74–91.

Bertini, I., H.B. Gray, S.J. Lippard and J.S. Valentine. 2007. Bioinorganic Chemistry. New Delhi, Viva Books Private Limited.

Bhatt, M., D. Maity, V. Hingu, E. Suresh, B. Ganguly and P. Paul. 2017. Functionalized calix[4]arene as a colorimetric dual sensor for Cu(II) and cysteine in aqueous media: Experimental and computational study. New J. Chem. 41(21): 12541–12553.

Bonaccorso, C., F. Nicoletta, V. Zito, G. Arena, D. Sciotto and C. Sgarlata. 2013. Tunable Zn^{2+} and Cu^{2+} calixarene complexes as polytopic building blocks for guest recognition. Supramol. Chem. 25(9-11): 615–625.

Bonal, C., Y. Israëli, J.-P. Morel and N. Morel-Desrosiers. 2001. Binding of inorganic and organic cations by p-sulfonatocalix[4]arene in water: A thermodynamic study. J. Chem. Soc., Perkin Trans. 2(7): 1075–1078.

Bonal, C., P. Malfreyt, J.-P. Morel and N. Morel-Desrosiers. 2006. Thermodynamics of Complexation of Tetraalkylammonium and Rare-earth Cations with Two Sulfonatocalix[n]arenes (n = 4 and 6) in Aqueous Solution. Supramol. Chem. 18(3): 183–190.

Casnati, A., S. Barboso, H. Rouquette, M.-J. Schwing-Weill, F. Arnaud-Neu, J.-F. Dozol and R. Ungaro. 2001. New efficient calixarene amide ionophores for the selective removal of strontium ion from nuclear waste: synthesis, complexation, and extraction properties. J. Am. Chem. Soc. 123(49): 12182–12190.

Chen, L., Y. Wang, X. Yuan, Y. Ren, N. Liu, L. Yuan and W. Feng. 2018. Highly selective extraction of uranium from nitric acid medium with phosphine oxide functionalized pillar[5]arenes in room temperature ionic liquid. Sep. Purif. Technol. 192: 152–159.

Christelle, P., B. Damien, K. Jan, K. Oleg, S. Miroshnichenko, R. Valentyn, B. Volker and F.D. Jean. 2008. CMPO-calix[4]arenes and the influence of structural modifications on the Eu(III), Am(III), Cm(III) separation. Radiochim. Acta. 96(4-5): 203–210.

Chung, T.D., J. Park, J. Kim, H. Lim, M.-J. Choi, J.R. Kim, S.-K. Chang and H. Kim. 2001. Self-assembled monolayer of a redox-active calix[4]arene: voltammetric recognition of the Ba^{2+} ion in aqueous media. Anal. Chem. 73(16): 3975–3980.

Corradini, R., A. Dossena, G. Galaverna, R. Marchelli, A. Panagia and G. Sartor. 1997. Fluorescent Chemosensor for Organic Guests and Copper(II) Ion Based on Dansyldiethylenetriamine-Modified β-Cyclodextrin. J. Org. Chem. 62(18): 6283–6289.

Crichton, R.R. 2008. Biological Inorganic Chemistry An Introduction. Amsterdam, Elsevier.

Dai, Y., K. Xu, C. Wang, X. Liu and P. Wang. 2017. Study on the synthesis of novel fluorescent macrocyclic sensors and their sensitive properties for Cu^{2+} and Fe^{3+} in aqueous solution. Supramol. Chem. 29(4): 315–322.

Dai, X., S. Wu and S. Li. 2018. Progress on electrochemical sensors for the determination of heavy metal ions from contaminated water. J. Chinese Adv. Mater. Soc. 6(2): 91–111.

Dalgarno, S.J., M.J. Hardie, J.E. Warren and C.L. Raston. 2004. Lanthanide crown ether complexes of p-sulfonatocalix[5]arene. Dalton Trans. (16): 2413–2416.

Danil de Namor, A.F. and O. Jafou. 2001. Complexation of calixarene derivatives and lanthanide cations in nonaqueous media. J. Phys. Chem. B. 105(33): 8018–8027.

David, O., S. Maisonneuve and J. Xie. 2007. Generation of new fluorophore by click chemistry: synthesis and properties of β-cyclodextrin substituted by 2-pyridyl triazole. Tetrahedron Lett. 48(37): 6527–6530.

Deng, X.Y., W.T. Xu, M. Liu, M.X. Yang, Q.J. Zhu, B. Lü and Z. Tao. 2019. Cucurbit[8] uril-improved recognition using a fluorescent sensor for different metal cations. Supramol. Chem. 31(9): 616–624.

Eddaif, L., A. Shaban and J. Telegdi. 2019. Sensitive detection of heavy metals ions based on the calixarene derivatives-modified piezoelectric resonators: a review. Int. J. Environ. Anal. Chem. 99(9): 824–853.

Ehala, S., E. Makrlík, P. Toman and V. Kašička. 2010. ACE applied to the quantitative characterization of benzo-18-crown-6-ether binding with alkali metal ions in a methanol–water solvent system. Electrophoresis. 31(4): 702–708.

Fang, Y., L. Wu, J. Liao, L. Chen, Y. Yang, N. Liu, L. He, S. Zou, W. Feng and L. Yuan. 2013. Pillar[5]arene-based phosphine oxides: novel ionophores for solvent extraction separation of f-block elements from acidic media. RSC Adv. 3(30): 12376–12383.

Fang, Y., X. Yuan, L. Wu, Z. Peng, W. Feng, N. Liu, D. Xu, S. Li, A. Sengupta, P.K. Mohapatra and L. Yuan. 2015a. Ditopic CMPO-pillar[5]arenes as unique receptors for efficient separation of americium(III) and europium(III). Chem. Commun. 51(20): 4263–4266.

Fang, Y., C. Li, L. Wu, B. Bai, X. Li, Y. Jia, W. Feng and L. Yuan. 2015b. A non-symmetric pillar[5]arene based on triazole-linked 8-oxyquinolines as a sequential sensor for thorium(IV) followed by fluoride ions. Dalton Trans. 44(33): 14584–14588.

Farruggia, G., S. Iotti, L. Prodi, M. Montalti, N. Zaccheroni, P.B. Savage, V. Trapani, P. Sale and F.I. Wolf. 2006. 8-Hydroxyquinoline derivatives as fluorescent sensors for magnesium in living cells. J. Am. Chem. Soc. 128(1): 344–350.

Feng, W.-X., Z. Sun, Y. Zhang, Y.-M. Legrand, E. Petit, C.-Y. Su and M. Barboiu. 2017. Bis-15-crown-5-ether-pillar[5]arene K+-Responsive Channels. Org. Lett. 19(6): 1438–1441.

Geng, Q.-X., H. Cong, Z. Tao, L.F. Lindoy and G. Wei. 2016. Cucurbit[7]uril-improved recognition by a fluorescent sensor for cadmium and zinc cations. Supramol. Chem. 28(9–10): 784–791.

Georghiou, P.E., S. Rahman, A. Alrawashdeh, A. Alodhayb, G. Valluru, K.S. Unikela and G.J. Bodwell. 2020. Synthesis, supramolecular complexation and DFTstudies of a bis(pyrene)-appended 'capped' triazole-linked calix[4]arene as Zn^{2+} and Cd^{2+} fluorescent chemosensors. Supramol. Chem. 32(5): 325–333.

Gokel, G.W., L.J. Barbour, R. Ferdani and J. Hu. 2002. Lariat Ether Receptor Systems Show Experimental Evidence for Alkali Metal Cation−π Interactions. Acc. Chem. Res. 35(10): 878–886.

Guler Akgemci, E., H. Bingol, M. Ersoz and I. Stibor. 2008. Facilitated transfer of alkali metal ions by a tetraester derivative of thiacalix[4]arene at the liquid–liquid interface. Electroanalysis. 20(12): 1354–1360.

Gumpu, M.B., S. Sethuraman, U.M. Krishnan and J.B.B. Rayappan. 2015. A review on detection of heavy metal ions in water–an electrochemical approach. Sens. Actuators B Chem. 213: 515–533.

Gupta, V.K., A.K. Jain, U. Khurana and L.P. Singh. 1999. PVC-based neutral carrier and organic exchanger membranes as sensors for the determination of Ba^{2+} and Sr^{2+}. Sens. Actuators B Chem. 55(2): 201–211.

Gupta, V.K., M.R. Ganjali, P. Norouzi, H. Khani, A. Nayak and S. Agarwal. 2011. Electrochemical analysis of some toxic metals by ion–selective electrodes. Crit. Rev. Anal. Chem. 41(4): 282–313.

Han, C., L. Zhang and H. Li. 2009. Highly selective and sensitive colorimetric probes for Yb^{3+} ions based on supramolecular aggregates assembled from β-cyclodextrin–4,4′-dipyridine inclusion complex modified silver nanoparticles. Chem. Commun. (24): 3545–3547.

Israëli, Y., C. Bonal, C. Detellier, J.-P. Morel and N. Morel-Desrosiers. 2002. Complexation of the La(III) cation by *p*-sulfonatocalix[4]arene—A [139]La NMR study. Can. J. Chem. 80(2): 163–168.

Izatt, R.M., J.S. Bradshaw, R.L. Bruening, B.J. Tarbet and M.L. Bruening. 1995. Solid phase extraction of ions using molecular recognition technology. Pure & Appl. Chem. 67(7): 1069–1074.

Jain, A.K., J. Raisoni and S. Jain. 2008. Calcium(II)-selective potentiometric sensor based on p-isopropylcalixarene in PVC matrix. Int. J. Environ. Anal. Chem. 88(3): 209–221.

Jaiswal, M.K., P.K. Muwal, S. Pandey and P.S. Pandey. 2017. A novel hybrid macrocyclic receptor based on bile acid and calix[4]arene frameworks for metal ion recognition. Tetrahedron Lett. 58(22): 2153–2156.

Ji, H.-F., R. Dabestani, G.M. Brown and R.A. Sachleben. 2000. A new highly selective calix[4]crown-6 fluorescent caesium probe. Chem. Commun. (10): 833–834.

Ji, H.-F., R. Dabestani, G.M. Brown and R.L. Hettich. 2001. Synthesis and sensing behavior of cyanoanthracene modified 1,3-alternate calix[4]benzocrown-6: a new class of Cs^+ selective optical sensors. J. Chem. Soc., Perkin Trans. 2. (4): 585–591.

Jia, Y., Y. Fang, Y. Li, L. He, W. Fan, W. Feng, Y. Yang, J. Liao, N. Liu and L. Yuan. 2014. Pillar[5]arenes bearing phosphine oxide pendents as Hg^{2+} selective receptors. Talanta. 125: 322–328.

Jobe, K., C.H. Brennan, M. Motevalli, S.M. Goldup and M. Watkinson. 2011. Modular 'click' sensors for zinc and their application *in vivo*. Chem. Commun. 47(21): 6036–6038.

Joseph, R. 2020. Selective detection of Fe^{3+}, F^-, and cysteine by a novel triazole-linked decaamine derivative of pillar[5]arene and its metal ion complex in water. ACS Omega. 5(11): 6215–6220.

Joseph, R., A. Asok and K. Joseph. 2020. Quinoline appended pillar[5]arene (QPA) as Fe^{3+} sensor and complex of Fe^{3+} (FeQPA) as a selective sensor for F^-, arginine and lysine in the aqueous medium. Spectrochim. Acta A Mol. Biomol. Spectrosc. 224: 117390.

Kaim, W., B. Schwederski and A. Klein. 2014. Bioinorganic Chemistry: Inorganic Elements in the Chemistry of Life: An Introduction and Guide. Germany, Wiley.

Kaur, N. and S. Kumar. 2011. Colorimetric metal ion sensors. Tetrahedron. 67(48): 9233–9264.

Kaur, B., N. Kaur and S. Kumar. 2018. Colorimetric metal ion sensors—a comprehensive review of the years 2011–2016. Coord. Chem. Rev. 358: 13–69.

Kim, J.S., O.J. Shon, J.W. Ko, M.H. Cho, I.Y. Yu and J. Vicens. 2000. Synthesis and metal ion complexation studies of proton-ionizable calix[4]azacrown ethers in the 1,3-alternate conformation. J. Org. Chem. 65(8): 2386–2392.

Kim, J.S., O.J. Shon, S.H. Yang, J.Y. Kim and M.J. Kim. 2002. Chromogenic indoaniline armed-calix[4]azacrowns. J. Org. Chem. 67(18): 6514–6518.

Kim, S.H., J.K. Choi, S.K. Kim, W. Sim and J.S. Kim. 2006a. On/off fluorescence switch of a calix[4]arene by metal ion exchange. Tetrahedron Lett. 47(22): 3737–3741.

Kim, S.H., J.S. Kim, S.M. Park and S.-K. Chang. 2006b. Hg^{2+}-Selective OFF−ON and Cu^{2+}-selective ON−OFF type fluoroionophore based upon cyclam. Org. Lett. 8(3): 371–374.

Kothur, R.R., F. Fucassi, G. Dichello, L. Doudet, W. Abdalaziz, B.A. Patel, G.W.V. Cave, I.A. Gass, D.K. Sarker, S.V. Mikhalovsky and P.J. Cragg. 2016. Synthesis and applications of copillar[5]arene dithiols. Supramol. Chem. 28(5-6): 436–443.

Kozlowski, C.A., T. Girek, W. Walkowiak and J.J. Koziol. 2005. Application of hydrophobic β-cyclodextrin polymer in separation of metal ions by plasticized membranes. Sep. Purif. Technol. 46(3): 136–144.

Lambert, B., V. Jacques, A. Shivanyuk, S.E. Matthews, A. Tunayar, M. Baaden, G. Wipff, V. Böhmer and J.F. Desreux. 2000. Calix[4]arenes as selective extracting agents. an NMR dynamic and conformational investigation of the Lanthanide(III) and Thorium(IV) complexes. Inorg. Chem. 39(10): 2033–2041.

Lau, Y.H., J.R. Price, M.H. Todd and P.J. Rutledge. 2011. A click fluorophore sensor that can distinguish CuII and HgII via selective anion-induced demetallation. Chem. Eur. J. 17(10): 2850–2858.

Leray, I., Z. Asfari, J. Vicens and B. Valeur. 2002. Synthesis and binding properties of calix[4]biscrown-based fluorescent molecular sensors for caesium or potassium ions. J. Chem. Soc., Perkin Trans. 2(8): 1429–1434.

Leydier, A., D. Lecerclé, S. Pellet-Rostaing, A. Favre-Réguillon, F. Taran and M. Lemaire. 2008. Sequestering agents for uranyl chelation: new calixarene ligands. Tetrahedron. 64(49): 11319–11324.

Li, C. and W.-T. Wong. 2002. Luminescent terbium(III) complexes with pendant crown ethers responding to alkali metal ions and aromatic antennae in aqueous solution. Chem. Commun. (18): 2034–2035.

Li, H., J. Zhan, M. Chen, D. Tian and Z. Zou. 2010. Metal ions recognition by 1,2,3-triazolium calix[4]arene esters synthesized via click chemistry. J. Incl. Phenom. Macrocycl. Chem. 66(1): 43–47.

Liang, Z., Z. Liu and Y. Gao. 2007. A selective colorimetric chemosensor based on calixarene framework for lanthanide ions-Dy^{3+} and Er^{3+}. Tetrahedron Lett. 48(20): 3587–3590.

Lippolis, V. and M. Shamsipur. 2006. Synthesis, coordination properties, and analytical applications of mixed donor macrocycles containing the 1,10-phenanthroline sub-unit. J. Iran. Chem. Soc. 3(2): 105–127.

Liu, Y., H. Wang, H.-Y. Zhang and P. Liang. 2004. A metallo-capped polyrotaxane containing calix[4]arenes and cyclodextrins and its highly selective binding for Ca^{2+}. Chem. Commun. 20: 2266–2267.

Ma, X.-Q., Y. Wang, T.-B. Wei, L.-H. Qi, X.-M. Jiang, J.-D. Ding, W.-B. Zhu, H. Yao, Y.-M. Zhang and Q. Lin. 2019. A novel AIE chemosensor based on quinoline functionalized Pillar[5]arene for highly selective and sensitive sequential detection of toxic Hg^{2+} and CN^-. Dyes Pigm. 164: 279–286.

Makrlík, E., P. Vaňura, P. Selucký and Z. Asfari. 2015. Calix[4]arene-bis(coumarin-crown-6) as an extraordinarily effective macrocyclic receptor for Cs^+, Ag^+, and Tl^+. J. Incl. Phenom. Macrocycl. Chem. 81(1): 169–174.

Malval, J.-P., I. Leray and B. Valeur. 2005. A highly selective fluorescent molecular sensor for potassium based on a calix[4]bisazacrown bearing boron-dipyrromethene fluorophores. New J. Chem. 29(8): 1089–1094.

Maniyazagan, M., C. Rameshwaran, R. Mariadasse, J. Jeyakanthan, K. Premkumar and T. Stalin. 2017. Fluorescence sensor for Hg^{2+} and Fe^{3+} ions using 3,3′–dihydroxybenzidine: α–cyclodextrin supramolecular complex: Characterization, in-silico and cell imaging study. Sens. Actuators B Chem. 242: 1227–1238.

Matthews, S.E., P. Schmitt, V. Felix, M.G.B. Drew and P.D. Beer. 2002. Calix[4]tubes: a new class of potassium-selective ionophore. J. Am. Chem. Soc. 124(7): 1341–1353.

Métivier, R., I. Leray and B. Valeur. 2004. Lead and mercury sensing by calixarene-based fluoroionophores bearing two or four dansyl fluorophores. Chem. Eur. J. 10(18): 4480–4490.

Mikulášek, L., B. Grüner, C. Dordea, V. Rudzevich, V. Böhmer, J. Haddaoui, V. Hubscher-Bruder, F. Arnaud-Neu, J. Čáslavský and P. Selucký. 2007. tert-Butyl-calix[4]arenes substituted at the narrow rim with cobalt bis(dicarbollide)(1–) and CMPO groups – new and efficient extractants for lanthanides and actinides. Eur. J. Org. Chem. 2007(28): 4772–4783.

Pearson, A.J. and W. Xiao. 2003. Fluorescent photoinduced electron transfer (PET) sensing molecules with p-Phenylenediamine as electron donor. J. Org. Chem. 68(13): 5361–5368.

Pereira Silva, M.J.J., J.M. Haider, R. Heck, M. Chavarot, A. Marsura and Z. Pikramenou. 2003. Ruthenium and osmium podate cyclodextrins with dual-function recognition sites for luminescent sensing. Supramol. Chem. 15(7-8): 563–571.

Podyachev, S.N., S.N. Sudakova, V.V. Syakaev, A.K. Galiev, R.R. Shagidullin and A.I. Konovalov. 2008. The preorganization effect of the calix[4]arene platform on the extraction properties of acetylhydrazide groups with transition metal ions. Supramol. Chem. 20(5): 479–486.

Ramírez, F.d.M., S. Varbanov, J. Padilla and J.-C.G. Bünzli. 2008. Physicochemical properties and theoretical modeling of actinide complexes with a para-tert-Butylcalix[6] arene bearing phosphinoyl pendants. Extraction capability of the calixarene toward f elements. J. Phys. Chem. B. 112(35): 10976–10988.

Root, H.D., G. Thiabaud and J.L. Sessler. 2020. Reduced texaphyrin: A ratiometric optical sensor for heavy metals in aqueous solution. Front. Chem. Sci. Eng. 14(1): 19–27.

Roper, E.D., V.S. Talanov, M.G. Gorbunova, R.A. Bartsch and G.G. Talanova. 2007. Optical determination of Thallium(I) and Cesium(I) with a fluorogenic calix[4]arenebis(crown-6 ether) containing one pendent dansyl group. Anal. Chem. 79(5): 1983–1989.

Roper, E.D., V.S. Talanov, R.J. Butcher and G.G. Talanova. 2008. Selective recognition of thallium(I) by 1,3-alternate calix[4]arene-bis(crown-6 ether): A new talent of the known ionophore. Supramol. Chem. 20(1-2): 217–229.

Rouis, A., R. Mlika, C. Dridi, J. Davenas, H. Ben Ouada, H. Halouani, I. Bonnamour and N. Jaffrezic. 2006. Optical spectroscopy studies of the complexation of chromogenic azo-calix[4]arene with Eu^{3+}, Ag^+ and Cu^{2+} ions. Mater. Sci. Eng. C. 26(2): 247–252.

Sansone, F., M. Fontanella, A. Casnati, R. Ungaro, V. Böhmer, M. Saadioui, K. Liger and J.-F. Dozol. 2006. CMPO-substituted calix[6]- and calix[8]arene extractants for the separation of An^{3+}/Ln^{3+} from radioactive waste. Tetrahedron. 62(29): 6749–6753.

Sengupta, A., X. Yuan, W. Feng, N.K. Gupta and L. Yuan. 2017. Highly efficient extraction of tetra- and hexavalent plutonium using DGA functionalized pillar[5]arene in RTIL: understanding speciation, thermodynamics and radiolytic stability. Sep. Sci. Technol. 52(17): 2767–2776.

Sessler, J.L., P.J. Melfi, D. Seidel, A.E.V. Gorden, D.K. Ford, P.D. Palmer and C.D. Tait. 2004. Hexaphyrin(1.0.1.0.0.0). A new colorimetric actinide sensor. Tetrahedron. 60(49): 11089–11097.

Shahid, M., H.M. Chawla and P. Bhatia. 2016. A calix[4]arene based turn off/turn on molecular receptor for Cu^{2+} and CN^- ions in aqueous medium. Sens. Actuators B Chem. 237: 470–478.

Singh, N. and D.O. Jang. 2009. Synthesis of calix[4]arene-based dipodal receptors: competitive solvent extraction and liquid bulk membrane transport for selective recovery of Cu^{2+} supramol. Chem. 21(5): 351–357.

Sirit, A., E. Kocabas, S. Memon, A. Karakucuk and M. Yilmaz. 2005. Synthesis and metal ion recognition properties of a novel chiral calix[4](azoxa)crown-7. Supramol. Chem. 17(3): 251–256.

Sliwa, W. and T. Girek. 2010. Calixarene complexes with metal ions. J. Incl. Phenom. Macrocycl. Chem. 66(1): 15–41.

Song, K.C., M.G. Choi, D.H. Ryu, K.N. Kim and S.-K. Chang. 2007. Ratiometric chemosensing of Mg^{2+} ions by a calix[4]arene diamide derivative. Tetrahedron Lett. 48(31): 5397–5400.

Souchon, V., I. Leray and B. Valeur. 2006. Selective detection of cesium by a water-soluble fluorescent molecular sensor based on a calix[4]arene-bis(crown-6-ether). Chem. Commun. (40): 4224–4226.

Suresh, P., I.A. Azath and K. Pitchumani. 2010. Naked-eye detection of Fe^{3+} and Ru^{3+} in water: Colorimetric and ratiometric sensor based on per-6-amino-β-cyclodextrin/p-nitrophenol. Sens. Actuators B Chem. 146(1): 273–277.

Talanova, G.G., E.D. Roper, N.M. Buie, M.G. Gorbunova, R.A. Bartsch and V.S. Talanov. 2005. Novel fluorogenic calix[4]arene-bis(crown-6-ether) for selective recognition of thallium(I). Chem. Commun. (45): 5673–5675.

Tamanini, E., A. Katewa, L.M. Sedger, M.H. Todd and M. Watkinson. 2009. A synthetically simple, click-generated cyclam-based Zinc(II) sensor. Inorg. Chem. 48(1): 319–324.

Tamanini, E., K. Flavin, M. Motevalli, S. Piperno, L.A. Gheber, M.H. Todd and M. Watkinson. 2010. Cyclam-based "Clickates": Homogeneous and heterogeneous fluorescent sensors for Zn(II). Inorg. Chem. 49(8): 3789–3800.

Tieke, B., A. El-Hashani, A. Toutianoush and A. Fendt. 2008. Multilayered films based on macrocyclic polyamines, calixarenes and cyclodextrins and transport properties of the corresponding membranes. Thin Solid Films. 516(24): 8814–8820.

Turshatov, A.A., N.B. Melnikova, Y.D. Semchikov, I.S. Ryzhkina, T.N. Pashirova, D. Möbius and S.Y. Zaitsev. 2004. Interaction of monolayers of calix[4]resorcinarene derivatives with copper ions in the aqueous subphase. Colloids Surf. A Physicochem. Eng. Asp. 240(1): 101–106.

Ueno, A., A. Ikeda, H. Ikeda, T. Ikeda and F. Toda. 1999. Fluorescent cyclodextrins responsive to molecules and metal ions. Fluorescence properties and inclusion phenomena of N^{α}-Dansyl-l-lysine-β-cyclodextrin and monensin-incorporated N^{α}-Dansyl-l-lysine-β-cyclodextrin. J. Org. Chem. 64(2): 382–387.

Wallace, K.J., A.D.G. Johnson, W.S. Jones and E. Manandhar. 2018. Chemodosimeters and chemoreactands for sensing ferric ions. Supramol. Chem. 30(5-6): 353–383.

Wang, P., M. Saadioui, C. Schmidt, V. Böhmer, V. Host, J.F. Desreux and J.-F. Dozol. 2004. Dendritic octa-CMPO derivatives of calix[4]arenes. Tetrahedron. 60(11): 2509–2515.

Wang, K., D. Guo, B. Jiang and Y. Liu. 2009. Highly selective fluorescent chemosensor for Na$^+$ based on pyrene-modified calix[4]arene derivative. Sci. China Chem. 52(4): 513–517.

Wei, P., D. Li, B. Shi, Q. Wang and F. Huang. 2015. An anthracene-appended 2: 3 copillar[5] arene: Synthesis, computational studies, and application in highly selective fluorescence sensing for Fe(III) ions. Chem. Commun. 51(82): 15169–15172.

Wei, T.-B., J.-F. Chen, X.-B. Cheng, H. Li, Q. Lin, H. Yao and Y.-M. Zhang. 2016a. A novel functionalized pillar[5]arene for forming a fluorescent switch and a molecular keypad. RSC Adv. 6(70): 65898–65901.

Wei, T.-B., X.-B. Cheng, H. Li, F. Zheng, Q. Lin, H. Yao and Y.-M. Zhang. 2016b. Novel functionalized pillar[5]arene: synthesis, assembly and application in sequential fluorescent sensing for Fe^{3+} and F$^-$ in aqueous media. RSC Adv. 6(25): 20987–20993.

Wong, J.K.H., S. Ast, M. Yu, R. Flehr, A.J. Counsell, P. Turner, P. Crisologo, M.H. Todd and P.J. Rutledge. 2016. Synthesis and evaluation of 1,8-disubstituted-cyclam/naphthalimide conjugates as probes for metal ions. ChemistryOpen. 5(4): 375–385.

Wu, L., Y. Fang, Y. Jia, Y. Yang, J. Liao, N. Liu, X. Yang, W. Feng, J. Ming and L. Yuan. 2014. Pillar[5]arene-based diglycolamides for highly efficient separation of americium(III) and europium(III). Dalton Trans. 43(10): 3835–3838.

Xia, W.-S., R.H. Schmehl, C.-J. Li, J.T. Mague, C.-P. Luo and D.M. Guldi. 2002. Chemosensors for lead (II) and alkali metal ions based on self-assembling fluorescence enhancement (SAFE). J. Phys. Chem. B. 106(4): 833–843.

Xin, P., Y. Sun, H. Kong, Y. Wang, S. Tan, J. Guo, T. Jiang, W. Dong and C.-P. Chen. 2017. A unimolecular channel formed by dual helical peptide modified pillar[5] arene: correlating transmembrane transport properties with antimicrobial activity and haemolytic toxicity. Chem. Commun. 53(83): 11492–11495.

Xin, P., H. Kong, Y. Sun, L. Zhao, H. Fang, H. Zhu, T. Jiang, J. Guo, Q. Zhang, W. Dong and C.-P. Chen. 2019. Artificial K$^+$ channels formed by pillararene-cyclodextrin hybrid molecules: tuning cation selectivity and generating membrane potential. Angew. Chem. Int. Ed. 58(9): 2779–2784.

Yakimova, L.S., D.N. Shurpik, A.R. Makhmutova and I.I. Stoikov. 2017. Pillar[5]arenes bearing amide and carboxylic groups as synthetic receptors for alkali metal ions. Macroheterocycles. 10(2): 226–232.

Yang, R., Y. Zhang, K.A. Li, F. Liu and W. Chan. 2004. Fluorescent ratioable recognition of Cu^{2+} in water using a pyrene-attached macrocycle/γ-cyclodextrin complex. Anal. Chim. Acta. 525(1): 97–103.

Yuan, X., Y. Cai, L. Chen, S. Lu, X. Xiao, L. Yuan and W. Feng. 2020. Phosphine oxides functionalized pillar[5]arenes for uranyl extraction: Solvent effect and thermodynamics. Sep. Purif. Technol. 230: 115843.

Yusof, N.N.M., T. Kobayashi and Y. Kikuchi. 2016. Ionic imprinting polymers using diallylaminomethyl-calix[4] resorcinarene host for the recognition of Pb(II) ions. Polym. Polym. Compos. 24(9): 687–694.

Zhang, X.X., K.E. Krakowiak, G. Xue, J.S. Bradshaw and R.M. Izatt. 2000. A highly selective compound for lead: Complexation studies of decamethylcucurbit[5]uril with metal ions. Ind. Eng. Chem. Res. 39(10): 3516–3520.

Zhao, J., H.-J. Kim, J. Oh, S.-Y. Kim, J.W. Lee, S. Sakamoto, K. Yamaguchi and K. Kim. 2001. Cucurbit[n]uril derivatives soluble in water and organic solvents. Angew. Chem. Int. Ed. 40(22): 4233–4235.

Zhu, W., H. Fang, J.-X. He, W.-H. Jia, H. Yao, T.-B. Wei, Q. Lin and Y.-M. Zhang. 2018. Novel 2-(hydroxy)-naphthyl imino functionalized pillar[5]arene: A highly efficient supramolecular sensor for tandem fluorescence detection of Fe^{3+} and F^- and the facile separation of Fe^{3+}. New J. Chem. 42(14): 11548–11554.

Zor, E., A. Saf and H. Bingol. 2013. Spectrophotometric and voltammetric characterization of a novel selective electroactive chemosensor for Mg^{2+}. Cent. Eur. J. Chem. 11(4): 554–560.

Chapter 3

Macrocyclic Receptors for Precious Metal Ions

INTRODUCTION

Precious metals also known as *noble metals* are rare, less reactive than most elements and usually have high economic value than common industrial metals. Noble metals such as gold and silver are the most popular precious metals. They are widely used not only in jewellery, coinage and art, but also in various industrial applications. Other precious metals include the platinum group of metals: platinum, palladium, rhodium, osmium, iridium and ruthenium. These metals usually occur together in nature due to the similarity in their physical and chemical properties. Among these group of metals, platinum is most well-known due to its major use in jewellery making. Owing to their remarkable mechanical, corrosion and oxidation-resistant properties, these metals are extensively used in electrical, electronics and medical components as well as high-performance catalysts (Zaghbani et al. 2007). Precious metals such as silver, gold, platinum and palladium are frequently used in the preparation of dental materials, anticancer drugs, catalysts and fuel cells. The demand for precious metals is driven not only by their practical use but also from the perspective of investment. For instance, there is a rising demand for precious metals in car catalytic converters and the jewellery industry.

However, contrary to their growing applications in industrial and jewellery sectors, precious metals are scarce due to their low natural abundance as well as the complexity of their extraction and refining processes (Zaghbani et al. 2007). Moreover, rapid technological advancements in recent decades have led to a generation of enormous e-waste which causes environmental pollution, health hazards and substantial loss of valuable metal resources. Consequently, the recovery of precious metals from e-waste and industrial waste solutions is of paramount importance to meet future demands and also to preserve ecology. This has resulted in an ever-growing interest among researchers to develop separation methods to retrieve precious metals as well as real-time monitoring methods to detect these metal ions in environmental and biological samples. These include the design and synthesis of chemosensors, ion exchange and chelating resins, precipitation methods, liquid and solid-phase extraction and liquid membrane transport techniques. Among these methods, the hydrometallurgical process of solvent extraction for the recovery of noble metals from acidic media is well-known. However, the selectivity of these methods is a very crucial parameter in the recovery of noble metals from waste mixtures.

In this regard, various research strategies and approaches to create tailor-made synthetic supramolecular assemblies for noble metal ions have been meticulously explored and reported in literature. A variety of macrocyclic receptors based on crown ethers, aza-crown ethers, thia-crown ethers, aza-thia-crown ethers, cyclam, cyclen, thiacalixarenes, pillarenes and their functionalized derivatives have been designed and exclusively investigated for a response towards precious metal ions. According to Pearson's HSAB theory, ligands bearing S as donor atoms are classified as soft bases and tend to have an affinity towards soft acids, such as silver(I), gold(III) and platinum group of metals. For instance, the presence of sulphur atoms in place of the usual methylene bridges in thiacalixarenes remarkably modifies its properties than that of the parent calixarenes. Consequently, thiacalixarenes and their derivatives have been reported to exhibit modified solubility and better conformational flexibility as well as complexation ability towards soft metal ions (Fontàs et al. 2007). Likewise, the substitution of oxygen by sulphur atoms in the crown and aza-crown ethers can lead to greater soft metal ion discrimination. For example, mixed N, O, S- donor crown ethers have been efficiently used as selective extractants for soft metal ions (Xu et al. 2010).

Though research literature offers a large number of supramolecular complexes with several noble metals, this chapter summarizes the application of macrocyclic receptors with respect to the three most researched precious metals ions, silver(I), gold(III) and palladium(II). The selected examples of macrocyclic host molecules for these cations are primarily considered and described on the basis of properties, such as complexation ability, solvent effect, binding affinity, selectivity, sensitivity, spectroscopic properties, extraction ability and ion transport.

MACROCYCLIC RECEPTORS FOR Ag$^+$ IONS

Silver is commonly produced during the refining of gold, copper, nickel and zinc. As a precious metal, it is widely used to make jewellery and other valuable articles. The compounds of silver(I) are extensively used in photography, electricals, electroplating and pharmaceutical industries. In the human body, silver is not known to have any biological role. However, ingestion of silver(I) compounds may lead to serious health conditions such as argyria, neuronal disorders, mental fatigue, gastroenteritis, rheumatism and knotting of the cartilage (Gumpu et al. 2015). Silver oxide is known to obliterate the environmental benign bacteria by obstructing their growth and reproductive ability (Liu et al. 2015).

As silver ions gain access to the environment through wastes, their bioaccumulation and toxicity become a major problem that has received considerable attention from researchers (Chen et al.2010). Hence, detection and monitoring of Ag^+ ions in soil, food and water are essential to preserving the environment and human health. Various methods such as UV-VIS spectrometry, atomic absorption spectrometry, inductively coupled plasma spectrometry and fluorescence spectrometry have been effectively utilized to detect Ag^+ ions. Several macrocyclic receptors containing N, O, S as binding sites have been designed and synthesized to complex with soft silver(I) ions.

Gherou and coworkers developed new fixed sites plasticized cellulose triacetate membranes containing macrocyclic polyethers, dibenzo-18-crown-6 **1**, hexathia-18-crown-6 **2**, diaza-18-crown-6 **3** and hexaaza-18-crown-6 **4, and** examined them for the transport of Cu^{2+}, Ag^+, and Au^{3+} ions. As compared to supported liquid membranes (SLM) containing the same polyethers, but lacking the membrane phase, fixed sites membranes (FSM) exhibited high transport efficiency for the Ag^+, Cu^{2+} and Au^{3+}. Thus, the incorporation of a selective macrocyclic polyether into the polymer matrix gave stability and durability (over 15 days) to the membranes for ion transport (Gherrou et al. 2004).

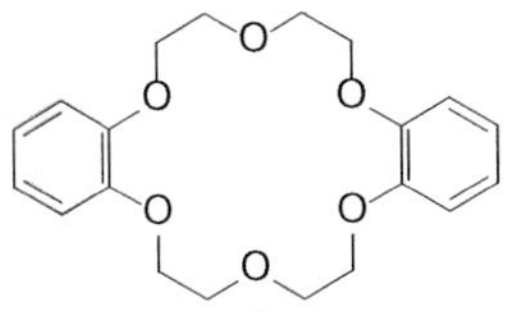

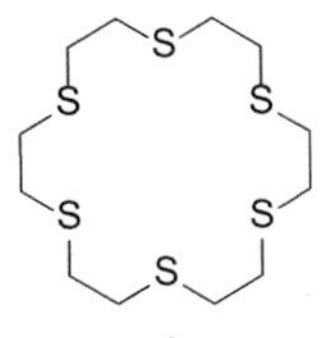

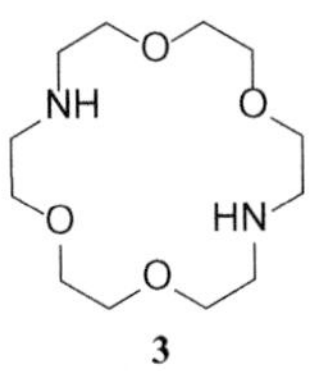

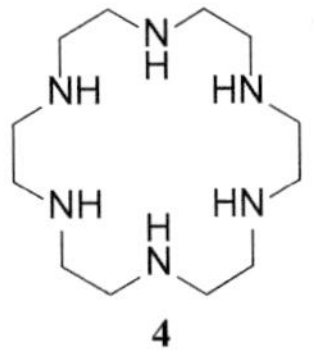

With the aim to study structure-function effects in terms of metal-ion discrimination behaviour, Price et al. synthesized compound **5-10** by progressive N-benzylation of the secondary amines in O_2N_4-donor 20-membered macrocyclic rings. A range of techniques including X-Ray diffraction, DFT computations, liquid-liquid extraction (aqueous/ chloroform) and competitive membrane transport experiments (water/chloroform/water) was used for this purpose. Collectively, the results indicated that with a progressive increase in N-benzylation of the secondary amine donor groups in the parent macrocycle the selectivity towards Ag^+ ions also enhanced as compared to the other six metal ions examined, with ligand 5 being the least selective. The observed behaviour served to exemplify the mechanism of *'selective detuning'* for metal ion discrimination where N-benzylation tends to significantly decrease the affinity of such ligands towards the tested Co^{2+}, Ni^{2+}, Zn^{2+}, Cu^{2+}, Cd^{2+}, Pb^{2+} metal ions, except Ag^+ (Price et al. 2004).

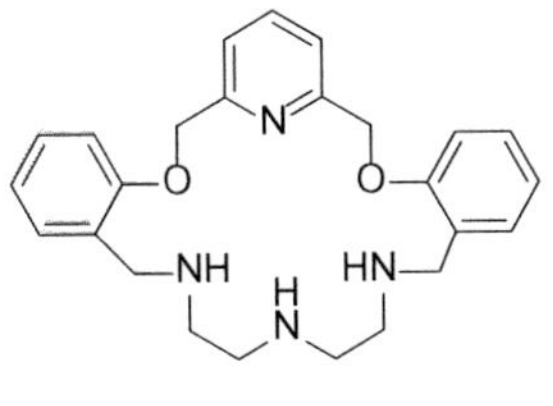

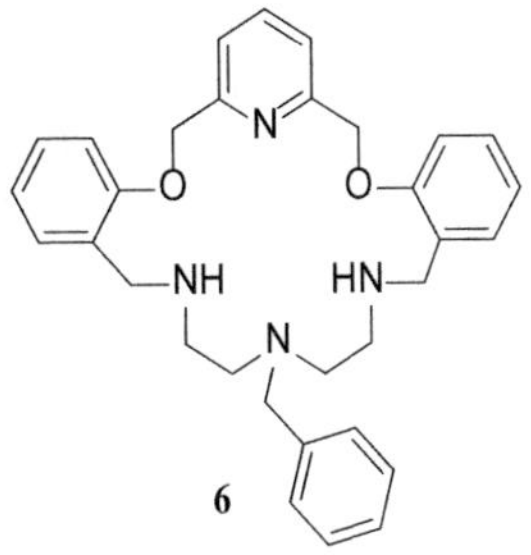

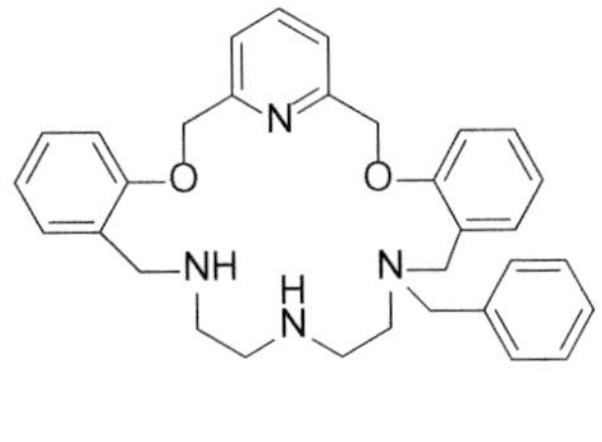

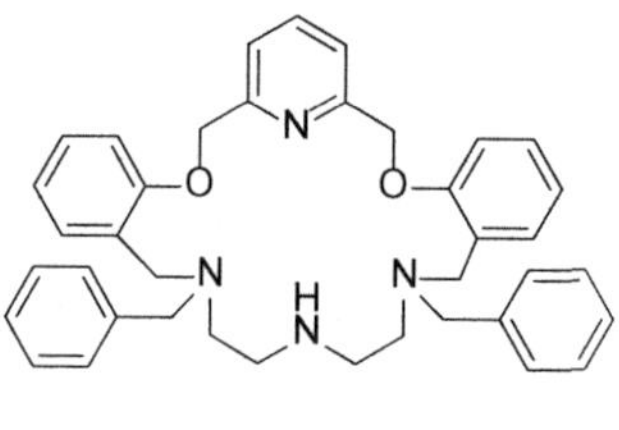

8

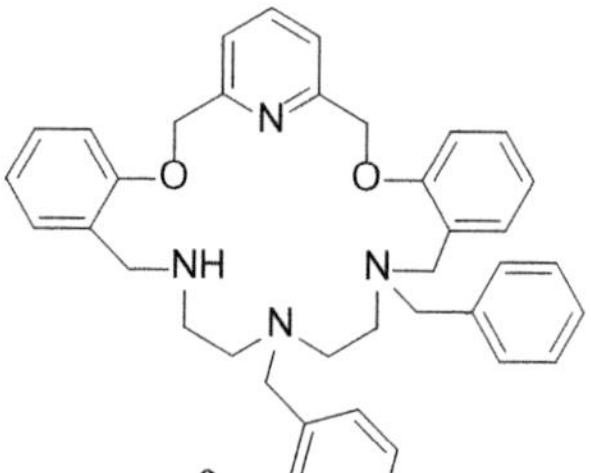

9

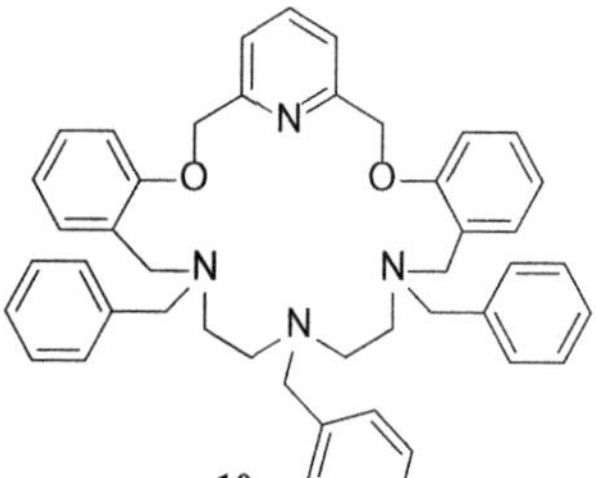

10

In a similar study, Dong and coworkers investigated the interaction of a series of successively N-benzylated derivatives of 1,4,8,11-tetraazacyclotetradecane (cyclam) with Co^{2+}, Ni^{2+}, Zn^{2+}, Cu^{2+}, Cd^{2+}, Pb^{2+} and Ag^+ ions. During the membrane transport experiments (water/chloroform/water) with each of the benzylated cyclam ligands in the chloroform phase, the tetra-N-benzyl cyclam derivative **11** exhibited remarkable sole transport selectivity for Ag^+ ions from the aqueous phase over the other competitive metal ions used in this study (Dong et al. 2003).

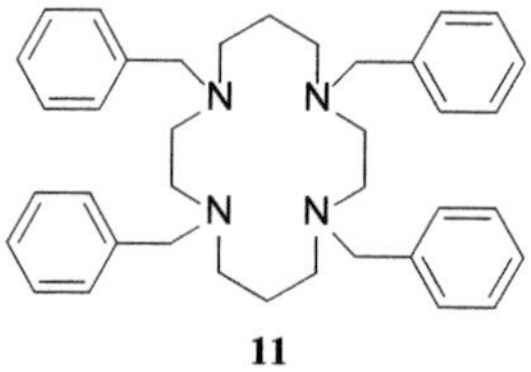

11

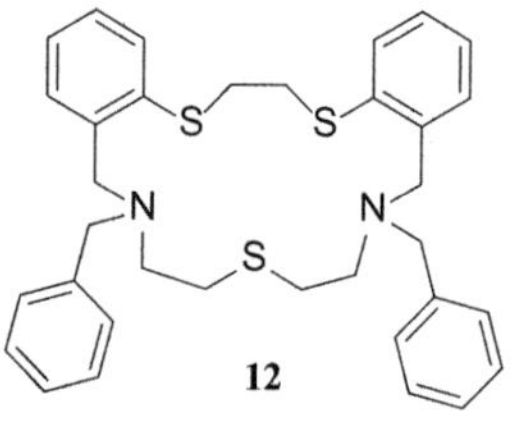

12

Vasilescu et al. designed and synthesized dibenzyl-N-substituted, 17-membered macrocyclic ring **12**, which contained an S_3N_2-donor set with the desired N–S–N donor atom sequence. Bulk membrane transport studies (water/chloroform/water) in the presence of six transition and post-transition metal ions were performed. The ligand was found to be highly selective towards Ag^+ ions over other metal ions. The stability constant for silver complex (1:1) was at least 10^5 times more stable than any other metal complexes examined. Thus, the ligand exhibited considerable discrimination for silver over the other six metal ions (Vasilescu et al. 2006).

Electrospray ionization mass spectrometry (ESI-MS) was used by Williams and coworkers to investigate the metal-binding selectivity of a series of novel caged aza-crown ether macrocycles. The results showed that the nitrogen-containing caged macrocycles **13**, **14**, **15**, **16** were found to be very selective for Ag^+ over a variety of alkali and other transition metal ions. The enhanced silver affinity and selectivity was attributed to a larger cavity size, nitrogen hetero atoms and the presence of aromatic and structurally stabilizing substituents on the ligand framework. The aromatic group may be involved in pi- interactions with the metal ions. Furthermore, the triaza-18-crown-6 bearing two phenyl groups and a cage group macrocycle 13 was far more superior than ligand 14 in selectively extracting Ag^+ ions from aqueous solutions over other transition and the most common alkali and alkaline earth metal ions (Williams et al. 2003).

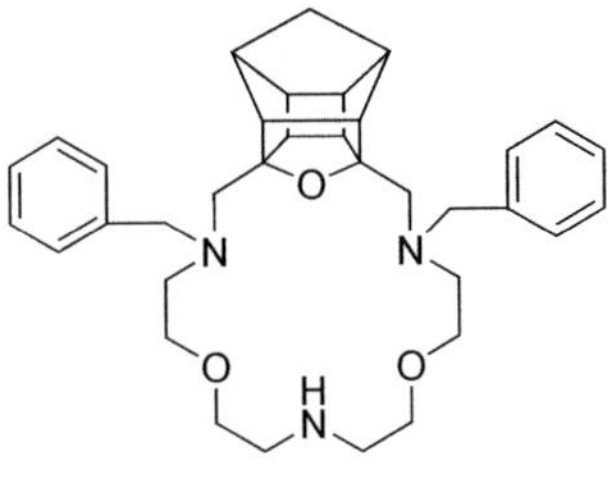

13

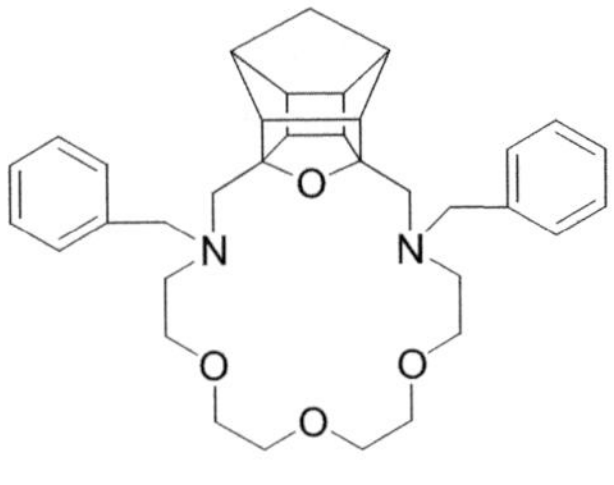

14

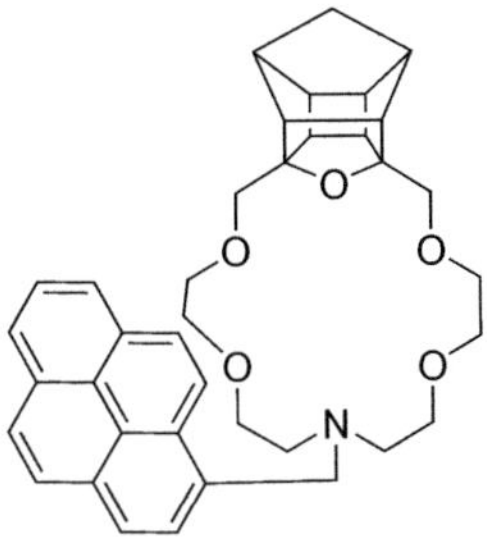

15

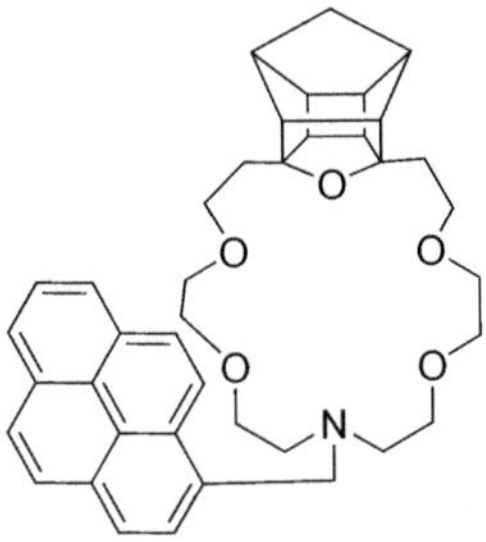

16

Intending to design N_2S_2-crown ether extractants for metal ions with small cavity size, Ocak et al. synthesized new N_2S_2-macrocyclic Schiff base ligands **17, 18**. The metal extractability and selectivity of the two ligands was explored using liquid-liquid extraction of metal picrates of K^+, Na^+, Hg^{2+}, Cd^{2+}, Zn^{2+}, Cu^{2+}, Ni^{2+} and Mn^{2+} from an aqueous phase to an organic phase. The effect of $CHCl_3$ and CH_2Cl_2 as organic solvents over the metal picrate extractions was examined by UV-Vis spectrometry. It was found that both 14-membered N_2S_2 macrocycle 17 and 16-membered N_2S_2 macrocycle 18 extracted Ag^+ ions selectively over other metal ions tested. The Ag^+ complexes with 2:1 (L:M) stoichiometry were formed for both the ligands. However, ligand 18 was able to extract Ag^+ from the aqueous phase to the dichloromethane phase most efficiently (Ocak et al. 2006).

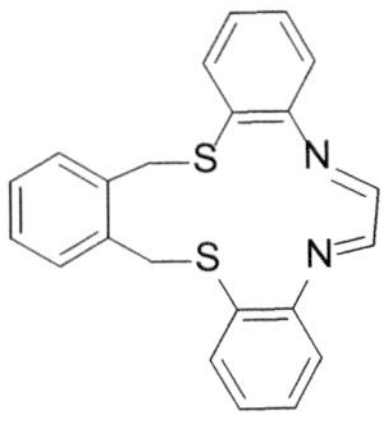

17

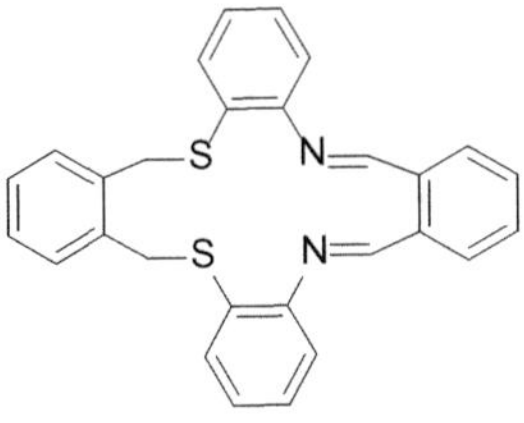

18

Due to its high absorption coefficient and high fluorescence quantum yield, the napththalimide group has been extensively used as a signalling unit in the design of functional supramolecules (Chen et al. 2010). Xu and coworkers synthesized two naphthalimide derivatives **19, 20** and examined the fluorescence responses of the two molecules for a wide range of metal ions in CH_3CN-H_2O (50:50, v/v; 0.5 M HEPES buffer at pH = 7.4). Probe 20 could detect Ag^+ ions with a selective fluorescent enhancement ($\sim$14-fold) and high association constant ($Ka = 1.64 \times 10^5$ M^{-1}). Moreover, the detection limit was as low as 1.0×10^{-8} M. On the contrary, the fluorescence response for reference compound 19 without the carbonyl group did not reflect strong binding with Ag^+, suggesting that the CO group positioned between 1,8-naphthalimide and [15]aneNO$_2$S$_2$ in probe 20, played a significant role in the selective fluorescent enhancement (Xu et al. 2010).

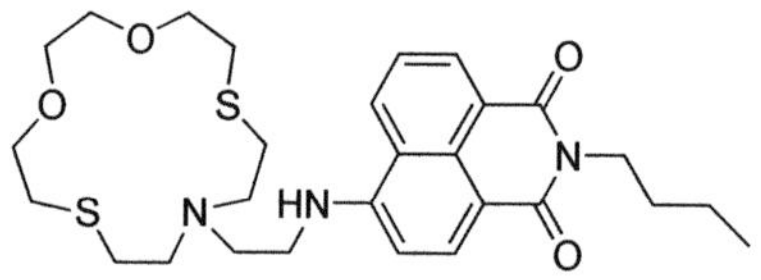

19 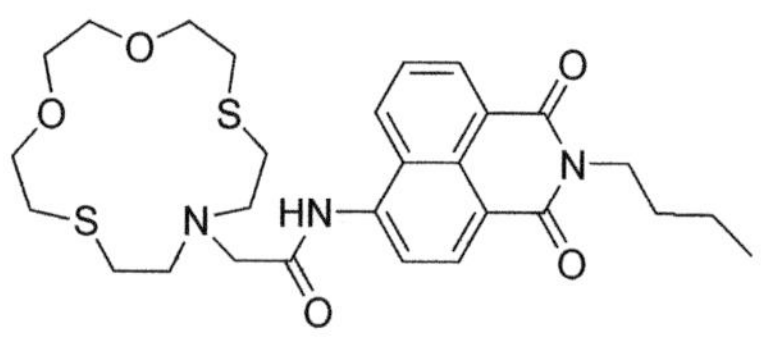   **20**

Chen et al. 2010 also reported the synthesis and fluoroionophoric properties of a novel fluorescent sensor **21** composed of an aminonaphthalimide fluorophore and thioether-rich NS$_4$ crown macrocycle attached through a space linker. The metal-recognition behaviour of the ligand was investigated in ethanol-water (20:80, v/v) solution towards different alkali, alkaline earth and transition metal ions. Only in the presence of Hg^{2+}, the compound showed remarkable fluorescent enhancement (5-fold) at 532 nm. Even the thiophilic metal ions such as Pb^{2+} and Cu^{2+} showed a negligible fluorescent effect on the ligand. Further addition of Ag^+, quenched the enhanced fluorescent signal for ligand-Hg^{2+} complex due to the intramolecular d–π interaction between the fluorophore and Ag^+. Hence, the ligand displayed highly selective and sensitive Hg^{2+}/Ag^+- induced OFF–ON–OFF type fluorescent signal control behaviour which could be used in designing a functional supramolecular switch.

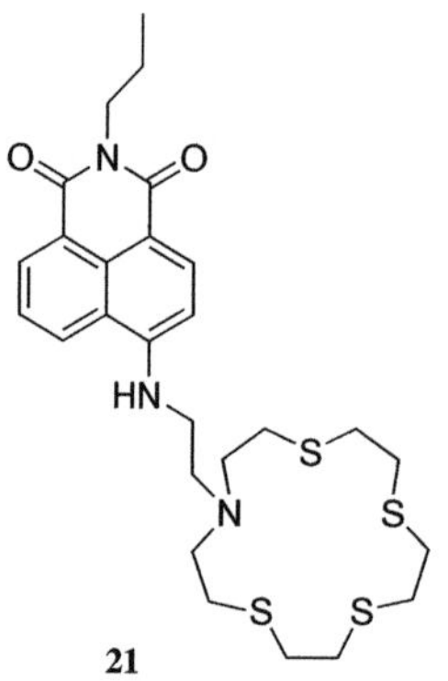

21 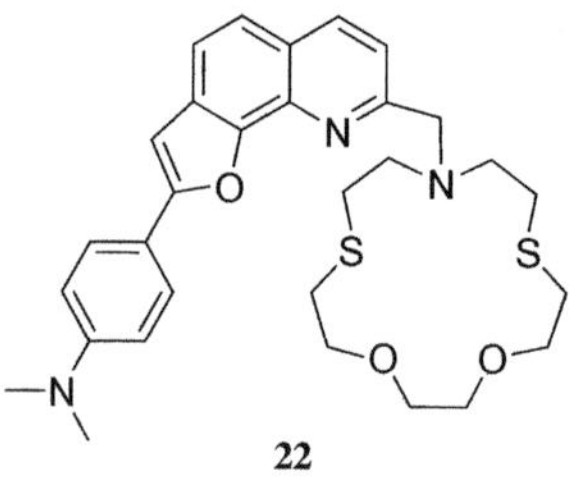   **22**

Wang and coworkers reported a new ratiometric fluorescent sensor **22** selective for Ag^+ ions by incorporating furoquinoline fluorophore into azacrown [N, S, O]. Upon addition of Ag^+ ions, the original UV-Vis absorption band at 346 nm decreased and a new significant absorption was observed at 405 nm. The ligand demonstrated a large Stokes shift of 173 nm and gave rise to a bathochromic shift up to 50 nm in the emission spectra. As compared to other competitive metal ions in ethanol, the sensor exhibited high selectivity and affinity for silver ions (log K = 7.21) with the formation of a 1:1 stoichiometric complex (Wang et al. 2010).

Schmittel and Lin designed and synthesized aza-dithia-dioxa crown-ether-appended iridium(III) and ruthenium(II) complexes **23, 24**. The metal binding ability of the two compounds was studied by UV-Vis absorption and photoluminescence measurements in the MeCN/H_2O (1:1, v/v) system towards a wide variety of competing metal ions as their perchlorate salts. The results showed that both complexes displayed selective binding properties by characteristic luminescence responses for Ag^+ and Hg^{2+}. In comparison to the analogous ruthenium complex 24, the iridium complex 23 was found to be a far superior sensor for Ag^+ ions. Complex 23 displayed a 10 times higher luminescence enhancement factor than 1, which was attributed to the dominance of emission due to di-aza-phen ligand which binds the Ag^+ ion. However, in the case of complex 24, the emission does not involve di-aza-phen but phen ligand, which has a remote effect upon the addition of Ag^+ ions (Schmittel and Lin 2007).

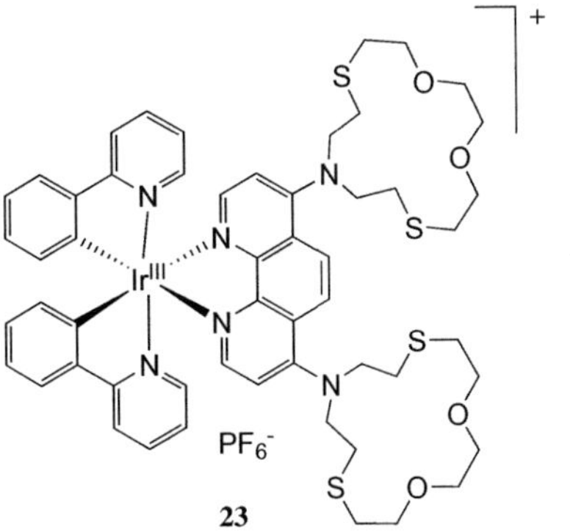

23

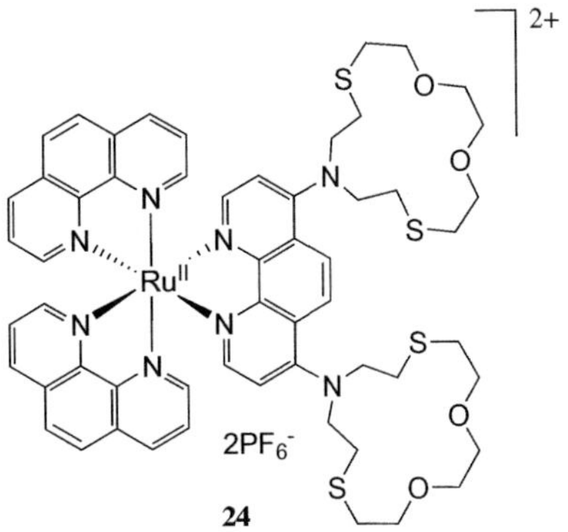

24

N-heterocyclic carbene complexes with many transition metals are quite stable because of the strong σ-donation ability of the ligand. Liu and coworkers synthesized N-heterocyclic carbene macrometallocycle **25** and examined the metal recognition and selectivity of the ligand with respect to a wide range of metal ions in CH_3OH using fluorescent and UV-Vis titrations. On addition of Ag^+ ions, the ligand exhibited a prominent fluorescent enhancement (10-fold) accompanied by a redshift of 14 nm. The ligand was found to be highly selective and sensitive towards Ag^+ ions with a detection limit of 1.7×10^{-8} M. ^1H NMR and IR spectra confirmed that the main binding force of the ligand for Ag^+ originates from the Ag–I interactions because of the strong affinity of iodine for silver ions. Once bound to iodine, the electron-withdrawing effect of Ag^+ leads to ICT (internal charge transfer) with fluorescent enhancement (Liu et al. 2015).

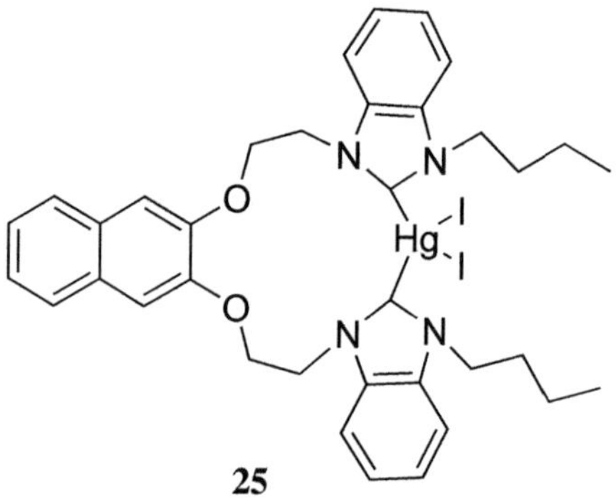

25 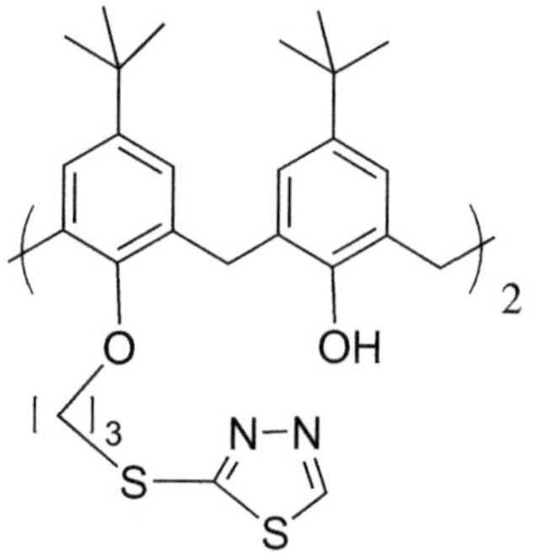   **26**

Wang et al. reported the synthesis of a new type of voltammetric sensor, a glassy carbon electrode modified by LB (Langmuir-Blodgett) film of a p-tertbutylcalix[4]arene derivative, that is 5,11,17,23-tetra-tertbutyl-25,27-di(3-thiadiazole-propanoxy)-26,28-dihydroxycalix[4]arene **26**. This derivative of p-tertbutylcalix[4]arene provided extra cavity dimensions and N, S atoms coordination sites of thiadiazole moiety. The modified glassy carbon electrode (GCE) was thoroughly investigated for its electrochemical properties and was found to exhibit good selectivity, sensitivity and recovery towards Ag^+ ions in aqueous solutions over a wide range of metal ions. The detection limit was as low as 8×10^{-9} M. When employed to analyze real samples such as lake water and tap water, the GCE gave satisfactory results (Wang et al. 2007).

In another study, Stankovic et al. 2008 exhaustively examined the extraction of silver from nitric acid solutions by calix[4]arene-thiotetramide **27**. The receptor was found to be highly efficient in extracting Ag^+ from acid as well as neutral solutions, with a very high degree of extraction (>99%). Even in the presence of Na^+ the ligand was found to be indifferent towards Na^+ and extracted Ag^+ selectively. Moreover, the ligand was found to have a high loading capacity since the distribution data showed that two silver ions were extracted per receptor molecule (Stankovic et al. 2008).

Fu and coworkers reported the synthesis of a new 1,3-alternate-thiacalix[4] arene derivative linked with two 4-chloro-7-nitrobenzofurazan groups at the lower rim of the thiacalixarene by amino groups **28**. On spectroscopic investigations, the receptor was found to have highly efficient colourimetric and fluorescence sensor for Ag$^+$. In the presence of Ag$^+$ ions, fluorescence was quenched dramatically, a redshift in the absorption spectrum and visible colour change was observed. The fluorescence quenching was attributed to the intramolecular electron transfer process between the Ag$^+$ ion and the two amine units (Fu et al. 2012).

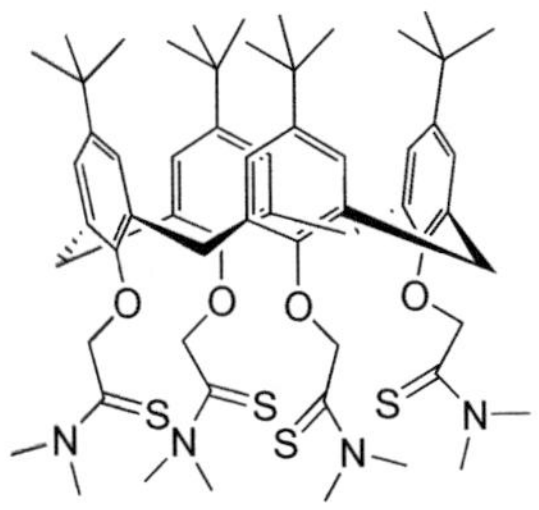

27 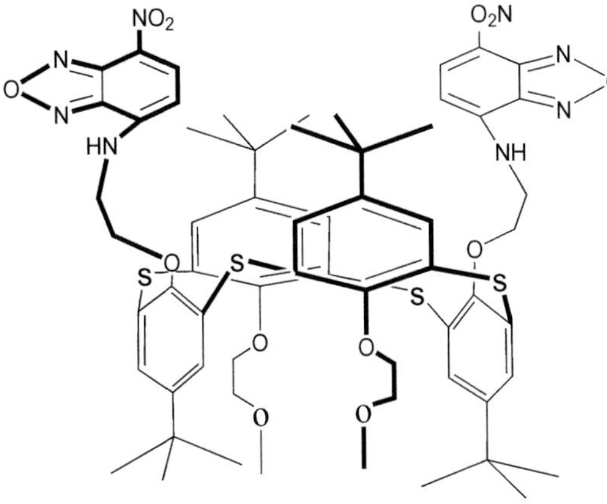   **28**

A bis-nitrobenzoxadiazole (NBD) derivative of calix[4]arene **29** was reported as a fluorescent sensor for Ag$^+$ and HCHO by Zhang and coworkers. The photophysical properties of ligand were tested for a wide range of metal ions using UV-Vis and fluorescence spectroscopy. Upon addition of Ag$^+$ ions to the ligand in THF/water, the fluorescence intensity at 527 nm decreased and a new emission band at 576 nm emerged with significant fluorescence enhancement. The binding ratio was determined to be 1:1 (M:L) with a detection limit of ~6.2 × 10^{-7} M. It was suggested that two 1,2,3-triazole units of the sensor combined with the silver ion in an intra-molecular manner. Furthermore, the introduction of HCHO to the ligand-Ag$^+$ complex reduced the emission band at 576 nm and fully re-generated the original emission profile of the free ligand (Zhang et al. 2016).

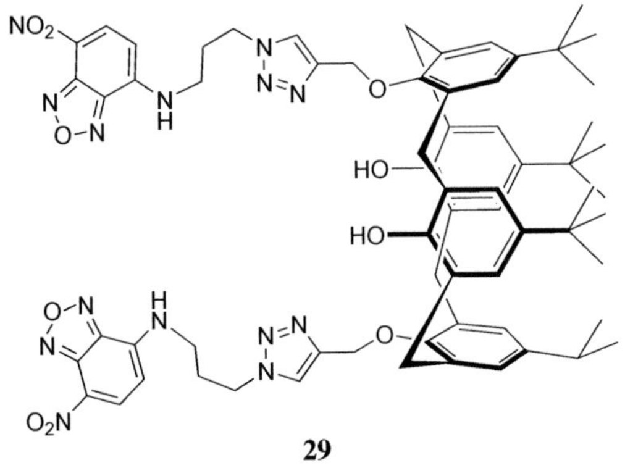

29

Table 3.1   Macrocyclic receptors for Ag^+ ions

Macrocyclic Receptor	Selectivity for	Application	Selectivity over	Reference
crown ethers	Ag^+, Cu^{2+}, Au^{3+}	Membrane Transport	---	(Gherrou et al. 2004)
N-benzylated, O_2N_4-donor macrocycle	Ag^+	Solvent Extraction & Membrane Transport	Co^{2+}, Ni^{2+}, Zn^{2+}, Cu^{2+}, Cd^{2+}, Pb^{2+}	(Price et al. 2004)
tetra-N-benzylated-cyclam derivative	Ag^+	Membrane Transport	Co^{2+}, Ni^{2+}, Zn^{2+}, Cu^{2+}, Cd^{2+}, Pb^{2+}	(Dong et al. 2003)
N-benzylated N_2S_3-donor macrocycle	Ag^+	Membrane Transport	Co^{2+}, Ni^{2+}, Zn^{2+}, Cu^{2+}, Cd^{2+}, Pb^{2+}	(Vasilescu et al. 2006)
caged aza-crown ethers	Ag^+	Receptor, Extractant	Group I, Cu^{2+}, Zn^{2+}, Mn^{2+}, Fe^{3+}, Ni^{2+}, Pb^{2+}, Au^{3+}, Co^{2+}, Mg^{2+}, Ca^{2+}	(Williams et al. 2003)
N_2S_2-Macrocyclic Schiff Base	Ag^+	Extractant	K^+, Na^+, Hg^{2+}, Cd^{2+}, Zn^{2+}, Cu^{2+}, Ni^{2+}, Mn^{2+}	(Ocak et al. 2006)
[15]aneNO_2S_2	Ag^+	Fluorescent probe	Li^+, Na^+, K^+, Mg^{2+}, Ca^{2+}, Co^{2+}, Ni^{2+}, Cu^{2+}, Zn^{2+}, Cd^{2+}, Fe^{2+}, Fe^{3+}, Cr^{3+}, Pb^{2+}, Hg^{2+}	(Xu et al. 2010)
Naphthalimide based thioether-rich crown	Hg^{2+}, Ag^+	Fluorescent sensor	Na^+, K^+, Cs^+, Ca^{2+}, Mg^{2+}, Ba^{2+}, Fe^{2+}, Fe^{3+}, Co^{2+}, Ni^{2+}, Zn^{2+}, Cd^{2+}, Cr^{3+}, Cu^{2+}, Pb^{2+}	(Chen et al. 2010)
Furoquinoline-azacrown[N,S,O]	Ag^+	Ratiometric fluorescent sensor	K^+, Mg^{2+}, Ca^{2+}, Cr^{3+}, Fe^{3+}, Cu^{2+}, Hg^{2+}, Zn^{2+}, Cd^{2+}, Pb^{2+}, Co^{2+}, Ni^{2+}, Mn^{2+}	(Wang et al. 2010)
iridium-phenanthroline crown ether complex	Ag^+	Luminescence sensor	Na^+, K^+, Mg^{2+}, Ca^{2+}, Ba^{2+}, Cr^{3+}, Co^{2+}, Ni^{2+}, Cu^{2+}, Ag^+, Zn^{2+}, Cd^{2+}, Hg^{2+}, Pb^{2+}	(Schmittel and Lin 2007)
N-heterocyclic carbene macrometallocycle	Ag^+	Fluorescent sensor	Li^+, Na^+, NH_4^+, Ag^+, Ca^{2+}, Co^{2+}, Ni^{2+}, Cu^{2+}, Zn^{2+}, Cd^{2+}, Cr^{3+}, Al^{3+}, Pb^{2+} and Hg^{2+}	(Liu et al. 2015)

Macrocyclic Receptor	Selectivity for	Application	Selectivity over	Reference
N-benzylated O_3N_2 & O_2N_3- donor macrocycles	Ag^+	Extractant, Membrane Transport	Co^{2+}, Ni^{2+}, Zn^{2+}, Cu^{2+}, Cd^{2+}, Pb^{2+}	(Kim et al. 2002)
p-tert-butylcalix[4]arene derivative	Ag^+	Voltammetric sensor	Group I, II, Fe^{2+}, Mn^{2+}, Co^{2+}, Ni^{2+}, Zn^{2+}, Cr^{3+}, Cd^{2+}, Pb^{2+}, Cu^{2+}, Hg^{2+}	(Wang et al. 2007)
calix[4]arene thiotetramide	Ag^+	Extractant	Na^+	(Stankovic et al. 2008)
4-chloro-7-nitrobenzofurazan-thiacalix[4]arene derivative	Ag^+	Colorimetric & Fluorescent sensor	Group I, Mg^{2+}, Ca^{2+}, Ba^{2+}, Co^{2+}, Ni^{2+}, Cu^{2+}, Zn^{2+}, Pb^{2+}, Cd^{2+}, Ru^+, Hg^{2+} Fe^{3+}	(Fu et al. 2012)
Calix[4]arene-bis-nitrobenzoxadiazole derivative	Ag^+	Fluorescent sensor	Na^+, K^+, Mg^{2+}, Ca^{2+}, Ba^{2+}, Fe^{3+}, Co^{2+}, Ni^{2+}, Cu^{2+}, Zn^{2+}, Cd^{2+}, Hg^{2+}	(Zhang et al. 2016)
N/O/S-donor macrocycles with 1,10-phenanthroline sub-unit	Ag^+	Membrane Transport	Mg^{2+}, Ca^{2+}, Sr^{2+}, Ba^{2+}, Co^{2+}, Ni^{2+}, Zn^{2+}, Cu^{2+}, Cd^{2+}, Pb^{2+}, Hg^{2+}	(Lippolis and Shamsipur 2006)

MACROCYCLIC RECEPTORS FOR Au^{3+} IONS

The noble metal gold is greatly in demand for its use in various fields, including the jewellery sector, electronics, catalysts, aerospace and spacecraft industry and medicines. The extensive use of gold in modern electronic circuitry has most promisingly transformed e-waste in worldwide assets. Due to limited natural resources of gold in nature and to meet the increasing demand for gold, it is imperative that the secondary sources of gold, such as e-waste be tapped for efficient recovery of gold. However, it is a challenge to selectively recycle gold ions from e-waste as gold is usually present along with other metals, for example, copper, zinc and nickel and its concentration in the leaching solutions are very low (Xu et al. 2019).

Highly toxic inorganic cyanides are frequently used during gold recovery to convert gold(0) into Au(CN)$_2$, a water-soluble coordination compound by means of a process called leaching (Liu et al. 2013). This is followed by isolation of the metal through solvent extraction, chemical precipitation, ion exchange or absorption. The use of cyanide solution for leaching gold ions is problematic as it often leads to contamination of the environment due to accidental leakages and exposures (Liu et al. 2013). Hence the development of eco-friendly technologies for the extraction and recovery of gold has emerged as an area of great research interest. In this regard, various macrocyclic host molecules have been designed and investigated for efficient recognition and isolation of gold ions from waste mixtures.

Mashahadizadeh and coworkers used tetrathia-12-crown-4 **30** as a specific ion carrier into a chloroform bulk liquid membrane to examine the transport of Au^{3+} ions. In the presence of a suitable stripping agent, 5.0×10^{-2} M KSCN at pH 6.5 in the receiving phase, the macrocyclic ligand with concentration as low as 5.0×10^{-4} M was able to selectively, efficiently and quantitatively transport Au^{3+} ions across the liquid membrane at a transport time period of 90 minutes. The excellent gold(III) ion selectivity of this system under optimum experimental conditions was investigated in the presence of a wide variety of alkali, alkaline earth, transition and heavy metal ions (Mashahadizadeh et al. 2004).

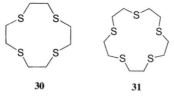

30   **31**

At the same time Bagheri et al. explored the solid-phase extraction of Au^{3+} by sorption on octadecyl silica membrane discs incorporated with pentathia-15-crown-5 **31** using flame atomic absorption spectrometry (FAAS). Various parameters such as the influence of flow rates of eluent and sample solution, types of eluents, amount of ligand with respect to gold elution from discs were investigated. This method was found to be simple, rapid, efficient and highly selective for the separation, concentration and determination of Au^{3+} ions. The detection limit with discs modified with 5 mg of the ligand was found to be 1.0 mg L^{-1}. When applied to pharmaceutical, synthetic and water samples, the method was found to be very successful in Au^{3+} determination and recovery (Bagheri et al. 2003).

Cyclam has been extensively used and examined as a ligand for a variety of metal ions including noble metals. Jermakowicz-Bartkowiak reported the synthesis, characterization and sorption properties of a new resin **32** based on vinylbenzyl chloride–divinylbenzene copolymer functionalized by 1,4,8,11-tetraazacyclotetradecane (cyclam). The sorption studies were conducted under static conditions with a mixture of precious and base metal ions. The resin bearing the cyclam moieties was found to be highly effective and selective for Au^{3+}, Pt^{4+} and Pd^{2+} from HCl solutions. Maximum sorption properties were achieved in 0.1M HCl solution. During a dynamic procedure, the sorption of Au^{3+}, Pt^{4+} and Pd^{2+} in the presence of 20-fold excess metals, i.e., Cu^{2+}, Ni^{2+}, Fe^{3+} resulted in the outcome of up to 400 mg of noble metals per gram of dry resin. Thus, the cyclam functionalized resin was found to be very suitable for the preconcentration and separation of Au, Pt and Pd in hydrometallurgical or separation procedures (Jermakowicz-Bartkowiak 2007).

32   33

Katagiri et al. investigated the metal-complexing ability of thiocalix[4]aniline **33**, a cyclic tetramer of p-tert-butylaniline with four sulphide bridges using the solvent extraction method. The ligand exhibited highly specific extractability towards Au^{3+} and Pd^{2+} ions over 41 interfering metal ions including soft metal ions Hg^{2+}, Cd^{2+}, Zn^{2+}, Pb^{2+} and Cu^{2+} in highly acidic nitrate media. The ligand was thought to have developed softer characteristics towards the soft metal ions, Au^{3+} and Pd^{2+} due to the introduction of an amino group and sulphur bridges. Through further scrutiny using ^1H NMR and X-Ray analysis, it was concluded that these two metal ions were specifically extracted by forming complexes with the ligand by coordination with the nitrogen atom of amino groups and bridging sulphur atoms (Katagiri et al. 2002).

A novel thiacalix[4]arene derivative **34** bearing three amide groups was investigated for the extraction of Au^{3+} and Pd^{2+} in chloride media by Zaghbani and coworkers. The liquid-liquid extraction experiments showed that the new macrocyclic compound could efficiently extract both Au and Pd from an aqueous solution into chloroform. Subsequently, a solid-supported liquid membrane system was developed for Au^{3+} and Pd^{2+} permeation by dissolving the ligand in 2-Nitrophenyl octyl ether (NPOE), 0.5 M NaSCN at pH 2 was used as the stripping agent. After evaluating several parameters affecting liquid membrane performance, the designed liquid membrane system was examined for its extractability towards various metal ions. The results showed that the ligand was highly effective and selective in transporting Au^{3+} and Pd^{2+} from the feed solutions containing other interfering metal ions. It was suggested that the formation of an ion-pair complex between the metal chloro-complexes and the ionophore was responsible for the extraction of the two noble metals (Zaghbani et al. 2007).

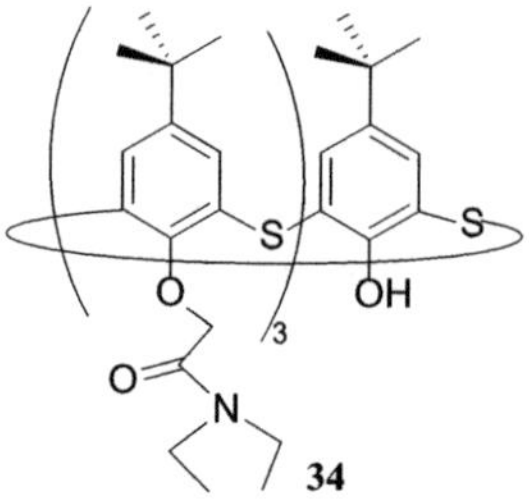

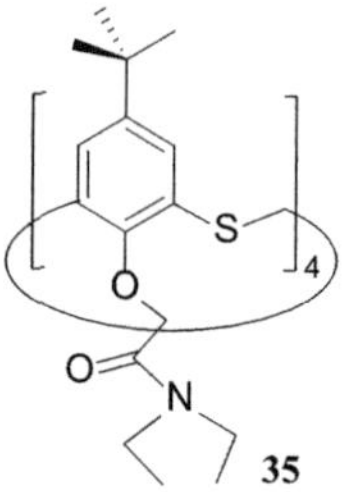

With an aim to ascertain the role of functional groups in the extraction of noble metals from chloride solution, Fontas et al. examined the behaviour of thiacalix[4] arenes **35, 36, 37, 38** with amide or ester groups as compared to the corresponding p-tert-butylthiacalix[4]arene and thiacalix[4]arene. The liquid-liquid extraction results showed that among the four thiacalixarene dissolved in $CHCl_3$, the macrocycle 35, bearing the amide groups was highly effective and efficient in extracting Au from chloride solution than Pd and Pt by an ion-exchange mechanism, and the recovery was almost complete. For calixarene 35, the order of affinity was ranked as Au > Pd > Pt. Also, the thiacalix[4]arene (macrocycle 38 selectively extracted Pd^{2+} via the formation of a $PdCl_2L_2$ complex, where the metal is coordinated through S atoms of the ligand (L). In the metal membrane transport experiments, Au was selectively transported over Pd and Pt in an SLM system (supported liquid membrane) containing calixarene 35, while Pd was recovered through a PPM (plasticized polymeric membrane) system containing thiacalix[4]arene 38. Based on the results, a new solid-phase extraction (SPE) support developed by the adsorption of calixarenes 35 or 38 on top of commercial polymeric cartridge supports. The extraction ability of these supports was greatly enhanced for Au^{3+} and Pd^{2+} due to the presence of the respective thiacalixarenes (Fontàs et al. 2007).

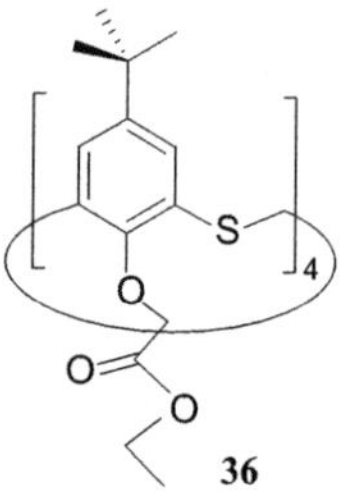

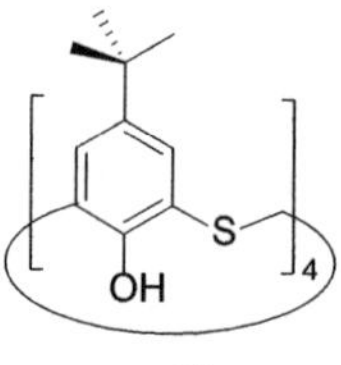

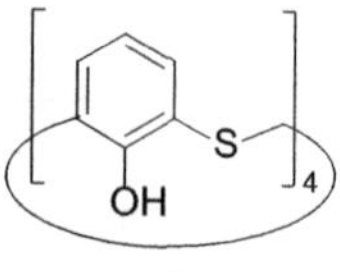

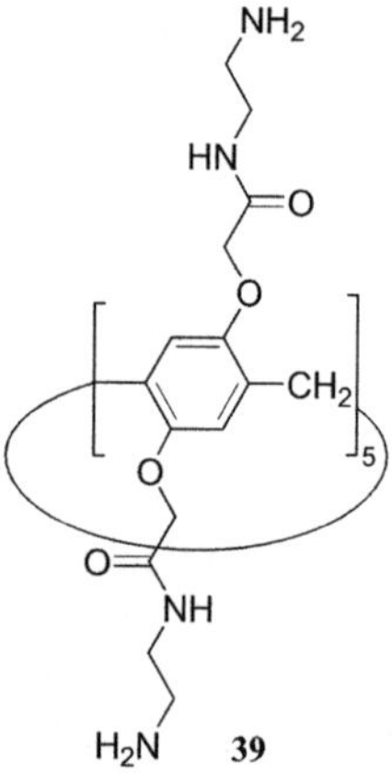

Recently, Yang and coworkers reported the synthesis of a novel fluorescent probe, amino-pillar[5]arene **39** for highly selective detection of Au^{3+} ion in aqueous solutions. The metal recognition characteristics of the sensor were examined using fluorescence and ^1H NMR spectroscopy towards 22 metal ions in an aqueous solution. The binding ratio of the ligand to Au^{3+} ion was $2:1$ with a detection limit of 7.59×10^{-8} M. In the presence of Au^{3+} ions, the probe demonstrated stable fluorescence characteristics within a wide pH range of 1- 13.5, with no significant interference from other coexisting metal ions (Yang et al. 2019).

Table 3.2 Macrocyclic receptors for Au^{3+} ions

Macrocyclic Receptor	*Selectivity*	*Application*	*Selectivity over*	*Reference*
tetrathia-12-crown-4	Au^{3+}	Membrane Transport	Na^+, K^+, Ca^{2+}, Mg^{2+}, Fe^{3+}, Co^{2+}, Ni^{2+}, Cu^{2+}, Zn^{2+}, Tl^+, Ag^+, Pb^{2+}	(Mashahadizadeh et al. 2004)
pentathia-15-crown-5	Au^{3+}	Extractant	K^+, Na^+, Cd^{2+}, Co^{2+}, Zn^{2+}, Ag^+, Mg^{2+}, Ca^{2+}, Cu^{2+}, Hg^{2+}	(Bagheri et al. 2003)
1,4,8,11-tetraazacyclo-tetradecane (cyclam)	Au^{3+}, Pt^{4+}, Pd^{2+}	Metal sorption resin	Fe^{3+}, Cu^{2+}, Ni^{2+}	(Jermakowicz-Bartkowiak 2007)
thiacalix[4]aniline	Au^{3+}	Extractant	41 metal ions including Hg^{2+}, Cd^{2+}, Zn^{2+}, Pb^{2+}, Cu^{2+}	(Katagiri et al. 2002)
thiacalix[4]arene bearing amide groups	Au^{3+}, Pd^{2+}	Extractant, Membrane Transport	Pb^{2+}, Cd^{2+}, Ni^{2+}, Zn^{2+}, Cu^{2+}	(Zaghbani et al. 2007)
thiacalix[4]arene derivatives	Au^{3+}, Pd^{2+}	Extractant, Membrane Transport	Pt^{2+}	(Fontàs et al. 2007)
amino-pillar[5]arene	Au^{3+}	Fluorescent probe	Group I, Group II, Al^{3+}, Sb^{3+}, Pb^{2+}, Sn^{2+}, various transition metal ion	(Yang et al. 2019)
calix[4] arene derivatives	Au^{3+}, Pd^{2+}, Hg^{2+}	Extractant	Hg_2^{2+}, Ag^+, Pt^{2+}, Ni^{2+}	(Yordanov et al. 1996), (Yordanov et al. 1997)
calix[4] arene derivatives	Au^{3+}, Pd^{2+}, Hg^{2+}, Ag^+	Extractant	Sn^{2+}, Hg^{2+}, Pb^{2+}, Cd^{2+}, Fe^{2+}, Mn^{2+}, Co^{2+}, Ni^{2+}, Zn^{2+}, Pt^{2+}, Pt^{4+}	(Yordanov et al. 1998)
calix[6]arene derivatives	Au^{3+}	Extractant	Zn^{2+}, Ni^{2+}, Pb^{2+}, Fe^{3+}, Co^{2+}	(Belhamel et al. 2003)

MACROCYCLIC RECEPTORS FOR Pd^{2+} IONS

Among the platinum group of metals, palladium is very valuable and useful in various fields, such as catalysis, electronics, dentistry, jewellery and fuel cells. Due to its high melting point, corrosion resistance and versatile catalytic properties, palladium is extensively used in catalytic converters in automotive exhaust systems. Consumption of palladium has increased considerably due to rapid advances of catalytic systems for the petrochemical industry and automotive exhausts (Bai et al. 2013). However, the limited terrestrial supply of palladium may not be able to meet its increasing industrial demands. Hence, palladium recovery from industrial wastes such as spent catalysts and electronic components has gained worldwide recognition (Leng et al. 2013). Moreover, recently, considerable attention is being given to developing methods for efficient recovery of large amounts of fission palladium from the radioactive high-level liquid waste (HLLW) which is produced during the reprocessing of nuclear-spent fuels (Bai et al. 2013, Leng et al. 2013, Wu et al. 2020). However, the complexity of HLLW and the separation of palladium from other interfering metal ions remains a daunting task (Bai et al. 2013).

Additionally, the release and accumulation of palladium from industrial and nuclear wastes in the environment may impact human health adversely. Palladium(II) ions are cytotoxic as they are capable of binding with proteins, DNA, RNA and can cause eye irritations, skin problems, asthma, etc. (Kim et al. 2011; Bai et al. 2013; Berhanu et al. 2019). Various methods such as solvent extraction, ion exchange, membrane transport and adsorption have been explored for palladium enrichment from nuclear wastes. Intensive efforts have been made to make use of the host-guest chemistry to design macrocyclic receptors for effective detection, separation and recovery of palladium.

Bai and coworkers reported the synthesis of a new silica-based adsorbent functionalized with macrocyclic polyether isomer cis-di(aminocyclohexyl)-18-crown-6 **40** for the selective recovery of palladium. The absorbent exhibited efficient adsorption and excellent selectivity towards Pd^{2+} in HNO$_3$ media and stimulated high-level liquid waste (HLLW) containing a large amount of interfering metal ions. A recovery rate higher than 90% for Pd^{2+} was obtained from these experiments. The proposed metal-ligand binding mechanism involved the formation of a complex ion-pair formation (Bai et al. 2013).

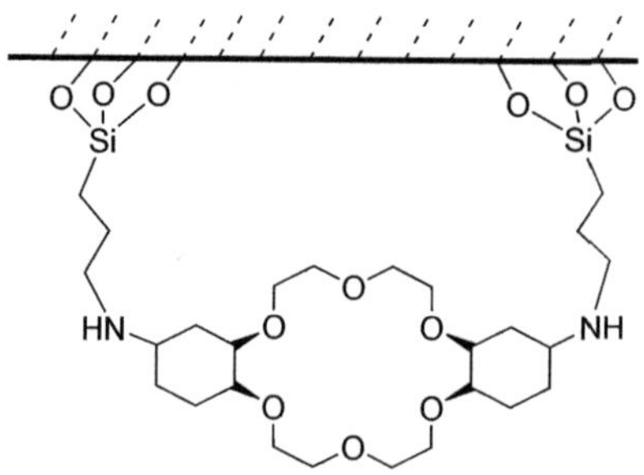

40 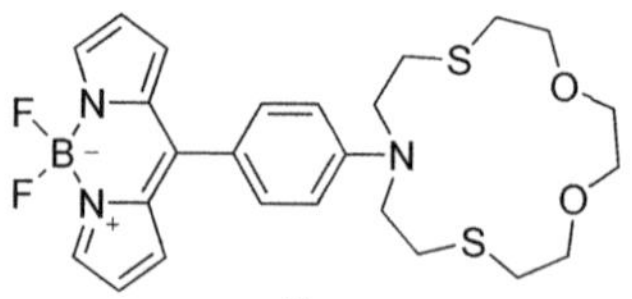   **41**

Kaur et al. developed a 'turn on' chemosensor **41** for Pd^{2+} detection based on dithia-dioxa-aza crown ether-linked to a fluorescent signalling handle, BODIPY. The metal-binding properties of the receptor towards 25 metal ions were investigated by spectrophotometric measurements and later results were correlated with DFT (density functional theory) and time-dependent DFT (TD-DFT) calculations. The ligand in CH_3CN exhibited an intense absorption band at 484 nm, however, on gradual addition of Pd^{2+} ions a bathochromic shift of this intense band to 498 nm and a visual colour change from pink to light orange was observed. The ligand was found to form a 1:1 complex with Pd^{2+}, and the detection limit of 1.18 ppb was achieved. According to the DFT/TD-DFT studies, the 'turn-on' behaviour of the fluorescent sensor on binding with Pd^{2+} could be attributed to restricted photoinduced electron transfer (PET) (Kaur et al. 2014).

Atkinson and coworkers reported the synthesis of two tri-linked, 16 membered N_2S_2-donor macrocycles **42, 43** incorporating a 1,3,5- 'tribenzyl' or a phloroglucinol core. Pd(II) and Pt(II) complexes of the ligands were synthesized and ^1H NMR, X-Ray structure determination and mass spectral studies, and spectrophotometric titrations were conducted. The collective results confirmed that within a particular complex a metal ion (Pd or Pt) was bound to each of the macrocyclic ring, spaced by the central phloroglucinol hub (Atkinson et al. 2001).

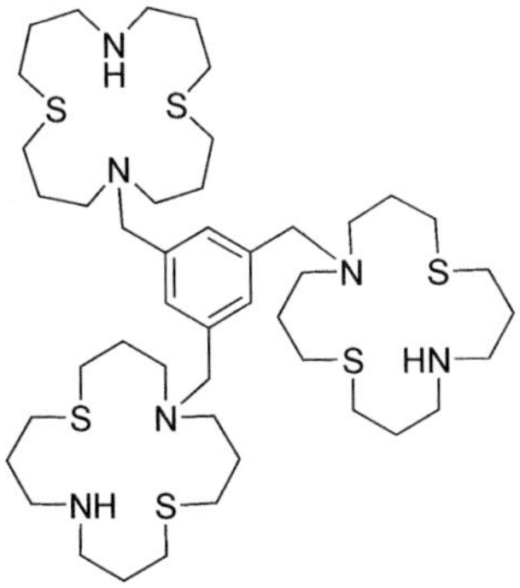

42

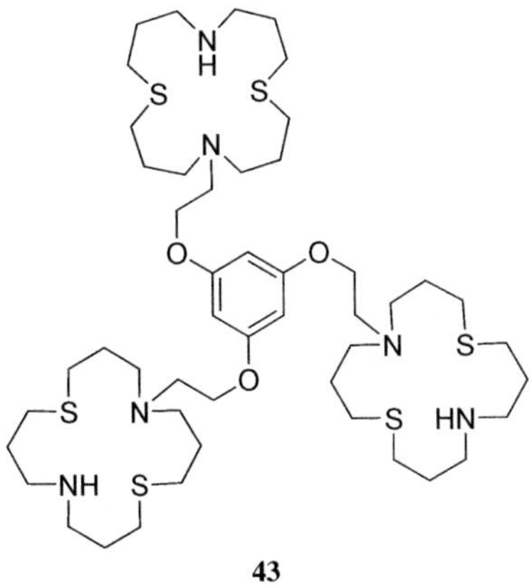

43

Extending their studies, Atkinson et al. reported the synthesis of a dendrimer **44** by incorporating nine S_2N_2-donor macrocycles. Pd(II) complex with the ligand was investigated using ^{1}H NMR and mass spectroscopy. Finally, spectroscopic titration of the ligand and Pd^{2+} in acetonitrile gave definitive evidence for the complexation of nine Pd^{2+} ions by the individual macrocyclic rings within the dendritic architecture of the ligand (Atkinson et al. 2002).

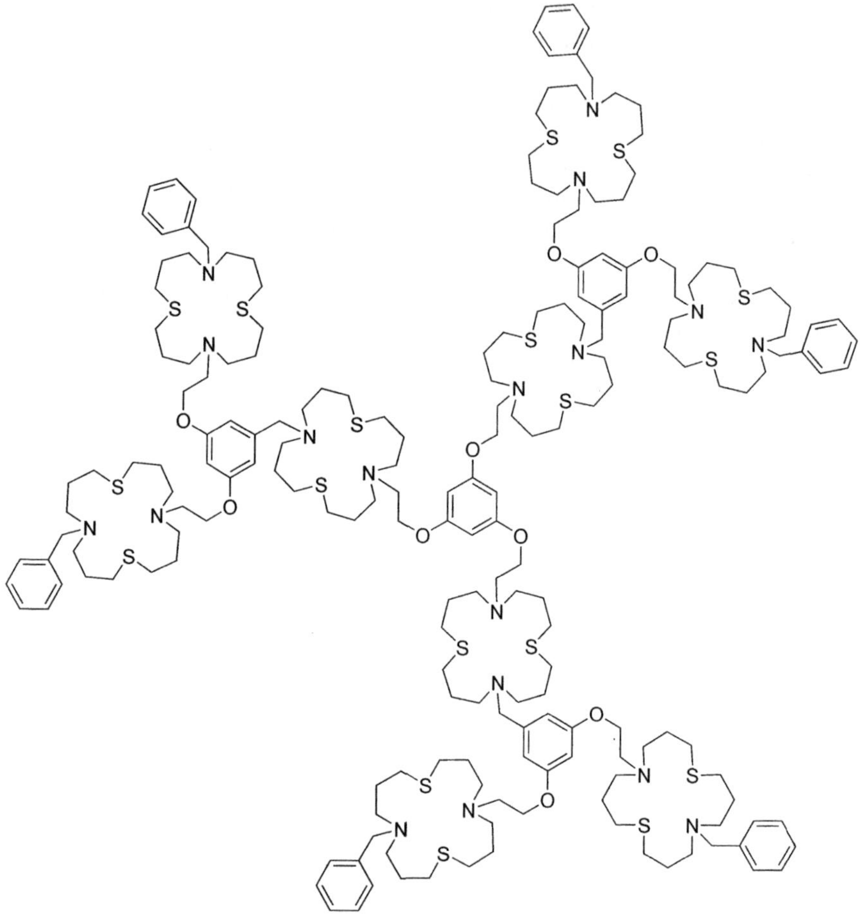

44

Cyclen derivatives are frequently used as ligands due to their strong complexing ability with various transition metal ions. Kim and coworkers developed a Pd^{2+}-specific fluorescent chemosensor **45** based on rhodamine hydroxamate with a cyclen-tri(tert-butyl ester) and a pyridine moiety as a binding unit. UV-Vis and fluorescence spectroscopy was used to evaluate metal ion recognition behaviour of the ligand towards Pd^{2+} ions over several other biologically relevant metal ions in water (DMSO 1%, v/v). The receptor was found to be highly sensitive and selective for Pd^{2+} and exerted a strong fluorescent signal at 581 nm, accompanied by a visual colour change of the solution from colourless to pink. Similarly, the UV-Vis absorption spectrum also displayed a distinctive absorption band at 480-610 nm, and the ligand formed a 1:2 complex with Pd^{2+} ions with a high binding constant (log K = 10.26). No other interfering metal ions tested caused any significant change in fluorescence intensity (Kim et al. 2011).

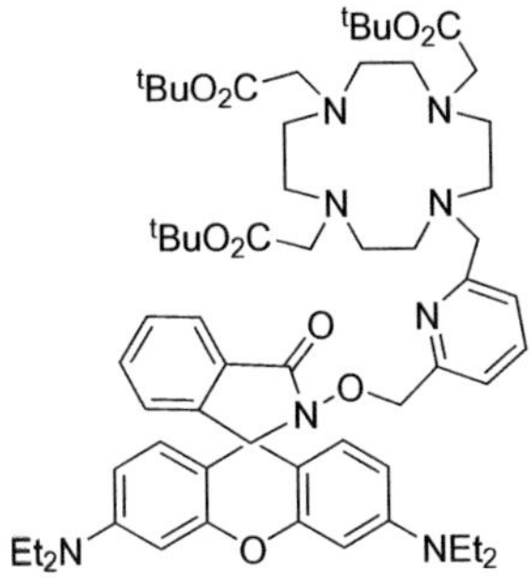

45	**46**	**47**

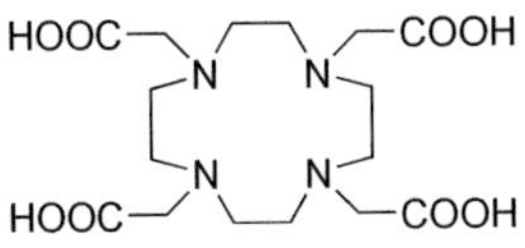

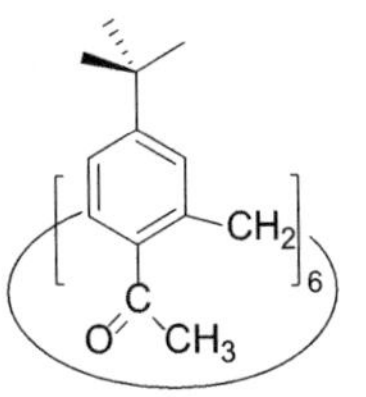

Wu et al. developed a novel polyazamacrocyclic receptor **46**, 1,4,7,10-teraazacyclododecane-tetraacetic acid decorated core-shell superparamagnetic microspheres. The receptor displayed highly selective binding towards Pd^{2+} over various other metal ions in HNO_3 media. Under the applied magnetic field, the microspheres possessed a high adsorption capacity which allowed an efficient enrichment and fast separation time (13 seconds) of Pd^{2+}. The host-guest coordination mechanism was investigated using XPS and FT-IR spectrophotometry and a synergistic contribution of the nitrogen donors in cyclic amines and the oxygen donors in carboxylic arms to the Pd^{2+} binding was proposed (Wu et al. 2016).

Subsequently, Wu and coworkers reported a novel composite resin C8-Cyclen/CG-71M, with high efficiency and selective binding ability to capture Pd^{2+} ions from HNO_3 solution. The resin was developed by impregnating the macroporous polymeric resins (Amberchrom CG-71M) with tetraoctyl-substituted 1,4,7,10-tetraazacyclododecane (C8-Cyclen) macrocyclic receptor. The resin exhibited favourable adsorption ability, rapid kinetics and high affinity towards Pd^{2+} in a wide concentration range of HNO_3, and was unaffected by the presence of other interfering metal ions (Wu et al. 2020).

With regards to conformational mobility, calix[6]arene occupies an intermediate position between calix[4]arene and calix[8]arene. Mathew and Khopkar examined the metal-chelating ability of hexaacetato calix[6]arene **47** in different organic diluents by solvent extraction method. Toluene was found to be the best diluent, and the ligand in low concentration in toluene was able to quantitatively extract microgram concentration of Pd^{2+} ions at pH 7.2 in the presence of a large number of diverse metal ions. Though Cu^{2+} and Cr^{3+} showed interference, the proposed method was found to be rapid and selective and allowed the separation of Pd^{2+} from commonly associated Fe^{3+}, Co^{2+}, Ni^{2+} and Pd^{2+} ions (Mathew and Khopkar 1997).

With the aim to add chromogens on the upper rim of thiacalixarene, Chakrabarti et al. synthesized several new tetrathiacalix[4]arenes by incorporating arylazo-, thiazoleazo- and β-naphthylazo- moieties into the tetrathiacalix[4] arene molecular framework through diazotization and coupling reactions. These macrocyclic structures were investigated for their metal recognition properties using UV-Vis spectroscopy towards various alkali, alkaline earth and transition metal ions in a chloroform/methanol solvent system. The results showed that out of the various tetrathiacalix[4]arene derivatives synthesized, the thiazoleazo coupled tetrathiacalix[4]arene **48** exhibited significant selective bathochromic shifts for Pd^{2+}, Cs^+ and Rb^+. In the case of Pd^{2+} ions, the marked bathochromic shift was accompanied by a visual colour change from yellow to bluish-green. The original absorption intensity of the ligand at 468 nm gradually decreased, a new absorption band was formed at 617 nm in the presence of Pd^{2+} ions due to d–d transitions. The ligand-metal binding ratio for Pd^{2+} was found to be 1:2 (Chakrabarti et al. 2006).

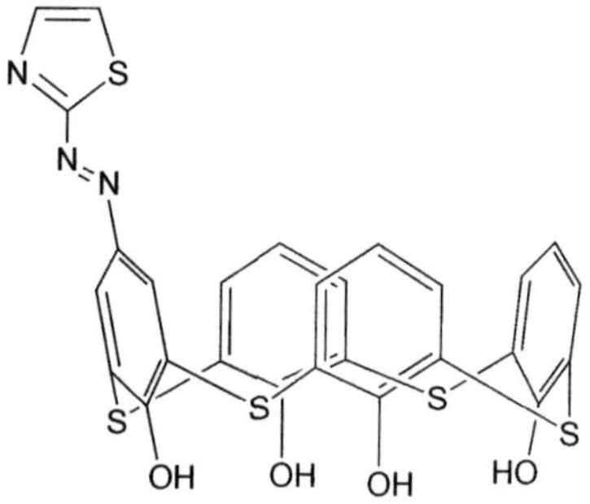

48

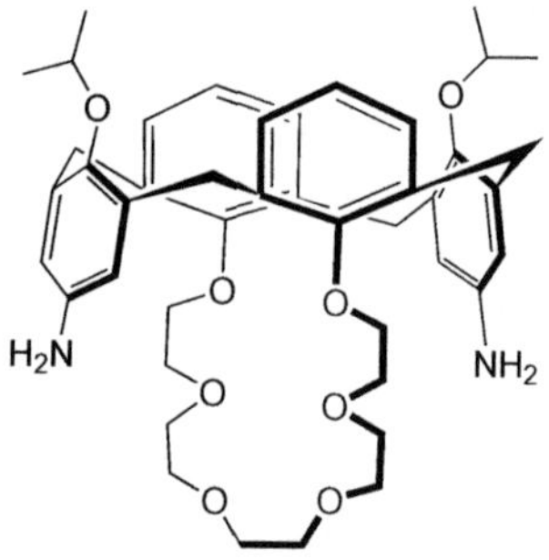

49

A new amino-bearing 1,3-alternate calix[4]arene-crown-6 receptor **49** was synthesized by Leng and coworkers for selective extraction of Pd^{2+} in HNO_3 media. Solvent extraction experiments resulted in a high distribution ratio (D) of 278.6 within 10 minutes. Furthermore, the receptor was covalently grafted on to micro-sized silica substrate to obtain a novel solid-phase extraction (SPE) material. Detailed investigations through batch and column experiments showed that the SPE material exhibited fast kinetics and favourable adsorption capacity towards Pd^{2+} in HNO_3 solutions. On this basis, a facile chromatographic process was proposed for selective enrichment and recovery of Pd^{2+} in stimulated HLLW (high-level liquid waste) and a successful recovery rate of 99.3% was obtained (Leng et al. 2013).

Table 3.3 Macrocyclic receptors for Pd^{2+} ions

Macrocyclic Receptor	Selectivity	Application	Selectivity over	Reference
polyether cis-di(aminocyclohexyl)-18-crown-6	Pd^{2+}	Absorbent	Group I, II, other transition metal ions	(Bai et al. 2013)
dithia-dioxa-aza crown ether with BODIPY	Pd^{2+}	Fluorescent sensor	Various Group I, II, transition and lanthanide metal ions	(Kaur et al. 2014)
tri-linked N_2S_2-donor macrocycle	Pd^{2+}, Pt^{2+}	Receptor	-----	(Atkinson et al. 2001)
dendrimer-nine S_2N_2-donor macrocycle	Pd^{2+}	Receptor	-----	(Atkinson et al. 2002)
cyclen-conjugated rhodamine hydroxamate	Pd^{2+}	Fluorescent sensor	Li^+, Na^+, K^+, Mg^{2+}, Ca^{2+}, Ba^{2+}, various other transition metal ions.	(Kim et al. 2011)
polyazamacrocyclic receptor	Pd^{2+}	Magnetic solid-phase extraction (MSPE)	K^+, Cs^+, Ba^{2+}, Sr^{2+}, Cd^{2+}, Ni^{2+}, Ru^{3+}, Fe^{3+}, Nd^{3+}, Cr^{3+}	(Wu et al. 2016)
polyazamacrocyclic receptor	Pd^{2+}	Polymeric resin, Solid-phase extraction	K^+, Na^+, Cs^+, Sr^{2+}, Ba^{2+}, Cd^{2+}, Ni^{2+}, Nd^{3+}, Cr^{3+}, Ru^{2+}, Rh^{2+}, Fe^{3+}, $Mo(VI)$, $Zr(VI)$, $U(VI)$, Am^{3+} and Pu^{4+}	(Wu et al. 2020)
calix[6]arene-acetyl derivative	Pd^{2+}	Extractant	Group I, II, various transition metal ions, Al^{3+}, In^{3+}, Se^{4+}, UO_2^{2+}, Th^{4+}	(Mathew and Khopkar 1997)
thiazoleazo coupled tetrathiacalix[4]arenes	Pd^{2+}, Cs^+, Rb^+	Chromoionophore	Other Group I cations, Ca^{2+}, Mg^{2+}, Ba^{2+}, Cr^{3+}, Fe^{2+}, Co^{2+}, Ni^{2+}, Cu^{2+}, Hg^{2+}, Pt^{2+}	(Chakrabarti et al. 2006)
1,3-alternate calix[4]arene-crown-6-amino derivative	Pd^{2+}	Extractant	---	(Leng et al. 2013)
15-membered azamacrocycle	Pd^{2+}	Membrane Transport	Pt (IV)	(Masllorens et al. 2006)

CONCLUSION

Many of the macrocyclic host molecules presented in this chapter displayed highly selective recognition for targeted precious metal ion species. The replacement of oxygen by sulphur atoms in the ligand macrocycle largely modified the solubility, conformational flexibility and complexation ability towards soft metal ions. Due to the high thiophilicity of soft acids, silver(I), gold(III) and palladium(II) were favourably responsive to such ligands. Hence, the basic consideration while designing probes for sensing these ions is the incorporation of soft centres such as N and S within the ligand framework. In most cases, the receptor must display strong binding affinity and selectivity for target metals in the presence of other interfering ionic species in waste solutions. Owing to the high sensitivity and easy operation, fluorescent probes have been found to be powerful tools to monitor and discriminate Ag+ ions in the presence of environmental interferences and competitive ions.

Among the commonly investigated separation methods, liquid and solid-phase extraction techniques have been recognized as the most favourable method for recovering precious metals. As compared to solvent extraction which suffers from low sensitivity, a tedious process, loss of high purity solvents and disposal of solvents, while the solid-phase extraction presents the advantages of convenient operation and minimal organic waste generation. Furthermore, the liquid membrane transport technique is considered to be a powerful tool for the concentration, separation and retrieval of these cations. However, the development of efficient macrocyclic receptors to selectively extract/ transport noble metals from waste mixtures and release them in a pure form remains a daunting challenge. It is hoped that in the future, more sophisticated, economically viable and greener technologies will be developed for the recovery of precious metal ions.

REFERENCES

Atkinson, I.M., J.D. Chartres, A.M. Groth, L.F. Lindoy, M.P. Lowe, G.V. Meehan, B.W. Skelton and A.H. White. 2001. New linked macrocyclic systems. Interaction of palladium(II) and platinum(II) with tri-linked N_2S_2-donor macrocycles and their single-ring analogues. J. Chem. Soc., Dalton Trans. 19: 2801–2806.

Atkinson, I.M., J.D. Chartres, A.M. Groth, L.F. Lindoy, M.P. Lowe and G.V. Meehan. 2002. A second generation dendrimer incorporating nine S_2N_2-donor macrocycles and its palladium(II) complex. Chem. Commun. 20: 2428–2429.

Bagheri, M., M.H. Mashhadizadeh and S. Razee. 2003. Solid phase extraction of gold by sorption on octadecyl silica membrane disks modified with pentathia-15-crown-5 and determination by AAS. Talanta. 60(4): 839–844.

Bai, F., G. Ye, G. Chen, J. Wei, J. Wang and J. Chen. 2013. Highly selective recovery of palladium by a new silica-based adsorbent functionalized with macrocyclic ligand. Sep. Purif. Technol. 106: 38–46.

Belhamel, K., T.K.D. Nguyen, M. Benamor and R. Ludwig. 2003. Design of calixarene-type ligands for second sphere complexation of noble metal ions. Eur. J. Inorg. Chem. 2003(22): 4110–4116.

Berhanu, A.L., Gaurav, I. Mohiuddin, A.K. Malik, J.S. Aulakh, V. Kumar and K.-H. Kim. 2019. A review of the applications of Schiff bases as optical chemical sensors. Trends Analyt. Chem. 116: 74–91.

Chakrabarti, A., H.M. Chawla, T. Francis, N. Pant and S. Upreti. 2006. Synthesis and cation binding properties of new arylazo- and heteroarylazotetrathiacalix[4]arenes. 62(6): 1150–1157.

Chen, T., W. Zhu, Y. Xu, S. Zhang, X. Zhang and X. Qian. 2010. A thioether-rich crown-based highly selective fluorescent sensor for Hg^{2+} and Ag^+ in aqueous solution. Dalton Trans. 39(5): 1316–1320.

Dong, Y., S. Farquhar, K. Gloe, L.F. Lindoy, B.R. Rumbel, P. Turner and K. Wichmann. 2003. Metal ion recognition. Interaction of a series of successively N-benzylated derivatives of 1,4,8,11-tetraazacyclotetradecane (cyclam) with selected transition and post-transition metal ions. Dalton Trans. (8): 1558–1566.

Fontàs, C., E. Anticó, F. Vocanson, R. Lamartine and P. Seta. 2007. Efficient thiacalix[4] arenes for the extraction and separation of Au(III), Pd(II) and Pt(IV) metal ions from acidic media incorporated in membranes and solid phases. Sep. Purif. Technol. 54(3): 322–328.

Fu, Y., X. Zeng, L. Mu, X.-K. Jiang, M. Deng, J.-X. Zhang and T. Yamato. 2012. Use of a new thiacalix[4]arene derivative bearing two 4-chloro-7-nitrobenzofurazan groups as a colorimetric and fluorescent chemosensor for Ag^+ and AcO^-. Sens. Actuators B Chem. 164(1): 69–75.

Gherrou, A., H. Kerdjoudj, R. Molinari, P. Seta and E. Drioli. 2004. Fixed sites plasticized cellulose triacetate membranes containing crown ethers for silver(I), copper(II) and gold(III) ions transport. J. Membr. Sci. 228(2): 149–157.

Gumpu, M.B., S. Sethuraman, U.M. Krishnan and J.B.B. Rayappan. 2015. A review on detection of heavy metal ions in water—an electrochemical approach. Sens. Actuators B Chem. 213: 515–533.

Jermakowicz-Bartkowiak, D. 2007. A preliminary evaluation on the use of the cyclam functionalized resin for the noble metals sorption. React. Funct. Polym. 67(12): 1505–1514.

Katagiri, H., N. Iki, Y. Matsunaga, C. Kabuto and S. Miyano. 2002. 'Thiacalix[4]aniline' as a highly specific extractant for Au(III) and Pd(II) ions. Chem. Commun. (18): 2080–2081.

Kaur, P., N. Kaur, M. Kaur, V. Dhuna, J. Singh and K. Singh. 2014. 'Turn-on' coordination based detection of Pd^{2+} and bioimaging applications. RSC Adv. 4(31): 16104–16108.

Kim, J., T.-H. Ahn, M. Lee, A.J. Leong, L.F. Lindoy, B.R. Rumbel, B.W. Skelton, T. Strixner, G. Wei and A.H. White. 2002. Metal ion recognition. The interaction of cobalt(II), nickel(II), copper(II), zinc(II), cadmium(II), silver(I) and lead(II) with N-benzylated macrocycles incorporating O_2N_2-, O_3N_2- and O_2N_3-donor sets. J. Chem. Soc., Dalton Trans.(21): 3993–3998.

Kim, H., K.-S. Moon, S. Shim and J. Tae. 2011. Cyclen-conjugated rhodamine hydroxamate as Pd^{2+}-specific fluorescent chemosensor. Chem. Asian J. 6(8): 1987–1991.

Leng, Y., J. Xu, J. Wei and G. Ye. 2013. Amino-bearing calixcrown receptor grafted to micro-sized silica particles for highly selective enrichment of palladium in HNO_3 media. Chem. Eng. J. 232: 319–326.

Lippolis, V. and M. Shamsipur. 2006. Synthesis, coordination properties, and analytical applications of mixed donor macrocycles containing the 1,10-phenanthroline sub-unit. J. Iran. Chem. Soc. 3(2): 105–127.

Liu, Z., M. Frasconi, J. Lei, Z.J. Brown, Z. Zhu, D. Cao, J. Iehl, G. Liu, A.C. Fahrenbach, Y.Y. Botros, O.K. Farha, J.T. Hupp, C.A. Mirkin and J. Fraser Stoddart. 2013. Selective isolation of gold facilitated by second-sphere coordination with α-cyclodextrin. Nat. Commun. 4(1): 1855.

Liu, Q.-X., Q. Wei, R. Liu, X.-J. Zhao and Z.-X. Zhao. 2015. NHC macrometallocycles of mercury(II) and silver(I): synthesis, structural studies and recognition of Hg(II) complex 4 for silver ion. RSC Adv. 5(36): 28435–28447.

Mashahadizadeh, M.H., R. Mohyaddini and M. Shamsipur. 2004. Selective and efficient liquid membrane transport of Au(III) by tetrathia-12-crown-4 as a specific carrier. Sep. Purif. Technol. 39(3): 161–166.

Masllorens, J., A. Roglans, E. Anticó and C. Fontàs. 2006. New applications of azamacrocyclic ligands in ion recognition, transport and preconcentration of palladium. Anal. Chim. Acta. 560(1): 77–83.

Mathew, V.J. and S.M. Khopkar. 1997. Hexaacetato calix(6)arene as the novel extractant for palladium. Talanta. 44(10): 1699–1703.

Ocak, Ü., H. Alp, P. Gökçe and M. Ocak. 2006. The synthesis of new N2S2-macrocyclic schiff base ligands and investigation of their ion extraction capability from aqueous media. Sep. Sci. Technol. 41(2): 391–401.

Price, J.R., M. Fainerman-Melnikova, R.R. Fenton, K. Gloe, L.F. Lindoy, T. Rambusch, B.W. Skelton, P. Turner, A.H. White and K. Wichmann. 2004. Macrocyclic ligand design. Structure–function relationships involving the interaction of pyridinyl-containing, oxygen–nitrogen donor macrocycles with selected transition and post transition metal ions on progressive N-benzylation of their secondary amines. Dalton Trans. (21): 3715–3726.

Schmittel, M. and H. Lin. 2007. Luminescent iridium phenanthroline crown ether complex for the detection of silver(I) ions in aqueous media. Inorg. Chem. 46(22): 9139–9145.

Stankovic, V., L. Outarra, F. Zonnevijlle and C. Comninellis. 2008. Solvent extraction of silver from nitric acid solutions by calix[4]arene amide derivatives. Sep. Purif. Technol. 61(3): 366–374.

Vasilescu, I.M., D.J. Bray, J.K. Clegg, L.F. Lindoy, G.V. Meehan and G. Wei. 2006. Rational ligand design for metal ion recognition. Synthesis of a N-benzylated N_2S_3-donor macrocycle for enhanced silver(I) discrimination. (43): 5115–5117.

Wang, L., B.-T. Zhao and B.-X. Ye. 2007. Electrochemical properties of electrode modified with langmuir-blodgett film of p-tert-Butylcalix[4]arene derivatives and its application in determining of silver. Electroanalysis. 19(9): 923–927.

Wang, H.-H., L. Xue, Y.-Y. Qian and H. Jiang. 2010. Novel ratiometric fluorescent sensor for silver ions. Org. Lett. 12(2): 292–295.

Williams, S.M., J.S. Brodbelt, Z. Huang, H. Lai and A.P. Marchand. 2003. Complexation of silver and co-recovered metals with novel aza-crown ether macrocycles by electrospray ionization mass spectrometry. Analyst. 128(11): 1352–1359.

Wu, F., G. Ye, R. Yi, T. Sun, C. Xu and J. Chen. 2016. Novel polyazamacrocyclic receptor decorated core–shell superparamagnetic microspheres for selective binding and magnetic enrichment of palladium: synthesis, adsorptive behavior and coordination mechanism. Dalton Trans. 45(23): 9553–9564.

Wu, F., C. Yang, Y. Liu, S. Hu, G. Ye and J. Chen. 2020. Novel polyazamacrocyclic receptor impregnated macroporous polymeric resins for highly efficient capture of palladium from nitric acid media. Sep. Purif. Technol. 233: 115953.

Xu, Z., S. Zheng, J. Yoon and D.R. Spring. 2010. Discovery of a highly selective turn-on fluorescent probe for Ag^+. Analyst. 135(10): 2554–2559.

Xu, W., X. Mo, S. Zhou, P. Zhang, B. Xiong, Y. Liu, Y. Huang, H. Li and K. Tang. 2019. Highly efficient and selective recovery of Au(III) by a new metal-organic polymer. J. Hazard. Mater. 380: 120844.

Yang, J.-L., Y.-H. Yang, Y.-P. Xun, K.-K. Wei, J. Gu, M. Chen and L.-J. Yang. 2019. Novel amino-pillar[5]arene as a fluorescent probe for highly selective detection of Au^{3+} ions. ACS Omega. 4(18): 17903–17909.

Yordanov, A.T., D. Max Roundhill and J.T. Mague. 1996. Extraction selectivities of lower rim substituted calix[4]arene hosts induced by variations in the upper rim substituents. Inorganica Chim. Acta. 250(1): 295–302.

Yordanov, A.T., O.M. Falana, H.F. Koch and D.M. Roundhill. 1997. (Methylthio)methyl and (N,N-Dimethylcarbamoyl)methyl upper-rim-substituted calix[4]arenes as potential extractants for Ag(I), Hg(II), Ni(II), Pd(II), Pt(II), and Au(III). Inorg. Chem. 36(27): 6468–6471.

Yordanov, A.T., B.R. Whittlesey and D.M. Roundhill. 1998. Calixarenes derivatized with sulfur-containing functionalities as selective extractants for heavy and precious metal ions. Inorg. Chem. 37(14): 3526–3531.

Zaghbani, A., R. Tayeb, M. Dhahbi, M. Hidalgo, F. Vocanson, I. Bonnamour, P. Seta and C. Fontàs. 2007. Selective thiacalix[4]arene bearing three amide groups as ionophore of binary Pd(II) and Au(III) extraction by a supported liquid membrane system. Sep. Purif. Technol. 57(2): 374–379.

Zhang, S., H. Yang, Y. Ma and Y. Fang. 2016. A fluorescent bis-NBD derivative of calix[4]arene: switchable response to Ag^+ and HCHO in solution phase. Sens. Actuators B Chem. 227: 271–276.

Chapter 4

Macrocyclic Anion Receptors

INTRODUCTION

Anions are ubiquitous in nature. It is a known fact that chloride anions exist/ are present in oceans, acid rain contains nitrate and sulfate anions and carbonates are crucial components of biomineralized materials. Phosphate and nitrates from agriculture runoff and other human activities cause major pollution hazards like eutrophication. Anions produced at industries, refineries, mining and chemical storage can leach into groundwater and cause hazards to health and the environment e.g., chlorite, bromate and fluoride can be leaked from water treatment operations, perchlorate from arms industries and cyanide from mining (Cheremisinoff 2002). The radioactive pertechnetate ions produced during nuclear fuel reprocessing is also a matter of environmental concern (Meena and Arai 2017).

Anions are also important and play a vital role in the maintenance of life. Chloride is a major contributor to the osmotic pressure gradient between the ICF (intracellular fluid) and ECF (extracellular fluid), maintaining the electrical neutrality in the ECF and also plays an important role in maintaining hydration. Disorder associated with high chloride anion level are cystic fibrosis, inherited kidney stone diseases, myotonia and epilepsy (Ashcroft 1999; Berend 2017).

The second most abundant anion in blood is bicarbonate (Wang et al. 2014). Its principal function is to maintain the body's acid-base balance. Part of calcium-phosphate salts are present in bone and teeth, phosphate is found in phospholipids in the cell membrane, ATP, nucleotides and buffers. Abnormally increased levels of phosphates in the blood are associated with decreased renal function (Soyoral et al. 2014; Vervloet et al. 2017).

Fluorine is essential for the maintenance and solidification of bones and to prevent dental decay. Its abnormal levels leads to teeth decay, osteoporosis and is harmful to the kidney, bones, reproductive organs, nerve and muscle; when in excess, it interfere with the thyroid gland function and disturbs blood glucose level (KONO 1994; Dey and Giri 2016).

The physiological role of iodine in the human body is in the synthesis of thyroid hormones by the thyroid gland. Deficiency of iodine leads to goiter and it's toxicity may lead to coma (Haldimann et al. 1998; Ahad and Ganie 2010). On a very different level, an inability to process or catabolize effectively xenobiotic anions, including simple anions such as cyanide, oxalate, arsenate or nitrite, can produce symptoms of chronic or acute toxicity.

Inorganic anions are an important class of analytes since they are involved in biological processes (Bianchi et al. 1997; Martinez-Manez and Sancenon 2003) but are also environmental pollutants (Okesola and Smith 2016; Kumar et al. 2017). Therefore, methods allowing rapid, selective and inexpensive determination of anions (important analytes) in different types of samples (environmental, industrial, biological fluid) are desirable for their differentiation, detection and quantification (Kaniansky et al. 1999; Fernández-Ramos et al. 2011; Gale et al. 2014). Various methods include optical detection, using an appropriately functionalized organic chromophore (Gale 2000; Gale and Gunnlargson 2010), which can sometimes also be used for differentiation of particular anions (Ali et al. 2018; Borissov et al. 2019; Men et al. 2019; Rifai et al. 2019), where detection and reporting might be based on variation in an inherent physical property (color, basicity, etc.) have been developed (Park and Simmons 1968; Graf and Lehn 1976; Duke et al. 2010).

Investigation of supramolecular systems contribute to the development of many fields of science such as modern analytical chemistry, medicine, technology and environmental protection.

Parameters that Affect Anion Complexation

The design of receptors for anions is particularly challenging when compared to the design of cation receptors due to number of difficulties in designing the receptors capable of anion binding. There are several reasons for this.

1. **Size of anions**: Anions are larger than the equivalent isoelectronic cations (Agmon 2017), which gives them a lower charge to radius ratio. The anion with diffuse nature reduces the contribution of electrostatic binding interactions with the host.

2. **Geometry of anions**: Anions have a wide range of geometries which includes simple spherical halide, linear azides, planer nitrate anion, sulfate, phosphate anions with tetrahedral geometries and cobalt hexafluoride (octahedral). A higher degree of design and complementarity is required to make receptors for a particular anionic guest, unlike most simple cations (Steed and Atwood 2013).

3. **pH of solution**: Anions may get protonated at low pH and lose their negative charge. Thus, receptors must be designed to function within the pH range of their stable target anion. This becomes a problem while designing protonated receptors (like ammonium moiety) for anions. Neutral receptors are less problematic or those containing permanent built-in charges, designed to operate in aprotic media.

4. **Nature of solvent**: The nature of the solvent plays an important role in regulating the anion-binding strength as well as selectivity. Electrostatic interactions not only stabilize anions but also dominate over other factors. Hydroxylic solvents have the ability to form a hydrogen bond with anions. Therefore, a potential anion receptor must compete effectively with the solvent environment in which anion-binding is to take place. On the other hand, a charged receptor can benefit from electrostatic effects and thus may compete more effectively with polar protic solvents. The anions reactivity is drastically increased in the solvents that are not attracted to the anion, leaving the anion naked and therefore, are able to bind easier with the anion acceptor. The anion receptor must not just compete with the solvent but also with the counter cation that is necessarily paired with the targeted anion. Ion-pairing can be very significant, particularly in non-polar solvents as the anion may be weakly solvated in non-polar solvents but there may be significant ion-pairing. It is imperative to remember that binding experiments in solution always comprise an element of competition either from solvent or from counter ion (Schmidtchen and Berger 1997).

5. **Hydrophobicity**: The selectivity of the receptor towards anions can be influenced by hydrophobicity. The Hofmeister series or lyotropic series (Shimizu et al. 2006) (Scheme 1), which was first established through studies on the effect of salts on the solubility of proteins, orders anions by their decreasing hydrophobicity (and therefore increasing degree of aqueous solvation). Hydrophobicity may therefore be used by chemists in the design of anion receptors to bias selectivity towards larger anions with low charge.

Organic anions $> ClO_4^- > SCN^- > I^- >$ salicylate $> NO_3^- > Br^- > Cl^- > HCO_3^-$
$> H_2PO_4^- > F^- > SO_4^{2-} > HPO_4^{2-}$

Scheme 1   The Hofmeister series

Receptors for Anions

Anionic receptors can be categorized into several groups depending on their interactions with ions. Hosts capable of interacting with an anion may have a positive charge and can bind with the substrate by electrostatic interactions or maybe a neutral anion receptor forming a complex through the hydrogen bond (C–H⋯anion interactions). Therefore, ligands are designed and synthesized in such a manner that they can either form a hydrogen bond or are capable of electrostatic interactions. For instance, derivatives of amide, urea, sulfonamide, thiourea or heterocycles bearing –NH group, the polarity of NH fragment (of the host) can be enhanced by introducing electron-withdrawing substituents or positively charged groups, which supports the anion binding; C–H⋯anion interactions can be strengthened by polarizing the C–H bond. Conversely, if polarization is too strong, it may result in deprotonation of ligand (proton transfer from the ligand to the anion) (Izatt et al. 1991; Choi and Hamilton 2003; Boiocchi et al. 2004; Amendola et al. 2010; Frontera Beccaria 2018; Sengupta et al. 2018; Tuo et al. 2018; Luo et al. 2019; Berry et al. 2020).

The design and synthesis of abiotic anion receptors are of substantial significance in many areas of science. Hydrophilic ligands can be used as enzyme models as they can form complexes in aqueous solution and most of the biochemical processes occur in water. They can also act as a selective ion transport agent through a hydrophobic membrane and can provide valuable information about cellular processes. They can act as drug-like molecules that increase the ion permeability through lipid bilayers in membranes containing defective channel proteins (Gunnlaugsson et al. 2005). Furthermore, suitable lipophilic anion receptors can act as ionophores in ion-selective electrodes. Potentiometric sensors of this kind are typically used in clinical or environmental analysis. They allow easy determination of ions (Pomecko et al. 2010). Receptors containing chromo- or fluorophore moiety may be used for the identification and determination of an anion using spectroscopic methods. Complexation of the anion with a ligand may result in a change of color or fluorescence which allow its simple identification or even semiquantitative assaying (the so-called 'naked-eye sensor') (Martinez-Manez and Sancenon 2003; Sessler et al. 2003; Martinez-Manez and Sancenón 2006; Santos-Figueroa et al. 2013).

Measurement Techniques

The role of a supramolecular chemist does not limit it to design and synthesize receptors but encompasses evaluation of their binding properties and selectivity. The first step towards an understanding of the Host (H) and its Guest (G) supramolecular chemistry is the accurate determination of the binding affinity between them. (Eq. (1)) (Li et al. 2009; Steed and Atwood 2013; Schalley 2012; Hirose 2007)

$$H + G \leftrightarrows GH \tag{1}$$

$$K_a = \frac{[H]_e [G]_e}{[G \subset H]_e} \tag{2}$$

To understand supramolecular effects at deeper levels, the thermodynamic parameters ΔH and ΔS are evaluated (Schmidtchen 2002; Moonen et al. 2005; Schmidtchen 2006; Sessler et al. 2006a). This information serves as the basis for both understanding and applications in various areas such as molecular transport biochemistry, template-directed synthesis, and sensing (Silverman and Lindskog 1988; Sato et al. 1998; Badr et al. 1999, 2000; Yamamoto et al. 2000; Accardi and Miller 2004; Kubik et al. 2005; Vickers and Beer 2007). Primary data must be resolved before the interpretation of thermodynamic values (Schneider and Yatsimirsky 2000) such as:

(i) Job's Plot method is generally used to determine the binding stoichiometry (Connors 1987)

(ii) Proposition of suitable chemical model for host-guest system (Long and Drago 1982; Cram 1988)

(iii) Estimation of the suitability of the titration experiment for predicting the thermodynamic values precisely (Connors 1987; Hirose 2007)

The area of anion receptor chemistry involves various techniques to measure the stability constant of the host with a guest. These techniques involve titration of a guest into a solution of the host to observe changes in the composition of the solution. The range of spectroscopic or calorimetric tools may be used for titration such as NMR, UV-Vis absorption spectroscopy, fluorescence emission spectroscopy or Isothermal Titration Calorimetry (ITC). Each of these techniques looks at a different part of the binding process and/or overall equilibrium. Titrations involving NMR spectroscopy provides a rich amount of structural information e.g., it can provide insight into the anion interaction with the hydrogen bond donor groups of the receptor. NMR is suitable for moderate binding strengths, $K_a = 1 - 10^3 \, M^{-1}$.

Isothermal Titration Calorimetry (ITC) measures the heat released or absorbed on interaction of the host with a guest by titrating one reactant into a second reactant under isothermal conditions. ITC can determine all the thermodynamic parameters (stoichiometry, ΔH, ΔS, ΔC_p, ΔG) in one experiment. The limitation of the technique lies in the fact that it is sensitive to equilibria, which are not involved with the host-guest interaction, but are associated with the absorption or release of heat.

Spectrophotometric techniques are complementary to these methods for the characterization of host-guest interactions with large binding constants ($K_a = 10^3 - 10^7 \, M^{-1}$) (Sessler et al. 2006a). UV-Vis and fluorescence spectroscopy reflect changes in the optical properties of the light-absorbing/emitting portions of the receptor.

Experiments must be conducted under identical conditions (e.g., temperature, solvent, concentration) while conducting the Host-Guest binding studies, or else any conclusions established may be invalid.

Subsequent sections focus on the topology and examples of anion receptors based on bile acids, Calix[4]pyrroles, Redox-active chemical species and Porphyrin cages.

BILE ACID-BASED MACROCYCLES AS ANION RECEPTORS

The design and synthesis of bile acid-based macrocycles have been a subject of considerable research interest because of their great potential applications in supramolecular chemistry. All bile acids consist of a rigid, chiral steroid skeleton and a short aliphatic side chain. The bile acid skeleton is curved because six membered rings A and B are in a cis connection (Fig. 4.1). The hydroxyl groups are directed toward the 'inner' cavity and form the hydrophilic face with the carboxylic group of the side chain. The 'outer' face is hydrophobic due to the presence of the three methyl groups. Bile acids are therefore amphiphilic compounds, are inexpensive, readily available natural steroids.

	R_1	R_2
Cholic acid	OH	OH
Deoxycholic acid	H	OH
Chenodeoxycholic acid	OH	H
Lithocholic acid	H	H
Ursodeoxycholic acid	β–OH	H

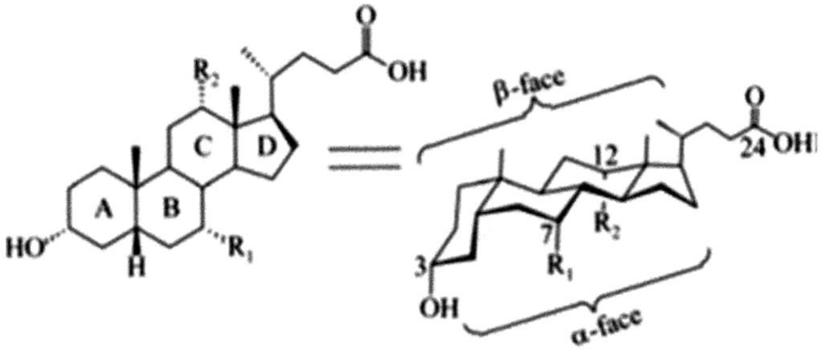

Figure 4.1 Chemical structure of bile acids.

Their concave hydrophilic faces with two or three hydroxyl groups are accessible to act as H-bond donors, but are incapable of forming intramolecular hydrogen bonding. Therefore, these can act as natural binding sites for anions. The strength of the bile acid receptor can be enhanced by the modification of hydroxyl groups. This can be achieved by simple derivatization e.g., conversion of OH to O–CO–NHAr (carbamate), great scope can be achieved by first replacing OH groups with amino groups and these may be further transformed into a variety of structures such as ureas or thioureas which can act as multiple H–bond donor centers. The resulting podands ('cholapods' derived from cholic acid) can fringe around a bound anion with H–bond donor units. (Brotherhood and Davis 2010). Bile acid based receptors can be synthesized by head-to-tail cyclization of the bile acids, by joining of two to four bile acid units together by different spacer groups, for example bile acid-based crown ethers can also be prepared (Tamminen and Kolehmainen 2001). By varying the bile acid and the bridging groups of the cyclic and cleft-type structures, it is possible to prepare many interesting bile acid-based molecular assemblies for recognizing anions.

Whitmarsh and co-workers converted cholic acid into cyclotrimeric and cyclotetrameric toroidal amphiphiles, enhanced the facial amphiphilicity by first converting inward-directed –OH to $–NH_3^+$ then cyclooligomerizing (Whitmarsh et al. 2008). The resulting systems **1** (Fig. 4.2) were toroidal facial amphiphiles with hydrophobic outer surfaces and strongly hydrophilic interiors. They exploited the potential of toroidal facial amphiphiles with hydrophobic exteriors for membrane transport and reported that cyclotrimer **1b** affects the transport of chloride ions across phospholipid bilayers. Ion Selective Electrode-based studies showed good activity for chloride nitrate exchange in vesicles while confirming that K^+ was not transported. Experiments designed to distinguish between the two modes of action (as a mobile carrier or by stacking within the phospholipid bilayer to form anion-selective pores **1b**) suggested that the mobile carrier mechanism was in operation.

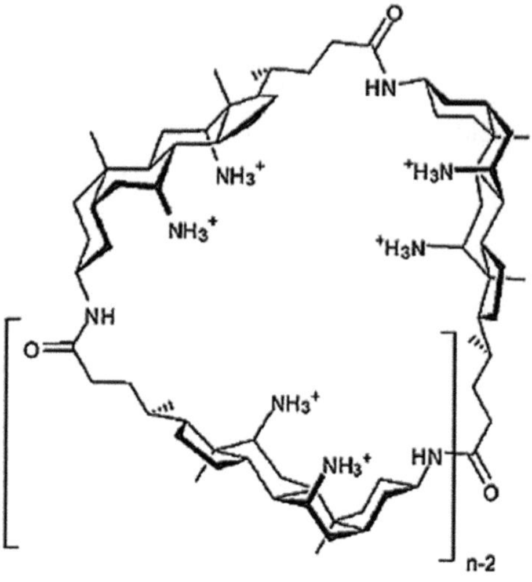

1a n = 2   1b n = 3   1c n = 4

Figure 4.2 Chemical structure of macrocyclic oligomers (cyclocholamides). Reprinted with permission from (Whitmarsh et al. 2008).

Anion exchange was studied in two-phase, water-chloroform system for 'cholaphane' (Sisson et al. 2005). In the absence of a receptor, the transfer of anions from the aqueous to organic phase is governed primarily by hydrophobicity, following the classical Hofmeister series. Compound **2** (Fig. 4.3) perturb the Hofmeister series by forming H-bonds with the more hydrophilic anions, promoting their extraction. Macrocycle **2** showed true selectivity for halides, extracting bromide and chloride in preference to both more and less hydrophilic anions.

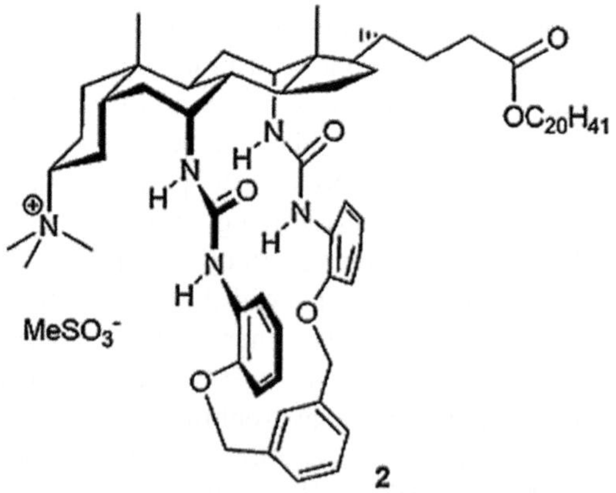

2

Figure 4.3 Chemical structure of macrocycle **2** reprinted with permission from (Sisson et al. 2005).

Cyclic (**3**) and acyclic (**4**) bile acid-based 1,2,3-triazolium receptors (Fig. 4.4) synthesized using click chemistry showed remarkable ability to recognize anions through C–H---X⁻ hydrogen bond interactions by Prof. Pandey's group (Kumar and Pandey 2008). Receptor **3a** showed a remarkable affinity for fluoride ion with an association constant of 560 M^{-1}. The selectivity trend was observed to be maximum for F^- followed by $Cl^- > Br^- > I^- > CH_3COO^-$. No significant binding was observed with $H_2PO_4^-$ ion though, receptor **3b** showed a considerable affinity for the $H_2PO_4^-$ ion with a binding constant of 1100 M^{-1}, which could be attributed to the larger cavity size of the receptor because of the para-substituted benzene ring. The selectivity trend observed toward anions was $H_2PO_4^- > Cl^- > Br^- > F^- > I^- > CH_3COO^-$. Remarkably, the acyclic receptor **4** showed much higher affinity and selectivity toward $H_2PO_4^-$ ion as compared to the cyclic receptor **3b**, having a binding constant of 1920 M^{-1}. The selectivity trend was $H_2PO_4^- > Cl^- > F^- > Br^- > I^- > CH_3COO^-$, which may be attributed to the greater flexibility of the acyclic receptor for adapting the suitable geometry required for the binding of a bulky anion (Fig. 4.4).

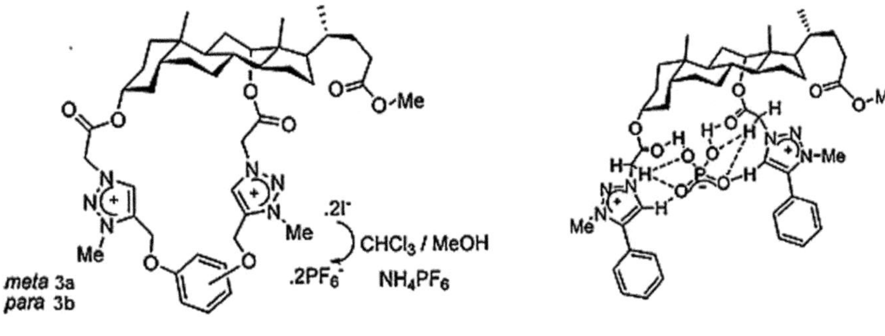

Figure 4.4 Cyclic (**3**) and acyclic (**4**) bile acid-based 1,2,3-triazolium receptors reprinted with permission from (Kumar and Pandey 2008).

The anion binding properties of bile acid-based cyclic bisbenzimidazolium receptors **5-7** bridged with m-xylene, p-xylene, and 2,6-dimethylpyridine (Fig. 4.5) studied by ^1H NMR titration (Khatri et al. 2007). Receptors **5** and **6** exhibit much greater binding affinity for fluoride (association constant of 3500 M^{-1}) and chloride ions (association constant of 20500 M^{-1}), respectively, association constant values are much higher as compared to the imidazolium receptors **8** and **9** for which the association constants obtained for fluoride and chloride ions, 2400 and 12000 M^{-1}, respectively. Receptor **7**, however shows high selectivity but very low binding affinity for anions (order of association constant of receptor **7** with different anions was $Cl^- > CH_3COO^- > Br^- > F^- > HSO_4^-$) which was attributed to the presence of the lone pair of electrons on nitrogen that prevents the anions from binding correctly in the cavity of the receptor. They also observed that all the receptors show very weak binding with HSO_4^- ion (K_a) 80-300 M^{-1}, Receptors **7** and **12** containing a pyridyl group have been found to have a better binding affinity than receptors **5** and **6** that may be attributed to the additional hydrogen bond interaction involving pyridyl nitrogen and the proton of the anion. None of the receptors show any significant binding with $H_2PO_4^-$ ion.

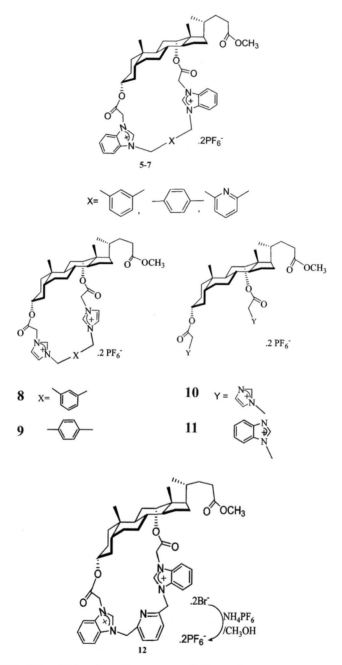

Figure 4.5 Bile acid-based cyclic receptors (5-12) reprinted with permission from (Khatri et al. 2007).

The same group (Chhatra et al. 2011) Pandey et al. synthesized bistriazolium derivative of the cholaphane (Fig. 4.6) which exhibited high selectivity for the chloride ion (K_a = 3700 M^{-1}) followed by HSO$_4^-$ > H$_2$PO$_4^-$ > F$^-$ > Br$^-$ > CH3COO$^-$ > I$^-$. The significant selectivity of this for chloride anion was due to the cavity size.

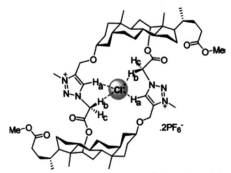

Figure 4.6 Chemical structure of bistriazolium derivative of the cholaphane receptor with encapsulated chloride ion; reprinted with permission from (Chhatra et al. 2011).

The Maitra group prepared the cyclodimeric structure (Fig. 4.7) and showed by NMR studies that it binds to Bu$_4$N$^+$F$^-$ in CDCl$_3$ with 1:2 stoichiometry (K_{a1} ≈2000M^{-1}, K_{a2} ≈250M^{-1}). Complexation was thought to take place through the formation of two OH$\cdots$F$^-$ and one CH$\cdots$F$^-$ hydrogen bonds, as shown in Fig 4.7. Direct evidence for the CH$\cdots$F$^-$ interaction was provided by ^{1}H NMR spectra, in which the chemical shift for one CH moves strongly downfield on titration with the substrate (Ghosh et al. 2005).

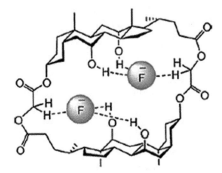

Figure 4.7 Fluoride ions encapsulated by bile acid-based cyclodimer receptor. Reprinted with permission from (Ghosh et al. 2005).

CALIX[4]PYRROLES BASED MACROCYCLES AS ANION RECEPTORS

Octamethylcalix[4]pyrrole (Baeyer 1886) was first synthesized by the cyclocondensation of pyrrole with acetone in the presence of hydrochloric acid by Baeyer in 1886, and in 1916 Chelintzev and Tronov characterized the structure of the compound which was called $\alpha,\beta,\gamma,\delta$-octamethylporphinogen by Fischer.

Calix[4]pyrroles originally named pyrrole-acetone and formally known as *meso-*octaalkylporphyrinogens consist of four pyrrole units linked together by four di-substituted *meso*-carbons at α positions. Figure 4.8 (OMCP) shows the basic skeleton of calix[4]pyrrole (where n = repeating pyrrolic units linked to each other by sp³ hybridized carbon atoms). Fully substituted *meso*-carbon makes calix[4]pyrroles resistant to oxidation. As a result, all the four pyrrole units are composed of neutral NH-bearing forms without any electron delocalization within the macrocycle.

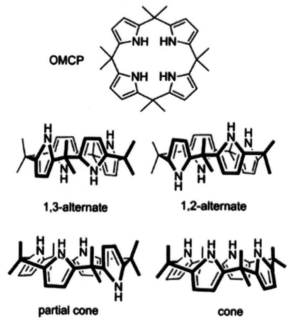

Figure 4.8   Structure of *meso*-octamethylcalix[4]pyrrole (OMCP) and its four different conformers. Reprinted with permission from (Saha et al. 2015).

Calix[4]pyrrole macrocycles can neither undergo oxidation nor are intrinsically effective ligands as a result they were only studied sporadically during most of the 20th century until the early 1990s when Floriani and co-workers explored their metal coordination (Benech et al. 1999; Kadish et al. 2000). In 1996, Sessler and co-workers uncovered calix[4]pyrroles as receptors for anions (Gale et al. 1996). On further investigation by the same group, it was observed that octamethylcalix[4]pyrrole on anion binding, undergoes a structural change from 1,2-alternate, 1,3-alternate, or partial cone to a cone conformation (Gale et. al 1996). This impelled the group to introduce the name calix[4]pyrrole to recognize octamethylcalix[4]pyrrole and its analogues.

Since the discovery of the anion recognition properties of calix[4]pyrrole, many modified calixpyrroles as anion receptors have been synthesized and studied by various groups worldwide and its importance can be judged from different reviews with calix[4]pyrrole are being published over the years and recently

(Rather et al. 2019; Shah and Bhatt 2019; Kohnke 2020; Peng et al. 2020). Two main strategies have been used to tailor new calix[4]pyrrole motifs for the recognition of anions. These include adding substituents on the sp³ carbon constituting the meso-bridges or modifying β-pyrrolic positions, or both (Lee et al. 2008; Saha et al. 2015)

Condensation of a dipyrromethane using acid as a catalyst, with an excess of acetone has been helpful in the functionalization of the meso-positions. For instance, one well-designed approach of this encompasses strapped calix[4] pyrroles. Such systems comprise of one or more bridging 'straps' that span the calix[4]pyrrole backbone to create a bi- or polycyclic system (Fig. 4.9). Strapped calix[4]pyrroles largely exhibit enhanced affinities and superior selectivity towards anionic species as compared to calix[4]pyrroles which may be assigned to various factors that include (i) the straps, in general which provides additional cooperative binding sites (ii) strapped (calix[4]pyrrole) assisted improvised preorganization of the host and substrate as such; (iii) multi-macrocyclic receptors including strap provides a better opportunity to encapsulate the targeted guests deeply within its cavity. Another important factor, the presence of straps reduces the freedom of the host making guest binding more entropically favorable. Recently, a review summarized the progress made in the field of strapped calix[4]pyrroles chemistry, with special emphasis on work over the last decade (Peng et al. 2020).

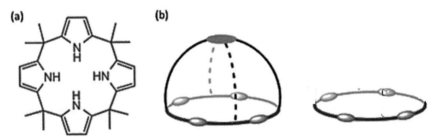

Figure 4.9 (a) Chemical structures of parent calix[4]pyrrole and (b) generalized representation of a strapped calix[4]pyrrole bearing at least one 'closed' strap and parent calix[4]pyrrole. Reprinted with permission from (Peng et al. 2020).

The addition of functional groups on meso-positions of calix[4]pyrroles result in the development of macrocycles with pendant arms, deep cavity systems and capsules, apart from strapped calix[4]pyrroles.

Few of the known β-substituted calix[4]pyrroles in literature were synthesized by a condensation strategy. Typically, a functionalized pyrrole bearing substituents on one or both of the β-pyrrolic positions is condensed with a ketone under acid-catalyzed conditions. However, a number of mono-substituted calix[4]pyrroles have been prepared by a post-synthetic strategy. Here a functional group, such as an ester, formyl or iodine substituent, is introduced into one of the β-pyrrolic positions within OMCP. In many cases, this is done by subjecting the calix[4] pyrrole to deprotonation with a strong base (e.g., n-butyl lithium), followed by the addition of the appropriate electrophile. On the other hand, iodination has typically been achieved by reacting calix[4]pyrrole with iodine-[bis(trifluoroacetoxy) iodo] benzene. These functionalized systems have been elaborated further to create, inter alia, anion sensors or materials that may be attached to solid or polymeric supports.

Very recently, reviews on calix[4]pyrrole have been published which demonstrates the importance of macrocycles as receptors (Rather et al. 2019; Shah and Bhatt 2019; Chen et al. 2020; Kohnke 2020; Peng et al. 2020). As most examples have been covered in these reviews, in this chapter the most recent work involving anion recognition by calix[4]pyrroles receptors are highlighted. The first example in this context is from the research article published in the year 2020 by Heo et al. (Heo et al. 2020) where they showed that anions selectivity and affinity can be tuned by adjusting the size and rigidity of the linkers which form a strap serving to connect two opposing meso carbons. This strategy further allows the introduction of hydrogen bonding donors and acceptors type of auxiliary recognition motif linkers. For this, they synthesized calix[4]pyrroles **13** and **14**, strapped with a phenanthroline via ester and amide linkages, respectively as anion receptors.

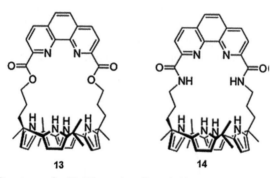

Figure 4.10   Structure of calix[4]pyrroles **13** and **14**, strapped with a phenanthroline by ester and amide linkages. Reprinted with permission from (Heo et al. 2020)

It was shown that both these receptors are effective in capturing bicarbonate anion. The bicarbonate anion is planar with pseudo-trigonal symmetry. The proton of bicarbonate, HCO_3^- has the potential to act as a hydrogen bond donor while phenanthroline subunits (lewis base) present in receptors **13** and **14** were intended to offer an additional benefit as a binding motif (hydrogen bond acceptor) for the bicarbonate anion. Amide linkers present in receptor **14** were introduced to act as hydrogen bonding donors and assist in anion capture. Along with this it was shown that their relative affinities toward anions can be tuned by changing the polarity of the solvent. ^1H NMR titrations of receptors **13** and **14** with various anions using $CDCl_3$ and 15% aqueous DMSO solution were conducted and it was established that $CDCl_3$ solution of both receptors were able to bind the bicarbonate anion with good selectiveity over other test anions and in a deuterated-chloroform solvent, receptor **13** was found to having higher binding affinity for most of the test anions as compared to receptor **14** while in 15% aqueous DMSO solution which is a more polar medium, the affinity towards anions were enhanced in case of receptor **14** while for receptor **13** affinity was observed to reduce to the point that no notable interaction between the receptor and the test anions was noticed. This solvent-based reversal in affinities and selectivities for test anions was evident from the association constant (M^{-1}) values of the host and guest. Receptor **13** showed maximum affinity towards HCO_3^- ($K_a = 1330$ M^{-1}) followed by SO_4^{2-} ($K_a = 565 \pm 86$ M^{-1}), F^- ($K_a = 580$ M^{-1}), $HP_2O_7^{3-}$ ($K_a = 447 \pm 24$ M^{-1}), HSO_4^- ($K_a = 364 \pm 47$ M^{-1}), Br^- ($K_a = 338 \pm 25$ M^{-1}), I^- ($K_a = 312 \pm 23$ M^{-1}), $H_2PO_4^-$ ($K_a = 174 \pm 17$ M^{-1}), Cl^- ($K_a = 121 \pm 11$ M^{-1}) while in 15% aqueous DMSO solution no binding was observed for SO_4^{2-}, HSO_4^-, Br^-, I^-, $H_2PO_4^-$, Cl^- and for $HP_2O_7^{3-}$, HCO_3^- binding constant values could not be calculated as during titration hydrolysis or precipitation occurred. In the case of receptor **14** the trend was reverse, it showed maximum binding for HCO_3^- in 15% aqueous DMSO solution ($K_a > 10\,000$ M^{-1}) in comparison to deuterated-chloroform solvent ($K_a = 4880$ M^{-1}) and F^-, $H_2PO_4^-$ also showed association constant value to be $> 10\,000$ M^{-1} in 15% aqueous DMSO solution and low K_a values in deuterated-chloroform solution. Chloride and bromide anions also showed higher affinity towards receptor **14** in 15% aqueous DMSO solution, while in iodide there was no binding affinity towards receptor **14** in either of the solvents. HSO_4^- and SO_4^{2-} showed no affinity towards receptor in 15% aqueous DMSO solution but their affinity (HSO_4^-, $K_a = 53 \pm 11$ M^{-1} and SO_4^{2-}, $K_a = 8$ M^{-1}) were shown in deuterated-chloroform solution. This result was attributed to the weakening of hydrogen bonding effects of amide NH-phenanthroline, allowing for an effective reversal in the anion affinities on moving from $CDCl_3$ to the more polar (and protic) medium comprised of 15% D_2O in DMSO-d_6. Receptor **14** was less effective than receptor **14** in $CDCl_3$ to bind most other anions than bicarbonate which was concluded to the presence of competing intramolecular amide NH-phenanthroline hydrogen bonding interactions in the case of receptor **14**. By changing the solvent from $CDCl_3$ to the polar protic solvent system, the affinity of receptor **13** to bind with anions decreased drastically whereas the affinity of receptor **14** improved significantly which was attributed to the weakening of the intramolecular amide NH-phenanthroline hydrogen bonding effects, which results in a reversal in the anion affinities on moving from $CDCl_3$ to the more polar (and protic) medium. They also reported DFT calculations and that both receptors **13** and **14** formed energetically favorable complexes with the bicarbonate anion where the phenanthroline group acts as a hydrogen bond acceptor.

Chahal and co-workers reported the successful synthesis of a highly conjugated calix[4]pyrrole (Chahal et al. 2019). This was achieved by a 'knock-on' regioselective reaction to introduce β-functionality at an appropriate position by first selective N-alkylation of the calix[4]pyrrole with subsequent N-alkylation at the opposing pyrrole group, this was done to promote stronger interactions of the oxoporphyrinogen 'OxP' with oxoanions. The resulting host molecule **15** (Fig. 4.11) exhibits anion binding of higher order with common chloride and nitrate anions enhanced by an order of magnitude stronger than its parent host conjugated calix[4]pyrrole Bn_2OxP. ^1H NMR and UV-Vis titrations (using $CDCl_3$ and CH_2Cl_2 as solvents in respective titrations) were followed to rationalize the interactions between the host **15** with the test anions (as their tetrabutylammonium (TBA) salts). Regarding Type A salts, initially, simple 1:1 host-guest binding model was applied in order to evaluate the binding affinity of **15** (host) to TBA salts (guest). However, the 1:1 binding model could not give consistent results of binding constants obtained from NMR (high concentration) and UV-Vis (low concentration) spectroscopic methods. They proposed that the possible reason for the discrepancy could be the assumption that in non-polar solvents, TBA salts are completely dissociated and all anions are free to bind with the host. This led to the analysis of the fraction of free anion as a function of salt dissociation constant K_d and total salt concentration. Thus, Host-Guest association constant values were found to be dependent on K_d. For reasonable determination of both of the important binding parameters, K_a and K_d, a method could be developed for simultaneous fitting of NMR and UV-Vis binding isotherms together.

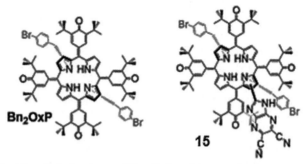

Figure 4.11   Chemical structures of Bn_2OxP and host molecules **15** reprinted with permission from (Chahal et al. 2019).

Gale and co-workers synthesized novel bis-triazole-functionalized calixpyrroles, intended for having transmembrane anion transport properties (Clarke et al. 2020). They reported the interaction between strap-extended calixpyrroles **16** and **17** with Cl^- through hydrogen bonding with NH and CH. Also they demonstrated improved Cl ion uniport activity due to extended alkyl linker or increased lipophilicity in comparison to shorter straps but compromised Cl^- over H^+/OH^- selectivity which was proposed due to the large size of binding cavity enabling interactions with deprotonated fatty acids.

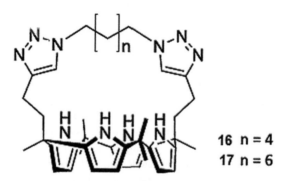

Figure 4.12   Structures of tripodal tristhiourea and strapped calixpyrroles **16**, **17** reprinted with permission from (Clarke et al. 2020).

Table 4.1 Derivatives of calix[4]pyrroles as host, their guest anions along with the outcomes of their interactions and techniques to study them are tabulated below:

S.No.	Host	Guest	Studies/techniques used	Outcome/results	References
1	Cyclic derivative of calix[4]pyrroles 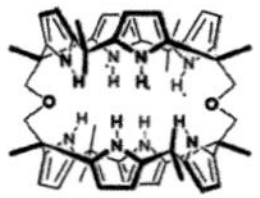	F⁻, Cl⁻, Br⁻, SCN⁻, NO₃⁻, H₂PO₄⁻, HSO₄⁻, SO₄²⁻, HP₂O₇³⁻	^{1}H NMR, X-ray diffraction, DFT calculations	Smallest bis-calix[4] pyrrole host capable of trapping fluoride anion in 1:2 manner but could not interact appreciably with other anions.	(Xiong et al. 2020)
2	a single fluorescent probe, i.e., the bis-calix[4]pyrrole-appended 9,14-diphenyl-9,14-dihydrodibenzo[a,c] phenazine (**DPAC-bisC4P**) 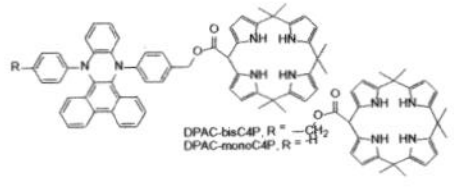 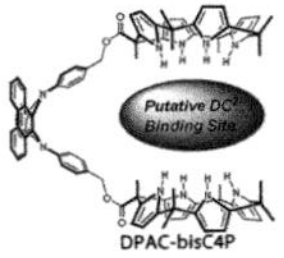	aromatic and linear saturated dicarboxylate anions (as their tetrabutylammonium salts) with different lengths and shapes in acetonitrile 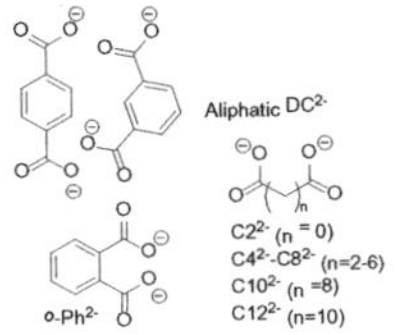	Fluorescence titration UV–Vis spectroscopic titrations semiempirical methods at the PM6 level ^{1}H NMR, ^{13}C NMR spectroscopy and high-resolution mass spectrometry (HRMS) for characterization	Fluorescent chemosensor and anions form pseudomacrocyclic host–guest complexes via multiple hydrogen bonding. the unique and distinct guest-dependent fluorescent signature was produced for linear saturated dicarboxylate anions as well as for three isomeric phthalate dianions. (see Table 4.2 for binding constant and LOD values)	(Chen et al. 2019)

Table 4.1 (Contd.)

Table 4.1 (Contd.) Derivatives of calix[4]pyrroles as host, their guest anions along with the outcomes of their interactions and techniques to study them are tabulated below:

S.No.	Host	Guest	Studies/techniques used	Outcome/results	References
3	*meso*-substituted calix[4]pyrroles containing a direct *meso*-ethynyl linker 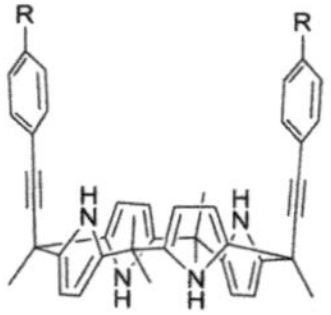 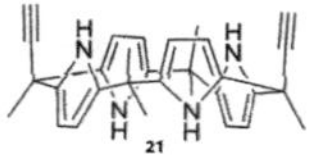 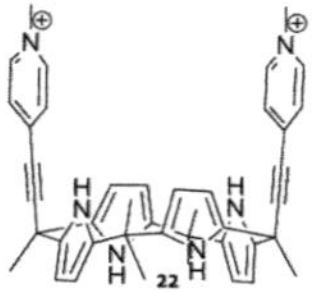 18 R = ethynyl, 19 R = H, 20 R = F Possible Binding Modes and Mechanism of Representative Receptor 19 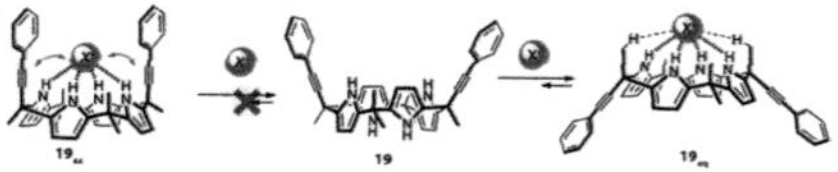	Halide anions F⁻, Cl⁻,Br⁻	¹H NMR in CD$_3$CN isothermal titration calorimetry (ITC) measurements	*meso*-Substituted calix[4]pyrroles showed pseudo-equatorial conformation of the *meso*-(aryl)ethynyl groups on complexation with halide anion, which was attributed to anion–alkyne repulsions and released steric strain it was also observed in the case of calix[4]pyrrole **22** bearing cationic *meso* components, which displayed the highest binding affinity for chloride anions ($K_a = 4.50 \times 10^6 \pm 1.01 \times 10^6$) followed by **18** ($K_a = 6.71 \times 10^5 \pm 3.30 \times 10^4$), **20** ($K_a = 5.73 \times 10^5$), **19** ($K_a = 4.25 \times 10^5$), **21** ($K_a = 1.16 \times 10^5 \pm 4.85 \times 10^3$) ITC studies in CH$_3$CN solvent showed higher affinity of all receptors for chloride as compared to bromide anion	(Dutta et al. 2019)

S.No.	Host	Guest	Studies/techniques used	Outcome/results	References
4	squaramide-substituted calix[4]pyrrole **23**	UV titration, ^{1}H NMR titrations, and computational DFT model NMR and HRMS for characterization	F^-, Cl^-, Br^-, I^- and $H_2PO_4^-$	anion-binding properties of squaramide-substituted calix[4]pyrrole **23** were investigated and were compared with parent calix[4]pyrrole. UV titrations of the receptor with anions in CH_2Cl_2 show 1:1 stoichiometry. Anions with larger radius show increased affinity (K_a(**23**)/K_a(**p**) = 2.47 × 10^5/97) for $H_2PO_4^-$ > (K_a(**23**)/K_a(**p**)= 2.38 × 10^5/3.50 × 10^2) for Cl^- > K_a(**23**)/K_a(**p**)= 2.51 × 10^5/1.72 × 10^4 for F^-). (where **p** is parent calix[4] pyrrole) while the affinity of **23** for Br^- and I was too low to be determined by the UV titration. The observation was attributed to the large distance between the calix[4]pyrrole moiety and squaramide moiety for all six NH protons to contribute strongly to the smaller	(He et al. 2019)

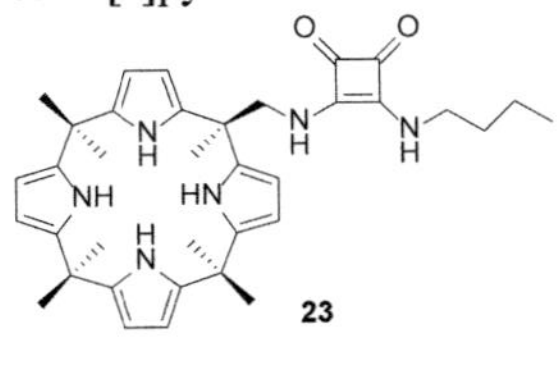

Table 4.1 (Contd.)

Table 4.1 (Contd.) Derivatives of calix[4]pyrroles as host, their guest anions along with the outcomes of their interactions and techniques to study them are tabulated below:

S.No.	Host	Guest	Studies/techniques used	Outcome/results	References
				anion binding and the fact that F^-, Cl^- and $H_2PO_4^-$ is far more strongly hydrogen-bonded to the squaramide moiety or calix [4]pyrrole moiety than Br^- and I^- which was supported by NMR titrations, where it was examined that $H_2PO_4^-$ with large radius can not only be strongly hydrogen-bonded to the calix[4] pyrrole framework but also strongly to squaramide unit. The hydrogen bonding between the squaramide unit and anion was also confirmed by the DFT calculations.	

S.No.	Host	Guest	Studies/techniques used	Outcome/results	References
5	triazole bridged bis-calix[4]pyrrole 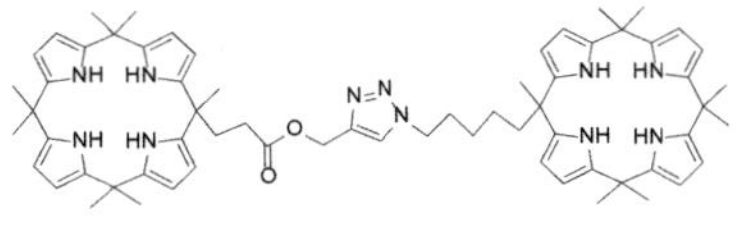	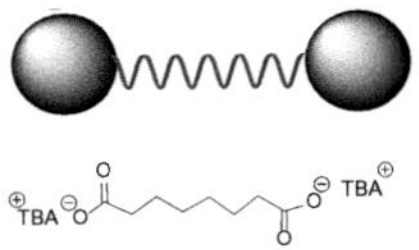	^{1}H-, DOSY NMR analyses, viscosity measurements, SEM	Binding affinity ($K_a = 1.65 \times 10^3 \pm 54.3$ M^{-1} in CDCl$_3$)laid the foundation to construct AABB type linear thermoresponsive supramolecular polymers due to the hydrogen bonding interactions between calix[4]pyrrole NHs and carboxylate units of homoditopic host and guest molecules. 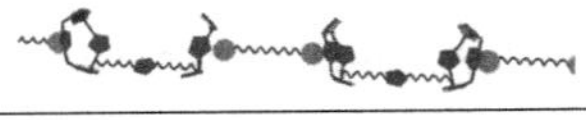	(Yuvayapan and Aydogan 2019)

Table 4.1 (Contd.)

Table 4.1 (Contd.) Derivatives of calix[4]pyrroles as host, their guest anions along with the outcomes of their interactions and techniques to study them are tabulated below:

S.No.	Host	Guest	Studies/techniques used	Outcome/results	References
6	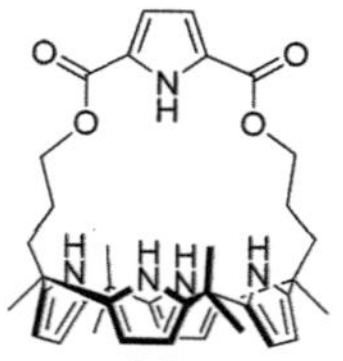 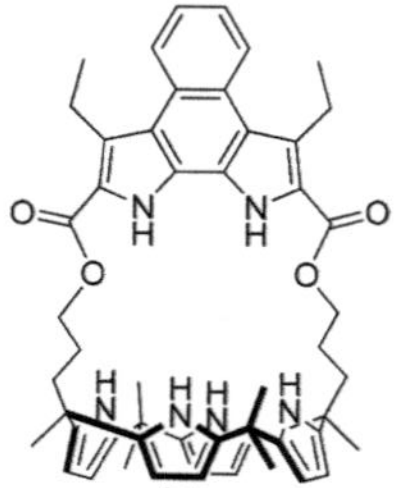	N_3^- (azide anion)	^{1}H NMR spectroscopic (20% CD$_3$OD in CDCl$_3$) and isothermal titration calorimetry Analyses Single crystal X-ray diffraction analyses and calculations using density functional theory	Higher affinities were observed for receptor **24** (K_a=9.14 × 10^5 M^{-1}), **25** (K_a=6.35 × 10^5 M^{-1}) with the azide anion in organic media as compared to the unfunctionalized calix[4] pyrrole (K_a=8.01 × 10^3 M^{-1} which was supported by Single-crystal X-ray diffraction analyses and calculations using density functional theory revealed that receptor **24** binds CsN$_3$ in two distinct binding modes, vertical [N=N=N]$^-$ or horizontal [N–N≡N]$^-$, that give rise to different N–N bond lengths within the bound azide anion. In contrast, azide anion is vertically bound to the NH protons of the calix[4] pyrrole core of receptor **25**. Coordination polymers were observed in a solid state depending on azide binding modes.	(Kim et al. 2018)

S.No.	Host	Guest	Studies/techniques used	Outcome/results	References
7	tetrathiafulvalene (TTF) annulated calix[4]pyrroles (C4Ps) **Benzo-TTF-C4P**	Chloride (tetra-butylammonium counter ion)	UV spectroscopy (molecular switches)	Complete vanishing of the wide charge-transfer bands were observed on adding tetrabutylammonium chloride to chloroform solutions containing nitroaromatic complexes of TTF-C4P host. This occurs by virtue of switching from the 1,3-alternate conformation to anion- cone conformation	(Park and Sessler 2018)

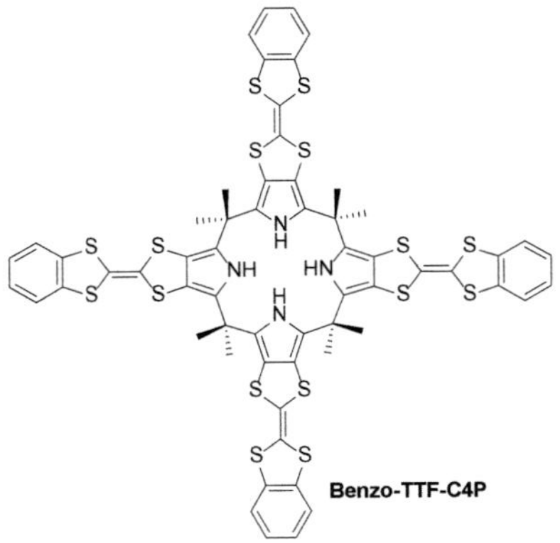

Table 4.1 (Contd.)

Table 4.1 (Contd.) Derivatives of calix[4]pyrroles as host, their guest anions along with the outcomes of their interactions and techniques to study them are tabulated below:

S.No.	Host	Guest	Studies/techniques used	Outcome/results	References
8	Dicationic meso-Bis(benzimidazolium) Calix[4]pyrrole 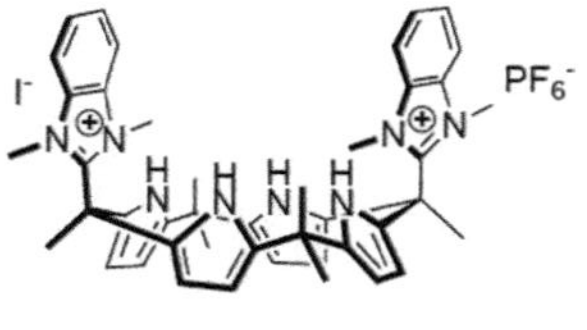	bicarbonate anion HCO_3^-	fluorescence dye displacement assay (FDDA) technique A UV-visible (UV-vis) spectral titration ^1H Nuclear Magnetic Resonance (NMR) spectroscopic titration, isothermal titration calorimetry solid-state structural analyses	The two benzimidazolium moieties in dicationic calix[4]pyrrole are tilted and are not completely parallel to one another which was attributed to an electrostatic effect of the counter anion. In conjunction with a fluorescent probe (chromenolate anion) receptor displays a high affinity for the bicarbonate anion. Under conditions of fluorescence displacement, it shown that the lowest detection limit at concentrations as low as 4 nM and on treatment with sodium ions, the sensor system can be regenerated. In the presence of alcoholic solvents, receptors promote the conversion of the HCO_3^- anion to a bound mono-carbonate ester.	(Mulugeta et al. 2017b)

S.No.	Host	Guest	Studies/techniques used	Outcome/results	References
9	large capsule-like biscalix[4]pyrrole 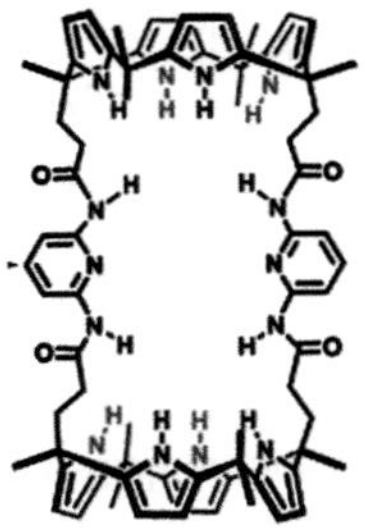	(Oxoanions; $H_2PO_4^-$, SO_4^{2-}, and $HP_2O_7^{3-}$) dihydrogen phosphate anions, dianionic sulfate anions, trianionic pyrophosphate anions	1H NMR and UV–vis spectroscopies and ITC titrations, single-crystal X-ray diffraction Analyses, DFT calculations	>$(H_2PO_4^-)_2 \subset$ receptor; the first binding constant ($K_1 = (1.67 \pm 0.15) \times 10^6$ M^{-1}) is slightly higher than that of the second one ($K_2 = (1.16 \pm 0.08) \times 10^5$ M^{-1}) > $SO_4^{2-})_2 \subset$ receptor log $K = 15.2 \pm 0.7$ > $(HP_2O_7^{3-})_2$ Cl ($K_1 = (6.52 \pm 0.56) \times 10^4$ M^{-1} and $K_2 = 49.0 \pm 0.56$ M^{-1}): two cobound anions (linked by water molecules were by multiple hydrogen-bonding interactions.) are bound concurrently within the large cavity	(He et al. 2017)

Table 4.1 (Contd.)

Table 4.1 (Contd.) Derivatives of calix[4]pyrroles as host, their guest anions along with the outcomes of their interactions and techniques to study them are tabulated below:

S.No.	Host	Guest	Studies/techniques used	Outcome/results	References
10	meso-(5,6-Dichlorobenzimidazole)-picket Calix[4]pyrrole;(calix[4] pyrrole anion receptor-bearing 5,6-dichlorobenzimidazole subunits at the diametrically opposed mesopositions) 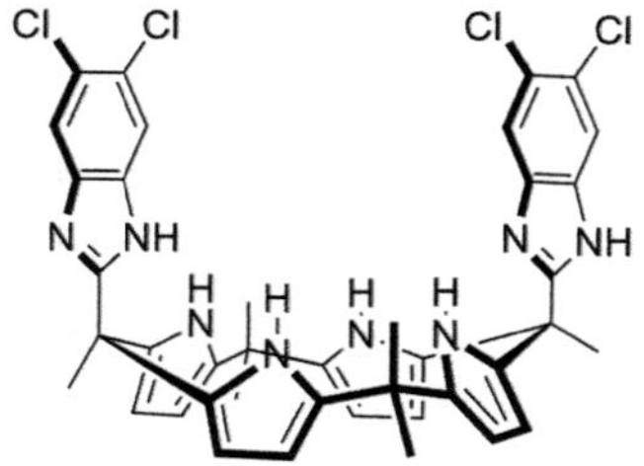	F⁻, Cl⁻, Br⁻, I⁻, AcO⁻, BzO⁻, H$_2$PO$_4^-$	¹H NMR spectroscopic and isothermal titration calorimetry (ITC) techniques	The largest binding affinity was found with fluoride anion, which benefits from a favorable entropic term ($K_a=1.81\ 10^6 \pm 4.82\ 10^5$; $T\Delta S=-5.16$ (kcal/mol)) followed by other test anions (Cl⁻, $K_a=4.12\ 10^4 \pm 1.27\ 10^3$, $T\Delta S = -2.17$ (kcal/mol); Br⁻, $K_a=1.10\ 10^3 \pm 3.12 \times 10^2$, 1.91; AcO⁻, $K_a=1.12\ 10^5 \pm 1.64\ 10^4$, $T\Delta S=-3.87$ (kcal/mol); BzO⁻, $K_a=4.29\ 10^5 \pm 6.51\ 10^4$, $T\Delta S=-3.13$ (kcal/mol); H$_2$PO$_4^-$, $K_a=2.29 \times 10^4 \pm 1.90\ 10^3$, $T\Delta S = -1.26$(kcal/mol)). Since the benzimidazole moieties are somewhat flexible in terms of both rotation and breathing-like motions, geometrical adaptation seems to be obviously depending on the size of the anion. It was thus expected that the receptor would be able to accommodate anionic guests larger than halide anions	(Mulugeta et al. 2017a)

S.No.	Host	Guest	Studies/techniques used	Outcome/results	References
11	**P1**(poly(Nisopropylacrylamide)-*b*-poly(calix[4]pyrrole)-methacrylate) or PNIPAM-*b*-P(C4P-*co*-MMA), Polymer **P2**, lacking the C4P receptor 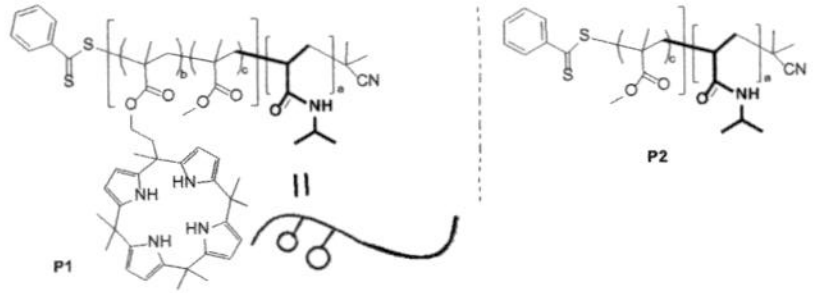	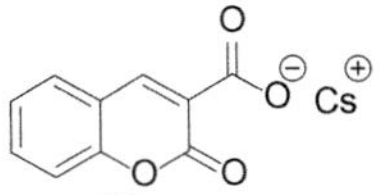 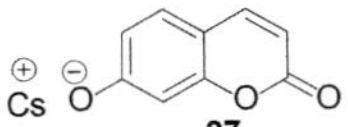 CsF, CsCl, CsBr, CsI, and CsNO₃	1H NMR spectroscopy and gel-permeation chromatography (GPC), dynamic light scattering (DLS), transmission electron microscopy (TEM), and turbidity measurements	The thermal-responsive amphiphilic polymer was designed to contain pendent calix[4]pyrrole receptors in the hydrophobic core. On adding to an anion-containing aqueous medium, it was found to self-assemble into micelle-like structures and anion gets captured within micelles. These micelles get precipitated from the aqueous medium on heating, which was removed by filtration allowing effective removal of anion-containing micelles from the medium. Polymer (1) was able to remove CsBr (39%) in preference over CsCl (32%) and CsF (16%) while CsI and CsNO3, were not removed appreciably from the aqueous source. Removal efficiencies for **26 and 27** displayed by **P1** were 62.5 and 48.9%,	(Ji et al. 2018)

Table 4.1 (Contd.)

Table 4.1 (Contd.) Derivatives of calix[4]pyrroles as host, their guest anions along with the outcomes of their interactions and techniques to study them are tabulated below:

S.No.	Host	Guest	Studies/techniques used	Outcome/results	References
				respectively. Polymeric material was regenerated by exposing the anion-trapped micelles to acidic aqueous solution leads to the competition-induced release of the bound anion.	

Table 4.2 Limits of Detection (LODs) and Binding Constants for **Receptor DPAC-bisC4P** and Anionic Dicarboxylate Guests in Acetonitrile at 293 K; permission from (Chen et al. 2019)

Anions	*LOD* (nM)	*binding constant* (M^{-1})	*Host:Guest*
C_2^{2-}	93.9	$(1.8 \pm 0.5) \times 10^6$	1:1
		$(1.2 \pm 0.2) \times 10^5$(stepwise)	1:2
C_4^{2-}	34.4	$(5.3 \pm 3.6) \times 10^5$	1:1
		$(2.1 \pm 0.9) \times 10^6$(stepwise)	1:2
C_5^{2-}	18.4	$(4.6 \pm 0.8) \times 10^6$	1:1
C_6^{2-}	14.2	$(8.2 \pm 0.6) \times 10^6$	1:1
C_7^{2-}	60.2	$(5.0 \pm 1.7) \times 10^7$	1:1
C_8^{2-}	79.5	$(1.9 \pm 0.4) \times 10^7$	1:1
C_{10}^{2-}	Not determined	$(3.1 \pm 0.8) \times 10^6$	1:1
C_{12}^{2-}	Not determined	$(1.8 \pm 0.4) \times 10^5$	1:1
p-Ph^{2-}	21.0	$(8.3 \pm 2.4) \times 10^6$	1:1
m-Ph^{2-}	7.55	$(5.7 \pm 1.4) \times 10^7$	1:1

REDOX-ACTIVE RECEPTORS AS ANION RECEPTORS

Redox-active molecular receptors are capable of sensing charged or neutral substrates and account for their presence by means of an electrochemical response. The receptors (Beer et al. 1999) generally consist of a guest binding site in close proximity to a redox-active moiety (Fig. 4.13) The electrochemical recognition of a guest by such a receptor requires that the guest binding site and the redox-active group can communicate optically and/or electrochemically. Electrochemical receptors for anions are expected to show cathodic shifts in their redox process when complexed to an anion as they are either easier to oxidize or harder to reduce than the free redox-active receptor.

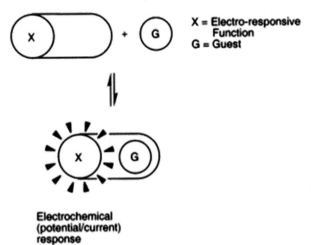

Figure 4.13 An electrochemical response is generated by the receptor upon guest binding; reprinted with permission from (Beer et al. 1999).

A novel ferrocene bis(triazole) macrocycle receptor was synthesized by intramolecular Eglinton coupling of an acyclic 1,1′-bis(triazolylalkyne) ferrocene precursor followed by alkylation (White and Beer 2012). The receptor was investigated for its anion binding properties by means of NMR titration in CD_3CN. WinEQNMR2 analysis showed that the cyclic receptor, **28**, binds strongly with benzoate ($K_a = 4.6(2) \times 10^3$ M^{-1}) and chloride benzoate ($K_a = 2.5(2) \times 10^3$ M^{-1}), while it binds weakly with iodide ($K_a = 2(2) \times 10^2$ M^{-1}); [NOTE: Estimated standard errors in K_a is given in parentheses] which was interpreted due to charge assisted C–H⋯anion interactions. Cyclic voltammetry analysis revealed that the redox-active macrocycle can sense chloride anion, causing a cathodic shift of the E_{pa} wave of the ferrocene/ferrocenium redox couple.

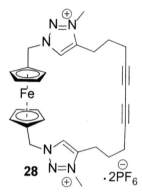

Figure 4.14   Structure of cyclic redox-active receptor **28**;
reprinted with permission from (White and Beer 2012).

Gregolińska et al. 2018 showed that fully conjugated [4]Chrysaorene system can be oxidized to well-defined radical cation and dication states and observed interactions between anions and different oxidation levels of a **29** which was attributed to the anion-binding ability along with reversible multi-redox behavior of [4]Chrysaorene. The cavity of the macrocycle acts as a receptor for halide ions and showed specific selectivity toward iodides ($K_a = 207 \pm 6$ M^{-1}) which was attributed to the exact match between the cavity size of the macrocycle and the ionic radius of iodide ion. NMR titration of **29** with a solution of diiodine assisted to establish a relationship between anion binding and redox chemistry and provided evidence for the binding of the anion inside the cavity of the macrocycle receptor. [29]•+ and **29** were observed to be in equilibrium and iodide formed a complex with both radial cation and neutral macrocycle. It was concluded that as compared to $I^- \subset$ **29**, the complex $I^- \subset [29]^{•+}$ was formed with a greater binding constant due to electrostatic interaction between ions. Excess of diiodine was inferred to react with the complexed iodide to produce a new ion pair, $I_3^- \subset [29]^{•+}$ and there was a fast exchange between $I^- \subset [29]^{•+}$ and $I_3^- \subset [29]^{•+}$. The radical cation–iodide adduct was generated in thin solid films of [4]Chrysaorene by simple exposure to diiodine vapor.

Figure 4.15 Structure of fully conjugated [4]Chrysaorene system **29**; reprinted with permission from (Gregolińska et al. 2018).

PORPHYRIN CAGES BASED MACROCYCLES AS ANION RECEPTORS

Porphyrin structures (Figueira et al. 2016) are planar, conjugated and rigid which favor interactions involving π-electrons with inherent optical and redox properties (Nappa and Valentine 1978; Hinman and Pavelich 1989). Expanded porphyrins in comparison to porphyrins exhibit various π-conjugation pathways, distinct optical features, better flexibility and a tendency to interact with anions. A variety of porphyrin-based supramolecular assemblies have been reported in literature, that contain nitrogen-based hydrogen-bond donor residues like integrated amide, ammonium, urea, pyrrole, pyrrole and imidazolium which can interact with different anions. This results in a change of redox potential or optical property that can be measured either using cyclic voltammeter or UV-Vis spectrophotometer etc., to get the desired information about the Host-Guest binding.

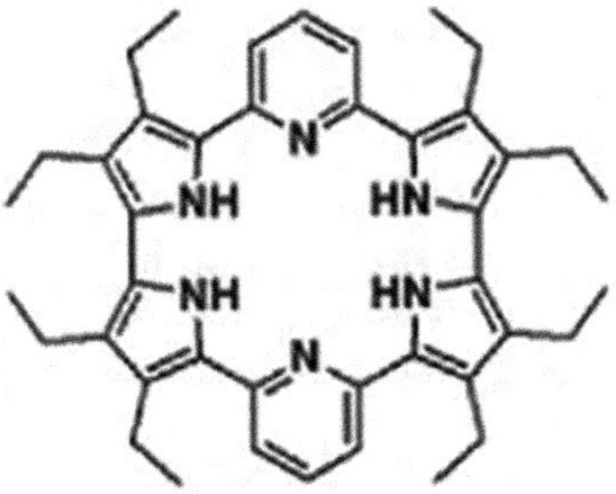

30

Figure 4.16 Structure of expanded porphyrin; reprinted with permission from (Zhang et al. 2015).

Zhang and co-workers demonstrated that one representative class of expanded porphyrin, the cyclo[m]pyridine[n]pyrroles (m + n = 6), may be used to stabilize a unique class of anion-induced self-assembly that can act as attractive chemical sensors (Zhang et al. 2015). For instance, expanded porphyrins(**30**) typically were used with diacids (like 4,4′-biphenyldisulfonic acid **DPDSA**) in CH_2Cl_2, to form anion-derived self-assembly which exhibited remarkable optical response with fluoride, chloride, bromide, iodide, nitrate, sulfate monobasic phosphate.

Zhang and co-workers reported the interaction of zinc porphyrin cage (based on a click-chemistry reaction) as a receptor (Fig. 4.17) towards tetrabutylammonium azide (TBAN$_3$) by titrating porphyrin cage with TBAN$_3$ (Zhang et al. 2012). The N$_3$ anion forms a Zn–NNN–Zn unit by interacting with both Zn atoms in the porphyrin cage, which was verified by variations in the NMR signals of the porphyrin cage **31**. Furthermore, the binding properties of cage **31** with N$_3$ anion were explored by UV-Vis spectroscopy in different solvents like acetone, CH$_2$Cl$_2$, and THF

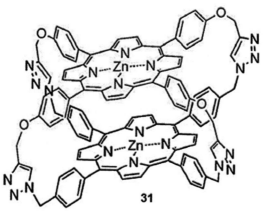

Figure 4.17   Structure of zinc porphyrin cage;
reprinted with permission from (Zhang et al. 2012).

Gilday et al. 2012 described novel triazole- and triazolium-containing zinc(II) metalloporphyrin-cage. These systems conglomerate hydrogen bonding (C–H⋯anion) and lewis acidity to act as an optical sensor for anions (Gregolińska et al. 2018). They investigated their characteristics using UV-Visible spectroscopy, especially determining the association constants for complex formation with various anions in polar organic and organic–aqueous solvent mixtures. The neutral metalloporphyrin cage host **32** binds strongly with anions as can be interpreted from the large association constant values for F$^-$ (~8 × 10^4 M^{-1}) and SO$_4^{2-}$ (1 × 10^5 M^{-1}) in acetone. The trend displayed by receptor **32** towards the singly-charged anions, AcO$^-$ ~H$_2$PO$_4^-$ ≫ Cl$^-$ > Br$^-$ > I$^-$ can be correlated with anion basicity.

The observed affinity trend in 5% water:acetone solvent system for Cage **33** was observed to be the highest for sulfate dianion (5.2 × 10^5 M^{-1}) followed by AcO$^-$ and H$_2$PO$_4^-$ with an association constant 1.3 × 10^5 M^{-1} and 1.2 × 10^5 M^{-1} respectively. The receptor also displayed an affinity for all the halides and the increasing order of affinity for halide ions was Br$^-$ ≈ I$^-$ (<5.0 × 10^1 M^{-1}) < Cl$^-$ (7.4 × 10^3 M^{-1}) < F$^-$ (8.7 × 10^3 M^{-1}). Receptor **33** demonstrated strong binding affinity (K$_a$= 2.5 × 10^5 M^{-1}) with sulfate in 15% water: acetone solvent system whereas the singly-charged anions were not bound with receptor **33** in the more aqueous solvent mixture.

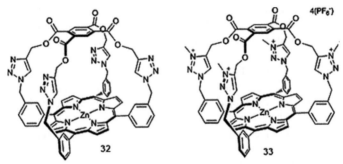

Figure 4.18 Structure of triazole- and triazolium-containing zinc(II) metalloporphyrin-cage systems. Reprinted with permission from (Gilday et al. 2012).

CONCLUSIONS

As illustrated by the examples provided in this chapter, bile acid, calix[4]pyrrole, redox-active and porphyrin cages-based macrocyclic anion receptors continue attracting a great deal of attention from research groups interested in the anion coordination chemistry.

REFERENCES

Accardi, A. and C. Miller. 2004. Secondary active transport mediated by a prokaryotic homologue of ClC Cl-channels. Nature 427(6977): 803–807.

Agmon, N. 2017. Isoelectronic theory for Cationic Radii. J. Am. Chem. Soc 139(42): 15068–15073.

Ahad, F. and S.A. Ganie. 2010. Iodine, iodine metabolism and iodine deficiency disorders revisited. Indian J. Endocrinol. Metab. 14(1): 13.

Ali, R., R.C. Gupta, S.K. Dwivedi and A. Misra. 2018. Excited state proton transfer (ESIPT) based molecular probe to sense F− and CN− anions through a fluorescence "turn-on" response. New J. Chem. 42(14): 11746–11754.

Amendola, V., L. Fabbrizzi and L. Mosca. 2010. Anion recognition by hydrogen bonding: urea-based receptors. Chem. Soc. Rev. 39(10): 3889–3915.

Ashcroft, F.M. (1999). Ion Channels and Disease. Academic Press.

Badr, I.H., M. Diaz, M.F. Hawthorne and L.G. Bachas. 1999. Mercuracarborand "anti-crown ether"-based chloride-sensitive liquid/polymeric membrane electrodes. Anal. Chem. 71(7): 1371–1377.

Badr, I.H., R.D. Johnson, M. Diaz, M.F. Hawthorne and L.G. Bachas. 2000. A selective optical sensor based on [9] mercuracarborand-3, a new type of ionophore with a chloride complexing cavity. Anal. Chem. 72(18): 4249–4254.

Baeyer, A. 1886. Ueber ein condensationsproduct von pyrrol mit aceton. Ber. Dtsch. Chem. Ges. 19(2): 2184–2185.

Beer, P.D., P.A. Gale and G.Z. Chen. 1999. Mechanisms of electrochemical recognition of cations, anions and neutral guest species by redox-active receptor molecules. Coord. Chem. Rev. 185: 3–36.

Benech, J.M., L. Bonomo, E. Solari, R. Scopelliti and C. Floriani. 1999. Porphomethenes and porphodimethenes synthesized by the two-and four-electron oxidation of the meso-octaethylporphyrinogen. Angew. Chem. Int. Ed. 38(13-14): 1957–1959.

Berend, K. 2017. Review of the diagnostic evaluation of normal anion gap metabolic acidosis. Kidney Dis. 3(4): 149–159.

Berry, S.N., L. Qin, W. Lewis and K.A. Jolliffe. 2020. Conformationally adaptable macrocyclic receptors for ditopic anions: analysis of chelate cooperativity in aqueous containing media. Chem. Sci. 11(27): 7015–7022.

Bianchi, A., K. Bowman-James and E. Garcìa-España. 1997. Supramolecular Chemistry of Anions, Vch Pub.

Boiocchi, M., L. Del Boca, D.E. Gómez, L. Fabbrizzi, M. Licchelli and E. Monzani. 2004. Nature of urea– fluoride interaction: incipient and definitive proton transfer. J. Am. Chem. Soc. 126(50): 16507–16514.

Borissov, A., I. Marques, J.Y. Lim, V.t. Félix, M.D. Smith and P.D. Beer. 2019. Anion recognition in water by charge-neutral halogen and chalcogen bonding foldamer receptors. J. Am. Chem. Soc. 141(9): 4119–4129.

Brotherhood, P.R. and A.P. Davis. 2010. Steroid-based anion receptors and transporters. Chem. Soc. Rev. 39(10): 3633–3647.

Chahal, M.K., J. Labuta, V. Březina, P.A. Karr, Y. Matsushita, W.A. Webre, D.T. Payne, K. Ariga, F. D'Souza and J.P. Hill. 2019. Knock-on synthesis of tritopic calix[4]pyrrole host for enhanced anion interactions. Dalton Trans. 48(41): 15583–15596.

Chen, W., C. Guo, Q. He, X. Chi, V.M. Lynch, Z. Zhang, J. Su, H. Tian and J.L. Sessler. 2019. Molecular cursor caliper: a fluorescent sensor for dicarboxylate dianions. J. Am. Chem. Soc. 141(37): 14798–14806.

Chen, L., S.N. Berry, X. Wu, E.N. Howe and P.A. Gale. 2020. Advances in anion receptor chemistry. Chem. 6(1): 61–141.

Cheremisinoff, N.P. 2002. Handbook of Water and Wastewater Treatment Technologies. Butterworth-Heinemann.

Chhatra, R.K., A. Kumar and P.S. Pandey. 2011. Synthesis of a bile acid-based click-macrocycle and its application in selective recognition of chloride ion. J. Org. Chem. 76(21): 9086–9089.

Choi, K. and A.D. Hamilton. 2003. Macrocyclic anion receptors based on directed hydrogen bonding interactions. Coord. Chem. Rev. 240(1-2): 101–110.

Clarke, H.J., X. Wu, M.E. Light and P.A. Gale. 2020. Selective anion transport mediated by strap-extended calixpyrroles. J. Porphyr. Phthalocyanines. 24(01n03): 473–479.

Connors, K.A. 1987. Binding Constants: The Measurement of Molecular Complex Stability. Wiley-Interscience.

Cram, D.J. 1988. The design of molecular hosts, guests, and their complexes (Nobel lecture). Angew. Chem. Int. Ed. Engl. 27(8): 1009–1020.

Dey, S. and B. Giri. 2016. Fluoride fact on human health and health problems: a review. Med. Clin. Rev. 2(1): 11.

Duke, R.M., E.B. Veale, F.M. Pfeffer, P.E. Kruger and T. Gunnlaugsson. 2010. Colorimetric and fluorescent anion sensors: an overview of recent developments in the use of 1, 8-naphthalimide-based chemosensors. Chem. Soc. Rev. 39(10): 3936–3953.

Dutta, R., S. Samala, H. Jo, K.M. Ok and C.-H. Lee. 2019. meso-Bis(ethynyl) versus meso-Bis(aryl) Calix[4]pyrroles: dimensionally well-modulated receptors that can regulate the anion binding domains. J. Org. Chem. 84(11): 6851–6857.

Fernández-Ramos, M., L. Cuadros-Rodríguez, E. Arroyo-Guerrero and L. Capitán-Vallvey. 2011. An IUPAC-based approach to estimate the detection limit in co-extraction-based optical sensors for anions with sigmoidal response calibration curves. Anal. Bioanal. Chem. 401(9): 2881.

Figueira, F., J.M. Rodrigues, A.A. Farinha, J.A. Cavaleiro and J.P. Tomé. 2016. Synthesis and anion binding properties of porphyrins and related compounds. J. Porphyrins Phthalocyanines. 20(08n11): 950–965.

Frontera Beccaria, A. 2018. Encapsulation of anions: macrocyclic receptors based on metal coordination and anion-pi interactions. Coord. Chem. Rev. 257: 1716–1727.

Gale, P.A., J.L. Sessler, V. Kral and V. Lynch. 1996. Calix[4]pyrroles: old yet new anion-binding agents. J. Am. Chem. Soc. 118(21): 5140–5141.

Gale, P.A. 2000. Anion coordination and anion-directed assembly: highlights from 1997 and 1998. Coord. Chem. Rev. 199(1): 181–233.

Gale, P.A. and T. Gunnlaugson. 2010. Themed issue: supramolecular chemistry of anionic species. Chem. Soc. Rev 39(10): 3581–4008.

Gale, P.A., N. Busschaert, C.J. Haynes, L.E. Karagiannidis and I.L. Kirby. 2014. Anion receptor chemistry: highlights from 2011 and 2012. Chem. Soc. Rev. 43(1): 205–241.

Ghosh, S., A.R. Choudhury, T.N. Guru Row and U. Maitra. 2005. Selective and unusual fluoride ion complexation by a steroidal receptor using OH... F-and CH... F-interactions: a new motif for anion coordination? Org. Lett. 7(8): 1441–1444.

Gilday, L.C., N.G. White and P.D. Beer. 2012. Triazole-and triazolium-containing porphyrin-cages for optical anion sensing. Dalton Trans. 41(23): 7092–7097.

Graf, E. and J.M. Lehn. 1976. Anion cryptates: highly stable and selective macrotricyclic anion inclusion complexes. J. Am. Chem. Soc. 98(20): 6403–6405.

Gregolińska, H., M. Majewski, P.J. Chmielewski, J. Gregoliński, A. Chien, J. Zhou, Y.-L. Wu, Y.J. Bae, M.R. Wasielewski and P.M. Zimmerman. 2018. Fully conjugated[4] chrysaorene. Redox-coupled anion binding in a tetraradicaloid macrocycle. J. Am. Chem. Soc. 140(43): 14474–14480.

Gunnlaugsson, T., H.D.P. Ali, M. Glynn, P.E. Kruger, G.M. Hussey, F.M. Pfeffer, C.M. dos Santos and J. Tierney. 2005. Fluorescent photoinduced electron transfer (PET) sensors for anions; from design to potential application. J. Fluoresc. 15(3): 287–299.

Haldimann, M., B. Zimmerli, C. Als and H. Gerber. 1998. Direct determination of urinary iodine by inductively coupled plasma mass spectrometry using isotope dilution with iodine-129. Clin. Chem. 44(4): 817–824.

He, Q., M. Kelliher, S. Bähring, V.M. Lynch and J.L. Sessler. 2017. A bis-calix[4]pyrrole enzyme mimic that constrains two oxoanions in close proximity. J. Am. Chem. Soc. 139(21): 7140–7143.

He, Y.-C., Z.-X. Ren, X.-F. Zhao, Y.-B. Zhang, J.-H. Wang, J.-B. Chao and M.-L. Wang. 2019. Squaramide-substituted calix[4]pyrrole: synthesis and ion binding studies. Tetrahedron. 75(36): 130491.

Heo, N.J., J.H. Oh, J.T. Lee, Q. He, J.L. Sessler and S.K. Kim. 2020. Phenanthroline-strapped calix[4]pyrroles: anion receptors displaying affinity reversal as a function of solvent polarity. Org. Chem. Front. 7(3): 548–556.

Hinman, A.S. and B.J. Pavelich. 1989. Influence of anion coordination on potentials for oxidation of the porphyrin ring in tetraphenylporphinatozinc. J. Electroanal. Chem. Interf. Electrochem. 269(1): 53–61.

Hirose, K. 2007. Determination of binding constants. pp. 17–54. *In*: C. Schalley (ed.). Analytical Methods in Supramolecular Chemistry. Wiley VCH Verlag GmbH & Co. KGaA

Izatt, R.M., K. Pawlak, J.S. Bradshaw and R.L. Bruening. 1991. Thermodynamic and kinetic data for macrocycle interactions with cations and anions. Chem. Rev. 91(8): 1721–2085.

Ji, X., C. Guo, W. Chen, L. Long, G. Zhang, N.M. Khashab and J.L. Sessler. 2018. Removal of anions from aqueous media by means of a thermoresponsive calix[4]pyrrole amphiphilic polymer. Chem. Eur. J. 24(59): 15791–15795.

Kadish, K., K.M. Smith and R. Guilard. 2000. The Porphyrin Handbook, Vol. 3. Elsevier.

Kaniansky, D., M. Masár, J. Marák and R. Bodor. 1999. Capillary electrophoresis of inorganic anions. J. Chromatogr. A. 834(1-2): 133–178.

Khatri, V.K., M. Chahar, K. Pavani and P.S. Pandey. 2007. Bile acid-based cyclic bisbenzimidazolium receptors for anion recognition: highly improved receptors for fluoride and chloride ions. J. Org. Chem. 72(26): 10224–10226.

Kim, S.H., J. Lee, G.I. Vargas-Zúñiga, V.M. Lynch, B.P. Hay, J.L. Sessler and S.K. Kim. 2018. Pyrrole-and naphthobipyrrole-strapped calix[4]pyrroles as azide anion receptors. J. Org. Chem. 83(5): 2686–2693.

Kohnke, F.H. 2020. Calixpyrroles: from anion ligands to potential anticancer drugs. Eur. J. Org. Chem. 2020(28): 4261–4272

KONO, K. 1994. Health effects of fluorine and its compounds. Nihon Eiseigaku Zasshi. 49(5): 852–860.

Kubik, S., C. Reyheller and S. Stüwe. 2005. Recognition of anions by synthetic receptors in aqueous solution. J. Incl. Phenom. Macrocycl. Chem. 52(3-4): 137–187.

Kumar, A. and P.S. Pandey. 2008. Anion recognition by 1, 2, 3-triazolium receptors: application of click chemistry in anion recognition. Org. Lett. 10(2): 165–168.

Kumar, P., A. Pournara, K.-H. Kim, V. Bansal, S. Rapti and M.J. Manos. 2017. Metal-organic frameworks: challenges and opportunities for ion-exchange/sorption applications. Prog. Mater. Sci. 86: 25–74.

Lee, C.-H., H. Miyaji, D.-W. Yoon and J.L. Sessler. 2008. Strapped and other topographically nonplanar calixpyrrole analogues. Improved anion receptors. Chem. Commun. (1): 24–34.

Li, Y., D.A.V. Griend and A.H. Flood. 2009. Modelling triazolophane–halide binding equilibria using Sivvu analysis of UV–vis titration data recorded under medium binding conditions. Supramol. Chem. 21(1-2): 111–117.

Long, J.R. and R.S. Drago. 1982. The rigorous evaluation of spectrophotometric data to obtain an equilibrium constant. J. Chem. Educ. 59(12): 1037.

Luo, J., J. Zhu, D.H. Tuo, Q. Yuan, L. Wang, X.B. Wang, Y.F. Ao, Q.Q. Wang and D.X. Wang. 2019. Macrocycle-directed construction of tetrahedral anion–π receptors for nesting anions with complementary geometry. Chem. Eur. J. 25(58): 13275–13279.

Martinez-Manez, R. and F. Sancenon. 2003. Fluorogenic and chromogenic chemosensors and reagents for anions. Chem. Rev. 103(11): 4419–4476.

Martinez-Manez, R. and F. Sancenón. 2006. Chemodosimeters and 3D inorganic functionalised hosts for the fluoro-chromogenic sensing of anions. Coord. Chem. Rev. 250(23–24): 3081–3093.

Meena, A.H. and Y. Arai. 2017. Environmental geochemistry of technetium. Environ. Chem. Lett. 15(2): 241–263.

Men, G., W. Han, C. Chen, C. Liang and S. Jiang. 2019. A cyanide-sensing detector in aqueous solution based on anion–π interaction-driven electron transfer. Analyst. 144(7): 2226–2230.

Moonen, N.N., A.H. Flood, J.M. Fernández and J.F. Stoddart. 2005. Towards a rational design of molecular switches and sensors from their basic building blocks. Molecular Machines, Springer: 99–132.

Mulugeta, E., R. Dutta, Q. He, V. Lynch, J. Sessler and C.H. Lee. 2017a. Anion-dependent binding-mode changes in meso-(5,6-dichlorobenzimidazole) picket calix[4]pyrrole. Eur. J. Org. Chem. 2017(33): 4891–4895.

Mulugeta, E., Q. He, D. Sareen, S.-J. Hong, J.H. Oh, V.M. Lynch, J.L. Sessler, S.K. Kim and C.-H. Lee. 2017b. Recognition, sensing, and trapping of bicarbonate anions with a dicationic meso-bis(benzimidazolium) calix[4]pyrrole. Chem 3(6): 1008–1020.

Nappa, M. and J.S. Valentine. 1978. The influence of axial ligands on metalloporphyrin visible absorption spectra. Complexes of tetraphenylporphinatozinc. J. Am. Chem. Soc. 100(16): 5075–5080.

Okesola, B.O. and D.K. Smith. 2016. Applying low-molecular weight supramolecular gelators in an environmental setting–self-assembled gels as smart materials for pollutant removal. Chem. Soc. Rev. 45(15): 4226–4251.

Park, C. and H. Simmons. 1968. Macrobicyclic amines. III. Encapsulation of halide ions by in, in-1,(k+ 2)-diazabicyclo [klm] alkane ammonium ions. J. Am. Chem. Soc. 90(9): 2431–2432.

Park, J.S. and J.L. Sessler. 2018. Tetrathiafulvalene (TTF)-annulated calix[4]pyrroles: chemically switchable systems with encodable allosteric recognition and logic gate functions. Acc. Chem. Res. 51(10): 2400–2410.

Peng, S., Q. He, G.I. Vargas-Zúñiga, L. Qin, I. Hwang, S.K. Kim, N.J. Heo, C.-H. Lee, R. Dutta and J.L. Sessler. 2020. Strapped calix[4]pyrroles: from syntheses to applications. Chem. Soc. Rev. 49(3): 865–907.

Pomecko, R., Z. Asfari, V. Hubscher-Bruder, M. Bochenska and F. Arnaud-Neu. 2010. Anion recognition by phosphonium calix[4]arenes: synthesis and physico-chemical studies. Supramol. Chem. 22(5): 275–288.

Rather, I.A., S.A. Wagay, M.S. Hasnain and R. Ali. 2019. New dimensions in calix[4] pyrrole: the land of opportunity in supramolecular chemistry. RSC Adv. 9(66): 38309-38344.

Rifai, A., N. AlHaddad, M. Noun, I. Abbas, M. Tabbal, R. Shatila, F. Cazier-Dennin and P.-E. Danjou. 2019. A click mediated route to a novel fluorescent pyridino-extended calix[4]pyrrole sensor: synthesis and binding studies. Org. Biomol. Chem. 17(23): 5818–5825.

Saha, I., J.T. Lee and C.H. Lee. 2015. Recent advancements in calix[4]pyrrole-based anion-receptor chemistry. Eur. J. Org. Chem. 2015(18): 3859–3885.

Santos-Figueroa, L.E., M.E. Moragues, E. Climent, A. Agostini, R. Martínez-Máñez and F. Sancenón. 2013. Chromogenic and fluorogenic chemosensors and reagents for anions. A comprehensive review of the years 2010–2011. Chem. Soc. Rev. 42(8): 3489–3613.

Sato, T., H. Konno, Y. Tanaka, T. Kataoka, K. Nagai, H.H. Wasserman and S. Ohkuma. 1998. Prodigiosins as a new group of H⁺/Cl⁻ symporters that uncouple proton translocators. J. Biol. Chem. 273(34): 21455–21462.

Schalley, C.A. 2012. Analytical Methods in Supramolecular Chemistry. John Wiley & Sons.

Schmidtchen, F.P. and M. Berger. 1997. Artificial organic host molecules for anions. Chem. Rev. 97(5): 1609–1646.

Schmidtchen, F.P. 2002. Surprises in the energetics of host– guest anion binding to calix[4] pyrrole. Org. Lett. 4(3): 431–434.

Schmidtchen, F.P. 2006. Reflections on the construction of anion receptors: is there a sign to resign from design? Coord. Chem. Rev. 250(23-24): 2918–2928.

Schneider, H.-J. and A.K. Yatsimirsky. 2000. Principles and Methods in Supramolecular Chemistry. John Wiley and Sons, New York.

Sengupta, A., Y. Liu, A.H. Flood and K. Raghavachari. 2018. Anion-binding macrocycles operate beyond the electrostatic regime: interaction distances matter. Chem. Eur. J. 24(54): 14409–14417.

Sessler, J., P. Gale and W.J.C.R. Cho. 2003. Anion Receptor Chemistry. Royal Society of Chemistry Cambridge.

Sessler, J.L., P.A. Gale and W.-S. Cho. 2006a. Anion Receptor Chemistry. Royal Society of Chemistry.

Sessler, J.L., D.E. Gross, W.-S. Cho, V.M. Lynch, F.P. Schmidtchen, G.W. Bates, M.E. Light and P.A. Gale. 2006b. Calix[4]pyrrole as a chloride anion receptor: solvent and countercation effects. J. Am. Chem. Soc. 128(37): 12281–12288.

Shah, H. and K.D. Bhatt. 2019. Review on calix[4]pyrrole: a versatile receptor. AOCL 6(1): 1–12.

Shimizu, S., W.M. McLaren and N. Matubayasi. 2006. The hofmeister series and protein-salt interactions. J. Chem. Phys. 124(23): 234905.

Silverman, D. and S. Lindskog. 1988. The catalytic mechanism of carbonic anhydrase: implications of a rate-limiting protolysis of water. Acc. Chem. Res. 21: 30–36.

Sisson, A.L., J.P. Clare and A.P. Davis. 2005. Contra-hofmeister anion extraction by cyclosteroidal receptors. Chem. Commun. (42): 5263–5265.

Soyoral, Y., M. Aslan, S. Ebinc, Y. Dirik and C. Demir. 2014. Life-threatening hypophosphatemia and/or phosphate depletion in a patient with acute lymphoblastic leukemia: a rare case report. Am. J. Emerg. Med. 32(11): 1437. e1433–1437. e1435.

Steed, J.W. and J.L. Atwood. 2013. Supramolecular Chemistry. John Wiley & Sons.

Tamminen, J. and E.J.M. Kolehmainen. 2001. Bile acids as building blocks of supramolecular hosts. 6(1): 21–46.

Tuo, D.-H., W. Liu, X.-Y. Wang, X.-D. Wang, Y.-F. Ao, Q.-Q. Wang, Z.-Y. Li and D.-X. Wang. 2018. Toward anion– π interactions directed self-assembly with predesigned dual macrocyclic receptors and dianions. J. Am. Chem. Soc. 141(2): 1118–1125.

Vervloet, M.G., S. Sezer, Z.A. Massy, L. Johansson, M. Cozzolino and D. Fouque. 2017. The role of phosphate in kidney disease. Nat. Rev. Nephrol. 13(1): 27–38.

Vickers, M.S. and P.D. Beer. 2007. Anion templated assembly of mechanically interlocked structures. Chem. Soc. Rev. 36(2): 211–225.

Wang, H.-S., Y. Chen, K. Vairamani and G.E. Shull. 2014. Critical role of bicarbonate and bicarbonate transporters in cardiac function. World J. Biol. Chem. 5(3): 334.

White, N.G. and P.D. Beer. 2012. A ferrocene redox-active triazolium macrocycle that binds and senses chloride. Beilstein J. Org. Chem. 8(1): 246–252.

Whitmarsh, S.D., A.P. Redmond, V. Sgarlata and A.P. Davis. 2008. Cationic cyclocholamides; toroidal facial amphiphiles with potential for anion transport. Chem. Commun. (31): 3669–3671.

Xiong, S., F. Chen, T. Zhao, A. Li, G. Xu, J.L. Sessler and Q. He. 2020. Selective inclusion of fluoride within the cavity of a two-wall bis-calix[4]pyrrole. Org. Lett. 22(11): 4451–4455.

Yamamoto, D., Y. Kiyozuka, Y. Uemura, C. Yamamoto, H. Takemoto, H. Hirata, K. Tanaka, K. Hioki and A. Tsubura. 2000. Cyploprodigiosin hydrochloride, a H^+/Cl^- symporter, induces apoptosis in human breast cancer cell lines. J. Cancer Res. Clin. Oncol. 126(4): 191–197.

Yuvayapan, S. and A. Aydogan. 2019. Supramolecular calix[4]pyrrole polymers from a complementary pair of homoditopic host–guest molecules. Chem. Commun. 55(60): 8800–8803.

Zhang, J., Y. Li, W. Yang, S.-W. Lai, C. Zhou, H. Liu, C.-M. Che and Y. Li. 2012. A smart porphyrin cage for recognizing azide anions. Chem. Commun. 48(30): 3602–3604.

Zhang, Z., D.S. Kim, C.-Y. Lin, H. Zhang, A.D. Lammer, V.M. Lynch, I. Popov, O.S. Miljanić, E.V. Anslyn and J.L. Sessler. 2015. Expanded porphyrin-anion supramolecular assemblies: environmentally responsive sensors for organic solvents and anions. J. Am. Chem. Soc. 137(24): 7769–7774.

Chapter 5

Macrocyclic Receptors for Sensing the Environmentally Important Gaseous Molecules

INTRODUCTION

Definition

A gas sensor is a device that detects the presence of a gaseous analyte simultaneously quantifying its concentration, in a given space and real-time condition, by converting it into a detectable/measurable response. It is usually a solid-state device that is exploited in sensing different gases or analytes in the vapor phase operating at the gas-solid or gas-liquid interfaces. However, a gaseous *molecular probe* is different from a gas sensor since it is not a solid-state device and it usually represents a group of atoms or molecules utilized in biology or chemistry. It's interaction with the gaseous analyte to detect them, quantifying their amount and also for the study of their properties including their structure, is by virtue of the alteration of the observable properties such as electrochemical, spectroscopic behavior (fluorescence, chemoluminescence) of the probe (Pinalli et al. 2018a). A gas sensors may involve molecular receptors in the monitoring process in the form of a thin film, a mono layer embedded in a polymer matrix or grafted on different surfaces, to improve the selectivity and sensitivity towards the gaseous analytes.

Gas sensing is relevant and important in monitoring of not only environmental pollutants (Kumar et al. 2017; Ray 2017; Pinalli et al. 2018b), point-of-care diagnostic including diagnosis of diseases (biomarkers) (Haick et al. 2014; Konvalina and Haick 2014), food testing (to find the freshness/staleness of different food items based on profiling of gases such as hydrogen sulfide and ammonia, etc., produced due to decomposition of carbohydrate, protein and fat present in it because of the action of bacteria and enzymes), but also in industrial quality control and safety which includes the field of hazard monitoring and fire warning, etc. (Arshak et al. 2004; Oh et al. 2011; Gubala et al. 2012; Ponzoni et al. 2012; Wilson 2013). It is estimated to become US$ 1 billion market by 2022. No doubt, the field has captured the attention as well as the interest of both scientists and engineers in the past few decades.

Challenges

The development of a gas sensor is one of the biggest challenges for the chemical community, particularly due to the stringent requirements of these sensors in terms of sensitivity, selectivity and ruggedness. The ability of a sensor to maintain its performance under the adverse operating conditions is known as ruggedness (Janata 2010). The key issues that need to be addressed are the necessity for high selectivity, the conflicting low parts per billion (ppb) sensitivity and combining these in a sensor (Hierlemann and Gutierrez-Osuna 2008). Therefore, the discovery of reliable gas sensors for rapid, sensitive and specific detection of a wide range of gaseous analytes necessitates innovative solutions and approaches that need to be different from the design of conventional sensing materials.

Although at present many gaseous monitoring techniques are available, such as gas chromatography, optical spectroscopy, mass spectroscopy, but the low time-consuming, sensitive and the rugged gas sensing technique is still a need of the hour.

The sensitivity of these available techniques can be increased by using principles of molecular recognition and molecular receptors. The exploitation of macrocyclic molecular receptors as sensing materials is particularly attractive to address the issue of selectivity. You may be aware that nature has provided several examples of exquisitely specific binding, such as enzyme-substrate, antigen-antibody and complementary DNA annealing. Biological systems also exploit the principle of molecular recognition to achieve uniquely high specificity among distinct chemical species which are complementary in terms of shape, length, width and charges. Although it is very well known that biological receptors function almost exclusively in aqueous conditions, where normal dispersion interactions are essentially canceled in transporting the substrate from water to the receptor site and also the entropic cost of the binding is compensated partially by releasing water molecules to bulk, but the same is not valid for man-made sensing methods where normal, non-specific dispersion interactions are crucial in transporting gaseous species to solids (Grate 1996).

Role of Macrocyclic Compounds in Gas Sensing

Chemists have designed and synthesized molecular receptors including those for a diverse pool of gaseous guests based on their inspiration and learning from the biological world. These synthetic receptors mimic the specificity of the biological receptors by the use of shape recognition concepts and binding site complementarity. This has become possible due to their ever-increasing understanding of supramolecular chemistry which offers a wide choice of synthetic receptors having different cavity size, shape, and directionality of the interactions that can be fine-tuned according to complementarity principles of structural recognition for having a specific affinity for a large number of analytes ranging from ions to complex organic molecules (Meyer 2003; Yang et al. 2014). The macrocyclic receptors can be designed and/or amended according to the analytical problem to be solved and can also be attached with various fluorescent, chemoluminescent or electrochemical tags to transform them into *molecular probes* with comparative ease since they are chemically stable, easy to functionalize, available in high purity and substantial quantities. Thanks to the several groups – who have, over the years utilized various macrocyclic molecular receptors, primarily comprising of cyclodextrins, calixarenes, cucurbiturils, cavitands, etc., in the development of gas sensing. Formative work of Stoddart revealed the field of cyclodextrins (Lai et al. 1988), which was followed by Gopel's work on Calixarenes (Schierbaum et al. 1992), subsequently, Dalcanale's work (Nelli et al. 1993), Dickart's (Dickert et al. 1993) and Gopel's work (Schierbaum et al. 1994) opened the field of other cavitands.

In this chapter, the emerging strategies and progress made in the gas and vapor sensing during the last three decades using different classes of macrocyclic molecular receptors and development of macrocyclic-based sensing materials have been summarized with the focus on the utilization of different cavities of macrocyclic molecules for the detection of gases.

Before moving any further, it is essential to briefly go through the different techniques that are generally exploited or have been used to measure or assess the interaction or association between a macrocyclic host and guest (gas or vapor molecules) for converting it into a detectable/measurable response. The interaction or association between the host and the gaseous guest in the gaseous state can be measured through several surface characterization techniques such as mass sensitive Quartz Micro Balance (QMB)/Quartz Crystal Microbalance (QCM), the Langmuir–Blodget (LB) films, Surface Acoustic Wave oscillator (SAW), etc. (Chawla et al. 2010).

Overview of a Few Different Types of Sensing Techniques
Quartz Micro Balance (QMB)

In a mass-sensitive Quartz Microbalance (QMB) also known as Quartz Crystal Microbalance (QCM), a piezoelectric effect is used which is referred to as an ability of piezoelectric material to produce oscillations when voltage (alternating) is applied to its surface. Such a sensor, when excited by an alternating voltage applied by the two electrodes result in mechanical oscillations of crystal which is measured as the crystal is put into oscillation circuit (García-Martinez et al. 2011). Analyte or any other mass bound on the surface of such crystals or more precisely on the surface of electrodes located on the crystal results in alteration of oscillation frequency (Pohanka 2017) and the change (decay) in frequency is proportional to mass bound of the crystal (Fig. 5.1). A typical example of a gas sensing system based on microbalance principle has been described by Auge et al. (Auge et al. 1995)

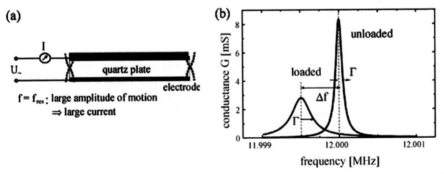

Figure 5.1 (a) Schematic diagram of the Quartz Crystal Microbalance (QCM) and (b) the electrical characteristic with (loaded) and without (unloaded) adsorbed mass. Reprinted with permission from (Johannsmann 2008).

The change in frequency, Δf of oscillating quartz working as a QCM sensor is a function of the mass change, Δm:

$$\Delta f = -\frac{f_0}{\rho_q Ad}\Delta m$$

Where f_0 is the frequency of an uncoated crystal the fundamental shear mode, d is the thickness of the quartz, A is the area of the coated surface and ρ_q is the density of the quartz crystal. So, it measures a mass variation per unit area by measuring the change in frequency of a quartz crystal. It can be used under vacuum or in the gas phase (King 1964). QCM sensors alone are non-selective transducers since they report any mass change on their surface, irrespective of the origin. Therefore coating should be selective/specific for a particular analyte gas. Nevertheless, due to the unavoidable presence of dispersion interactions between the analyte and the coated layer of receptors, the non- specific detection also comes into the picture, and usually gold or platinum electrodes are used on to a thin quartz disc.

Langmuir–Blodgett Method (LB)

The Langmuir–Blodgett (LB) technique is known to be an excellent method for the deposition of gas-sensitive materials on solid supports. Precisely defined thickness, well-ordered structure and high homogeneity of LB films make them especially suitable for gas sensing. Good sensitivity and selectivity of the gas sensors using LB films have been widely demonstrated. (Clemendot et al. 1992, Nabok et al. 1995)

Surface Acoustic Wave Oscillator (SAW)

The principle of the operation of this transducer is conceptually really simple. An acoustic wave is generated which is confined to the surface of some substrate material and is allowed to propagate. Properties of this acoustic wave e.g., amplitude, phase, harmonic content, etc., are altered if some matter (in monolayers) is present on the surface. This is due to the interaction of the wave and the matter and the measure of such variation in the surface wave characteristics properties is a sensitive gauge of the properties of the material existing on the surface of the device. The convenient generation of the acoustic wave on the surface requires a substrate material which is a piezoelectric transducer. (Wholtjen and Dessy 1979) In the case of thin, non-conductive monolayers having no lateral connectivity (i.e., covalent bonds) on piezoelectric substrate, electrical and viscoelastic effects are minimal and the resonant frequency shift (Δf) (Grate and Klusty 1991) depends mainly on the mass loading per unit area ($\Delta m/A$) of the sensing monolayer (Grate et al. 1992) or any adsorbed vapors according to the following equation

$$\Delta f = -Kf^2 \frac{\Delta m}{A}$$

where K is a material constant for the piezoelectric substrate and f is the fundamental resonant frequency of the oscillator (Fig. 5.2).

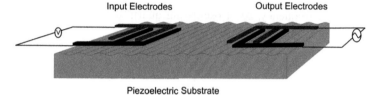

Figure 5.2   Typical set-up of a Surface Acoustic Wave (SAW) sensor. An acoustic wave propagates from a sender to a receiver passing the active sensor area where its amplitude and velocity is influenced by the sensor surrounding (i.e., liquid or adsorbed mass) reprinted with permission from (El Gowini and Moussa 2009).

Thin-film Bulk Acoustic Resonator (FBAR or TFBAR)

It is a device comprising of a piezoelectric material manufactured by thin-film methods sandwiched between two electrodes and acoustically secluded from the surrounding medium. FBAR devices exploit piezoelectric films with thicknesses ranging from several micrometers down to a tenth of micrometers resonating in the frequency range of roughly 100 MHz to 20 GHz (Fig. 5.3).

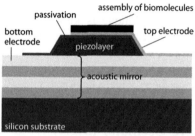

Figure 5.3   Schematic diagram Film bulk acoustic resonators (FBARs) consisting of a piezoelectric layer sandwiched between two electrodes over which the piezo layer is acoustically excited. The FBAR is isolated from the substrate by an acoustic mirror. Reproduced from (Nirschl et al. 2009).

Surface Plasmon Resonance

SPR sensors rely on the propagation of surface waves along the noble metals. Refractive index changes as a result of the binding of an analyte with the receptor coated at the sensing surface (Patching 2014) makes it a useful tool in the characterization of thin films of metal surfaces. SPR technique is also appropriate for the dynamic/kinetic measurements to investigate the binding constant of the analyte with the receptor. SPR technique was used by (Kretschmann and Raether 1968; Feresenbet et al. 2004a; Homola 2008) as a transduction scheme to diminish the effect of non-specific interactions since it offers higher sensitivity over QCM. SPR can detect vapor interactions with just monolayers of molecular receptors with a drastic reduction in dispersion interactions among the analytes and the sensing layer. SPR can be also exploited for gas sensing applications, adapting the standard Kretschmann configuration (Kretschmann and Raether 1968) (Fig. 5.4).

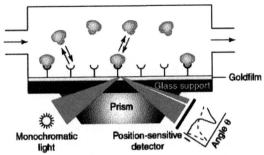

Figure 5.4 Schematic diagram of the Surface Plasmon Resonance (SPR). Monochromatic light is reflected on a gold surface. At a certain angle, where the surface plasmons are excited, the reflected light has a minimum intensity, which is continuously measured. This angle is directly connected with the analyte bound to the surface reproduced from (Zimmermann et al. 2002).

Solid-Phase Micro Extraction (SPME)

Solid-Phase Micro Extraction (SPME) is an important solvent-free, solid-phase extraction method integrating sampling, extraction and preconcentration in a single step and also the standard analytical method for environmental gases detection (Pawliszyn 1997) that typically comprises of the utilization of a fiber coated with an extracting phase such as liquid (organic polymers) or a solid (sorbent), which extracts different kinds of analytes, both volatile and non-volatile, from different kinds of media, that can be in liquid or gas phase (Pawliszyn 2011). The typical SPME process comprises of two basic steps (Fig. 5.5), first the partitioning of analyte gases between the extraction phase and the sample matrix, and the second, desorption of concentrated extracts into an analytical instrument such as GC-MS. However, the materials commonly used as sorbent material (activated carbon, Tenax TA®, graphitized carbon blacks such as Carbotrap 100® and Carbotrap 300®, etc.) are both not selective in the adsorption and in the desorption processes and also (the typically utilized) standard polymeric coatings of SPME fibers are less selective towards the different analytes that need to be adsorbed.

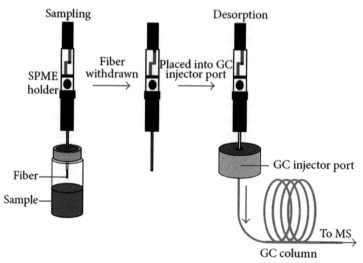

Figure 5.5 Diagram of analysis with solid phase micro extraction-gas chromatography-mass spectrometry (SPME-GC-MS). Reprinted with permission from (Schmidt and Podmore 2015)

Therefore the usual requirements of a sorbent material used for pre-concentrating gases/vapors including VOCs are should have large breakthrough volumes for the enrichment of the compounds, and allow complete desorption of the analytes at moderate temperatures, and it should possess good thermal stability and high selectivity for the enrichment of the desired organic compounds.

Evanescent Field Wave Sensor

It is a powerful well-established technique for the detection of analytes in water. In this technique, when an electromagnetic wave (light) in a dielectric medium of a particular refractive index is incident on the boundary between the optically denser medium and optically rarer medium at an angle greater than the critical angle so that total internal reflection takes place and the beam is reflected into a denser medium. But, if the incident light is not reflected instantaneously when it reaches the boundary, but super imposes in the incident and reflects beams takes place resulting in the formation of a standing electromagnetic wave then the amplitude of the electric field is maximum at the interface and decays exponentially in the outward direction of the rarer medium. This decaying field is called the evanescent field. If a part of fiber optic cladding is removed and is replaced by an absorbing fluid then a part of the evanescent field will be absorbed and hence the power transmitted by the fiber will decrease. This is the basis of the Fiber Optics Evanescent Wave Sensor (Fig. 5.6).

There are typically two approaches that have been adopted in these sensors. In the first, the wavelength of the light propagating coincides with the absorption band of the analyte and the evanescent waves interact directly with the analyte. Such sensors are called direct spectroscopic evanescent wave sensors. In the second approach, an intermediate reagent that responds optically to the analyte is attached to the core of the fiber. Such sensors are called reagent-mediated evanescent wave sensors (Chachlani and Chattopadhyay 2016).

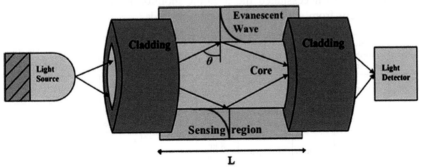

Figure 5.6   Schematic diagram of evanescent wave-based fiber optic sensor source (Sharma et al. 2019).

SENSING OF ENVIRONMENTALLY IMPORTANT GASES

The rapid growth of human population during the last century is one of the factors that have resulted in the swift surge in urbanization leading to the sharp rise in industrialization as well as a steep increase in the number of vehicles. All this has given rise to widespread entry of substances and energy into the environment causing undesirable changes in the environment (Francová et al. 2017; Kabashima et al. 2017) and global environmental issues such as ozone depletion, acid rain and global warming (Jacobson 2009; Ramanathan and Fang 2009). Further, extensive production and use of chemicals has also resulted in large-scale environmental pollution and over 300 major chemical accidents (Abbaspour and Mansouri 2005; Duan et al. 2011). Atmospheric pollution is responsible for more than 1.24 million premature deaths in India (Balakrishnan et al. 2019) apart from being accountable for damage to structures, ecosystems and for various disorders, illnesses and diseases in humans, plants and animals (Appelo and Postma 2005; Smith 2012). It, along with water pollution, can cause major disasters in a short period (Gifford and Hanna 1973; Elsom 1992; Tan et al. 2000; Molina and Molina 2004; Stedman 2004). Therefore, in view of the extreme dangers posed by their presence in minute quantities and to regulate them to prevent damage to our environment, the detection and monitoring of these dangerous pollutant gases as well as Volatile Organic Compounds (VOCs) both in the gaseous state or dissolved in water is of great importance. Out of six major pollutants i.e., Nitrogen dioxide (NO_2), Ozone (O_3), Sulfur dioxide (SO_2), Particulate Matter (PM), Carbon monoxide (CO), and Lead (Pb) that are of major concern for maintaining the quality of air in urban regions, few pollutants, like CO, Pb, NO_2, PM and SO_2, are emitted directly from a variety of sources including industries and increased traffic congestion. Ozone (O_3), present in stratosphere, acts as a protective shield for the life on earth since it has ability to absorb ultraviolet (UV) radiation thus acting as a filter from damaging ultraviolet radiation. VOCs are responsible for the formation of O_3 in troposphere near the earth surface when they react with NO_2, SO_2, ammonia (NH_3), and other gases emitted from various sources. Stringent domestic and international standards for emission control necessitated the accurate monitoring of these pollutants to protect the society from adverse public health hazards.

Pollutant	UK threshold Concentration	Standard measured	WHO threshold Concentration	Standard measured	USA threshold Concentration	Standard measured	Indian threshold Concentration	Standard measured
Benzene	5 ppb	Annual mean	Data not available	Data not available	Data not available	Data not available	5.0 µg/m3	Annual mean
3-Butadiene	1 ppb	Annual mean	Data not available	Data not available	Data not available	Data not available	-	-
Carbon monoxide (CO)	8:6 ppm	8 hr mean	8:6 ppm	8 hr mean	9 ppm	8 hr mean	2.0 mg/m^3	8 hr mean
Lead (Pb)	0:25 µg/m^3	Annual mean	0.5 µg/m^3	Annual mean	1.5 µg/m^3	Quarterly mean	1.0 µg/m^3	24 hr mean
Nitrogen dioxide (NO_2)	21 ppb	Annual mean	105 ppb	1 hr mean	0.05 ppm	Annual mean	80.0 µg/m^3	24 hr mean
Ozone (O_3)	50 ppb	8 hr mean	60 ppb	8 hr mean	0.08 ppm	8 hr mean	60.0 µg/m^3	24 hr mean
Particles (PM10)	50 µg/m^3	24 hr mean	70 µg/m^3	24 hr mean	50 µg/m^3	Annual mean	100.0 µg/m^3	24 hr mean
Particulate matter (size less than 2.5 µm	25 µg/m^3	Annual mean	-	-	-	-	60.0 µg/m^3	24 hr mean
Sulfur dioxide (SO_2)	100 ppb	15 min mean	188 ppb	10 min mean	0.14 ppm	24 hr mean	80.0 µg/m^3	24 hr mean
NH_3	-	-	-	-	-	-	100.0 µg/m^3	Annual mean

Data from (Ray 2017) and from (NAAQS CPCB, 2009)

Volatile Organic Compounds

Volatile Organic Compounds (VOCs) are organic compounds that possess high vapor pressure and have low water solubility. According to the European Union, a VOC is "any organic compound having an initial boiling point less than or equal to 250°C measured at a standard atmospheric pressure of 101.3 kPa" or "it is any organic compound, as well as the fraction of creosote, having at 293.15 K vapor pressure of 0.01 kPa or more or having corresponding volatility under the particular conditions of use" (Directive 2010/75/EU, 2010; Directive 2004/42/CE, 2004). VOCs are reported to bind with ozone O_3 at ground-level and form a photochemical smog which is a cause for breathing difficulty for asthma patients (Ray 2017). VOCs not only comprise of anthropogenic forms that are often unhealthy chemicals making them relevant for environmental monitoring but also include naturally occurring scents, pheromones, essential oils, etc. The most important human-made category is, as they harmful ones, the halogenated and aromatic hydrocarbons.

BTEX

Benzene, Toluene, Ethylbenzene and Xylenes (BTEX) are considered as hazardous because of their toxic effects on human health. Adverse effects include impact on respiratory, cardiovascular and endocrine systems. These health hazards are associated with chronic exposure to BTEX (Bolden et al. 2015).

Although BTEX can bound to soils and sediments for a small time or spill in sea water before eventually released being into the atmosphere where they can react with other pollutants leading to the formation of photochemical smog. These compounds, in a manner, are very similar to other VOCs and also play a role in the formation of ground-level ozone.

One of the main emission sources of BTEX compounds is the petrochemical industry. Manufacturing of rubber, plastics, cosmetics and pharmaceutical products is also a source of BTEX.

Benzene is listed among the most harmful VOCs because it is recognized as a human carcinogen by the US Environmental Protection Agency and by the European Commission. (USEPA DWC, 2009). (EC AQS 2015). Long-term exposure to higher concentrations of benzene can be toxic to the liver, kidneys, central nervous system and eyes. However, exposure to BTEX, at normal atmospheric concentrations, or slightly higher concentrations over a short period, is unlikely to cause significant health damage.

The primary goal of all these studies has been the monitoring of aromatic, aliphatic and chlorinated VOCs due to their importance for environmental..

Table 5.1 Exposures level to BTEX compounds as indicated by OSHA[a] and NIOSH[b], January 2018 (Data from (Clément and Llobet 2020)

Compound	*TWA[c]* **(ppm)**	*STEL[d]* **(ppm)**	*IDLH[e]* **(ppm)**
Benzene	1	5	500
Toluene	100	150	500
Ethyl benzene	100	125	800
Xylene	100	150	900

[a]OSHA: Occupational Safety and Health Administration (USA)
[b]NIOSH: The National Institute for Occupational Safety and Health (USA)
[c]TWA: (time-weighted average)
[d]STEL: (Short term exposure limit)
[e]IDLH: Immediately dangerous to life and health exposure level

Cyclodextrins

Cyclodextrins (CDs) are natural products, which are obtained from starch. They are available in three different forms i.e., cyclic hexamers (α), heptamers (β (Fig.5.7)) and octamers (γ) of glucose having different pre organized rigid cavity sizes and are well-known for displaying interesting host-guest chemistry with organic molecules especially in aqueous media (Saenger 1984) CDs possessing hydroxyl groups on the rims of their cavity on which various chemical transformations can be done to alter the binding affinity of the host.

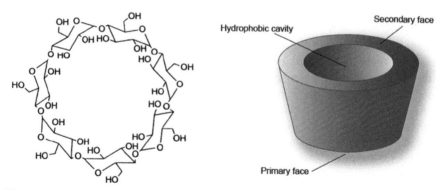

Figure 5.7   Schematic representations of β-cyclodextrin reprinted from (Becuwe et al. 2008) Derivatized cyclodextrins immobilized on capillary columns have been exploited for separations of a variety of organic species in gas phases at raised temperatures (>50°C) (Konig et al. 1989; Schurig et al. 1991).

In an inspirational work of Stoddart et al. (Lai et al. 1988) three chemically modified cyclodextrins (CDs) namely 2,6-Per-*O*-(*t*-butyldimethylsilyl)-α-cyclodextrin (**1**), 2,6-per-*O*-allyl-α-cyclodextrin (**2**), and 2,6-per-*O*-methyl-β-cyclodextrin-3-perbenzoate (**3**) containing nonpolar groups to support selectivity towards benzene over the other interfering molecules like methane, propane, butane, pentane, acetylene, ammonia, nitrobenzene and toluene were synthesized and their thin films were coated on piezoelectric transducers to create a gas sensor and were exposed to the analyte vapors to study the frequency change as the function of the absorbed mass of the analyte. The performance of **1** was found to be the best for the 0.08-400 mg dm^{-3} of benzene in the air. Toluene was found to be the most interfering molecule. But, the decrease in frequency observed upon exposure to toluene was only <20% of the response with a corresponding amount of benzene for **1** sensor.

Modified asymmetric cyclodextrins having a chemical environment to form a complex with volatile organic compounds namely α-cyclodextrindodeca(2O,3O)benzoate (**4**), β-cyclodextrintetradeca (2O,3O)-benzoate (**5**), β–cyclodextrintetradeca(2O,3O)acetate (**6**), were placed on oxide surfaces, terminated with hydroxyl groups, exploiting stronger bis(1,6-trichlorosilylhexane) coupling reagents for the genesis of covalently bound molecular Self-Assembled Monolayers (SAMs) with tailored selectivity (Moore et al. 1995). This is to ensure that most of the surface attached nanometer-sized host molecules, cyclodextrins are aligned upward to reduce non-specific extra cavity interactions. These sensors demonstrated low ppm sensitivity towards BTX. Moreover the sensors also displayed responsiveness to many interfering molecules like carbon tetrachloride, perchloroethylene, 1,1,1-trichloroethane, chloroform, n-hexane and tetrahydrofuran.

SAW devices have been created by *in-situ* formation of polysiloxane cyclodextrin **7**, (Fig. 5.8) films on the transducer's surface by immersing the devices in a DMF solution of trimethoxysilane functionalized CDs, achieved two-fold benefits, first, it provides a simple and easily reproducible approach to multilayer films for optimization of sensitivity. Second, the use of a siloxane intervoid network, to minimize adsorption of the target VOCs into interstitial sites. (Swanson et al. 1998).

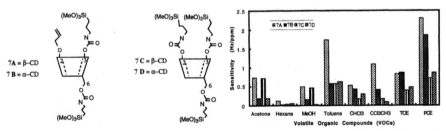

Figure 5.8 Schematic diagram of Siloxane CD polymer precursors (7A, B, C and D) and Sensor array response patterns to different VOCs with four 250 MHz SAW devices coated with 7 (A, B, C, and D) reprinted with permission form (Swanson et al. 1998).

These CD derivative coated SAW devices are sensitive and selective toward several VOCs, found to be stable and could detect analytes (perchloroethylene (PCE), trichloroethylene (TCE), trichloroethane, etc.) in tens of ppm. An α-CD coated SAW sensor is more selective to toluene and hexane than β-CD coated sensors. On the other hand, a β-CD coated sensor is more selective to halogenated hydrocarbons.

Different fluorescent indolizine modified β-cyclodextrin (Fig. 5.9) have been investigated for their sensing ability as potential molecular chemosensors (molecular probe) towards 1-adamantanol, phenol, *p*-cresol (Surpateanu et al. 2007). Multi conformational search by molecular modeling tools, MM3 and AM1 in the gaseous state and water respectively suggested 'open cavity' structure. Indolizine derivative with a perfluoro aromatic fragment showed fluorescent properties and sensitivity factor which indicated the potential of a sensor for the detection of volatile organic compounds (VOCs) (Surpateanu et al. 2007) 2D ROESY NMR spectrum recorded at neutral pH suggested that strong dipolar interactions exist between the protons localized inside the β-CD core and the aromatic protons of a fluorophore. However, these through space interactions disappeared at acidic pH indicating that inside-outside molecular motion of fluorescent moiety, controlled by the protonation of free pyridyl nitrogen (Fig. 5.9). To study the effect of placement of indolizine derivatives on β-cyclodextrin the sensing ability of these derivatives for benzene, toluene, phenol and *p*-cresol and 1-adamantanol was compared with that of β-cyclodextrin by computing formation constant of such complexes realized by means of a spectrophotometric spectral displacement with Methyl Orange (MO) which is known to form inclusion complex due to encapsulation of its diazo part and stability constant value of the inclusion compounds is found to be very similar to that of β-cyclodextrin indicating insignificant effect of the placement of indolizine derivatives apart from steric interactions that occurs between the various guests and the indolizine moiety of the sensor. To determine the sensing activity of the fluorescent sensors, the sensitivity factor $\Delta I/I0$, a measure of quenching of the fluorescence intensity of the sensor on interaction with the guest mentioned above was used. Results largely indicated the potential of these CD sensors by immobilizing on a solid support, for the detection of VOCs.

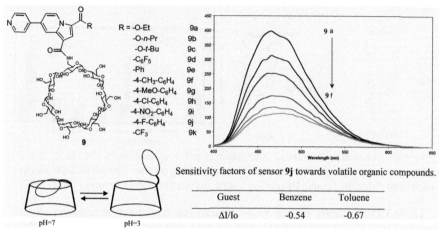

Sensitivity factors of sensor **9j** towards volatile organic compounds.

Guest	Benzene	Toluene
$\Delta I/I_0$	-0.54	-0.67

Figure 5.9 Pyridin-4-yl indolizine modified β-cyclodextrin derivatives (top left), structures of pyridin-4-yl indolizine β-cyclodextrin **9a** at pH 3 and pH 7 (bottom left), fluorescence spectra of (a) the sensor **9j** in aqueous solution (0.1 mM, 25°C), at various concentrations of toluene (b) 2.7 mM, (c) 5.4 mM, (d) 13.5 mM, (e) 18.9 mM, (f) 27 mM (top right) an sensitivity factor of 9j towards benzene and toluene (bottom right) reprinted from (Becuwe et al. 2008).

In 2009, solid-state fluorescence molecular probe comprising of a self-assembled supramolecular aggregate of perylene bisimide-bridged bis-(permethyl-β-cyclodextrins (**10**) formed by $\pi-\pi$ stacking interactions between perylene bisimide was reported (Liu et al. 2009) for VOCs detection, utilizing fluorescent indicator–displacement assay (Nguyen and Anslyn 2006) as the basis of detection and study. Perylene bisimides was chosen and incorporated as a fluorescence probe due to their strong solid-state fluorescence, photochemical stability and low quantum yield of intersystem crossing. Aggregates of **10** was embedded in the poly(vinylidene fluoride) (PVDF) membrane to study the vapor sensing and on exposure of this membrane to analyte vapors, gaseous guest molecules get captured inside permethyl-β-cyclodextrins cavity because of the host-guest interactions leading to the quenching of fluorescence of the membrane due to perylene bisimide chromophores, which is attributed to analyte inclusion and associated changes in the packing of the dyes in addition to the photoinduced electron transfer. This membrane system exhibited fluorescence quenching to several saturated vapors of volatile organic compounds with a response time of 10 seconds including not only organic amines and nitro-based compounds (o-methyl-aniline, aniline, benzylamine, butylamine, triethylamine, hydrazine, nitroethane, nitromethane and nitrobenzene) but also general organic solvents (chloroform, toluene, chlorobenzene, methanol and acetonitrile) and fluorescence alteration. Further, PVDF membrane-embedded system was found to be much more sensitive to aniline than to toluene as the fluorescence was quenched 30% by aniline with a vapor pressure of 33 ppm (Fig. 5.10) and 15% by toluene with a vapor pressure of 4100 ppm. The higher sensitivity to aniline was attributed to photoinduced electron transfer from the reducing reagents to the excited states of chromophores

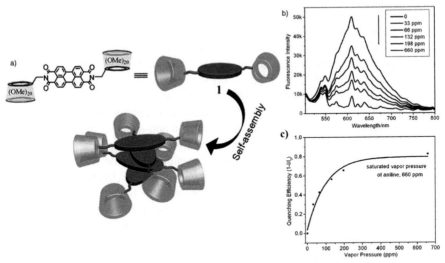

Figure 5.10 Structural illustration of 10 and its self-assembly mode (left), fluorescence spectra of the PVDF membrane-embedded on exposure to the vapor of aniline at different vapor pressures, λ_{max} = 490 nm (top right) and the fluorescence quenching efficiency as a function of the vapor pressure of aniline (bottom left) (Liu et al. 2009).

However replacing the spacer that has been used between the perylene core and the two 6-deoxy-6-amino-permethyl-β-CD units in the above unit in **10** with elongated junction amidic spacer resulted in device with excellent sensing reversibility, quick recovery as well as high selectivity of the resultant molecular fluorescence probe system **11** (Fig. 5.11) towards aniline vapors (Jiang et al. 2011). The quenched fluorescence could be recovered completely by blowing the membrane with a gas blower for 150 seconds

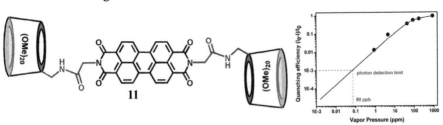

Figure 5.11 Structural illustration of **11** (left) and fluorescence quenching efficiency $(I-I/I_0)$ as a function of the vapor pressure of aniline (data error (6%) fitted with the Langmuir equation (right). Reprinted from (Jiang et al. 2011).

Sensing studies of benzene/toluene analyte with a composite of pristine polyaniline, a conducting polymeric material and β-cyclodextrin (CD) in both undoped as well as doped form was reported (Subramanian et al 2016). The undoped materials with non-polar/hydrophobic environment could detect the analyte and performed well by displaying a 39-57% sensor efficiency. On the other hand, doped composite known to display increased conductivity, showed insignificant sensitivity and non-linear sensor behavior towards the analytes owing to the hydrophilic character imparted through the doping process. The introduction of cyclodextrin in the polyaniline polymeric material leads to the formation of composite inclusion complex with CD having greater hydrophobicity and resulted in the enhanced sensor response of polyaniline towards analyte.

The analysis of VOCs in exhaled breath samples possess a potential for diagnosis of illnesses including cancer, multiple sclerosis, Parkinson's disease, tuberculosis, diabetes and more by breath tests and therefore represent a new developing frontier in medical diagnostics which is a non-invasive and a potentially inexpensive way to detect illnesses (Haick et al. 2014; Konvalina and Haick 2014). Cyclodextrin-based materials are found to be beneficial in the detection of various VOCs in the exhaled breath which also include aromatic hydrocarbons such as benzene and toluene that are prominent biomarkers for the early detection of cancer. Development of a novel electronic nose system comprising of a functionalized β-cyclodextrin (CD) wrapped Reduced Graphene Oxide (RGO) conductive CD nanocomposite (CCDC) sensors have been reported by Nag and co-workers (Nag et al. 2014; Duarte et al. 2016) by creating a synergistic effect between RGO possessing high conductivity as well as high surface area and a CD which is known for its host-guest recognition properties towards the VOCs. Various supramolecular assemblies were constructed.

The one which is abbreviated as CD functionalized RGO@PYAD, was created by the non-covalent coupling of chemically functionalized CDs with RGO utilizing pyrene attached to adamantene thorough amide spacer (PYAD) as 'glue' where on one side the adamantene terminal of the spacer was acting as a guest to functionalized CDs while the pyrene side of the linker was linked to RGO via $\pi-\pi$ interactions. Another was constructed by immobilization of perbenzylated CD on RGO (Fig. 5.12). An Array of sensors was constructed by exploiting a variety of functionalized CDs such as native CD (providing OH polar groups susceptible of interaction with polar analytes), mannose functionalized CD i.e., MCD (providing OH polar groups with a longer spacer capable of easier disconnection) perbenzylated CD i.e., PBCD (providing a strong non-polar character to the junction due to its numerous aromatic rings) and amino-functionalized CD i.e., NCD (bringing reversible covalent bonding of its free amino group with acetone, for instance) immobilized on RGO directly or through PYAD. The sensitivity/selectivity of the different CD functionalized RGO@PYAD sensors were then investigated, after successive exposures to rectangular pulses of 5 minutes of saturated VOC, followed by 5 minutes of pure nitrogen, as the function of the modification of junctions physico-chemical properties of these conductive CD nanocomposites. Exposure to a set of 10 VOCs (methanol, ethanol, propanol, isopropanol, benzene, toluene, p-xylene, acetone and formaldehyde each one of which belongs to cancer biomarkers) along with water present in a breath at more than 90%, was analyzed. The resulting CCDC-quantum chemo-resistive transducers (QRS) have demonstrated high sensitivity, evidenced by signal to noise ratio (SNR) up to 88 at a concentration as low as 400 ppb for benzene (Fig. 5.12 (bottom right)). An array of three CCDC sensors, namely RGO@ PYAD-MCD MCD, RGO@PYAD-NCD and RGO@PBCD was assembled in an e-nose and exposed to the same set of 10 VOC biomarkers. For each analyte, five VOC/nitrogen successive sequences were recorded and all signals maxima were collected into an m by n matrix (m being the number of measurements and n the number of sensors), and subsequently treated with a principal component analysis (PCA) algorithm (Fig. 5.12). The separation of clusters of points corresponding to different analytes is a clear indication of the effectiveness of this array, based on only three sensors, to discriminate between the 10 biomarkers.

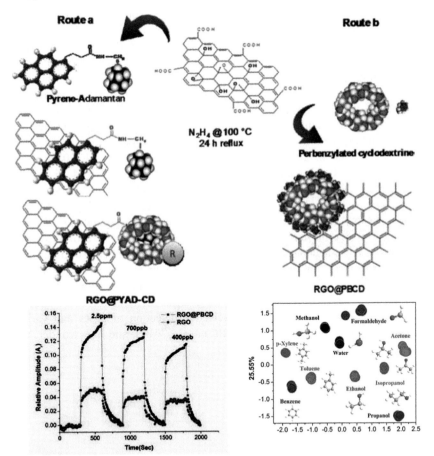

Figure 5.12 (Top) Different routes of synthesis of functionalized cyclodextrin linked to reduced graphene oxide by supramolecular assembly, (Route a) Route for the synthesis of RGO@PYAD-CD RGO@PYAD-MCD & RGO@PYADNCD (R is a hydroxyl group for CD, an amino group for NCD, and a mannose sugar for MCD), (Route b) Route for the synthesis of RGO@PBCD. (Bottom left) Responses of RGO@PBCD and RGO sensors to benzene vapor in the ppm-ppb concentration range. (Bottom right) Pattern recognition of the electronic nose exposed to a set of lung cancer biomarker VOC after principal component analysis (PCA) reprinted from (Nag et al. 2014).

Calixarenes

Calixarenes where n = 4, 6 or 8 are macrocycles that comprise of phenolic units appended to each other at an ortho position through methylene bridges. The cyclic structure of calix[n]arenes is similar to other macrocycles like cyclodextrin (Liu et al. 2003; Del Valle 2004; Laza-Knoerr et al. 2010) and crown ethers (Lambet al. 1980; Gokel 1991; Gokel et al. 2004) and has created a large interest among the researchers and is one of the most studied macrocyclic molecule around the world for developing molecular receptors. Calix[n]arene family consists of different members having varying cavity size/dimensions and these members adopt several different conformers. Native calix[4]arene possess the smallest cavity dimension and can also adopt several different conformers including the most stable rigid cone conformation, where all the phenolic OH groups are in the same direction to form strong hydrogen bonds. Calixarenes derivatives as a host are important since their recognition properties can be tuned by its functionalization, at either the upper and/or lower rim, as a function of the class of analytes to be detected and also due to their ability to control the rigidity and flexibility which is used to encapsulate different small guest molecules including gaseous molecules through multiple non-covalent interactions such as hydrogen bonding, cation-π and CH-π interactions, π–π stacking (Atwood et al. 2004; Chawla et al. 2010; Chawla et al. 2011; Chawla et al. 2012a, b, c; Pinalli et al. 2018b). In summary, these Calix[n]arenes (n = 4, 6 and 8) molecules offer different structural features such as different cavity sizes and functional groups for molecular recognition, pre organization ability in different conformations to accommodate guests of various sizes, ease of functionalization at both the rims for the easy modification of the shape and size of the cavity as well as for the attachment of the host molecules to different solid surfaces, porous structure and simultaneously these macrocycles can even detect/bind organic molecule at low concentration and can discriminate between different guests based on better host-guest fit making them an attractive and important choice as a host for the sensitive and selective detection of gases and various gases and vapor monitorings. (Farrukh et al. 2014) and no wonder that

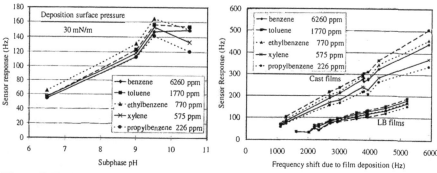

Figure 5.13 Steady-state sensor response as a function of the subphase pH for a film deposited at 30 mN/m (left) and comparison between the sensor responses of calixarene LB and cast sensing films (right). Reprinted with permission from (Munoz et al. 1999).

these molecules have been extensively investigated for gas and vapor sensing in combination with different transducers as well as optical transaction-based sensors.

Langmuir–Blodgett (LB) and cast methods were employed by Munoz et al. (Munoz et al. 1999) for coating thin films of calix[6]arenes at different pH for the development of odor sensing QCM sensors array which was used for testing of the environmentally important vapors of the aromatic compounds with different side chains such as benzene, toluene, ethylbenzene, xylene and propylbenzene and to study the effect of pH on sensors response. The odor sensing systems are also important from the quality control perspective, i.e., in the inspection of food, beverages, cosmetics and other products. The comparative results of Langmuir–Blodgett (LB) and cast methods were used to determine the relationship between the structure of the sensing film and the characteristics of the sensor (Fig. 5.13).

The results also indicated that at a high pH, the stability of the monolayers having complexation of calixarenes with alkali ions improves at the water-air interface. The sensor responses of LB films were lower than that of cast film.

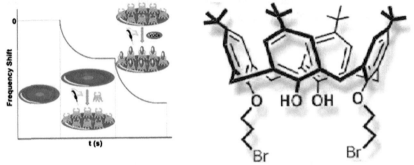

Figure 5.14 Graphical presentation of frequency shift upon deposition of a film up to sensing of VOCs. (left) and structural representation of **12** Reprinted with permission from (Temel et al. 2017).

Sensing studies of QCM quartz crystal sensors coated with a thin film of calix[4]arene derivatives bearing different functional groups selective towards the volatile organic compounds (VOCs) such as benzene, carbon tetrachloride, chloroform, ethyl acetate, n-hexane, methylene chloride and toluene were done in aqueous media (Temel et al. 2017). Experiments showed that calix[4]arenes derivative bearing bromopropyl functionalities (**12**) (Fig. 5.14) coated QCM sensor was a most useful sensor for toluene emissions among all the calix[4]arene derivatives studies. Results were corroborated with the help of sensitivity studies, host: guest stochiometry, partition coefficient, detection limit, time constant, drawing of Langmuir and Freundlich isotherms and testing of different coating techniques to demonstrate that the coating of QCM sensor surface with calixarenes was a good approach for sensing of the VOCs (Fig. 5.14).

Another sensor based on a piezoelectric quartz resonator modified with the amphiphilic aminomethylated calix[4]resorcinarene (**13**) films coated using Langmuir–Blodgett (LB) method has been developed. Various parameters of sensor material such as effects of the pH, copper metal ion content of the subphase on the behavior of the Langmuir monolayers of calix[4]resorcinarene, and the effects of the number of monolayers in a sensor coating on the response of the resulting piezoelectric quartz sensors to the vapors of various volatile organic compounds were studied and parameters for the formation of ordered LB films have been optimized. It demonstrated short response time and reproducibility of detection of volatile organic compounds (ethanol, benzene, toluene, ethylbenzene, ethyl acetate, acetone, hexane, and cumene) (Rusanova et al. 2009)

The spin-coating method (at 2000 rpm) was utilized by Hassan et al. (Hassan et al. 2001) to deposit 12 nm thin films of calix[4]resorcinarene derivatives having four azo-dye groups as substituents on quartz slides. The calix[4]resorcinarene derivatives (14 and 15) possessing an azo-calixresorcinarene molecular cavity have been used as a host to form a host-guest interaction with benzene vapors, at low concentration. Surface Plasmon Resonance (SPR) studies of these sensors showed fast and reversible adsorption on exposure to benzene vapor, with a linear response up to 400 ppm concentration of benzene and a response time of a few seconds. 15 displayed pronounced response to benzene in comparison to 14 which was attributed to its comparatively large cavity size where interaction with benzene vapors may have taken place (Fig. 5.15).

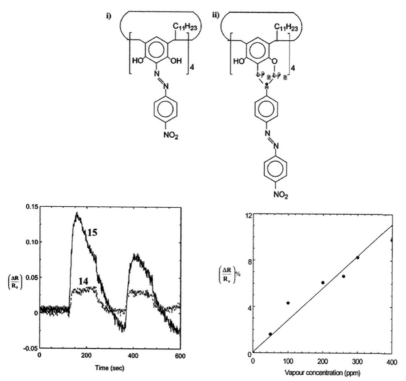

Figure 5.15. Chemical structure of i) tetra-undecyl-tetra-p-nitro-phenylazocalix[4] resorcinarene (14) ii) tetra-undecyl-tetra-(4-aminomethyl-4′-nitroazobenzyl)calix[4] resorcinarene (15) (left), the kinetic response of an overlayer of 15 and 14 compounds on exposure to 400 ppm of vapors of benzene (middle) and dependence of 15 films response to vapor concentration of benzene (right). Reprinted with permission from (Hassan et al. 2001).

A report of almost decade earlier of the sensing study of the films with volatile organic vapors using SPR was done by Erdogan et al. (Erdoğan et al. 2010) in which active layer/thin film of calixarene derivatives (**16**) (Fig. 5.16 left) was deposited using the LB film technique. Fast and reversible adsorption of organic vapors was observed and that the penetration of organic vapors into calix[4]amine LB film was rapid and diffusion coefficients were dependent on the organic vapors used. Diffusion coefficients for benzene were faster than other organic vapors such as toluene and xylene since benzene has the lowest molar volume and relatively high viscosity parameter which indicates that benzene molecules are more mobile than the other organic molecules and penetrate easily into the calix[4]amine LB film.

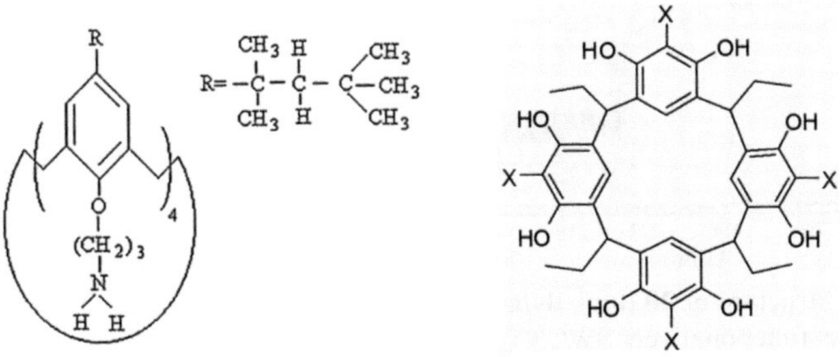

Figure 5.16 Molecular structures of **16** (left) reprinted with permission from (Erdoğan et al. 2010) and **17** (X = H), **18** (X = OH), **19** (X = CH₃) (right).

Different calixarene derivatives like resorcinol cyclic tetramer (**17**), 2-methyl resorcinol cyclic tetramer (**18**), pyrogallol cyclic tetramer (**19**), and 4-tert-butylcalix[4]arene were coated (deposited using the Langmuir–Blodgett technique) and Quartz Crystal Microbalance (QCM) sensing devices were constructed and exposed to 16 different organic vapors, such as isopropanol, ethanol, methanol, ethyl ether, n-hexane, cyclohexane, chloroform, carbon tetrachloride, dichloromethane, benzene, toluene, o-xylene, m-xylene, p-xylene, ethylbenzene and chlorobenzene, showing that **17** was found to be the most efficient actively coating material for sensing of isopropanol molecule (Zou et al. 2011).

Since the limited conductivity of calixarenes is a major obstruction in the way of development of calixarene-only chemiresistive sensors, Tapan et al. (Sarkar et al. 2018) reported development of calixarene-based chemiresistor sensing device by exploiting to non-covalent functionalization of high conductivity SWCNTs with calixarene derivative (**20**) (Fig. 5.17) having excellent affinity for certain analytes, to improve sensitivity. The hybrid material was fabricated by solvent casting. These sensors were exposed to BTEX group vapors such as benzene, toluene, ethylbenzene and xylenes with successive increments in concentration from 50 to 250 ppm, and limits of detection were found to be 25, 7.5, 6.5, and 4 ppm respectively which is well below their Occupational Safety and Health Administration (OSHA) Permissible Exposure Limit (PEL) except for benzene. Mechanistic study via field-effect transistor measurements suggested that the sensing mechanism for such sensing is dominated by an electrostatic gating effect. These SWCNT-calixarene hybrid sensor arrays have the potential for the realization of electronic nose applications.

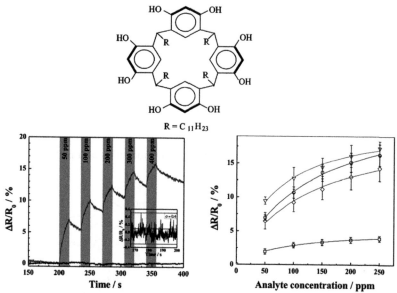

Figure 5.17 Structure of **20** (top), dynamic responses ($\Delta R/R_o$) of pristine SWCNTs (blue) and calixarene functionalized SWCNTs (red) towards 50 to 250 ppm of ethylbenzene vapor; (inset) noise (s) of the SWCNT calixarene hybrid sensor (bottom left) and calibration curves of SWCNT-Calixarene hybrid sensors to 50–250 ppm of benzene (black/□), toluene (red/◊), ethylbenzene (blue/o) and Xylenes (olive/Δ) vapor at room temperature (bottom right) (number of the device; n = 5). Reprinted with permission from (Sarkar et al. 2018).

Kalchenko et al. (Kalchenko et al. 2002) used different calixarene derivatives as coating material which was deposited by by either of the methods like thin and thick drop, LB thermal evaporation, etc., to prepare QCM sensors. It was observed that sensors coated with *p-tert*-butylcalix[6] and *p-tert*-butylcalix[8] arene, **21**, **22**, and **23** provide the largest response with toluene vapors and that QCM-based sensors and sensor arrays are promising devices for the detection of organic volatile molecules in ambient air (Fig. 5.18).

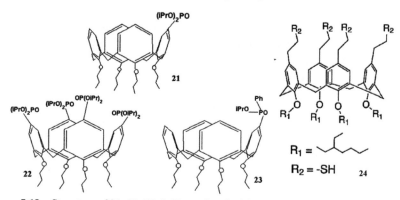

Figure 5.18 Structure of **21, 22, 23** (left) reprinted with permission from (Kalchenko et al. 2002) and calixarene ligand **24** (right). Reprinted with permission from (Kimura et al. 2011).

Similarly, Kimura et al. have prepared a quartz crystal microbalances (QCMs) mass sensor coated with gold nanoparticles capped with calix[4]arene ligand **24** (Fig. 5.18) prepared via direct synthesis approach and they were found to be highly sensitive and selective for sensing volatile organic compounds. The sensor displayed a good response toward toluene vapors due to the interaction of aromatic vapor with an electron-rich calixarene cavity (Kimura et al. 2011) Attachment of calixarene ligands to the surface of nanoparticles resulted in the enhancement of the sensor response utilizing sensing films with large film thicknesses through the good resonance propagation of the nanoparticle films and limited swelling during VOC sorption and the creation of porous structures by packing of nanoparticles

A fast, sensitive and reversible sensor for toluene with a sensitivity of 231 parts per million by volume (ppmv) has been developed by coating calixarene functional material that exhibits sensitivity to the aromatic compounds toluene and benzene while being relatively insensitive to the aliphatic hydrocarbon hexane on an optical fibre's Long-Period Grating (LPG). A nanoscale cavity of calixarene molecules entraps the VOC molecule, thus altering the refractive index of the coating and influencing the transmission spectrum of the LPG (Topliss et al. 2010).

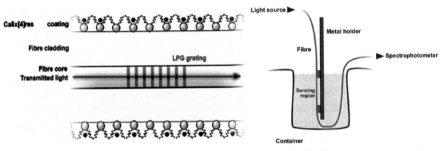

Figure 5.19 Schematic of the optical fiber sensor with Calix[4]resorcinarene coating and experimental setup reproduced with permission from (Partridge et al. 2014).

Calix[4]resorcinarene coated optical fiber long-period grating sensor for the detection of toluene in water with detection range up to ~100 ppm with semi-selectivity and low (< 10 ppm) variation was reported (Partridge et al. 2014) where water quality monitoring can be carried out without the requirement of water sampling or pre-concentration (Fig. 5.19).

An array of three Long-Period Gratings (LPG) out of which one was modified with a mesoporous film. The array was comprised of SiO_2 nanoparticles (NPs) along with poly(allylamine hydrochloride) polycation PAH and calixarene derivative (*p*-sulphanato calix[4]arene, CA[4] or *p*-sulphanato calix[8]arene, CA[8]) as a functional compound. It was used for the detection of individual Volatile Organic Compounds (VOCs) and their mixtures (Hromadka et al. 2017). The sensing mechanism was based on the measurement of the Refractive Index (RI) change induced by complexation of VOCs with calixarene. The LPG was exposed to VOCs like chloroform, benzene, toluene and acetone vapors as well as their mixture evaporating from freshly painted surfaces. At high VOC concentrations, a small but a measurable shift of the resonance band was observed (Fig. 5.20).

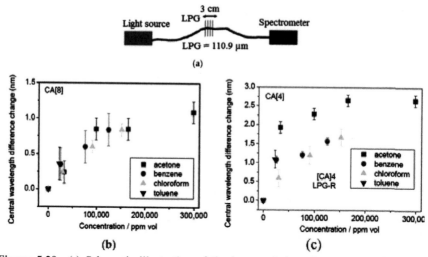

Figure 5.20 (a) Schematic illustration of the long period grating sensor (left), (b) 26, and (c) CA[4] Sensor's response related towards selected VOCs at different concentration levels (the error bars represent the standard deviation). Reprinted with permission from (Hromadka et al. 2017).

Other Cavitands

The word cavitand is usually used for container-shaped molecules possessing a cavity, which allows them to engage in host-guest chemistry with guest molecules of a complementary shape and size. Due to their outstanding molecular recognition properties, they are perfectly suitable for supramolecular sensing. The rational design of these synthetic receptors is achieved based on the analyte gas to be detected through proper functionalization of a host such as resorcinarenes, quinoxaline arenes, controlling weak host-guest interactions. Bridging groups often play an important role in the design of cavitands for the detection of different gases.

Early attempts to monitor VOCs in the air were made by placing a coat of quinoxaline bridged cavitands on Quartz Crystal Microbalances (QCM) transducers due to the capability of quinoxaline cavitands to selectively complex various aromatic vapors such benzene and toluene both in the gas phase as well as in solid phase (Vincenti et al. 1990; Soncini et al. 1992; Dickert et al. 1993; Nelli et al. 1993; Dalcanale and Hartmann 1995; Vincenti and Dalcanale 1995; Hartmann et al. 1996; Dickert et al. 1997).

However, above-mentioned QCM sensors coated with cavitands have an inherent disadvantage that they are unable to distinguish between specific intra-cavity complexation from extra cavity adsorption since these sensors account for any mass change on their surface through frequency alterations, regardless of the origin. Therefore, the potential of these cavitands could not be tapped fully (Pirondini and Dalcanale 2007; Ryvlin et al. 2017).

Keeping in view the above-mentioned disadvantage, the selectivity of quinoxaline bridged cavitands toward the aromatic vapors such as benzene and toluene has been studied by SPR by depositing a thin layer of it onto the gold substrate start (Feresenbet et al. 2004a). This selectivity was further enhanced on depositing the quinoxaline cavitand, **25** monolayers on the gold SPR surface (Feresenbet et al. 2004b) by installing four thioether legs at the lower rim of the cavitand (van Velzen et al. 1995).

SPME sampling using sorbent material of airborne aromatic hydrocarbon vapors, benzene, toluene and xylene was done using methylene-bridged and quinoxaline-bridged resorcinarene cavitands as molecular receptors grafted on to silica through sol-gel technology with the help of the hydrosilylation on the lower rim of the cavitands. The resulting sorbent material selectively retains BTX at ppb levels and was developed for the selective determination of benzene and chlorobenzenes at ultra-trace levels in air and water samples. This SPME coated fiber also fulfilled the requirement of the sorbent material i.e., having excellent thermal (400°C) and chemical stability as well as a very good fiber-to-fiber and batch-to-batch repeatability (Bianchi et al. 2003; Bianchi et al. 2008) (Fig. 5.21).

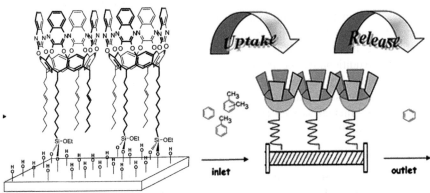

Figure 5.21 Quinoxaline-bridged cavitands immobilized on silica gel (left) schematic representation of gas sorption Quinoxaline-bridged cavitands immobilized on silica gel. Reprinted with permission from (Bianchi et al. 2003).

Triptycene containing quinoxaline cavitands, **25** possessing cavity roofing of two triptycene moieties at the upper rim of the tetraquinoxaline walls in order to impede the uptake of the bulkier substituted aromatics such as toluene, ethyl benzene and xylenes and to increase the selectivity of the sorbent material towards the benzene was reported by Bertani et al. for supramolecular detection of benzene, toluene, ethylbenzene, and xylenes (BTEX) chlorinated hydrocarbons in the air (Bertani et al. 2016). The inclusion properties in the solid-state of the cavitands were tested by Solid-Phase Micro-Extraction (SPME) sampling of BTEX at trace levels in the air.

In another report (Riboni et al. 2016), tetraquinoxaline cavitands, **26** with methylenoxy bridges at the upper rim of the quinoxaline walls to bring in both the conformational rigidity as well as the reduction of the cavity opening. The cavitands were used as selective solid-phase microextraction (SPME) coating for the determination of BTEX at trace levels in air which allowed higher enrichment factors and very low detection limits and increased selectivity toward benzene and also coating is found to be insensitive to most common interfering agents present in the air, i.e., aliphatic hydrocarbons and water (Fig. 5.22).

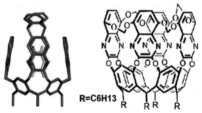

R=C6H13

Figure 5.22   Molecular structure of triptycene containing quinoxaline cavitands, **25** (left) and QxCav, **26** with methylenoxy bridges (right). Reprinted with permission from (Bertani et al. 2016) and (Riboni et al. 2016) respectively.

Nitrogen-containing basic gases and volatile compounds

The presence of ammonia and other volatile organic derivatives of ammonia such as amines in the environment beyond certain concentrations have been termed as dangerous to both humans and animals. For example, the presence of ammonia in concentration beyond 25 ppm in the air is considered to be unsafe for humans. Also pyridine is considered as a toxic nitrogenous aromatic compound with a heteroatom. Additionally, the monitoring of various organic amines is also of importance in quality control of food in food industries, for diagnostics in medical/ clinical fields and safety at work, whereas that of gaseous ammonia is crucial in clinical, environmental as well as chemical and automotive industrial processes. (Timmer et al. 2005).

Cyclodextrin

Azophenol dyes possessing both permethylated cyclodextrins as well as crown moieties **27**, **28**, and **29** have been reported to provide a critical method for discriminating 1° and 3° amines with unique color changes. An addition of 1° and 2° amines to **27** shifted the absorbance maximum from 380 to 580 and 530 nm, respectively, but no change was observed with 3° amines. The high selectivity of **27** is mainly due to H-bonding between the ammonium H atoms of the amine and oxygen atoms of the crown-6 (Fig. 5.22).

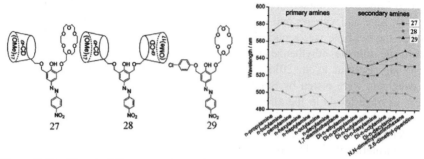

Figure 5.23   Chemical Structure of **27**, **28**, and **29** and plot of the absorption maxima of different amines having different substitutions in chloroform. Reprinted with permission from (June et. al 2006).

Calixarenes

A lithium complex of calix[4]arene-bearing nitrophenylazophenol chromogenic moiety **30** immobilized onto filter paper and was utilized as a visual indicator for the presence of gaseous amines such as trimethylamine (TMA) (McCarrick et al. 1994). The complexed calixarene has been found to undergo a dramatic color change from yellow to red in the presence of gaseous TMA at concentrations above 20 ppm with UV absorption max changed from 490 nm to 376 nm (Fig. 5.24).

Figure 5.24 (Left) The absorption spectra of **30** (2.5 mL, 5.0×10^{-5} M with 1.0 M LiClO$_4$) in butanol on exposure to the following concentrations of the gaseous TMA (*a*) 0, (*b*) 0.45, (*c*) 2.25, (*d*) 4.88, (*e*) 22.50, (*f*) 45.00 ppm (reprinted with permission from (McCarrick et al. 1994), (middle) Structure of the nitrophenylazophenol calix[4]arene **30** (right) optical responses of the lithium complex to gaseous ammonia and c) A, 100; B, 50; C, 20; D, 10; E,5; and F, 2 ppm; G, N$_2$. A scan of the coated fibre in nitrogen was used as a reference reprinted with permission from (MacCraith and McKervey 1997).

An early report of optical detection/ sensing of ammonia was through the utilization of optical evanescent wave sensor coated with PVC film containing lithium complex of the above mentioned chromogenic calix[4]arene moiety, **30** (MacCraith and McKervey 1997) On exposure to various concentration of ammonia both as a free ligand or in the form of lithium complex the absorption maximum of the chromogenic group shift to the 500 nm due to deprotonation of chromogenic calixarene.

Coating of the mixtures of the polyaniline, PA (emeraldine base) and phosphorylated resorcinarene derivative **31** (20 wt%), doped with HCl, prepared by LB method were used for conductometric gas sensors. The films are found to be quite stable at the air-water interface. The films were found to be highly sensitive to NH$_3$ (Lavrik et al. 1996) (Fig. 5.25).

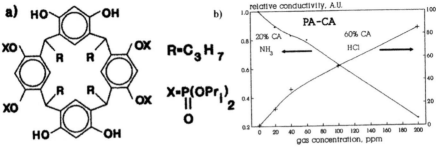

Figure 5.25 a) The chemical structures and space-filled molecular models of phosphorylated calix[4]resorcinolarene derivative **31** and calibration curves for NH$_3$ and HCl of the sensors containing PA–**31** 20 and PA–**31** 60 films, respectively. Reprinted with permission from (Lavrik et al. 1996).

Monitoring/detection of organic amines has been done by utilizing various techniques such as electrochemical devices, ion-selective electrodes and other chemical sensors. However colorimetric/Fluorometric-based sensing methods or molecular probes offers few distinct advantages over other techniques due to their portability, involvement of simple detection process, large-signal output and comparatively less expensive and in many cases, the output signal can even be perceived with the naked eye.

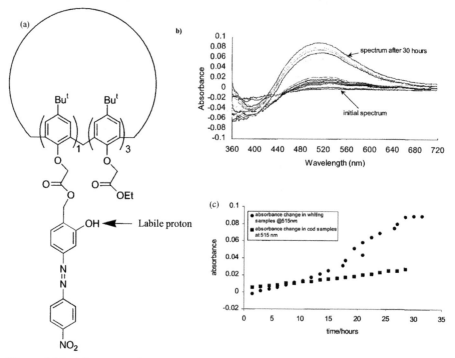

Figure 5.26. a) Structure of mononitrophenolazophenol calix[4]arene tetraester derivative, **32** used as the acidochromic dye (left), b) differential reflectance spectra of 3:1 w/w LiClO$_4$-dye disk in contact with fresh whiting (24 hours on ice since catch), 2.0 g of fish at room temperature (22±1°C), for 30 hours and c) comparison of the change in absorbance observed throughout 30 hours at room temperature obtained from fresh whiting and cod samples at 515 nm. Reprinted with permission from (Loughran and Diamond 2000).

The detection and monitoring of volatile amines such as trimethylamine, dimethylamine and ammonia from fish samples (headspace of cod and whiting samples) done through changes in the color of a sensitive calix[4]arene-based acidochromic dye, **32** (Fig. 5.26) immobilized on test paper disks employing UV-Vis reflectance spectroscopy with an increase in absorbance max centered at 500-510 nm. The sensitivity of the dye response was tuned by varying the ratio of LiClO$_4$ to dye in the dye solution due to the formation of more acidic Li$^+$ dye complex. Volatile components absorbing at lower wavelengths (below 410 nm) were detected in cod samples (Loughran and Diamond 2000). Fibre optical sensors based on nitrophenylazo-derivatives of calix[4]arene was exposed to different primary, secondary and tertiary amine vapors, the chromogenic effect was the largest for the primary amine and it was in parallel with the alkyl chain length, however, the effect was slightly less for the secondary and least for the tertiary amine. These results could be understood in terms of the basicity of the amines and the steric-hindrance effect due to the substituents (Liu et al. 2004).

Spun films of nitrophenylazo-calix[4]arenes, **33** (Fig. 5.27) were reported as gas-phase optical sensors for n-hexylamine and other amines with weak basicity, e.g., n-pentylamine, n-butylamine, and aniline, and amines with branched alkyl groups, e.g., dipropylamine, triethylamine, *iso*-pentylamine and *tert*-pentylamine. The redshift of the visible absorption bands was observed when these sensors were exposed to different amine vapors. The highest sensitivity was observed for n-hexylamine better than other primary amines such as n-pentylamine, n-butylamine and weak sensitivity was observed for branched amines. Further, the change in observed absorbance max towards n-hexylamine was also found to be dependent on substitutions present on the rims of calix[4]arene (Liu et al. 2005).

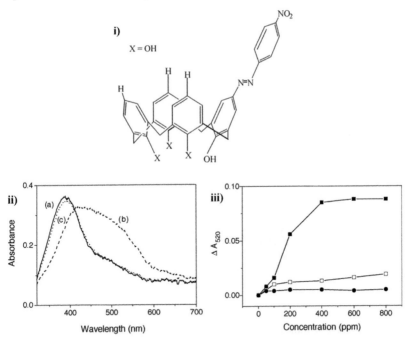

Figure 5.27 i) Nitrophenylazo-calix[4]arenes, **33** used in this study ii) Absorption spectra of the DQ-OH film: (a) before exposure, (b) after exposure to 600 ppm n-hexylamine, (c) after N_2 purge at 150°C. iii) ΔA_{520} vs. amine concentration curves for the DQ-OH film: (■) n-hexylamine, (□) n-pentylamine, (●) n-butylamine, (○) aniline reprinted with permission from (Liu et al. 2005).

Calix[4]arene derivatives (**34** and **35**) involving either one or two amino acids of a particular type, at the 1,3-bridge have been employed as a coating of QCM sensor for the detection of organic amines through covalent attachment between bi-sulfur atoms and the gold. The presence of two phenylalanine and two cysteine moieties resulted in the formation of a larger bridge at the lower rim of the calix[4arenes derivative which assisted the assembly of macrocyclic sensing material onto the surface of the gold electrode in a better manner as a comparison to the calixarene derivative with just two cysteine units resulting in a smaller bridge (**34**) (Fig. 5.28). The sensor with (**35**) exhibited better sensitivity for n-butylamine as compared to *iso*-butylamine or *tert*-butylamine vapors owing to the steric hindrance which a branched amine may offer during the formation of a complex with the host molecules (Li et al. 2004; Yuan–Yuan et al. 2005).

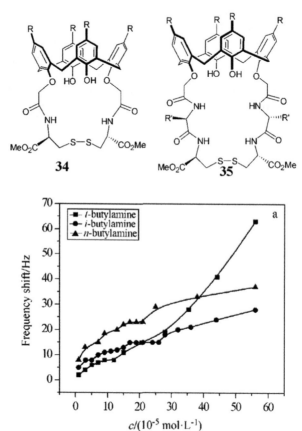

Figure 5.28 Structure of the compounds **34** (left) and **35** (middle) used for self-assembled coatings and adsorption curves of **35** QCM for butylamine isomers in gas (25°C). Reprinted with permission from (Yuan-Yuan et al. 2005).

The LB films of the blend of 5,10,15,20-tetrakis[3,4-bis(2-ethylhexyloxy) phenyl]-porphinato zinc(II) (**36**) and Mn(II) (**37**) and calix[8]arene derivative (**38**) were utilized to have solid films of improved quality, better lifetime and sensitivity towards amine vapors. The films with zinc complexed porphyrin (**36**) on exposure to amine vapors exhibited a spectral red shift, whereas Mn complexed porphyrin (**37**) displayed a higher frequency shift (Brittle et al. 2008) (Fig. 5.29).

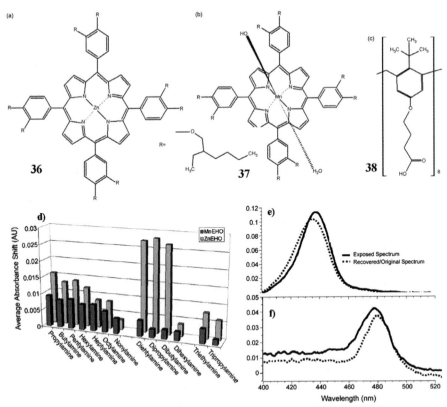

Figure 5.29 (a) Molecular structure of **36**, (b) **37**, and (c) (**38**). reprinted with permission from (Brittle et al. 2008) (d) average absorbance shift in **36** and **37** with exposure to different amines, (e) The red-shift of the Soret peak in spectra of **36** due to the exposure to different amine vapors (f) likewise the blue-shift of the Soret peak of **37** spectrum with exposure to amines. Reprinted with permission from (Brittle et al. 2008).

C-undecylcalix[4]resorcinarene (**38**) has been used as a coating material by Cao et al. (Cao et al. 1998) for constructing Thickness-Shear-Mode (TSM) acoustic wave sensors which were utilized in the detection and monitoring of pyridine vapors since their frequency shift response characteristics indicated that they as the most efficient host-guest adsorptive material for the vapors of pyridine. These sensors exhibited good reproducibility, high stability, short response time and recovery rate of 98.10–103.92% in the determination of pyridine vapors.

Acidic Vapors

Hydrochloric Acid

Monitoring of HCl vapors is considered important not only because it is a known corrosive environmental pollutant, but also because its presence in the atmosphere severely affects human health even at low concentration and can cause irritation of the respiratory tract including the throat (Bolstad-Johnson et al. 2000), eyes and nose in healthy individuals and exacerbate symptoms associated with conditions such as asthma and emphysema. Its presence also interferes in various atmospheric processes and is one of main factors in the production of photochemical smog. Since the dissolution of HCl in water produces a strong and corrosive acid (Bastidas et al. 2000) it also contributes to acid rain causing damage to building materials and reduce crop yields. HCl release is associated with combustion of fossil fuels like coal and heavy oils, and with several manufacturing processes, including cement production. Although it is not considered a combustible gas, it may react or form combustible compounds with alcohol and hydrogen cyanide or with metal catalysts such as aluminum-titanium alloys (Canakci and Van Gerpen 1999; Kamata et al. 2009).

A conductometric gas sensor was developed by Lavrik et al. for monitoring of HCl vapors by utilizing a Langmuir–Blodgett composite film made from HCl-doped polyaniline (emeraldine base) and phosphorylated calix[4]resorcinarene, **31** (Lavrik et al. 1996).

Nitrogen Oxides NO_x

NO_x gases comprises of NO, NO_2, N_2O_3 ($NO \cdot NO_2$), N_2O_4 ($NO_2 \cdot NO_2$), and N_2O_5. Environmental pollution caused due to the presence of nitrogen oxides (NO), a colorless and odorless gas, has gained attention from the past few decades owing to its responsibility in contribution to acid rain as well as to the photochemical smog particularly in urban areas. This has necessitated the development of efficient, selective and sensitive sensors for detecting/monitoring NO. These sensors are important to be used for monitoring of exhaust gases from combustion equipment like boilers, combustion furnaces, and diesel engines which utilizes different nitrogen-containing fuels (Satake et al. 1994). Its clinical monitoring is important for the diagnosis of specific types of airway inflammation (Maniscalco et al. 2016) due to its role in the regulation of important biological functions in the human body such as respiratory, cardiovascular and nervous systems through control of blood circulation (Pfeiffer et al. 1999; Ignarro 2000; Butler and Nicholson 2003).

Calix[4]arenes

Calix[4]arenes offer an important class of supramolecules that possess the ability to complex with NO^+ reversibly with the different conformation of the host. The colorimetric sensors that are capable of detecting micromolar/ppm level of NO in gaseous have been reported (Rathore et al. 2000) which were based on formation of colored charge transfer calix[4]arenenitrosonium complex due to the entrapment of NO gas upon its exposure to the host solution through the interaction of NO+ to the π-electron-rich cavity of the host. The cation-π interaction was found to be present between the calixarene aromatic cavity and NO^+, due to which the calixarene unit was able to bind to the cationic moiety (Botta et al. 2007). Calixarene-based nanotubes for storing NO^+ ions were also reported (Organo et al. 2005; Organo and Rudkevich 2007) Gambert et al. designed a device comprising of an absorber made of a carbon-alloy carrier that conducts electricity and calixarene as a pre-concentrator for sensing NO. Sensing has been achieved by breathing through the device to get NO^+ absorbed. Subsequently, NO was desorbed by heating, detected and measured by sensor signal proportional to the partial pressure in the breath (Gambert 2001).

N_2O Nitrous oxide is a colorless, sweet-smelling gas and is widely used in hospitals as a general anesthetic and analgesic. It is known for causing amusing effects, hence the name 'laughing gas' is associated with it. As an environmental pollutant, it does contribute to global warming since it has a high 'global warming potential' (310 times that of carbon dioxide). Nitrous oxide damages the ozone layer and it is involved in the formation of ozone in the troposphere near ground level. Almost 40% of N_2O emissions are due to human activities at a global level.

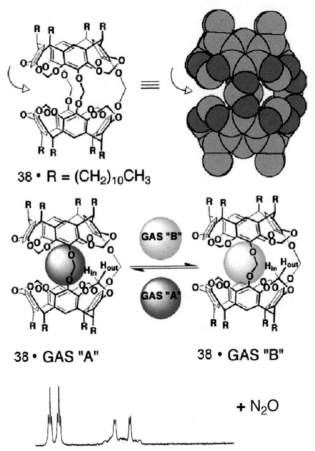

Figure 5.30 A macromodel representation of gas exchange in hemicarcerand **38** occurs both in solution and in the solid-state. The depicted OCH_2O inward-pointed (Hin) hydrogens are most sensitive to the presence of gas in the cavity.

Hemicarcerand, **38** (Fig. 5.30) a molecular container comprising of two resorcinarene hemispheres, connected by three methylene bridges, and having an inner cavity volume of ~ 110 Å^3 was used for entrapment, storage and release of different gases like helium, hydrogen carbon dioxide, etc. by Rudkevich et al (Leontiev and Rudkevich 2004) and the same was also used for making complex with N_2O on bubbling the gas through $CDCl_3$ or benzene solution of the molecular container and also in solid state through a slow process of gas exchange. The N_2O complex was studied and distinguished from other gas complexes by NMR spectroscopy.

NO_2 Nitrogen dioxide, a highly reactive reddish-brown suffocating gas, whose major source of release is the burning of fossil fuel, is a major atmospheric pollutant that forms nitric oxide and toxic organic nitrates in the air (Rudkevich 2005). It is a major component of NO_x gases and is known for its role in the production of smog and various nitrosation processes with biomolecules, causing cancer and other diseases. (Lerdau et al. 2000; Kirsch et al. 2002) NO_2 gas exists as dimer N_2O_4 at a higher concentration which disproportionate to $NO^+NO_3^-$. According to the European Commission air quality standards, the NO_2 concentration should not exceed the limit of $40\,\mu g/m^3$ at the averaging period of one year.

Calix[4]arene derivative **39** (Fig. 5.31) has been used for the chemoselective colorimetric molecular probe of NO_2/N_2O_4 gases both in solution as well as in the solid-state by Kang and Rudkevich 2004, since NO^+ form complex by interaction with the π-electron-rich inner part of the calix[4]arene cavity. The UV–vis spectra of **39** a and b exhibited broad charge-transfer (Rathore et al. 2000; Borodkin and Shubin 2001) band at λ_{max} 560-580 nm.

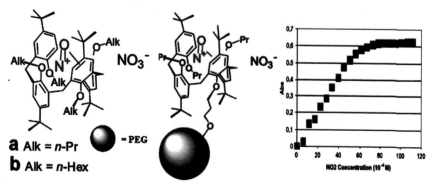

Figure 5.31 Molecular structure of calixarene **39**–nitrosonium complexes (left and middle) and UV–Vis spectroscopic titration experiments (CH_2Cl_2, 295 K) with calix[4] arene derivative and NO_2/N_2O reprinted with permission from (Kang and Rudkevich 2004).

Various supramolecular calix[4]arene based nanotubes (Fig. 5.32) and other systems for sensing, entrapment, fixation and conversion of NO_2/N_2O_4 (NO_x) gases have been reported by Zyranov et al. where chemical fixation has been demonstrated through the transformation into calixarene – NO^+ complex which is based on their study of the host-guest chemistry of the stable deeply colored. In calix[4]arene-nitrosonium (NO^+) complexes of tetra-o-alkylated calix[4]arenes **40** (Fig. 5.32) NO^+ get encapsulated within the calixarene cavity and forms stable charge-transfer complexes. Attachment of calix[4]arenes to silica gel, **41** has been demonstrated (Fig. 5.33), which allowed a solid material for visual detection and entrapment of NO_2/N_2O_4 (Zyryanov et al. 2002; Zyryanov et al. 2003; Organo et al. 2004; Rudkevich 2004; Rudkevich 2005).

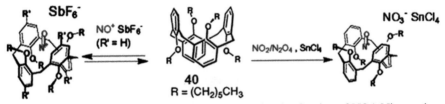

Figure 5.32 (Schematic representation of supramolecular fixation of NO₂) Nitrosonium complexes of calix[4]arene **40** (two methods of synthesis). Reprinted with permission from (Zyryanov et al. 2003).

Supramolecular fixation of NO_2/N_2O_4 gases in the synthetic containers and reaction vessels which are formed by caged, calixarene-based compounds has been discussed in detail by Rudkevich (Rudkevich 2004) On treatment with pure air, the calixarene's nitrosonium complex dissociates, enabling the parent calixarenes derivative molecule to be recovered (Rudkevich 2003). Calixarene based hemicarcerands derivatives have been used as nitrosonium storing and releasing reagents for organic synthesis (Rudkevich et al. 2005).

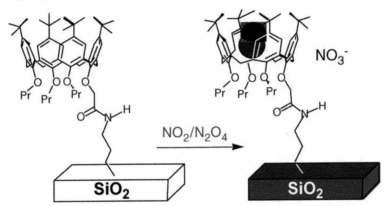

Figure 5.33 Schematic representation of sensing and entrapment of NO_x by **41** materials based on calixarene immobilized on silica. Reprinted with permission from (Rudkevich et al. 2005).

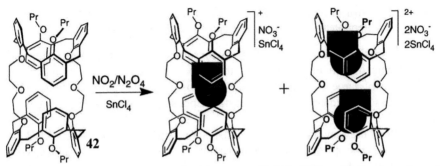

Figure 5.34 Calix[4]arene-based tube **42** reacts with NO_2/N_2O_4 with the formation of nitrosonium complexes [20]. Multiple-guest complexes are expected with tube elongation. Reprinted with permission from (Rudkevich et al. 2005).

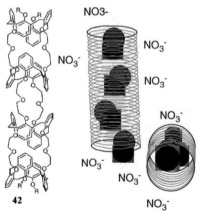

Figure 5.35   Molecular nanotubes **42** for NO_2/N_2O_4 fixation. On exposure to polyaromatic surfaces of the tubes, N_2O_4 disproportionates to $NO^+NO_3^-$. Nanotubes encapsulate NO^+ within their interiors. An in-and-out exchange of trapped NO^+ species is possible, and this can be applied for nitrosation reactions. Reprinted with permission from (Rudkevich et al. 2005)

LB film consisting of two components i.e., calix[8]arene derivative **43** and EHO porphyrin derivative **44** (Fig. 5.36) as a NO_2 gas sensing material was studied by Richardson et al via optical absorbance detection method (Richardson et al. 2006).

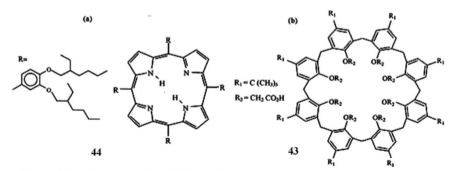

Figure 5.36   Structure of (a) EHO porphyrin **44** and (b) calix[8]arene derivative **43**. Reprinted with permission from (Richardson et al. 2006).

These sensing material exhibited up to sub-ppm detection range and fast response time due to the porosity of the calixarene domains originating from the sufficiently large cavities allowing transport of small gas molecules. The concentration dependence of the sensing response followed simple Langmuir kinetics.

The use of the Langmuir–Blodgett (LB) mixed films of 5,10,15,20-tetrakis[3,4-bis(2-ethylhexyloxy)phenyl]-21H,23H-porphine (EHO), **44** and p-tert-butylcalix[8] arene derivative, **43** have exhibited better-sensing capability towards the gaseous NO_2 since the Calix-8 matrix reportedly improved the sensing properties of the porphyrin molecules in the solid-state (Roales et al. 2011).

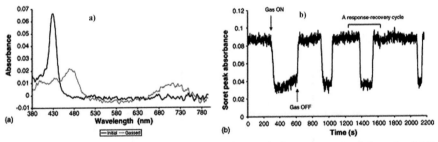

Figure 5.37 a) The absorbance spectrum before and after exposure of the mixed LB film to 4.6 ppm NO_2 (13 layers) and (b) the Soret band absorbance versus time for four exposure-recovery cycles (13 layers). Reprinted with permission from (Roales et al. 2011).

Use of Langmuir–Schaefer films (LS) of 5,10,15,20-tetrakis[3,4-bis(2-ethylhexyloxy)phenyl]- 21H,23H-porphine (EHO) (**44**) in conjunction with carboxylic acid substituted calix[8]arene (**45**) (Fig. 5.38) and Polymethyl methacrylate (PMMA) as a selective barrier layer on the top of EHO layer to differentiate between large and small analyte molecules and to further improve the selectivity towards NO_2 vapors was also reported recently (Evyapan and Dunbar 2015). The degree of selectivity is determined by the relative quantity of calix[8]arene molecules in the selective layer. Although the sensors have shown the sensitivity towards the vapors of four different analytes including NO_2, acetic acid, butyric acid and hexanoic acid, but the sensors exhibited comparatively better sensitivity and thermal reversibility towards NO_2 gas.

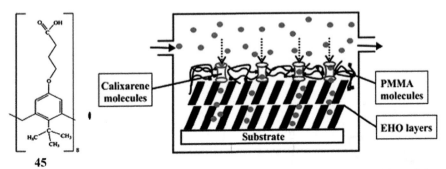

45

Figure 5.38 Structure of calix[8]arene derivative, **45** and schematic illustration of the structure of barrier layer. Reprinted with permission from (Evyapan and Dunbar 2015).

NO_2 gas sensing or storage have been studied by o-tetraalkylated calix[4] arene, a molecule that has shown to undergo a rapid color change when it complexes with the reactive nitrosonium ion in soluble state in hydrophobic Ionic Liquid (IL), 1-butyl-3-methylimidazolium bis(perfluoroethylsulfonyl)imide ([C4mim]-[BETI]) as color change in solution was also observed on bubbling of NO_2 which was suitable for colorimetric determination (Poplin et al. 2007)

Similarly immobilized calix[4]arene derivatives **46** and **47** (Fig. 5.39) made by co-dissolution with cellulose and followed by reconstitution into a solid form was also found to be suitable for reversible colorimetric detection of NO_x (Rogers et al. 2007)

The solid sensor was directly exposed to NO_x gas for it quick colorimetric determination via calixarene- NO+ complex formation. Subsequently, the process has been rapidly reversed with humidity (Hines et al. 2008)

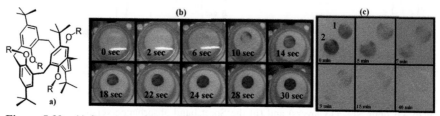

Figure 5.39 (a) Structure of the calix[4]arenes used in the study: **46** (R = $CH_2)_5CH_3$) and Calix **47** (R = $(CH_2)_2CH_3$). (b) Displaying reaction times for Calix **46** films, (c) reversal of 1% (w/w) Calix **46** and Calix **47** overtime in 80% relative humidity. Reprinted with permission from (Hines et al. 2008).

In another report, Rudkevich et al developed a fiber optic optical sensor with 1,3-alternate o-hexylcalix[4]arene **48** immobilized on fine mesh silica-gel which in turn is coated Thin Layer Chromatography (TLC) plate for selective determination of gaseous NO_2 at sub-ppm level by impinging the gaseous NO_2 on the surface of the sensing plate exploiting the formation of a deep purple colored complex on reaction with gaseous NO_2 (Fig. 5.40) (Ohira et al. 2009).

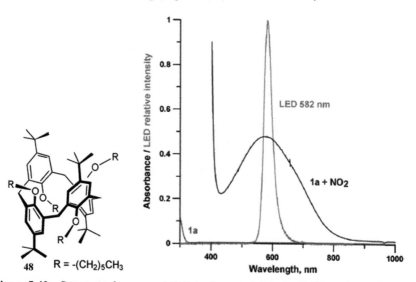

Figure 5.40 Structure of compound **48** (1,3- alternate O-hexyl calix[4]arene) and absorption spectrum obtained with **48** and NO_2 reprinted with permission from (Ohira et al. 2009).

Amir et al. studied the thermal stability of the intensively colored Charge Transfer Complex (CTC) of 1,3-alternate conformer of *tert*-butylcalix[4]arene with four *n*-propoxy substituents (**49**) (Fig. 5.41) with gaseous NO_2/ N_2O_4 and used it for the detection of nitrogen dioxide (NO_2) to assess its thermal reversibility (Khabibullin et al. 2012).

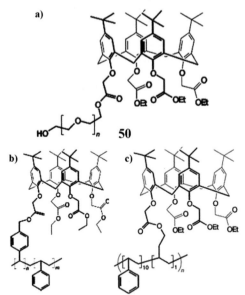

Figure 5.41 Molecular structure of **49** reprinted with permission from (Khabibullin et al. 2012)

5,11,17,23-tetrakis(*tert*-butyl)-25-carboxymethoxy-26,27,28-tris (ethoxycarbonylmethoxy) calix[4]arene (**50**) immobilized on the PEG and Merrifield′s resin polymer support or in the copolymerized state with styrene monomer (Fig. 5.42) was used in the quantitative formation of a stable and brightly colored nitrosonium complex (Gusak et al. 2014) The presence of ester groups in calixarene does not prevent the formation of a stable colored complex with nitrous gases.

Figure 5.42 a) Structure of PEG polymer immobilized with substituted calixarenes **50**, b) copolymerized with styrene, and c) on Merrifield′s resin. Reprinted with permission from (Gusak et al. 2014).

Cucurbituril

It has been reported that gases of intermediate sizes including N_2O and NO, can be absorbed and released repeatedly on decamethylcucurbit[5]uril (Miyahara et al. 2002).

Carbon Dioxide (CO_2)

It is a colorless acidic gas. It is a part of the earth's carbon cycle, and a primary greenhouse gas whose level has significantly increased during the past century due to human activities such as industrialization (in the production of methanol, urea, metal carbonates, metal bicarbonates, carbonated soft drinks, as a propellant, as refrigerant, as an acidity regulator in the food and beverage industry, etc.) and fossil fuel burning in industry, households as well as automobiles which have not only altered the carbon cycle through emission but also by altering natural carbon sinks (National Research Council 2010) Although at low concentration carbon dioxide is regarded as harmless, but is an asphyxiant gas and can cause adverse health effects at large concentration and is responsible for global warming during the past century.

Calixarenes Daschbach et al. studied the free energies of absorption of carbon dioxide by a low-density structure *p*-tert-butylalix[4]arene with loadings up to 2:1 of CO_2:calix[4]arene by molecular dynamics simulations method and found that although loading of CO_2 up to 1:1 is favored, but for the loading of 2:1 of CO_2, the structure is primarily dimeric and the retention of CO_2 is comparatively lower since free energy of inclusion above 1:1 is comparatively less favorable (Daschbach et al. 2009).

Functionalized calix[4]arene **51** (Fig. 5.43) modified carbon nanotube thin films based sensors have been used for very sensitive and selective detection and monitoring of CO_2 which were fabricated using the drop-casting method on a quartz crystal microbalance gold electrode (Mermer et al. 2012).

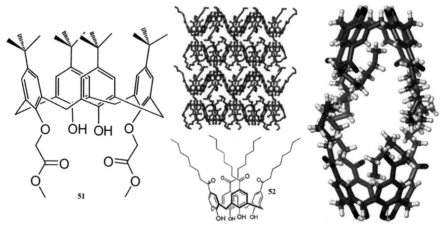

Figure 5.43 Structure of *p*-tert-butylcalix[4] diester **51** reproduced from (Mermer et al. 2012) and para-octanoylcalix[4]arene, **52** (middle) and its crystalline lattice, and capsular structure (right). Reprinted with permission from (Ananchenko et al. 2008).

A two-stage process for the capture of post-combustion CO_2 based on cellulose acetate hybrid membrane incorporating silica functionalized with p-tetranitrocalix[4] arene (dinitrogen selective) and CO_2 selective polyvinylamine cast on porous polysulfone was studied via simulation (Hussain et al. 2015). Crystalline capsular structure of calix[8]arene derivative para-octanoylcalix[4]arene, **52** was reported to be a good adsorbent for carbon dioxide as it permits gas adsorption even in the absence of permanent channels and by exploiting the adsorption characteristics it is possible to separate gases from one another (Ananchenko et al. 2008).

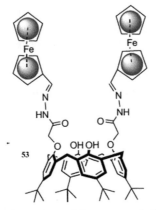

Figure 5.44   Chemical structure of ferrocene-substituted calix[4]arene **53**. Reprinted with permission from (Sayin et al. 2014).

QCM based sensor with ferrocene-substituted calix[4]arene derivative **53** (Fig. 5.44) coated on quartz crystal microbalance gold electrode prepared via the drop-casting method has been reported as a very effective gas sensor for monitoring of both carbon dioxide and carbon monoxide gases due to their complexing ability with the macrocyclic host (Sayin et al. 2014).

Cucurbituril The organic porous material based on cucurbit[6]uril obtained by recrystallization with HCl exhibited a high CO_2 sorption capacity at 298 K, 1 bar (Kim et al. 2010) The high selectivity of CO_2 in comparison to CO was attributed to the high enthalpy of CO_2 adsorption (Fig. 5.45).

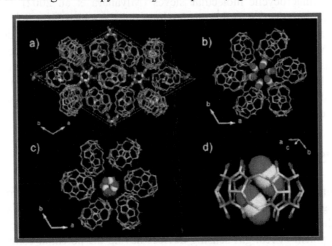

Figure 5.45   (a) X-ray crystal structure of CO_2 adsorbed cucurbit[6]uril, (b) sorption site **A**, (c) sorption site **B**, and (d) sorption site **C** (Kim et al. 2010).

Organic Carboxylic Acid Vapors

Enhancement in the vapor sensing selectivity of Langmuir–Schaefer (LS) films of free-base porphyrin 5,10,15,20-tetrakis[3,4-bis(2-ethylhexyloxy)phenyl]-21H,23H-porphine (EHO), **44** was achieved by the addition of selective barrier layers of 4-tert-Butylcalix[4]arene, 4-tert-butylcalix[6]arene and 4-tert-butylcalix[8] arene embedded in PMMA (Poly(methylmethacrylate)) on top of the porphyrin (Evyapan and Dunbar 2016). The films were deposited onto the surface of glass and silicon substrates using the Langmuir-Schaefer deposition method. These sensors responded by changing color on adsorption of the analyte gas to the sensor surface and were used for differentiating between acetic, butyric and hexanoic acids (Fig. 5.46).

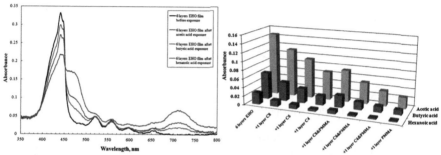

Figure 5.46 The UV absorption spectra of 6 layer EHO, **44** LS film on exposure of 855 ppm acetic acid, 846 ppm butyric acid, and 822 ppm hexanoic acid and the magnitude of the responses EHO, **44** LS film and of different barrier layers coated EHO films against the carboxylic acids. Reprinted with permission from (Evyapan and Dunbar 2016).

Neutral Gaseous Molecules

The report of macrocyclic molecular sieves (Miyahara et al. 2002) and *p-tert*-butylcalix[4]arene crystal polymorph (low density) (Brouwer et al. 2003; Ripmeester et al. 2006) that contains lattice voids formed in van der Waals assemblies (MacGillivray and Atwood. 1999a, b) large enough to accommodate small gaseous guest molecules under the ambient or mild condition of temperature and pressure has attracted researchers and regenerated their interest in gas sorption materials and separation media based on molecular recognition by macrocyclic host molecules such as calixarenes and cucurbiturils.

CONCLUSION

Although significant work has been done in the field of gas sensing involving the use of pre-organized macrocyclic molecular receptors (from a large available pool) to increase the sensitivity and selectivity of the techniques and the concept has led to the development of real gas sensing devices having the potential comparable to that of the other state-of-the-art technologies, but the field has further scope for improvement. Ways to achieve better selectivity and sensitivity lie in the synthetic modularity, chemical purity and diverse complexation properties of the macrocyclic receptor which offer unparalleled opportunities and their true potential can be tapped by tailoring the number and type of interactions between the macrocyclic host and gaseous guest molecule(s). As these interactions are generally weak. These receptors promise complete reversibility of the respective sensors.

REFERENCES

Abbaspour, M. and N. Mansouri. 2005. City hazardous gas monitoring network. J. Loss Prev. Process Ind. 18(4-6): 481–487.

Ananchenko, G.S., I.L. Moudrakovski, A.W. Coleman and J.A. Ripmeester. 2008. A channel-free soft-walled capsular calixarene solid for gas adsorption. Angew. Chem. Int. Ed. 120(30): 5698–5700.

Appelo, C. and D. Postma. 2005. Geochemistry, Groundwater and Pollution. CRC.

Arshak, K., E. Moore, G. Lyons, J. Harris and S. Clifford. 2004. A review of gas sensors employed in electronic nose applications. Sens. Rev. 24: 181–198.

Atwood, J.L., L.J. Barbour and A.J.A.C. Jerga. 2004. A new type of material for the recovery of hydrogen from gas mixtures. Angew. Chem. Int. Ed. 116(22): 3008–3010.

Auge, J., P. Hauptmann, J. Hartmann, S. Rösler and R. Lucklum. 1995. Versatile microcontrolled gas sensor array system using the quartz microbalance principle and pattern recognition methods. Sens. Act. B. 26(1-3): 181–186.

Balakrishnan, K., S. Dey, T. Gupta, R.S. Dhaliwal, M. Brauer and A.J. Cohen, et al. 2019. The impact of air pollution on deaths, disease burden, and life expectancy across the states of India: the Global Burden of Disease Study 2017. Lancet Planet. Health. 3(1): e26–e39.

Bastidas, J., J. Polo and E. Cano. 2000. Substitutional inhibition mechanism of mild steel hydrochloric acid corrosion by hexylamine and dodecylamine. J. Appl. Electrochem. 30(10): 1173–1177.

Becuwe, M., D. Landy, F. Delattre, F. Cazier and S. Fourmentin. 2008. Fluorescent indolizine -b-cyclodextrin derivatives for the detection of volatile organic compounds. Sensors. 8(6): 3689–3705

Bertani, F., N. Riboni, F. Bianchi, G. Brancatelli, E.S. Sterner, R. Pinalli, S. Geremia, T.M. Swager and E. Dalcanale. 2016. Triptycene-roofed quinoxaline cavitands for the supramolecular detection of BTEX in air. Chem. Eur. J. 22(10): 3312–3319.

Bianchi, F., R. Pinalli, F. Ugozzoli, S. Spera, M. Careri and E. Dalcanale. 2003. Cavitands as superior sorbents for benzene detection at trace level. New J. Chem. 27(3): 502–509.

Bianchi, F., M. Mattarozzi, P. Betti, F. Bisceglie, M. Careri, A. Mangia, L. Sidisky, S. Ongarato and E.J.A.c. Dalcanale. 2008. Innovative cavitand-based sol-gel coatings for the environmental monitoring of benzene and chlorobenzenes via solid-phase microextraction. Anal. Chem. 80(16): 6423–6430.

Bianchi, F., A. Bedini, N. Riboni, R. Pinalli, A. Gregori, L. Sidisky, E. Dalcanale and M. Careri. 2014. Cavitand-based solid-phase microextraction coating for the selective detection of nitroaromatic explosives in air and soil. Anal. Chem. 86(21): 10646–10652.

Bolden, A.L., C.F. Kwiatkowski and T. Colborn. 2015. New look at BTEX: are ambient levels a problem? Environ. Sci. Technol. 49(9): 5261–5276.

Bolstad-Johnson, D.M., J.L. Burgess, C.D. Crutchfield, S. Storment, R. Gerkin and J.R. Wilson. 2000. Characterization of firefighter exposures during fire overhaul. Am. Ind. Hyg. Assoc. J. 61(5): 636–641.

Borodkin, G.I. and V.G. Shubin. 2001. Nitrosonium complexes of organic compounds. Structure and reactivity. Russ. Chem. Rev. 70(3): 211–230.

Botta, B., I. D'Acquarica, G.D. Monache, L. Nevola, D. Tullo, F. Ugozzoli and M. Pierini. 2007. Nitrosonium complexes of resorc[4]arenes: spectral, kinetic, and theoretical studies. J. Am. Chem. Soc. 129(36): 11202–11212.

Brittle, S., T. Richardson, J. Hutchinson and C. Hunter. 2008. Comparing zinc and manganese porphyrin LB films as amine vapour sensing materials. Colloid. Surface A. 321(1-3): 29–33.

Brouwer, E.B., G.D. Enright, K.A. Udachin, S. Lang, K.J. Ooms, P.A. Halchuk and J.A. Ripmeester. 2003. The complex relationship between guest-free polymorphic products and desolvation of p-tert-butylcalix[4]arene inclusion compounds. J. Chem. Soc., Chem. Commun. (12): 1416–1417.

Butler, A.R. and R. Nicholson. 2003. Life, Death and Nitric Oxide. Royal Society of Chemistry, Cambridge. 33.

Canakci, M. and J. Van Gerpen. 1999. Biodiesel production viaacid catalysis. Trans ASAE 42(5): 1203.

Cao, Z., T. Li, X.H. Yang, K.M. Wang, H.G. Lin and R.Q. Yu. 1998. Studies on thickness-shear-mode acoustic wave pyridine sensor coated with calix[4]arenes. Chem. J. Chin. Univ.-Chin. 19(6): 882–884.

Chachlani, R. and M. Chattopadhyay 2016. Fiber optics evanescent wave sensor. International IJIRSET. 5(9): 17130-17134.

Chawla, H., N. Pant, S. Kumar, N. Kumar and D.S.C. Black 2010. Calixarene-based material for chemical sensors. pp. 117-181. *In*: G. Korotcenkov [ed.]. Chemical Sensors: Polymers and Other Materials, Vol. 3. Momentum Press, New York.

Chawla, H., S. Kumar, N. Pant, A. Santra, K. Sriniwas, N. Kumar and D.S. Black. 2011. Synthesis and evaluation of deep cavity imidazolyl calix[n]arenes. J. Incl. Phenom. Macrocycl. Chem. 71(1-2): 169–178.

Chawla, H., R. Srivastava, S.N. Sahu, S. Kumar and S. Upreti. 2012a. Synthesis and evaluation of neutral anion receptors based on acylhydrazide-appended calix[4]arenes. Supramol. Chem. 24(9): 672–683.

Chawla, H.M., G. Hundal, S. Kumar and P. Singh. 2012b. Synthesis and evaluation of novel 1, 3-bridged calix[4]arene-crown ethers for selective interaction with Na$^+$/K$^+$ cations. J. Incl. Phenom. Macrocycl. Chem. 72(3-4): 323–330.

Chawla, H.M., S.N. Sahu, R. Shrivastava and S. Kumar. 2012c. Calix[4]arene-based ditopic receptors for simultaneous recognition of fluoride and cobalt (II) ions. Tetrahedron Lett. 53(17): 2244–2247.

Clemendot, S., J. Bourgoin, G. Derost, M. Vandevyver, A. Ruaudel-Teixier and A. Barraud. 1992. Conducting Langmuir-Blodgett films of tetracyanoquinodimethane and ethylenedithio-tetrathiafulvalene derivatives applied to gas sensing: gas diffusion and detection ranges. Thin Solid Films. 210: 430–433.

Clément, P. and E. Llobet. 2020. Carbon nanomaterials functionalized with macrocyclic compounds for sensing vapors of aromatic VOCs. pp. 223–237. *In*: R. Jaaniso and O.K. Tan [eds]. Semiconductor Gas Sensors, 2nd Ed. Woodhead Publishing Series in Electronic and Optical Materials, Elsevier.

Dalcanale, E. and J. Hartmann. 1995. Selective detection of organic compounds by means of cavitand-coated QCM transducers. Sens. Acutator B. 24(1-3): 39–42.

Daschbach, J.L., X. Sun, T.-M. Chang, P.K. Thallapally, B.P. McGrail and L.X. Dang. 2009. Computational studies of load-dependent guest dynamics and free energies of inclusion for CO_2 in low-density p-tert-butylcalix[4]arene at loadings up to 2:1. J. Phys. Chem. A. 113(14): 3369–3374.

Del Valle, E.M. 2004. Cyclodextrins and their uses: a review. Process Biochem. 39(9): 1033–1046.

Dickert, F., U. Bäumler and G.J.S.m. Zwissler 1993. Supramolecular structures and chemical sensing. Synth. Met. 61(1-2): 47–52.

Dickert, F.L., U.P. Bäumler and H. Stathopulos. 1997. Mass-sensitive solvent vapor detection with calix[4]resorcinarenes: tuning sensitivity and predicting sensor effects. Anal. Chem. 69(6): 1000–1005.

Directive 2004/42/CE of the European Parliament and of the Council of 21 April 2004 on the Limitation of Emissions of Volatile Organic Compounds Due to the Use of Organic Solvents in Certain Paints and Varnishes and Vehicle Refinishing Products and Amending Directive 1999/13/EC. Available online:https://eur-lex.europa.eu/legal-content/EN/TXT/?uri=celex%3A32004L0042 (accessed on 19 November 2020).

Directive 2010/75/EU of the European Parliament and of the Council of 24 November 2010 on Industrial Emissions (Integrated Pollution Prevention and Control). Available online: https://eur-lex.europa.eu/LexUriServ/LexUriServ.do?uri=OJ:L:2010:334:0017:0119: en:PDF (accessed on 19 November 2020).

Duan, W., G. Chen, Q. Ye and Q. Chen. 2011. The situation of hazardous chemical accidents in China between 2000 and 2006. J. Hazard. Mater. 186(2-3): 1489–1494.

Duarte, L., S. Nag, M. Castro, E. Zaborova, M. Ménand, M. Sollogoub, V. Bennevault, J.F. Feller and P. Guégan. 2016. Chemical sensors based on new polyamides biobased on (Z) octadec-9-enedioic acid and β-cyclodextrin. Macromol. Chem. Phys. 217(14): 1620–1628.

European Commission. Air Quality Standards. 2015. http://ec.europa.eu/environment/air/quality/standards.htm

El Gowini, M.M. and W.A. Moussa 2009. A reduced three dimensional model for SAW sensors using finite element analysis. Sensors. 9(12): 9945–9964.

Elsom, D.M. 1992. Atmospheric Pollution: A global Problem, Blackwell Oxford.

Erdoğan, M., R. Çapan and F. Davis. 2010. Swelling behaviour of calixarene film exposed to various organic vapours by surface plasmon resonance technique. Sens. Actuators B Chem. 145(1): 66–70.

Evyapan, M. and A.D.F. Dunbar. 2015. Improving the selectivity of a free base tetraphenylporphyrin based gas sensor for NO_2 and carboxylic acid vapors. Sens. Actuators B Chem. 206: 74–83.

Evyapan, M. and A.D.F. Dunbar. 2016. Controlling surface adsorption to enhance the selectivity of porphyrin based gas sensors. Appl. Surf. Sci. 362: 191–201.

Farrukh, S., F.T. Minhas, A. Hussain, S. Memon, M. Bhanger and M. Mujahid. 2014. Preparation, characterization, and applicability of novel calix[4]arene-based cellulose acetate membranes in gas permeation. J. Appl. Polym. Sci. 131(6): 1–9.

Feresenbet, E.B., M. Busi, F. Ugozzoli, E. Dalcanale and D.K. Shenoy. 2004a. Influence of cavity depth on the responses of SPR sensors coated with self-assembled monolayers of cavitands. Sensor Lett. 2(3-4): 186–193.

Feresenbet, E.B., E. Dalcanale, C. Dulcey and D.K. Shenoy. 2004b. Optical sensing of the selective interaction of aromatic vapors with cavitands. Sens. & Act. B. 97(2-3): 211–220.

Francová, A., V. Chrastný, H. Šillerová, M. Vítková, J. Kocourková and M. Komárek. 2017. Evaluating the suitability of different environmental samples for tracing atmospheric pollution in industrial areas. Environ. Pollut. 220: 286–297.

Gambert, R. 2001. Device for the measurement of low partial pressures of nitrogen monoxide in breath by preconcentration on a calixarene layer. Application: DE, DE Patent 10130296: 10130296.

García-Martinez, G., E.A. Bustabad, H. Perrot, C. Gabrielli, B. Bucur, M. Lazerges, D. Rose, L. Rodriguez-Pardo, J. Fariña and C. Compère. 2011. Development of a mass sensitive quartz crystal microbalance (QCM)-based DNA biosensor using a 50 MHz electronic oscillator circuit. Sensors 11(8): 7656–7664.

Gifford, F. and S. Hanna. 1973. Modelling urban air pollution. Atmos. Environ. 7(1): 131–136.

Gokel, G.W., 1991. Crown Ethers and Cryptands. The Royal society of Chemistry, London.

Gokel, G.W., W.M. Leevy and M.E. Weber. 2004. Crown ethers: sensors for ions and molecular scaffolds for materials and biological models. Chem. Rev. 104(5): 2723–2750.

Grate, J.W. and M. Klusty. 1991. Surface acoustic wave vapor sensors based on resonator devices. Anal. Chem. 63(17): 1719–1727.

Grate, J.W., M. Klusty, R.A. McGill, M.H. Abraham, G. Whiting and J. Andonian-Haftvan 1992. The predominant role of swelling-induced modulus changes of the sorbent phase in determining the responses of polymer-coated surface acoustic wave vapor sensors. Anal. Chem. 64(6): 610–624.

Grate, J. 1996. GC Frye in Sensors Update, Vol. 2. Wiley-VCH, Weinheim.

Gubala, V., L.F. Harris, A.J. Ricco, M.X. Tan and D.E. Williams. 2012. Point of care diagnostics: status and future. Anal. Chem. 84(2): 487–515.

Gusak, A., E. Ivanova, P. Prokhorova, G. Rusinov, E. Verbitskiy and Y.Y. Morzherin. 2014. Synthesis and use of polymer-immobilized calix[4]arene derivatives as molecular containers for nitrous gases. Russ. Chem. Bull. 63(6): 1395–1398.

Haick, H., Y.Y. Broza, P. Mochalski, V. Ruzsanyi and A. Amann. 2014. Assessment, origin, and implementation of breath volatile cancer markers. Chem. Soc. Rev. 43(5): 1423–1449.

Hartmann, J., P. Hauptmann, S. Levi and E. Dalcanale. 1996. Chemical sensing with cavitands: influence of cavity shape and dimensions on the detection of solvent vapors. Sens. Actuators B Chem. 35(1-3): 154–157.

Hassan, A., A. Ray, A. Nabok and F. Davis. 2001. Spun films of novel calix[4]resorcinarene derivatives for benzene vapour sensing. Sens. Actuators B Chem. 77(3): 638–641.

Hierlemann, A. and R. Gutierrez-Osuna. 2008. Higher-order chemical sensing. Chem. Rev. 108(2): 563–613.

Hines, J.H., E. Wanigasekara, D.M. Rudkevich and R.D. Rogers. 2008. Calix[4]arenes immobilized in a cellulose-based platform for entrapment and detection of NOx gases. J. Mater. Chem. 18(34): 4050–4055.

Homola, J. 2008. Surface plasmon resonance sensors for detection of chemical and biological species. Chem. Rev. 108(2): 462–493.

Hromadka, J., S. Korposh, M. Partridge, S.W. James, F. Davis, D. Crump and R.P. Tatam. 2017. Volatile organic compounds sensing using optical fibre long period grating with mesoporous nano-scale coating. Sensors 17(2): 205.

Hunt, N.K. and B.J. Mariñas. 1997. Kinetics of *Escherichia coli* inactivation with ozone. Water Res. 31(6): 1355–1362.

Hussain, A., S. Farrukh and F.T. Minhas. 2015. Two-stage membrane system for post-combustion CO_2 capture application. Energy Fuels. 29(10): 6664–6669.

Ignarro, L.J. 2000. Nitric Oxide: Biology and Pathobiology. Academic Press, Cambridge.

Jacobson, M.Z. 2009. Review of solutions to global warming, air pollution, and energy security. Energy Environ. Sci. 2(2): 148–173.

Janata, J. 2010. Principles of Chemical Sensors. Springer Science & Business Media.

Jiang, B.-P., D.-S. Guo and Y. Liu. 2011. Reversible and selective sensing of aniline vapor by perylene-bridged bis (cyclodextrins) assembly. J. Org. Chem. 76(15): 6101–6107.

Johannsmann, D. 2008. Viscoelastic, mechanical, and dielectric measurements on complex samples with the quartz crystal microbalance. Phys. Chem. Chem. Phys. 10(31): 4516–4534.

Jung, J.H., S.J. Lee, J.S. Kim, W.S. Lee, Y. Sakata and T. Kaneda. 2006. α-CD/crown-appended diazophenol for selective sensing of amines. Org. Lett. 8(14): 3009–3012.

Kabashima, K., A. Otsuka and T. Nomura. 2017. Linking air pollution to atopic dermatitis. Nat. Immunol. 18(1): 5–6.

Kalchenko, V., I. Koshets, E. Matsas, O. Kopylov, A. Solovyov, Z. Kazantseva and Y.M. Shirshov. 2002. Calixarene-based QCM sensors array and its response to volatile organic vapours. Mater. Sci. 20(3): 73–88.

Kamata, H., S.-i. Ueno, N. Sato and T. Naito. 2009. Mercury oxidation by hydrochloric acid over TiO_2 supported metal oxide catalysts in coal combustion flue gas. Fuel Process. Technol. 90(7–8): 947–951.

Kang, Y. and D.M. Rudkevich. 2004. Polymer-supported calix[4]arenes for sensing and conversion of NO_2/N_2O_4. Tetrahedron. 60(49): 11219–11225.

Khabibullin, A.A., G.D. Safina, M.A. Ziganshin and V.V. Gorbatchuk. 2012. Thermal analysis of charge-transfer complex formed by nitrogen dioxide and substituted calix[4] arene: characterization of complexation reversibility. J. Therm. Anal. Calorim. 110(3): 1309–1313.

Kim, H., Y. Kim, M. Yoon, S. Lim, S.M. Park, G. Seo and K. Kim. 2010. Highly selective carbon dioxide sorption in an organic molecular porous material. J. Am. Chem. Soc. 132(35): 12200–12202.

Kimura, M., M. Yokokawa, S. Sato, T. Fukawa and T. Mihara. 2011. Volatile organic compound sensing by gold nanoparticles capped with calix[4]arene ligand. Chem. Lett. 40(12): 1402–1404.

King, W.H. 1964. Piezoelectric sorption detector. Anal. Chem. 36(9): 1735–1739.

Kirsch, M., H.-G. Korth, R. Sustmann and H. de Groot. 2002. The pathobiochemistry of nitrogen dioxide. Biol. Chem. 383(3-4): 389–399.

Konig, W., S. Lutz, M. Hagen, R. Krebber, G. Wenz, K. Baldenius, J. Ehlers and H. Dieck. 1989. Cyclodextrins as chiral stationary phases in capillary gas-chromatography. 4. heptakis (2, 3, 6-tri-o-pentyl)-beta-cyclodextrin. HRC J. High Resolut. Chromatogr. 12(1): 35–39.

Konvalina, G. and H. Haick. 2014. Sensors for breath testing: from nanomaterials to comprehensive disease detection. Acc. Chem. Res. 47(1): 66–76.

Kretschmann, E. and H. Raether. 1968. Radiative decay of non-radiative surface plasmons excited by light. Z. Naturforsch. A. 23(12): 2135–2136.

Krupa, A., M. Descamps, J.-F. Willart, B. Strach, E.B. Wyska, R. Jachowicz and F. Danede. 2016. High-energy ball milling as green process to vitrify tadalafil and improve bioavailability. Mol. Pharmaceutics. 13(11): 3891–3902.

Kumar, S., H. Chawla and R.J.T. Varadarajan. 2003. A convenient single step synthesis of p-thiomethylmethylcalixarenes and metal ion extraction studies. Tetrahedron. 59(38): 7481–7484.

Kumar, S., S. Chawla and M.C. Zou. 2017. Calixarenes based materials for gas sensing applications: a review. J. Incl. Phenom. Macrocycl. Chem. 88(3-4): 129–158.

Lai, C.S., G.J. Moody, J.R. Thomas, D.C. Mulligan, J.F. Stoddart and R. Zarzycki. 1988. Piezoelectric quartz crystal detection of benzene vapour using chemically modified cyclodextrins. J. Chem. Soc., Perkin Trans. 2(3): 319–324.

Lamb, J., J. Christensen, S. Izatt, K. Bedke, M. Astin and R. Izatt. 1980. Effects of salt concentration and anion on the rate of carrier-facilitated transport of metal cations through bulk liquid membranes containing crown ethers. J. Am. Chem. Soc. 102(10): 3399–3403.

Lavrik, N., D. De Rossi, Z. Kazantseva, A. Nabok, B. Nesterenko, S. Piletsky, V. Kalchenko, A. Shivaniuk and L. Markovskiy. 1996. Composite polyaniline/calixarene langmuir-blodgett films for gas sensing. Nanotechnology. 7(4): 315.

Laza-Knoerr, A., R. Gref and P. Couvreur. 2010. Cyclodextrins for drug delivery. J. Drug Targeting. 18(9): 645–656.

Leontiev, A.V. and D.M. Rudkevich. 2004. Encapsulation of gases in the solid state. Chem. Commun. (13): 1468–1469.

Lerdau, M.T., J.W. Munger and D.J. Jacob. 2000. The NO_2 flux conundrum. Sens. Actuators, A. 289(5488): 2291–2293.

Li, Y.Y., X.W. He, G.Z. Zhang, J.Q. He, and J.P. Cheng. 2004. Recognition of organic amines and alcohols using quartz crystal microbalance coated with novel amino acid-bearing 1,3-bridged calix[4]arenes. Acta Chim. Sinica. 62(2): 194–198.

Liu, Y., H. Wang, H.-Y. Zhang, L.-H. Wang and Y. Song. 2003. Compactness of linear aggregation controlled by molecular selective binding of bridged bis (β-cyclodextrin)s. Chem. Lett. 32(10): 884–885.

Liu, C.J., J.T. Lin, S.H. Wang and L.G. Lin 2004. Chromogenic calixarene sensors for amine detection. Chem. Sens. 20 (Suppl. B). 362–363.

Liu, C., J. Lin, S. Wang, J. Jiang and L. Lin. 2005. Chromogenic calixarene sensors for amine detection. Sens. Actuator B–Chem. 108(1-2): 521–527.

Liu, Y., K.R. Wang, D.S. Guo and B.P. Jiang. 2009. Supramolecular assembly of perylene bisimide with β-cyclodextrin grafts as a solid-state fluorescence sensor for vapor detection. Adv. Funct. Mater. 19(14): 2230–2235.

Loughran, M. and D. Diamond. 2000. Monitoring of volatile bases in fish sample headspace using an acidochromic dye. Food Chem. 69(1): 97–103.

MacCraith, B. and M.A. McKervey. 1997. Optical sensor for gaseous ammonia with tuneable sensitivity. Analyst 122(8): 803–806.

MacGillivray, L.R. and J.L. Atwood. 1999a. Strukturelle Klassifizierung von sphärischen molekularen Wirten und allgemeine Prinzipien für ihren Entwurf. Angew Chem. 111(8): 1080–1096.

MacGillivray, L.R. and J.L. Atwood. 1999b. Structural classification and general principles for the design of spherical molecular hosts. Angew. Chem. Int. Ed. 38(8): 1018–1033.

Maniscalco, M., C. Vitale, A. Vatrella, A. Molino, A. Bianco and G. Mazzarella. 2016. Fractional exhaled nitric oxide-measuring devices: technology update. Med. Devices (Auckl) 9: 151.

McCarrick, M., S.J. Harris and D. Diamond. 1994. Assessment of a chromogenic calix[4] arene for the rapid colorimetric detection of trimethylamine. J. Mater. Chem. 4(2): 217–221.

Mermer, Ö., S. Okur, F. Sümer, C. Özbek, S. Sayın and M. Yılmaz. 2012. Gas sensing properties of carbon nanotubes modified with calixarene molecules measured by QCM techniques. Acta Phys. Pol. A. 121(1): 240–242.

Meyer, E.A., R.K. Castellano and F. Diederich. 2003. Interactions with aromatic rings in chemical and biological recognition. Angew. Chem. Int. Ed. 42(11): 1210–1250.

Miyahara, Y., K. Abe and T. Inazu. 2002. "Molecular" molecular sieves: lid-free decamethylcucurbit[5]uril absorbs and desorbs gases selectively. Angew. Chem. 114(16): 3146–3149.

Molina, M.J. and L.T. Molina. 2004. Megacities and atmospheric pollution. J. Air Waste Manag. Assoc. 54(6): 644–680.

Moore, L.W., K.N. Sprjnger, J.-X. Shi, X. Yang, B.I. Swanson and D. Li. 1995. Surface acoustic wave chemical microsensors based on covalently bound self-assembled host monolayers. Adv. Mater. 7(8): 729–731.

Munoz, S., T. Nakamoto and T. Moriizumi. 1999. Comparisons between calixarene Langmuir-Blodgett and cast films in odor sensing systems. Sens. Mater. 11(7): 427–435.

NAAQ CPCB, Central poolution Control Board. 2009. Revised National Ambient Air Quality Standards (2009). NAAQS Notification Dated 18th November. (https://scclmines.com/env/DOCS/NAAQS-2009.pdf) accessed 10–01–2021.

Nabok, A., Z. Kazantseva, N. Lavrik and B. Nesterenko. 1995. Nitrogen oxide gas sensor based on tetra-tertbutyl copper phthalocyanine Langmuir-Blodgett films. Int. J. Electron. 78(1): 129–133.

Nag, S., L. Duarte, E. Bertrand, V. Celton, M. Castro, V. Choudhary, P. Guegan and J.-F. Feller. 2014. Ultrasensitive QRS made by supramolecular assembly of functionalized cyclodextrins and graphene for the detection of lung cancer VOC biomarkers. J. Mater. Chem. B. 2(38): 6571–6579.

Nelli, P., E. Dalcanale, G. Faglia, G. Sberveglieri and P. Soncini. 1993. Cavitands as selective materials for QMB sensors for nitrobenzene and other aromatic vapours. Sens. Act. B. 13(1-3): 302–304.

National Reseach Council. 2010. Advancing the Acience of Climate Change. National Research Council, The National Academies Press, Washington, DC, USA.

Nguyen, B.T. and E.V. Anslyn. 2006. Indicator–displacement assays. Coord. Chem. Rev. 250(23-24): 3118–3127.

Nirschl, M., A. Blüher, C. Erler, B. Katzschner, I. Vikholm-Lundin, S. Auer et al. M. Mertig. 2009. Film bulk acoustic resonators for DNA and protein detection and investigation of in vitro bacterial S-layer formation. Sens. Actuat. A. 156(1): 180–184.

Oh, E.H., H.S. Song and T.H. Park. 2011. Recent advances in electronic and bioelectronic noses and their biomedical applications. Enzyme Microb. Technol. 48(6-7): 427–437.

Ohira, S.-I., E. Wanigasekara, D.M. Rudkevich and P.K. Dasgupta. 2009. Sensing parts per million levels of gaseous NO_2 by a optical fiber transducer based on calix[4]arenes. Talanta. 77(5): 1814–1820.

Ohshima, T., K. Sato, H. Terauchi and M. Sato. 1997. Physical and chemical modifications of high-voltage pulse sterilization. J. Electrostat. 42(1-2): 159–166.

Organo, V., G. Zyryanov and D. Rudkevich. 2004. Synthetic nanotubes: new molecular containers for NO_2/N_2O_4 fixation. Abstracts of papers of the American Chemical Society, Amer. Chemical Soc. 1155 16TH ST, NW, Washington, DC 20036 USA.

Organo, V.G., A.V. Leontiev, V. Sgarlata, H.R. Dias and D.M. Rudkevich. 2005. Supramolecular features of calixarene-based synthetic nanotubes. Angew. Chem. 117(20): 3103–3107.

Organo, V.G. and D.M. Rudkevich. 2007. Emerging host–guest chemistry of synthetic nanotubes. Chem. Commun. (38): 3891–3899.

Partridge, M., R. Wong, S.W. James, F. Davis, S.P. Higson and R.P. Tatam. 2014. Long period grating based toluene sensor for use with water contamination. Sens. Actuators. B Chem. 203: 621–625.

Patching, S.G. 2014. Surface plasmon resonance spectroscopy for characterisation of membrane protein–ligand interactions and its potential for drug discovery. Biochim. Biophys. Acta. 1838(1): 43–55.

Pawliszyn, J. 1997. Solid Phase Microextraction: Theory and Practice. John Wiley & Sons.

Pawliszyn, J. 2011. Handbook of Solid Phase Microextraction. Elsevier.

Pfeiffer, S., B. Mayer and B. Hemmens. 1999. Nitric oxide: chemical puzzles posed by a biological messenger. Angew. Chem. Int. Ed. Engl. 38(12): 1714–1731.

Pinalli, R., A. Pedrini and E. Dalcanale. 2018a. Biochemical sensing with macrocyclic receptors. Chem. Soc. Rev. 47(18): 7006–7026.

Pinalli, R., A. Pedrini and E. Dalcanale. 2018b. Environmental gas sensing with cavitands. Chem. Eur. J. 24(5): 1010–1019.

Pirondini, L. and E. Dalcanale. 2007. Molecular recognition at the gas–solid interface: a powerful tool for chemical sensing. Chem. Soc. Rev. 36(5): 695–706.

Pohanka, M. 2017. The piezoelectric biosensors: principles and applications. Int. J. Electrochem. Sci. 12: 496–506.

Ponzoni, A., E. Comini, I. Concina, M. Ferroni, M. Falasconi, E. Gobbi et al. G. Sberveglieri. 2012. Nanostructured metal oxide gas sensors, a survey of applications carried out at sensor lab, Brescia (Italy) in the security and food quality fields. Sensors. 12(12): 17023–17045.

Poplin, J.H., D. Rudkevich, R. Swatloski and R.D. Rogers. 2007. Development of ionic liquid membranes for NOx gas detection and storage utilizing calix[4]arenes. ECS Trans. 3(38): 105.

Ramanathan, V. and Y. Feng. 2009. Air pollution, greenhouse gases and climate change: global and regional perspectives. Atmos. Environ. 43(1): 37–50.

Rathore, R., S.V. Lindeman, K.S. Rao, D. Sun and J.K. Kochi. 2000. Guest penetration deep within the cavity of calix[4]arene hosts: the tight binding of nitric oxide to distal (cofacial) aromatic groups. Angew. Chem. Int. Ed. 39(12): 2123–2127.

Ray, A. 2017. Organic materials for chemical sensing. pp. 1241–1266. *In*: S. Kasap and P. Capper [ed.]. Springer Handbook of Electronic and Photonic Materials. Boston, MA, Springer US.

Riboni, N., J.W. Trzcinski, F. Bianchi, C. Massera, R. Pinalli, L. Sidisky et al. M. Careri. 2016. Conformationally blocked quinoxaline cavitand as solid-phase microextraction coating for the selective detection of BTEX in air. Anal. Chim. Acta. 905: 79–84.

Richardson, T., R. Brook, F. Davis and C. Hunter. 2006. The NO_2 gas sensing properties of calixarene/porphyrin mixed LB films. Colloids Surf. A Physicochem. Eng. Asp. 284: 320–325.

Ripmeester, J., G.D. Enright, C.I. Ratcliffe, K.A. Udachin and I.L. Moudrakovski. 2006. What we have learned from the study of solid *p-tert*-butylcalix[4]arene compounds. Chem. Commun. (48): 4986–4996.

Roales, J., J.M. Pedrosa, P. Castillero, M. Cano and T.H. Richardson. 2011. Optimization of mixed Langmuir–Blodgett films of a water insoluble porphyrin in a calixarene matrix for optical gas sensing. Thin Solid Films. 519(6): 2025–2030.

Rogers, R., J.H. Poplin and D.M. Rudkevich. 2007. New platforms for immobilization of calixarenes for gas-sensing and trapping. Abstracts of papers of The American Chemical Society, Amer. Chemical Soc. 1155 16TH ST, NW, Washington, DC 20036 USA.

Rudkevich, D. 2004. Supramolecular fixation of NOx gases. Abstracts of papers of the American Chemical Society, Amer. Chemical. Soc. 1155 16TH ST, NW, Washington, DC 20036 USA.

Rudkevich, D. 2005. Sensing and fixation of NO_2 by calixarenes. J. Kem. Ind. 54(2): 57–63.

Rudkevich, D.M., Y. Kang, A.V. Leontiev, V.G. Organo and G.V. Zyryanov. 2005. Molecular containers for NO X gases. Supramol. Chem. 17(1-2): 93–99.

Rusanova, T.Y., A. Kalach, S. Rumyantseva, S. Shtykov and I. Ryzhkina. 2009. Determination of volatile organic compounds using piezosensors modified with the Langmuir-Blodgett films of calix[4]resorcinarene. J. Anal. Chem. 64(12): 1270.

Ryvlin, D., O. Dumele, A. Linke, D. Fankhauser, W.B. Schweizer, F. Diederich and S.R. Waldvogel. 2017. Systematic investigation of resorcin[4]arene-based cavitands as affinity materials on quartz crystal microbalances. Chempluschem. 82(3): 493–497.

Saenger, W. 1984. Inclusion compounds. pp. 231-260. *In*: J.L. Atwood, J.E. Davis and D.D. MacNicol [eds]. Inclusion Compounds. Academic Press, London.

Sarkar, T., S. Srinives, A. Rodriquez and A. Mulchandani. 2018. Single-walled carbon nanotube-calixarene based chemiresistor for volatile organic compounds. Electroanalysis. 30(9): 2077–2084.

Satake, K., A. Katayama, H. Ohkoshi, T. Nakahara and T. Takeuchi. 1994. Titania NOx sensors for exhaust monitoring. Sens. Actuators B Chem. 20(2-3): 111–117.

Sayin, S., C. Ozbek, S. Okur and M. Yilmaz. 2014. Preparation of the ferrocene-substituted 1, 3-distal p-tert-butylcalix[4]arene based QCM sensors array and utilization of its gas-sensing affinities. J. Organomet. Chem. 771: 9–13.

Schierbaum, K., A. Gerlach, M. Haug and W. Göpel. 1992. Selective detection of organic molecules with polymers and supramolecular compounds: application of capacitance, quartz microbalance and calorimetric transducers. Sens. Actuators, A. 31(1-3): 130–137.

Schierbaum, K., T. Weiss, E.T. Van Veizen, J. Engbersen, D. Reinhoudt and W. Göpel. 1994. Molecular recognition by self-assembled monolayers of cavitand receptors. Sens. Actuators, A. 265(5177): 1413–1415.

Schmidt, K. and I. Podmore. 2015. Current challenges in volatile organic compounds analysis as potential biomarkers of cancer. J. Biomark. 2015: Article ID 981458.

Schurig, V., D. Schmalzing and M. Schleimer. 1991. Enantiomer separation on immobilized Chirasil-Metal and Chirasil-Dex by gas chromatography and supercritical fluid chromatography. Angew. Chem. Int. Ed. Engl. 30(8): 987–989.

Sharma, A.K., J. Gupta and I. Sharma. 2019. Fiber optic evanescent wave absorption-based sensors: a detailed review of advancements in the last decade (2007-18). Optik. 183: 1008–1025.

Smith, W.H. 2012. Air Pollution and Forests: Interactions between Air Contaminants and Forest Ecosystems. Springer Science & Business Media.

Soncini, P., S. Bonsignore, E. Dalcanale and F. Ugozzoli. 1992. Cavitands as versatile molecular receptors. J. Org. Chem. 57(17): 4608–4612.

Stedman, J.R. 2004. The predicted number of air pollution related deaths in the UK during the August 2003 heatwave. Atmos. Environ. 38(8): 1087–1090.

Subramanian, E., B.M.L. Jeyarani, C. Murugan and D.P. Padiyan. 2016. Crucial role of undoped/doped state of polyaniline-β-cyclodextrin composite materials in determining sensor functionality toward benzene/toluene toxic vapor. J. Chem. Mater. Res. 5(6): 129–134.

Surpateanu, G.G., M. Becuwe, N.C. Lungu, P.I. Dron, S. Fourmentin, D. Landy and G. Surpateanu. 2007. Photochemical behaviour upon the inclusion for some volatile organic compounds in new fluorescent indolizine β-cyclodextrin sensors. J. Photochem. Photobiol. A Chem. 185(2-3): 312–320.

Swanson, B., S. Johnson, J. Shi and X. Yang. 1998. Cyclodextrin-based microsensors for volatile organic compounds. pp. 130–138. *In*: N. Akmal and A.M. Usmani [eds]. Polymers in Sensors: Theory and Practice. ACS Symposium Series, Vol. 690, American Chemical Society.

Tan, W.C., D. Qiu, B.L. Liam, T.P. Ng, S.H. Lee, S.F. van Eeden, Y. D'Yachkova and J.C. Hogg. 2000. The human bone marrow response to acute air pollution caused by forest fires. Am. J. Respir. Crit. Care Med. 161(4): 1213–1217.

Temel, F., E. Ozcelik, A.G. Ture and M. Tabakci. 2017. Sensing abilities of functionalized calix[4]arene coated QCM sensors towards volatile organic compounds in aqueous media. Appl. Surf. Sci. 412: 238–251.

Timmer, B., W. Olthuis and A. vaan den Berg. 2005. Ammonia sensors and their applications—a review. Sens. Actuators, B. 107: 666.

Topliss, S.M., S.W. James, F. Davis, S.P. Higson and R.P. Tatam. 2010. Optical fibre long period grating based selective vapour sensing of volatile organic compounds. Sens. Actuators, B. 143(2): 629–634.

United States Environmental Protection Agency. Drinking water contaminants. 2009. https://www.epa.gov/ground-water-and-drinking-water/national-primary-drinking-water-regulations (accessed 11-01-2021).

van Velzen, E.T., J.F. Engbersen and D.N. Reinhoudt. 1995. Synthesis of self-assembling resorcin[4]arene tetrasulfide adsorbates. Synthesis. 1995(08): 989–997.

Vincenti, M., E. Dalcanale, P. Soncini and G. Guglielmetti. 1990. Host-guest complexation in the gas phase by desorption chemical ionization mass spectrometry. J. Am. Chem. Soc. 112(1): 445–447.

Vincenti, M. and E. Dalcanale. 1995. Host–guest complexation in the gas phase. Investigation of the mechanism of interaction between cavitands and neutral guest molecules. J. Chem. Soc., Perkin Trans. 2(6): 1069–1076.

Wholtjen, H. and R. Dessy. 1979. Surface acoustic wave probe for chemical analysis. Anal. Chem. 51: 1458–1464.

Wilson, A. 2013. Diverse applications of electronic-nose technologies in agriculture and forestry. Sensors. 13(2): 2295–2348.

Yang, H., B. Yuan, X. Zhang and O.A. Scherman. 2014. Supramolecular chemistry at interfaces: host-guest interactions for fabricating multifunctional biointerfaces. Acc. Chem. Res. 47(7): 2106–2115.

Yuan-Yuan, L., Y. Hong-Zong, H. Xi-Wen, C. Lang-Xing, Z. Guo-Zhu and H. Jia-Qi. 2005. QCM coated with self-assembled cystine-bearing 1, 3-bridged calix[4]arenes for recognizing gas-phase butylamines. Chin. J. Chem. 23(5): 571–575.

Zimmermann, B., C. Hahnefeld and F.W. Herberg. 2002. Applications of biomolecular interaction analysis in drug development. Targets. 1(2): 66–73.

Zou, R.F., Z. Cao, J.L. Zeng, Y.L. Dai and L.X. Sun 2011. Characteristics and Mechanism for Host–Guest Recognition of Isopropanol Vapor Based on Calixarene Supramolecules. Advanced Materials Research, Trans. Tech. Publ.

Zyryanov, G.V., Y. Kang, S.P. Stampp and D.M. Rudkevich. 2002. Supramolecular fixation of NO_2 with calix[4]arenes. Chem. Commun. (23): 2792–2793.

Zyryanov, G.V., Y. Kang and D.M. Rudkevich. 2003. Sensing and fixation of NO_2/N_2O_4 by calix[4]arenes. J. Am. Chem. Soc. 125(10): 2997–3007.

pH-Responsive Pseudorotaxanes and Rotaxanes

INTRODUCTION

Supramolecular chemistry (Lehn 2007) dealing with a macrocyclic host and a guest has played and is playing a pivotal role in the development of chemistry (Dietrich et al. 1969a, b; Kyba et al. 1977; Shinkai and Feringa 2001; Lehn 2002; Gil and Hudson 2004; Kay et al. 2007; Lehn 2007; Mendes 2008; Lehn 2009; Leung et al. 2009; Browne and Feringa 2010; Feringa and Browne 2011; Lehn 2013; Mura et al. 2013; Zhuang et al. 2013; Blanco et al. 2015; Bleger and Hecht 2015; Coudert 2015; Fihey et al. 2015; Frisch and Besenius 2015; Lehn 2015; Marti-Centelles et al. 2015; Lutz et al. 2016; Lehn 2017; Liu et al. 2017; Murray et al. 2017). There is a well-established synergy between host-guest and supramolecular chemistry, inclination in the field (especially mechanically interlocked molecules) is not only on non-responsive systems but also on the development of stimuli-responsive host-guest systems and their practical applications, (Shinkai and Feringa 2001; Leung et al. 2009; Zhuang et al. 2013; Frisch and Besenius 2015; Marti-Centelles et al. 2015). Multi-responsive, Mechanically Interlocked Molecules (MIMs), exhibit a change in size and physicochemical properties in response to external stimuli (Stuart et al. 2010; Sanchez et al. 2011; Torres et al. 2014; Karimi 2016).

MIMs switchable assemblies (for more details refer to Chapter 9) such as, rotaxanes, pseudorotaxanes and catenanes are designed to be capable of mechanical motion induced by an external stimulus such as electrochemical (Bruns et al. 2014a,b; Chen et al. 2016), photochemical (Gokel et al. 2004; Baroncini et al. 2012; Ragazzon et al. 2015; Qu et al. 2015; Baroncini et al. 2018), thermal (Fujita et al. 1996), chemical (Mandl et al. 2004), solvent (Mateo-Alonso et al. 2006; Jena and Murugan 2013), metal–ion (Crowley et al. 2010), and pH (Coutrot et al. 2008b) which results in a change in the system's co-conformation. Among various other external stimuli, pH stimuli are one of the most efficient and commonly used methods for controlling the motions by acids and bases (Leung et al. 2009; Guang-Huan 2015). On applying external stimuli, the noncovalent interactions between the subunits of the assembly can be interrupted and formed reversibly, leading to mechanical motions. Chemical interactions such as hydrogen bonding, metal-ligand interactions, and C–H...π and π–π stacking interactions which exists between subcomponents of switchable assembly decide the type of stimulus that will induce mechanical motion (Schwarz et al. 2004) and the type of interaction between subunits in a switchable assembly decides the mechanism of switching. Consequently, the type of interactions between subunits also regulates the potential application(s) of the supramolecular assembly.

pH-responsive, switchable MIMs systems have two evident features: (i) subcomponents, host or guest (containing recognition site) typically contain reactive donor atoms (e.g., oxygen and/or nitrogen), which can be easily protonated and deprotonated with acids and bases; (ii) non-covalent interactions like hydrogen bonding, electrostatic interactions and hydrophobic effects are very sensitive to the variation of pH. Therefore, acid/base has been one of the most useful stimuli for controllable motions in MIMs. Switchable MIMAs, responsive to changes in pH not only requires the addition of an appropriate acid or base in order to protonate/deprotonate one or more of the binding sites in the system to obtain a specific co-conformation but also requires to sustain a co-conformation for which, the pH of the environment of the switchable assembly must be maintained. Therefore, these types of assemblies are relevant for applications in which maintaining an environmental pH is practically possible.

Macrocyclic receptors that can have non-covalent interacts with a guest, can respond to acid/base, and are water-soluble (according to *Lowry–Brønsted theory*, pH is an aqueous property) are discussed below:

Pillar[n]arenes, or simply pill[n]arenes (Ogoshi et al. 2008; Cao et al. 2009; Han et al. 2010; Li et al. 2010; Ogoshi et al. 2010; Hu et al. 2011; Li et al. 2011; Ma et al. 2011; Strutt et al. 2011; Zhang et al. 2011a, b; Cragg and Sharma 2012; Duan 2012; Guan et al. 2012; Han et al. 2012; Nierengarten et al. 2012; Ogoshi 2012; Ogoshi et al. 2012a, b, c; Strutt et al. 2012; Yao et al. 2012; Yu et al. 2012a, b, c, d; Xue et al. 2012). (PA[n]s), are a relatively new class of macrocyclic hosts discovered by Ogoshi et al. (Ogoshi et al. 2008) have gained growing attention due to their intrinsic unique rigid and symmetrical pillar-shaped architecture, tunable cavity size, easy modification and superior host-guest properties. Its structure is composed of hydroquinone or dialkoxybenzene units linked by methylene bridges at para-positions forming a unique rigid pillar-like architecture (Fig. 6.1). PA[n]s (n = 5–15) are symmetrical and rigid, easy to mono-/di-/per-functionalize (Strutt et al. 2014) and their derivatives are appropriately soluble in aqueous or organic solvents, where they display a very rich host-guest chemistry (Xue et al. 2012; Ogoshi et al. 2016). Due to the hydroquinone rings, native PA[n]s display an Electron-Donor (ED) core and ionophoric rims, which make these macrocycles appropriate hosts for Electron Acceptor (EA) organic dications, such as MVs, in organic media. A series of water-soluble pillararenes have been synthesized and demonstrated to act as scaffolding hosts to various guests (Ogoshi et al. 2012a, b; Hu et al. 2016; Yakimova et al. 2016). Among these water-soluble pillararenes, the pH-responsive ones have been reported in the construction of plenty of supramolecular systems (Yu et al. 2012b, d; Cao et al. 2014; Hu et al. 2016; Xiao et al. 2019). Furthermore, PA[n]s have as well the ability to complex a wide range of neutral organic guests by dipole-dipole, $\pi...\pi$, C–H...π, and hydrogen bonding interactions, as exemplified by the complexation ability of PA[5] with aliphatic nitriles, alcohols, esters, aldehydes, and ketones or haloalkanes (Wang and Ping et al. 2016).

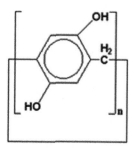

Figure 6.1   Chemical structure of pillar[n]arene.

Cucurbiturils (Lee et al. 2003; Isaacs 2014) (CB[n]s or Q[n], n = 5–11) are a family of pumpkin-shaped macrocycles prepared by condensation of n glycoluril units joined together by 2n methylene bridges, producing hollow molecules with inner hydrophobic cavities accessible through two identical carbonyl-laced portals (Fig. 6.2). Due to these structural features, CB[n]s display rich host-guest chemistry that has been extensively reviewed (Ko et al. 2007; Hu et al. 2016). Regarding the stimuli-responsiveness of CB[n]s, they can be considered as intrinsic pH-based MSs. For instance CB[7], with pKa = 2.2, reversibly forms

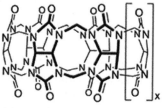

CB5: x = 0 **CB6:** x =1 **CB7:** x = 2
CB8: x = 3 **CB10:** x = 5

Figure 6.2   Chemical structures of the CBn macrocycles reprinted with permission from (Sinn and Biedermann 2018).

hydrogels in acidic aqueous media (Hwang et al. 2007). They are well known for their extraordinary affinity towards alkane diammoniums in water driven by the hydrophobic effect and ion-dipole interactions between the ammoniums and the polar carbonyl groups at the portals of CB[n]. Accordingly, the host-guest binding between CBs and ammoniums is very pH-dependent, which paves the way for constructing pH-controllable molecular motions. In this respect, pioneering works of pH-driven shuttling were performed in 1990 by Mock (Mock and Pierpont 1990) and Kim (Im Jun et al. 2000).

Crown ethers and their derivatives have been continuosly popular for the fabrication of pH-responsive MIMs. They are of versatile binding ability towards an enormous variety of organic guests, especially dialkylammonium ions, based on hydrogen bond interactions and electrostatic interactions. These interactions can be destroyed simply by removing the hydrogen linked to the nitrogen in the guest molecules with an appropriate base. Hence, lots of pH-responsive crown ether-based MIMs have been reported in the last few decades. Among the crown ethers and their derivatives, dibenzo-24-crown-8 (DB24C8; Fig. 6.3) is the most commonly designed and synthesized pH-responsive molecular switch and machine. Different recognition sites for DB24C8 have been explored to construct pH-switchable rotaxanes. DB24C8 is a well-known π-donor macrocycle present in many MIMs containing complementary acceptors, such as 4,4-bipyridinium, (Garaudée et al. 2005; Braunschweig et al. 2006) 1,2-bis(pyridinium)ethane, (Loeb and Wisner 1998b; Mercer and Loeb 2011; Zhu et al. 2016) or benz-imidazolium (Zhu et al. 2012; Farahani and Loeb 2016). DB24C8 was also widely used in the synthesis of rotaxanes incorporating dialkylammonium ions hosted in the crown ether cavity by the N^+–$H\cdots O$ interactions, which are manifested by short $H\cdots O$ contacts (Thibeault and Morin 2010). Dialkylammomium is usually employed as a reversible control unit and another permanent cation as a second recognition site. This strategy has generated various examples owing to the discovery of novel recognition stations for DB24C8. Early research was founded by Stoddart that [2] rotaxane containing a dibenzylammonium unit (R_2NH^{2+}) and bipyridinium unit ($BIPY^{2+}$) can be switched by acid/base (Martínez-Díaz et al. 1997; Ashton 1998a).

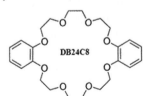

Figure 6.3 Chemical structure of Dibenzo-24-crown-8.

Cyclodextrins: With more than one century of history since their discovery by Villiers, (Crini 2014) naturally occurring cyclodextrins (CDs) are probably the most widely used family of macrocyclic hosts (Sollogoub 2013). Structurally, the water-soluble, non-toxic and commercially available α-, β-, and γ-CDs are bucket-shaped cyclic oligosaccharides consisting of six, seven or eight glucopyranose units attached by 1,4-linkages. Due to the oligomerized glucose units, CDs exhibit hydroxyl groups on the larger primary face and hydroxymethyl functions on the secondary, making these derivatives quite easy to functionalize (Croft and Bartsch 1983; Khan et al. 1998). Due to these features, an extensive number of CD derivatives have been reported to have a wide range of binding abilities. For instance, native CDs can complex a variety of organic guests in an aqueous media mostly because of the hydrophobic effect. Due to the different cavity diameters of CDs (Fig. 6.4), α-CD can accommodate linear alkanes and monocyclic aromatic molecules as appropriate guests, β-CD binds, in turn, bulky hydrocarbons, such as adamantane or polyaromatic compounds, such as naphthalene and anthracene derivatives and γ-CD can incorporate even larger guests, e.g., two g-CD molecules can sandwich a fullerene or two aromatic guests can be included in a single cavity.

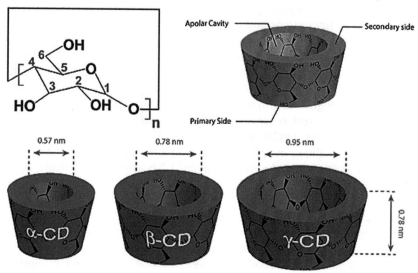

Figure 6.4   Top: Functional structural scheme of α-CD (n = 6), β-CD (n = 7) and γ-CD (n = 8). Bottom: Geometric dimensions of cyclodextrins with permission from (Crini 2014).

Benzylic amide macrocycles: Amide-based macrocycles also bind neutral guests through hydrogen bonding, which have been effectively used for pH-triggered molecular shuttles. The first example of pH-switchable shuttling involved anion recognition of a benzylic amide macrocycle by Leigh's group (Keaveney and Leigh 2004). [2]Rotaxane with a benzylic amide ring threaded onto an axle incorporating two potential hydrogen bonding stations (Fig. 6.5).

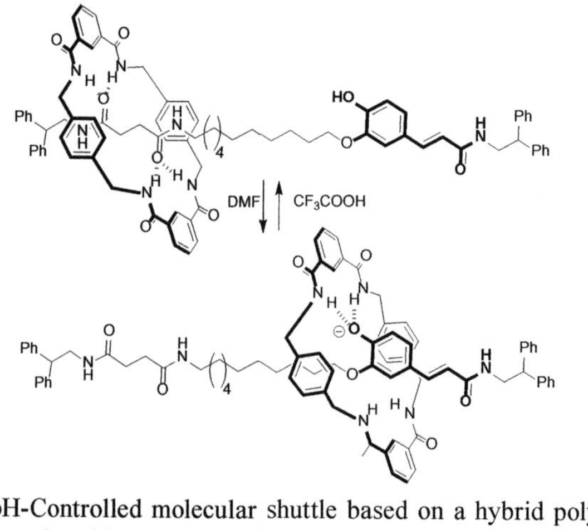

Figure 6.5   pH-Controlled molecular shuttle based on a hybrid polyether/amide macrocycle with permission from (Keaveney and Leigh 2004).

"...When we get to the very, very small world..............we have a lot of new things that would happen that represent completely new opportunities for design. ....as we go down and friddle around with the atoms down there, we are working with different laws, and we can expect to do different things. We can manufacture in different ways...............At the atomic level, we have new kinds of forces and new kinds of possibilities, new kinds of effects. The problems of manufacture and reproduction of materials will be quite different............., inspired by the biological phenomena in which chemical forces are used in repetitious fashion to produce all kinds of weird effects (one of which is the author)...."

—Richard P. Feynman "Plenty of Room at the Bottom" (Dec. 1959)

Rotaxane and pseudo rotaxane assemblies in which mechanical motion is induced by changes in pH have been reported in this chapter. Subsequently the topology along with the recent examples of pH-responsive pseudorotaxanes and rotaxanes are described.

PSEUDOROTAXANE

Pseudorotaxanes are of interest to scientists not only due to their topological importance but also due to their applications in the development of artificial molecular machines, which can respond to suitable external stimuli. Topologically pseudorotaxane (Sun et al. 2017) comprises a thread-like linear species, axle (acts as an internal guest; without bulky groups/stoppers at the ends) inserted through the cavity of a macrocycle (Fig. 6.6). The threading of a linear component through the cavity of the macrocycle is due to noncovalent bonding interactions to create a stable species in solution. The association/dissociation (threading/dethreading) [Fig. 6.6(a)] process between the axle and macrocycle components can be readily and reversibly attained by modulating the attractive/repulsive forces between the components. In addition to the main association/dissociation mobility, shuttling of the wheel along the (bistable) axle and rotation (Fig. 6.6(b,c)) are another type of mobilities for pseudorotaxanes. The threading/dethreading, shuttling or rotation of macrocycle along axle of a pseudorotaxane can be induced either by the change of solvent and/or temperature or by chemical stimuli, such as anion control, (Huang et al. 2003; Zhu et al. 2010; Gong et al. 2011; Rambo et al. 2012) pH control, (Han et al. 2006; Zhang et al. 2009; Chen et al. 2010; Zhang et al. 2011; Li et al. 2012; Ji et al. 2013) cation control, (Gibson et al. 2007) or addition of other competitive hosts/guest molecules (Wang et al. 2003; Ding et al. 2011). Based on the dynamic property of association and dissociation, pseudorotaxanes can be endowed with the functions of control/release and lock/key. It allows them to be applied in the construction of various molecular devices such as sensors, (Yamauchi et al. 2006) switches, (Liu et al. 2005; Share et al. 2010) logic gates (Credi et al. 1997), and nanovalves (Saha et al. 2007), developing drug delivery and release system, (Xu et al. 2017; Li et al. 2020) molecular machines (Jeong et al. 2003) and also for pseudorotaxane-type ligands with tunable structural dynamics offer an opportunity in the exploration of new actinide hybrid materials (Li et al. 2019). Thus different kinds of water-soluble macrocyclic hosts like pillararenes, crown ethers, etc., have been employed to construct pH-responsive pseudorotaxane (Xue et al. 2015; Yang et al. 2019; Yu et al. 2019).

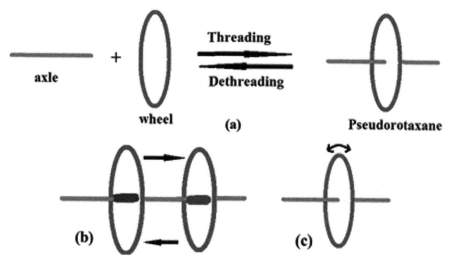

Figure 6.6 Cartoon representation of different types of mobility in pseudorotaxane (a) threading/dethreading (b) shuttling of the wheel along with the axle (c) rotation.

Here the focus on the recent examples of acid/base-controlled translocation, association and dissociation in pseudorotaxanes as well as different recognition motifs based on various macrocyclic hosts and their operating mechanism is discussed.

Cucurbiturils CB[n] Based pH Responsive Pseudorotaxanes

Lin et al. (Lin et al. 2020) took the advantage of different structural features and binding properties of cucurbit[n]uril and their derivatives to construct pseudorotaxanes. The group demonstrated two novel heterowheel [4]pseudorotaxanes (Fig. 6.7) comprising of cucurbit[7]uril (Q[7]) and symmetrical-tetramethyl-cucurbit[6]uril (TMeQ[6]) constructed via multi recognition mechanism, in which cucurbit[7]uril can rotate freely around the horizontal axis, while TMeQ[6] cannot as benzimidazolium group was complemented to the ellipsoidal cavity of the TMeQ[6]. It was observed that the formation and dissociation of [4]pseudorotaxanes could be controlled reversibly by regulating acid/base. At pH 10 [4]pseudorotaxanes $1^{4+} \subset Q[7] \cdot TMeQ[6]2$ and $2^{4+} \subset Q[7] \cdot TMeQ[6]2$ precipitated which indicates that the deprotonated axles 1^{2+} and 2^{2+} were insoluble in water, which lead to the disassociation of the [4]pseudorotaxanes into free wheels and axle while at pH 2 precipitates disappeared indicating the formation of [4]pseudorotaxanes $1^{4+} \subset Q[7] \cdot TMeQ[6]2$ and $2^{4+} \subset Q[7] \cdot TMeQ[6]2$.

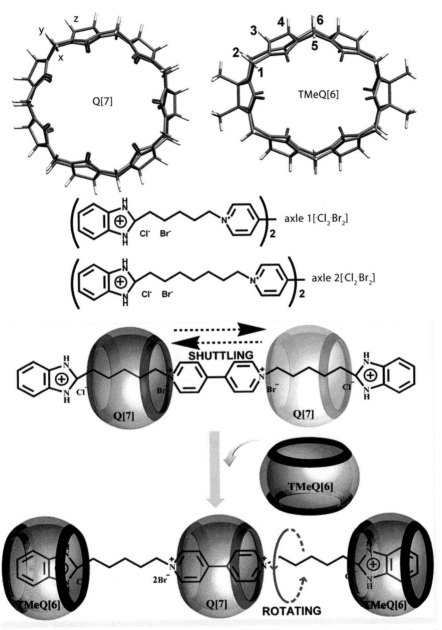

Figure 6.7 Molecular structures of Q[n], TMeQ[6] and the axles 1[·Cl$_2$·Br$_2$] and 2[·Cl$_2$·Br$_2$], along with [4]pseudorotaxanes 1^{4+}⊂Q[7]·TMeQ[6]$_2$ with permission from (Lin et al. 2020).

Li et al. (Li et al. 2020b) have developed pH-sensitive pseudorotaxanes to mimic the catalytic properties of Glutathione Peroxidase (GPx) which was based on the organoselenium compounds **3** (a diselenide catalytic center) and **4** (a selenide catalytic center) (Fig. 6.8a) modified with two primary amine groups, as GPx models and cucurbit[6]uril. It was shown that the GPx mimics form 1:1 host-guest pseudorotaxane complexes and the mimics were not able to show GPx activity because the active sites were encapsulated into the hydrophobic cavity of cucurbit[6] uril following the di-protonation of the amine groups. (Fig. 6.8b). But the increase of pH of the solution resulted in an increase of GPx activity, the reason for this was indorsed to the gradual deprotonation of primary amine groups and decrease in the binding affinity of cucurbit[6]uril with **compounds 3 and 4**. thus, it was clinched that the catalytic activity of the GPx model can be switched on/off by changing the pH. It was observed that reversible pH control switching can be attained at pH between 7 and 9 for **compound 3** or between 7 and 10 for **compound 4**.

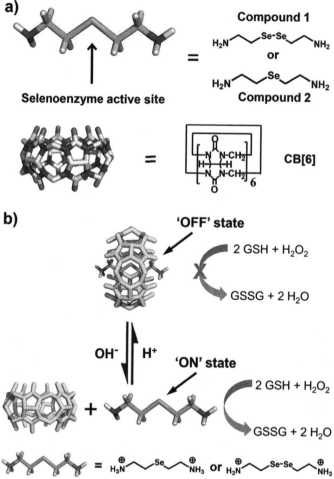

Figure 6.8   (a) Structures of organoselenium compounds 3, 4, and CB[6]; (b) schematic representation of the pH-sensitive smart GPx mimic based on the formation and dissociation of pH-responsive pseudorotaxanes formed by CB[6] and compounds 1 and 2 with permission from (Li et al. 2020b).

Seco et al. (Seco et al. 2020) synthesized pH-, light- and redox-responsive flavylium-bipyridinium molecular dyad as an axle to develop a pseudorotaxane with the macrocycle cucurbit[7]uril in an aqueous solution. Various spectroscopic and electrochemical techniques were devised to demonstrate the formation of stable pseudorotaxane under acidic and neutral conditions. It is noteworthy that the flavylium-bipyridinium tricationic (**5AH⁺**) dyad (axle) was only stable in highly acidic media, while it underwent a reversible hydration reaction at slightly acidic or neutral pH to give a *trans*-chalcone-bipyridinium dication(**6Ct**). ¹H NMR experiments showed that in this last species (*trans*-chalcone-bipyridinium dication) the cucurbit[7]uril was bonded to the bipyridinium unit of *trans*-chalcone-bipyridinium dication while in the tricationic species the macrocycle was positioned between the flavylium and the bipyridinium moieties. Displacement of the cucurbit[7]uril wheel from the peripheric bipyridinium station to the center of the axle allowed the control of the shuttling movement (Fig. 6.9) using pH as well as light stimuli to trigger the interconversion between these two species.

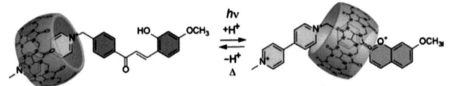

Figure 6.9 Structure for the 6Ct-CB[7] pseudorotaxane (left). This species can be converted into 5AH⁺-CB[7] pseudorotaxane (right) by pH and/or light stimuli inducing translocation of the CB[7] wheel with permission from (Seco et al. 2020).

Zubillaga et al. (Zubillaga et al. 2018) devised pH-gated photoresponsive ring-shuttling pseudorotaxanes using **7** and **8** (axle) to form inclusion complexes with cucurbit[7]uril (Fig. 6.10) which demonstrated water-soluble properties along with long-lived states. UV-Vis titration of **7** with CB[7] showed the formation of 1:1 of the host-guest complex with a binding constant of $K_{11} = 9.8 \times 10^5$ M⁻¹ (in H_2O at 25°C) and 2:1 host-guest complex ($K_{21} = 2.0 \times 10^2$ M⁻¹ in D_2O at 25°C) in the presence of an excess of the host which was proposed due to hydrophobic effects that result in the inclusion of the phenol group inside the cavity of a second cucurbit[7]uril molecule. Axle **7** on irradiation photochemically generated flavylium species **8** whose interaction with cucurbit[7]uril was investigated by UV-Vis and fluorescence titrations at pH = 2 in order to avoid the thermal back reaction to form **7** during the titration. Formation of 1:1 **8**:cucurbit[7]uril host-guest complex (cucurbit[7]uril molecule sitting around the phenyl moiety) and on the excess of cucurbit[7]uril 2:1 host-guest complex (first cucurbit[7]uril molecule pushed to the pyrylium moiety in order to allow the space for the second cucurbit[7]uril on the ammonium group) was observed with binding constants $K_{11} = 1 \times 10^7$ M⁻¹ and $K_{21} = 1.2 \times 10^4$ M⁻¹ (in H_2O, pH = 2, at 25°C), translocation was observed in an acidic medium while in neutral or basic conditions no translocation was observed. In the dark, the system was recognized to be reversible but a more acidic pH values it can be made irreversible (locked) and unlocked by neutralization.

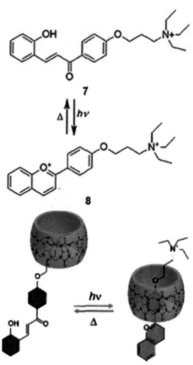

Figure 6.10 Structures of *trans*-chalcone **7**, flavylium cation **8**, and CB[7] molecules and Illustration showing the reversible displacement of the cucurbit[7]uril, CB[7] wheel after photoinduced interconversion of **1**:CB7 into **2**:CB7 complexes and thermal back-reaction with permission from (Zubillaga et al. 2018).

Diniz et al. (Diniz et al. 2015) synthesized a multistate molecular dyad containing flavylium and viologen units incorporated into the molecular axle to devise cucurbit[7]uril-based pseudorotaxanes. The flavylium cation was observed to be stable at extreme acidic conditions. As the pH increases above pH≈5 the hydration of the flavylium resulted in the formation of the hemiketal trailed by ring-opening tautomerization to give the *cis*-chalcone subsequently, *cis*-chalcone isomerizes to *trans*-chalcone. The flavylium cation and the *trans*-chalcone species were identified as thermodynamically stable and/or metastable species being the hemiketal and the *cis*-chalcone kinetic intermediates. All stable species present in the network were shown to form 1:1 and 2:1 inclusion complexes with cucurbit[7] uril 1:1 complexes showed higher values of binding constant as compared to 2:1. The 1:1 complexes behave as pH-responsive bistable pseudorotaxanes. The pH-dependent kinetics studies of the chalcone/flavylium molecular switching abilities showed that the rate of the co-conformational switch can be tuned from hours to seconds depending on the pH (Fig. 6.11). Kinetic studies showed the interconversion of flavylium cation into deprotonated *trans*-chalcone at pH≈12 the flavylium cation in a few minutes and under these conditions cucurbit[7] uril was located around the viologen unit. Decreasing the value of pH to ~1, resulted in the regeneration of the flavylium cation in seconds and translocation of the macrocycle to the middle of the axis. When the pH was decreased to 6 the deprotonated *trans*-chalcone was neutralized to metastable species that evolved thermodynamically stable flavylium cation in approximately 20 hours. The group proposed that spatiotemporal control of the molecular organization in pseudorotaxane could be achieved by taking the advantage of the pH-dependent kinetics of the *trans*-chalcone/flavylium interconversion.

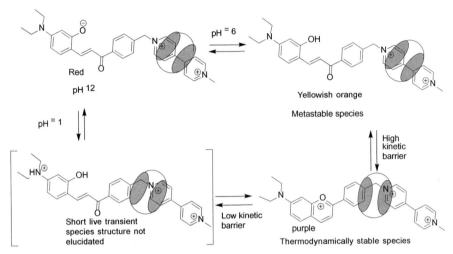

Figure 6.11 pH-dependent kinetics of wheel translocation in flavylium-based pseudorotaxanes with permission from (Diniz et al. 2015).

Shi et al (Shi et al. 2019) designed and synthesized cucurbit[7]uril based stimuli-responsive pseudorotaxane with linear rigid axle molecule in which two identical *p*-toluic acid units were located at its two ends and one 4,4′-bipyridinium (viologen nucleus) group at the middle of the axle 9^{2+}. ^{1}H NMR titration aided in establishing the sequential formation of a molecular shuttle and a [3]pseudorotaxane. Firstly, the axle 9^{2+} interacts with wheel cucurbit[7]uril to form a molecular shuttle with the cucurbit[7]uril oscillating back and forth along with the 9^{2+} (Fig. 6.12a); secondly, the addition of 2.0 equivalent of cucurbit[7]uril into the solution of the molecular shuttle leads to the formation of a [3]pseudorotaxane, in which two cucurbit[7]uril wheels reside over two sidearms of the 9^{2+} (Fig. 6.12b) which was further confirmed by MALDI-TOF mass spectra. At pH 9.0, terminal COOH groups of the axle 9^{2+} were deprotonated to their anionic carboxylate forms. Release of one of the cucurbit[7]uril wheel from the assembly into the solution takes place due to the repulsive electrostatic interactions between the negative charge of the carboxylate of the axle 1^{2+} and that of the carbonyl oxygen rims of the cucurbit[7]uril portals and was the driving factor for the other cucurbit[7]uril wheel movement to the viologen nucleus station, which leads to the formation of the [2]pseudorotaxane (Fig. 6.12c). On changing pH to 2 [2]pseudorotaxane structure was restored. So, it was established that the interconversion between [3]pseudorotaxane and [2]pseudorotaxane can be controlled by changing the pH of the solution.

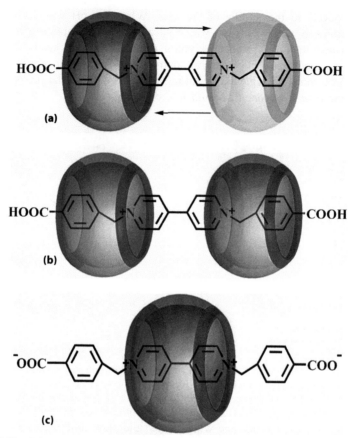

Figure 6.12   Schematic representation of (a) the molecular shuttle of 9^{2+} with cucurbit[7] uril (b) [3]pseudorotaxane of 9^{2+} with 2.0 equiv. of cucurbit[7]uril (c) [2]pseudorotaxane of 9^{2+} with cucurbit[7]uril with permission from (Shi et al. 2019).

He et al. (He et al. 2018) designed [2]/[3]pseudorotaxane (SD⊂CB[7]) based on Stilbene Dye (SD) with two recognition sites (stilbene site and a hexanoic acid site) for CB[7]. As shown in Fig. 6.13, the SD could be included by CB[7] with 1:1 and 1:2 stoichiometry to form [2] and [3]pseudorotaxanes respectively. Fluorescence spectra and ^{1}H NMR spectra confirmed the formation of 1:1 host-guest complex, [2]pseudorotaxane (inclusion constant of SD⊂CB[7] was estimated to be 2.93×10^4 L/mol) in neutral or weak alkaline conditions, with only one CB[7] molecule being encapsulated inside the stilbene site of SD. When the pH changes to the acidic condition, the carboxylate gets protonated, the hydrogen bonding between the carboxyl group and the carbonyl of CB[7] makes another CB[7] encapsulated inside the hexanoic acid site to form [3]pseudorotaxane of the 1:2 (SD⊂CB[7]) host-guest complex (shown in Fig. 6.13), in this way, the pH-controlled [2] or [3]pseudorotaxane can be achieved. The pH-controlled alternation between [2] and [3]pseudorotaxane based on stilbene dye SD⊂CB[7] is similar to 'on' and "off" in molecular switches.

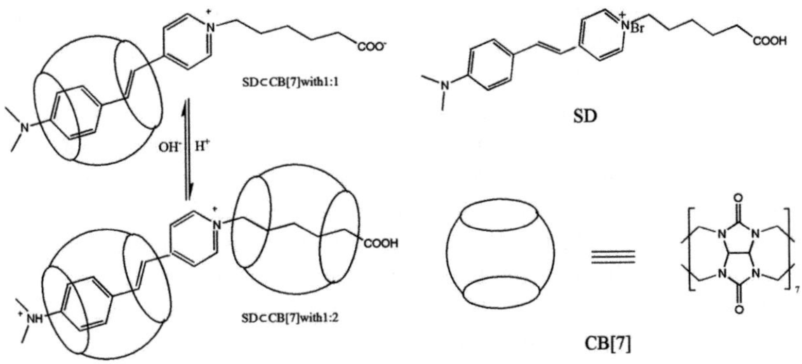

Figure 6.13 pH control [2] or [3]pseudorotaxanes based on SD⊂CB[7] with permission from (He et al. 2018).

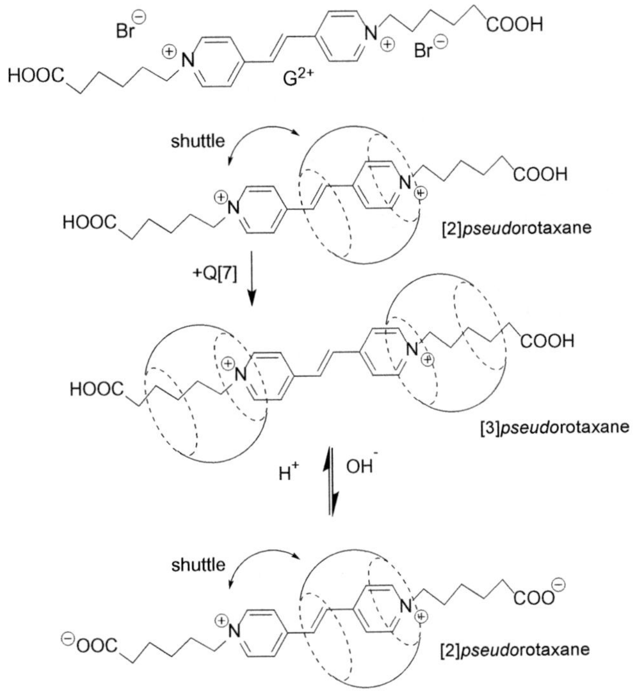

Figure 6.14 Chemical structures of G^{2+} and schematic representation of the shuttling and pseudorotaxane modes of the Q[7] wheel with permission from (Wu et al. 2018).

Wu et al. (Wu et al. 2018) studied host-guest triggered pseudorotaxane modes between Cucurbit[7]urils Q[7] and a 1,2-di(4-pyridyl)ethylenyl derived dicationic fluorescent axle guest (G^{2+}) containing hexanoic acid group as the end units, which not only increases the aqueous solubility of the guest but also reversibly switches between neutral and anionic states, depending on the pH of the media. ^{1}H NMR studies suggested that [2]pseudorotaxane (below 1.0 equivalent) under acidic conditions, behaves like a fast molecular shuttle on the bispyridinium ethylene axle and in the presence of 2.0 equivalent of Q[7] gives rise to the 2:1 complex of Q[7] with G^{2+}([3]pseudorotaxane). At pH 10, it was established that in the deprotonated state of an axle, [2]pseudorotaxane behaves like a fast molecular shuttle on the bispyridinium ethylene axle, and at acidic pH [3]pseudorotaxanes were recovered. Figure 6.14 demonstrates that the Q[7]-based [2]pseudorotaxane behaves like a molecular shuttle on the axle and can survive in both states (protonation/deprotonation) presented by the carboxylates. Additionally, bispyridinium ethylene moiety showed fluorescent response behavior which indicated that it not only acts as an axle component for the pseudorotaxane system but also behaves as an optical signaling unit during the host-guest complexation. The results were supported by density functional theory (DFT), ^{1}H NMR spectroscopic, UV-vis absorption and fluorescence spectroscopy techniques.

Pillar[5]arene Based pH Responsive Pseudorotaxanes

Xia et al. 2018 synthesized a novel pH-sensitive pillar[5]arene for [2]pseudorotaxane. The assembly included a morpholine group per-substituted pillar[5]arene as the Host (H) and an azastilbenzene derivative, *trans*-4,4-vinylenedipyridine, a photosensitive guest (*trans*-G). 1:1 complexation, *trans*-G $\subset$ H was observed. On addition of the solution of the [2]pseudorotaxane with trifluoroacetic acid (TFA), both H and *trans*-G changed into protonated products, A-*trans*-G and A-H, respectively. ^{1}H studies showed that the complexation between the H and *trans*-G was destroyed by TFA while on the addition of triethylamine, the H and *trans*-G complex was formed again. pH- and photo-dual stimuli-responsiveness was displayed by the [2]pseudorotaxane (Fig. 6.15) which was further used to fabricate a pH-, photo and cyanide-triple stimuli-responsive metallosupramolecular polypseudorotaxane by Cu(II) coordination (copper's strong coordination with pyridine groups).

With a view that single-stimulus responsive nanocontainers could not meet the complex demands Ding et al. 2016 designed multiple-stimuli responsive nanocontainers, which can be tuned to precisely release cargoes upon each stimulus (Alberti et al. 2015; Zhang et al. 2015). The group designed and fabricated acid/alkali/Zn^{2+} stimuli-responsive controlled release system by installing carboxylate substituted pillar[5]arenes (WP5, water-soluble) based bistable [2]pseudorotaxanes onto the exterior surface of mesoporous silica nanoparticles (MSNs; are extensively studied as they have biocompatibility, high loading efficiency, easy to up-scale and the diversity of surface functionalization enables MSNs to link different kinds of nanovalves (Yang et al. 2014; Noureddine et al. 2015; Song et al. 2015), regulating the flow of model cargoes, rhodamine B (RhB). WP5 form complex with 1,6-hexanediammonium (HAD) recognition stations, served as stoppers to block pore orifice and prevent RhB (entrapped cargoes) from leaking under neutral solution and get dissociated from HDA stations after being exposed to alkaline range (at pH 12, 78% total amount of RhB was released) as WP5 transfer to the 1-(6-aminohexyl)-pyridinium (APy) recognition sites which were far away from the orifice thus, opened doors for diffusion out of RhB whereas acidic stimulus (pH 2) facilitates WP5 to precipitate and results in the release of entrapped RhB (due to protonation of the terminal carboxylate groups of WP5). The strong chelation between Zn^{2+} and carboxylate groups of WP5 facilitates the unlocking of pores, realizing Zn^{2+}-triggered controlled release (Fig. 6.16). Stimuli-controlled release experiments were conducted using the real-time continuous monitoring UV-Vis spectroscopic measurement. The binding affinity of WP5 with N, N′-dimethyl-1,6-hexanediamine K_a, under neutral solution was found to be $1.29 \times 10^6 \, M^{-1}$; K_b, under alkaline solution, was $1.21 \times 10^4 \, M^{-1}$ and with N-hexylpyridinium bromide was found to be $K_c = 6.50 \times 10^5 \, M^{-1}$ using Isotherm Titration Calorimetry (ITC) experiment. T-SRNs, triple-stimuli-responsive nanocontainers were also tested for drug delivery vehicles owning acid/Zn^{2+}-triggered release.

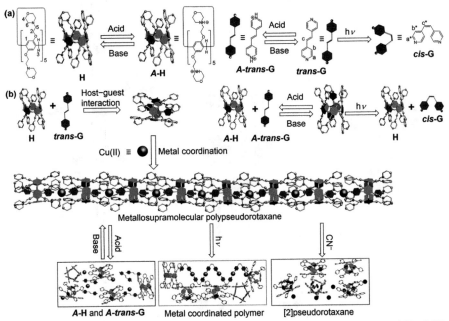

Figure 6.15 (a) Chemical Structures and Cartoon Presentation of Compound H, A-H, *trans*-G, A-*trans*-G, and cis-G and (b) Cartoon Representation of the Dual-Responsive [2] Pseudorotaxane Based on H ⊃ *trans*-G and the Fabrication of the Triple Stimuli-Responsive metallosupramolecular Polypseudorotaxane with permission from (Xia et al. 2018).

Zhao et al. 2019 synthesized ethylene glycol bridged pyridine and pillar[5] arene-based mechanically self-locked pseudo[1]rotaxane. MeP5A unit of pseudo[1] rotaxane (Fig. 6.17) could shuttle between ethylene glycol chain and pyridine unit as an acid/base-controllable molecular shuttle (Fig. 6.17a). The shuttling properties of pseudo[1]rotaxane were investigated by ^{1}H NMR. It was found that in diluted solution the pseudo[1]rotaxane could be operated as an acid/base-controllable switchable assembly between ethylene glycol chain and pyridine unit. Further, DOSY experiments showed that pseudo[1]rotaxane (Fig. 6.17) performed as self-locked-monomer in low concentration (10 mM) and interlocked-dimer (100 mM) in high concentration. After the addition of TFA to the concentrated solution (100 mM) of interlocked-dimer, it could change from a shrinking state to an extension state (Fig. 6.17b).

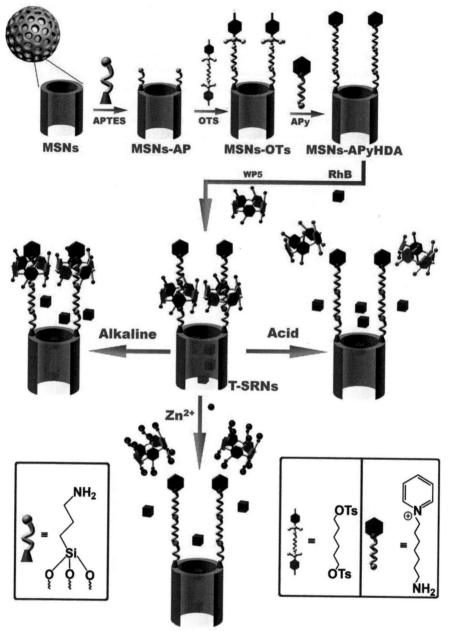

Figure 6.16 Schematic representation of the preparation of T-SRNs. MSNs were functionalized with [2]pseudorotaxanes on the surface and the chemical structure of the [2]pseudorotaxane containing WP5 ring and the functional stalk is shown. T-SRNs can be activated by acid, alkali, or Zn^{2+} stimuli to regulate the release of a model drug, rhodamine B (RhB) with permission from (Ding et al. 2016).

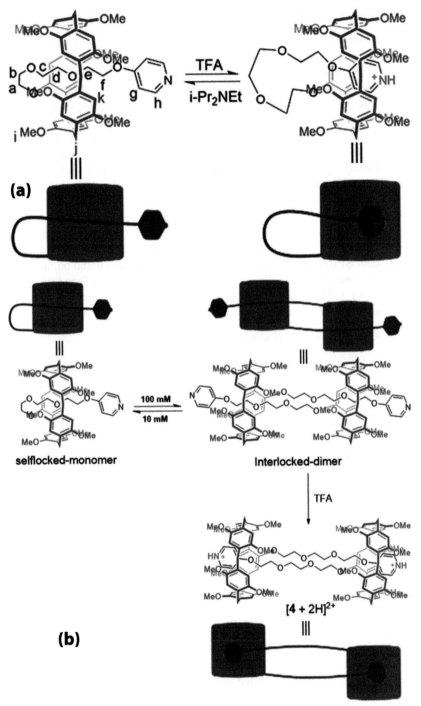

Figure 6.17 (a) Shuttling process of pseudo[1]rotaxane and (b) Assemble process of in different condition with permission from (Zhao et al. 2019).

Cyclodextrin Based pH Responsive Pseudorotaxanes

To disentangle the problem of amino acid metabolic abnormalities which can cause serious diseases due to various biological effect of enantiomers, Zhang et al. 2019 developed bioinspired γ-Cyclodextrin pseudorotaxane assembly nanochannel for the transportation of selective amino acid, for which they constructed a biomimetic artificial amino acid transporter by assembling γ-cyclodextrin (γ-CD, used as a host) into the N-(1-naphthyl) ethylenediamine(NEDA, acts as a guest)-functionalized biomimetic nanochannel (Fig. 6.18) which can easily actualize the reversible switching under the influence of a pH change. Patch–clamp technology aided in studying the individual translocation events of nanochannel system in the presence of different aromatic amino acids, comprising phenylalanine enantiomers (L-Phe and D-Phe), L-histidine (L-His), L-tryptophan (L-Trp) and L-tyrosine (L-Tyr), it was observed that nanochannel selectively transports L-Phe more than D-Phe and other amino acids. The Langmuir model demonstrated that the nanochannel produced stronger adsorption to L-Phe than D-Phe and other amino acids which was evidenced from the binding constant values calculated based on the Langmuir model that was 4.9 for L-Phe, 1.05 for D-Phe, 0.96 for L-Trp, 1.01 for L-Tyr, and 0.97 for L-His. The results suggested the use of pseudorotaxane assembly nanochannel for chiral detection and drug delivery.

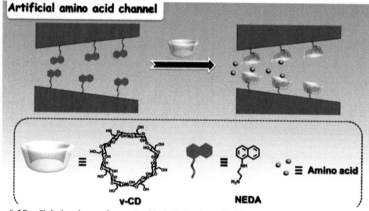

Figure 6.18   Fabrication of an Artificial Amino Acid Channel with permission from (Zhang et al. 2019).

ROTAXANES

'Rotaxane' (Xue et al. 2015; Bazargan et al. 2018; Blanco-Gómez et al. 2020; Zhou et al. 2020) is derived from the Latin words for 'wheel' and 'axle' , and describes a compound that consists of a linear dumbbell-shaped chain component referred to as the 'shaft' (sometimes called the rodlike part or guest) that threads through the cavity of one or more macrocyclic rings (cyclic species sometimes called the beadlike part or host) (Bruns and Stoddart 2014) bound together in a threaded structure by noncovalent forces. To prevent the ring(s) from dethreading, large end groups (so-called stoppers) are added to the ends of the shaft/axle. Although the cyclic and axle components of the rotaxane are not covalently bonded, rotaxanes are stable assemblies as a large amount of free energy of activation ($\Delta G^{\ddagger}$), is required to overcome the dissociation of a ring from the axis of a rotaxane (Fig. 6.19) (Wenz et al. 2006). Nomenclature for rotaxanes is given the form [n]rotaxane where the number n in the square bracket indicates the sum of the numbers of rings and axles in a rotaxane (Safarowsky et al. 2000; Wenz et al. 2006).

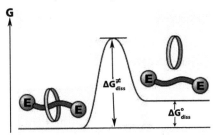

Figure 6.19 Energy diagram for a [2]-rotaxane and its constituents: E represents end group reprinted with permission from (Wenz et al. 2006).

As discussed earlier, an assembly without bulky stoppers is referred to as a pseudorotaxane. 'Pseudo' means false, so 'pseudorotaxane' without bulky end groups of the axle means false rotaxane, which is a supramolecular complex but not a compound, and the system is in a dynamic equilibrium between free and assembled molecular components. Sometimes a pseudorotaxane with only one stopper is called semirotaxane. Rotaxanes, covalently linked together form oligomeric or polymeric species called polyrotaxanes. Daisy chain rotaxanes (so-called 'molecular muscles') consist in the double thread of two symmetric macrocycles that are covalently linked to an axle bearing switchable stations (each station represents recognition site for macrocycle) and a bulky stopper to avoid unthreading (Jiménez et al. 2000; Bruns and Stoddart 2014; Kay and Leigh 2015; Stoddart 2009; Wu 2019). [1]-rotaxanes (Xue and Mayer 2010) or [2]rotaxanes (Holler et al. 2019) to daisy chains, oligorotaxanes (Belowich et al. 2012), to polyrotaxanes (Fig. 6.20) (Harada et al. 2009).

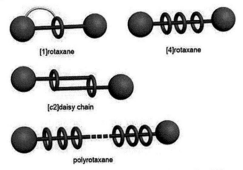

Figure 6.20 Shows a schematic cartoon representation of a [2]rotaxane containing one ring and one shaft along with an analogous pseudorotaxane and semirotaxane along with [1]rotaxane, [4]rotaxane, [c2]daisy chain, polyrotaxane reprinted with permission from (Xue et al. 2015).

The recent examples of the following types of pH-responsive rotaxanes are described next:

1. [2]-rotaxane
2. [3]-rotaxane
3. [4]-rotaxane
4. [c2]daisy chain
5. Polyrotaxane
6. Interconversion between pseudorotaxane and rotaxanes

pH-responsive [2]rotaxane

Corra et al. 2019 designed a mechanically planar chiral [2]rotaxane system where ring shuttling along the axle leads to interconversion of the two enantiomers by passing through an achiral conformation in which the ring is located in the center of the axle i.e., a three-station molecular shuttle that can be switched reversibly between symmetric prochiral and desymmetrized mechanically planar chiral states. (Figure 6.21a) Their rotaxane design was fabricated with a crown ether macrocycle, and dibenzylammonium and triazolium recognition sites located along the axle (Fig. 6.21b) that can be reversibly switched between prochiral and chiral states upon acid-base stimulation. It was based on the preference of a dibenzo[24]crown-8 (**DB24C8**)-type ring to encircle the ammonium center due to strong hydrogen bonding, while on deprotonation of the ammonium the ring can move on the triazolium station. Further, it was demonstrated that the supramolecular interaction of the positively charged rotaxane with optically active anions causes an imbalance in the population of the two enantiomeric which is an excellent example of chiral molecular recognition.

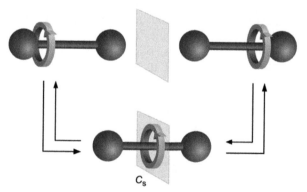

C_s

Figure 6.21a Two enantiomers of a coconformationally mechanically planar chiral rotaxane and their interconversion by ring shuttling through an achiral coconformation that features a mirror plane.

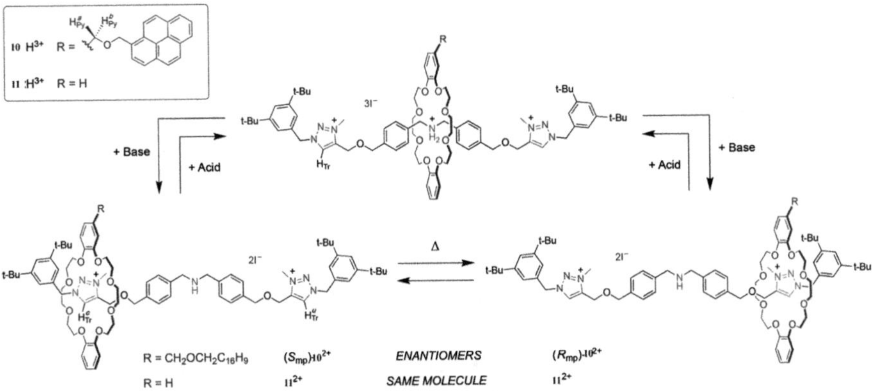

Figure 6.21b   Rotaxanes 10H^{3+} and 11H^{3+} (top), and Their Base-Triggered Switching to 10^{2+} and 11^{2+} reprinted with permission from (Corra et al. 2019).

Arumugaperumal et al. (Arumugaperumal et al. 2018) developed a class of a novel fluorescent acid-base switchable [2]rotaxane **NIR4** composed of two different molecular stations (dibenzylammonium (DBA) and triazolium unit) and rotaxane arms terminated with far-red boron-dipyrromethene (BODIPY) fluorophores. The molecular shuttling motion of Mechanically Interlocked Molecules (MIMs) was observed by the fluorescence signal transduction via distance-dependent photo-induced electron transfer process of [2]rotaxane **NIR4** triggered by external chemical stimuli. On the addition of acid the macrocycle (DB24C8, Dibenzo-24-crown-8 as a wheel) was stationed at protonated DBA site state I, whereas on the addition of base in [2]rotaxane NIR4 (subjected to deprotonation) the triazolium moiety acted as a binding site for the macrocycle in state II (Fig. 6.22). So, the far-red fluorescence of [2]rotaxane **NIR4** could be reversibly switched between 'ON' and 'OFF' via acid-base control, which is correlated to the position of the macrocycle either DBA (ON) or triazolium station (OFF). Additionally, the flexible arms of triazolium moiety in [2]rotaxane **NIR4** and its axle exhibited remarkable

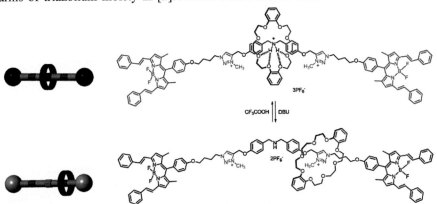

Figure 6.22   Conformation switching by external chemical stimuli of acid/base with different macrocycle positions in switchable [2]rotaxane NIR4 reprinted with permission from (Arumugaperumal et al. 2018).

selectivity and sensitivity towards $H_2PO_4^-$, where the specific mechanical molecular motion was supported by quantum mechanical calculations and could be applied for *in-vitro* imaging and clarify the distribution of $H_2PO_4^-$ at subcellular levels.

Kimura et al. (Kimura et al. 2017) synthesized [2]rotaxanes with dibenzo[24] crown-8 (DB24C8) as their macrocyclic component and dumbbell-like axle component having three recognition sites a dialkylammonium, an alkylarylamine and a tetra(ethylene glycol). [2]rotaxanes (Fig. 6.23 state 2) under different pH conditions act as four-state molecular shuttles (Fig. 6.23): (i) under acidic conditions, the DB24C8 unit encircled both ammonium stations; (ii) under neutral conditions, the dialkylammonium moiety was the predominant station (iii) under basic conditions and in the absence of a metal cation, when both ammonium centers were deprotonated, the alkylarylamine station mainly recognized the DB24C8 component; and (iv) under basic conditions in the presence of an alkali-metal cation, the tetra(ethylene glycol) unit recognized the DB24C8 component through cooperative binding of the alkali-metal ion. Under basic conditions, the deprotonated [2]rotaxanes recognized an alkali-metal cation in the cavity formed from the oligo(ethylene glycol) units, with the association of a Na^+ cation being stronger than that of a K^+ cation. The shuttling motions of the macrocyclic component proceeded reversibly.

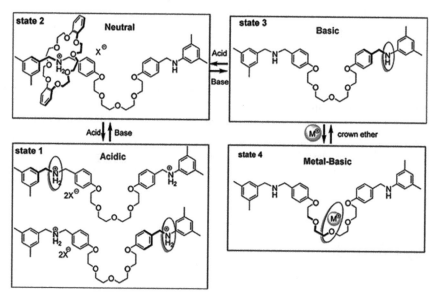

Figure 6.23   Cartoon representation of the four-state molecular shuttling of the [2]rotaxane $1H_1X_1$ reprinted with permission from (Kimura et al. 2017).

Riss-Yaw group (Riss-Yaw et al. 2016) reported the synthesis and the study of four pH-sensitive mannosyl molecular shuttles that consisted of dibenzo-24-crown-8 (DB24C8) as the macrocycle, anilinium and pyridinium amide as molecular stations. At acidic pH, the DB24C8 resides around the anilinium station for all the protonated [2]rotaxanes, while under basic conditions, it shuttles around the pyridinium site (due to hydrogen bonding) for every molecular shuttle.

The mannopyranose moiety lies under a $^1C_4$ chair-like conformation for all the molecular shuttles at the protonated state (**13a, 13b, 16a,** and **16b**) was observed in all the non-interlocked molecules because the anomeric carbon of the mannose is linked to the positively charged nitrogen of the pyridinium unit (Fig. 6.24); it underwent the Reverse Anomeric Effect (RAE) irrespective of the electron-withdrawing or -releasing inductive effect of the amide groups on the pyridinium ring. On addition of base (deprotonation of the anilinium) the DB24C8 shuttles

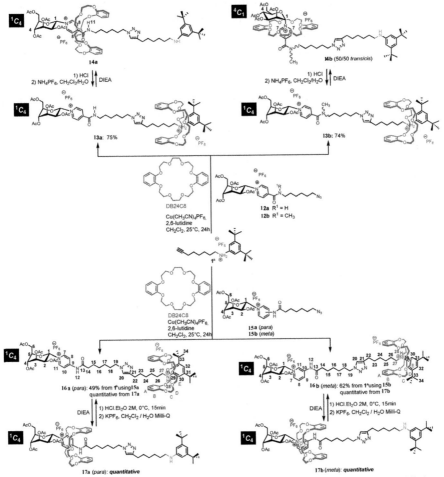

Figure 6.24 Synthesis of [2]rotaxane molecular shuttles containing various pyridinium amides (13, 14) and 3- and 4-alcanamidopyridin-1-ium molecular stations reprinted with permission from (Riss-Yaw et al. 2016).

with a large amplitude motion toward the pyridinium amide stations. In a single case, molecular shuttle **14b**, flipping off the chair-like conformation of mannose from $^1C_4$ and $^4C_1$ through the concealing of the positive charge was observed due to deprotonation of the di-tert-butylanilinium end of the encircled axle which triggered the large-amplitude motion of the DB24C8 along the thread toward the mannosylpyridinium site (i.e., the other end of the encircled molecular axle) while in all other deprotonated [2]rotaxane molecular shuttles, the interactions between the DB24C8 and the pyridinium station did not allow the efficient concealing of the positive charge, resulting in no switching of the RAE. Further, it was proposed that the possibility to switch between $^1C_4$ and $^4C_1$ conformations of a mannose moiety in response to pH stimulus should find interesting applications in the domain of molecules with changing properties.

Ueda et al. (Ueda et al. 2016) have reported the construction of pair of [2]rotaxanes, possessing different types of chromophores that form a five-state molecular shuttling system that operates based on the independent protonation/ deprotonation of four different amino groups in response to the addition of the acid and base; their five distinct and simple absorptive outputs were readily monitored using UV/Vis spectroscopy (Fig. 6.25b). A pair of acid-/base-responsive [2]rotaxanes, featuring dialkylamine (dialkylammonium) and arylamine (arylammonium) stations in their rodlike components as binding sites for DB24C8 and DB25C8 (24- and 25-membered crown ethers, Fig. 6.25a) as macrocyclic components and cyanodiazobenzene and coumarin as chromophores, respectively. In each [2]rotaxane, under basic conditions the macrocyclic component was located at its arylamine station and its alkylammonium station under neutral conditions; under acidic conditions, the alkylammonium station acted predominantly as the station for the azo-[2]rotaxane, whereas both ammonium stations were encircled by the macrocyclic component in the coumarin-[2]rotaxane. ^1H NMR and UV/Vis studies showed three states in terms of position of the macrocyclic component and the degree of protonation of the arylamine in each [2]rotaxane. The mixed [2]rotaxane system displayed stepwise and independent molecular shuttling behavior based on the degree of protonation of the amino groups in response to both the amount and strength of added acids or bases; as such, the system provided five different absorption signals as outputs that were studied using UV/ Vis spectroscopy (Fig. 6.25).

Franchi et al. (Franchi et al. 2019) demonstrated back and forth motions of [2]-rotaxane under an acid–base stimulus employing EPR spectroscopy for the first time. The paramagnetic rotaxane architecture included a dialkyl nitroxide functionality embedded in a crown ether-like ring interlocked with a dumb bell axle component that contains a dialkylammonium (NH_2^+) and a 4,4-bipyridinium (BPY^{2+}) unit as two different recognition sites. Diisopropylethylamine (iPr_2EtN) triggered the shuttling of the paramagnetic macrocycle toward the secondary BPY^{2+} station (Fig. 6.26a) due to deprotonate the NH_2^+ center while the addition of trifluoroacetic acid causes the return back movement of the ring onto the favored ammonium site. The group studied one-only stimulus supplied back and forth motions of the paramagnetic wheel in rotaxane $18H \cdot (PF_6)_3$, using acid **19** and derivatives **21-23** (Fig. 6.26a) to drive the switchable molecular shuttle in $18H \cdot (PF_6)_3$. Figure 6.26b shows fast and quantitative proton transfer (step A → B′) causes the movement of the wheel onto the ammonium site of the dumb bell. Loss of CO_2 (step B′ → B″) followed by slow back proton transfer from $[18H \cdot (PF_6)_2]^+$ to its counter anion (R^-) restores the position of the ring to the bipyridinium station. Back and forth motions were accelerated using the p-Cl derivative (electron-withdrawing) of the acid while employing p-CH$_3$ and p-OCH$_3$ derivatives (electron-donating), the back motion is strongly inhibited by the insurgence of collateral radical reactions.

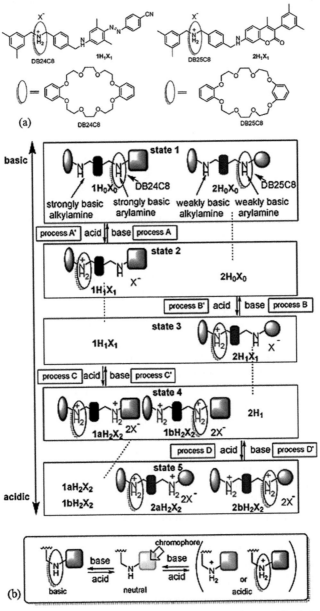

Figure 6.25. (a) Chemical structures of the [2]rotaxanes along with their macrocycle component (b) Cartoon representation of five-state molecular shuttling using a pair of [2]rotaxanes (their mono-ammonium forms) and the environments of their chromophore units under basic, neutral, and acidic conditions with permission from reprinted with permission from (Ueda et al. 2016).

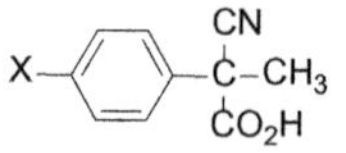

Figure 6.26a   Switching Process between 18H·(PF$_6$)$_3$ and 18·(PF$_6$)$_2$.

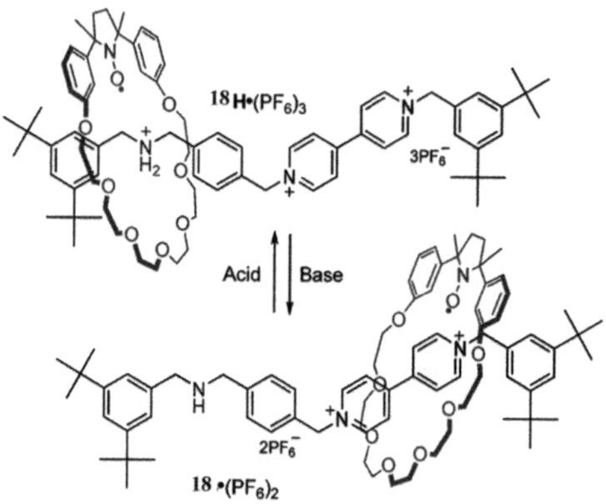

Figure 6.26b   Switching Motions of Rotaxane 18·(PF$_6$)$_2$ Triggered by Acid **19** reprinted with permission from (Franchi et al. 2019).

Shinohara et al. (Shinohara et al. 2020) investigated generation of $^1O_2$ on visible-light irradiation, by mechanically interlocked porphyrins appended to gold nanoparticles (AuNP), which can effectively quench the photosensitization via energy transfer from the excited porphyrin photosensitizer to the gold nanoparticle. For the study, mechanically-interlocked photosensitizer–quencher systems were synthesized based on free-base tetraphenylporphyrin (H$_2$TPP)–gold nanoparticle (AuNP) composites (as the size of gold nanoparticles was larger than the cavity of the 24-crown-8, hence it acted both as the quencher and as the capping agent) by utilizing a rotaxane architecture comprised of secondary ammonium and crown ether subunit. The H$_2$TPP-substituted 24-crown-8, **24** (Fig. 6.27) was observed to shuttle along the alkanethiolate axle, triggered by deprotonation/ protonation at the ammonium station, altering the H$_2$TPP–AuNP distance and the photoexcitation energy transfer efficiency. On switching, quantum yields for photosensitized singlet oxygen ($^1O_2$) generation and fluorescence after deprotonation were quenched by 46 and 42%, respectively.

Klein and Beer (Klein and Beer 2019) demonstrated that a bistable [2]rotaxane containing a halogen bonding benzimidazole-iodotriazole station directly conjugated to a naphthalimide station axle component undergoes macrocycle, **26** (Fig. 6.28b) shuttling translocation from the naphthalimide station to the benzimidazolium-iodotriazole station only upon both protonations of the benzimidazole moiety and addition of chloride anion (Fig. 6.28). A naked-eye detectable color response was found due to the co-conformational change of the host structure. In the neutral state, **26**, the macrocycle was found to reside at the naphthalimide station; neither addition of chloride nor protonation with HBF$_4$ caused any observable spectroscopic changes and it was observed only on addition of aqueous HCl solution or chloride to **26·HBF$_4$** that co-conformational modulation occurred.

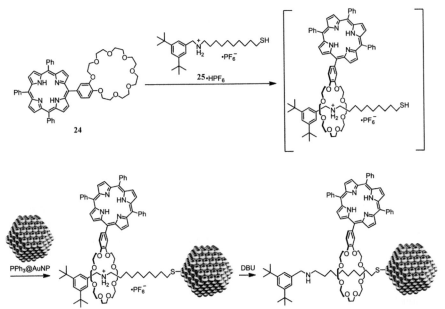

Figure 6.27 Synthetic route for porphyrin-appended gold nanoparticles via capping of a mechanically-bonded pseudorotaxane intermediate (diazabicycloundecene (DBU) reprinted with permission from (Shinohara et al. 2020).

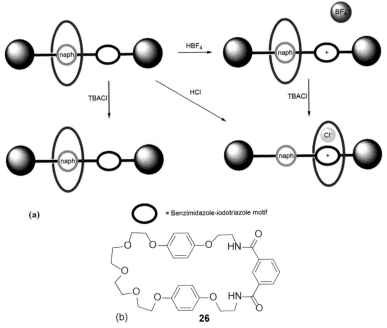

Figure 6.28 (a) Schematic outline of molecular switching exhibited by [2]rotaxane. Both protonation and chloride addition are required to induce macrocycle translocation (b) Chemical structure of macrocycle reprinted with permission from (Klein and Beer 2019).

(Li et al. 2018) reported the rarely-exploited neutral halogen bonding (XB) [2] rotaxane hosts (XB bis-iodotriazole-secondary amine functional motif) capable of orthosteric control of guest binding, where a chemical stimulus (acid) interacting with the interlocked host binding site switches the host's native guest preference from metal cations to anions. Rotaxane in neutral form was capable to bind strongly with metal cation whereas anions bound relatively weakly. However, protonation (using acid) of rotaxane attenuates the rotaxanes' ability to coordinate cations while it greatly enhances halide affinity through charge-assisted halogen bonding and hydrogen bonding interactions in competitive aqueous solvent media. This is due to the protonation of the secondary amine functionalities, concomitant with the conformational inversion of the XB iodotriazole motifs, activates the binding cavity for strong charge assisted XB and HB (hydrogen bonding HB receptors in competitive solvent media) anion coordination in competitive aqueous solvent media whilst concurrently attenuating metal cation binding (Fig. 6.29). In aqueous media, the cationic protonated rotaxanes also resulted in the unprecedented observation of anti-Hofmeister anion binding bias for Cl⁻ by XB [2]rotaxanes. Furthermore, attachment of a fluorescent anthracene reporter group to the rotaxane framework also enabled diagnostic sensory responses to cation/anion binding.

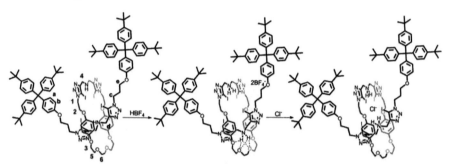

Figure 6.29 Schematic illustration showing capability of rotaxane to bind anions reprinted with permission from (Li et al. 2018).

NIR (near-infrared) photothermal agents can enhance photoacoustic imaging, can change absorbance cross-section due to triggered self-aggregation but has an inherent drawback of dependence on local concentration which can be difficult to control (Nam et al. 2009; Dragulescu-Andrasi et al. 2013; Guha et al. 2016) pH-sensitive croconaine dye and doped stealth liposomes with Croc dye (Croc-SL) for this purpose was studied. Dynamic light scattering measurements showed that Croc-SL does not change size in acidic pH, but the absorption spectrum slowly broadened with time. Further, they trapped the Croc dye within a tetralactam macrocycle generating an interlocked croconaine rotaxane (CrocRot) and prepared CrocRot doped stealth liposomes (CrocRot-SL) and observed that the spectral stability of CrocRot-SL was greatly improved. They demonstrated that CrocRot-SL is a robust biocompatible nanoparticle that can be switched reversibly between acid and base forms (Fig. 6.30) and that they survive strong and repeated laser heating. A stealth liposome composition not only permitted acid-activated, photothermal heating but also acted as an effective nanoparticle probe for ratiometric photoacoustic imaging of acidic pH in deep sample locations, including a living mouse.

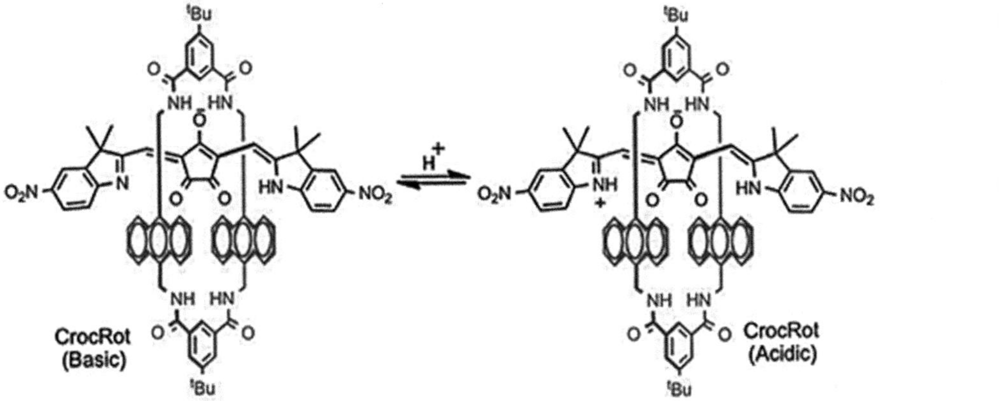

Figure 6.30 pH-sensitive croconaine rotaxane (CrocRot) reprinted with permission from (Guha et al. 2016).

Zhou et al. (Zhou et al. 2019) reported a 2,6-helic[6]arene-based [2]rotaxane **RH·BArF**, in which a protonated tertiary ammonium salt was a reversible control unit and hexamethylene chain was a second recognition site. It represented the first triply operable molecular switch based on helic[6]arenes where the states of macrocycle on different sites could be reversibly and efficiently switched under three different chemical stimuli including anions, acid/base and solvent polarity. Fluoride ions were chosen to drive the macrocycle move to the hexamethylene chain site. ^{1}H NMR studies showed the formation of a translational isomer **RH·F** on the addition of 2:2 equivalents of tetrabutylammonium fluoride (TBAF) to a CD_2Cl_2 solution of **RH·BArF**. As the protonated tertiary ammonium salt had acid/base responsiveness, the similar translational motion was readily actuated through alternately adding acid and base. 1,8-diazabicyclo[5.4.0]undec-7-ene (DBU) was used as a base to deprotonate the ammonium salt, which resulted in occupancy of hexamethylene chain site by the host while on the addition of acid the oxonium acid $[H(Et_2O)_2]^+[[3,5-(CF_3)_2C_6H_3]4B]^-$ (HBArF) resulted in protonation of the tertiary amine and helic[6]arene ring resided selectively on the ammonium unit under acidic condition (Fig. 6.31). Translational motion of **RH·BArF** was observed by altering solvent polarity, i.e., by adjusting the volume ratio of CD_2Cl_2/DMSO-d_6 which was studied for proton NMR. In low polarity solvent (CH_2Cl_2), the helic[6] arene ring resided at the ammonium site. While in the solvent of high polarity (DMSO), the macrocycle migrated to the hexamethylene chain site as the polar solvent molecules could destroy the host-guest interaction between helicarene and protonated ammonium salt; while the hexamethylene chain was inclined to be encircled by macrocycle because of their solvophobic nature.

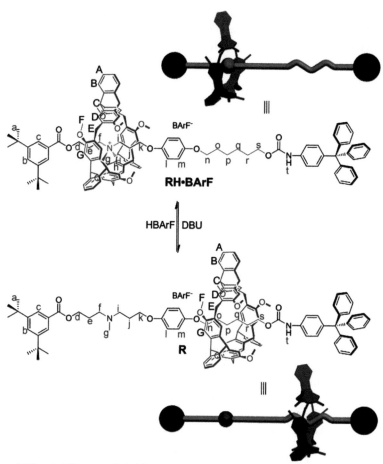

Figure 6.31   Acid/base-switchable motion of RH·BArF reprinted with permission from (Zhou et al. 2019).

pH-responsive [3]rotaxane

Altmann and Pöthig 2017 reported the first organometallic scaffolds, the so-called pillarplexes (Ag₈pillarplex) as the host and synthesized novel organometallic [2] rotaxane. The pillarplex's reactivity towards transmetallation agents such as (tht) AuCl was fully maintained within the mechanically interlocked architecture. The system demonstrated an acidic transformation to a purely organic [3]rotaxane which was reversible on addition of an adequate base – rendering the assembly a pH-dependent switch (Fig. 6.32). During the acidic cleavage of $[\mathbf{Ag_8(L^{Me})_2}]$ $[\mathbf{2}]\mathbf{Rot(OTf)_4}$ eight Ag+ ions per pillarplex were liberated and on treatment of an acetonitrile solution containing the newly formed $(\mathbf{H_6L^{Me}})_2[\mathbf{3}]\mathbf{Rot(OTf)_8}$ and the released silver(I) ions with N, N-diisopropylethylamine, a quantitative back-conversion to $[\mathbf{Ag_8(L^{Me})_2}][\mathbf{2}]\mathbf{Rot(OTf)_4}$ over 60 minutes at 70°C was observed (Fig. 6.32). This process was reversible for multiple cycles without significant loss of material.

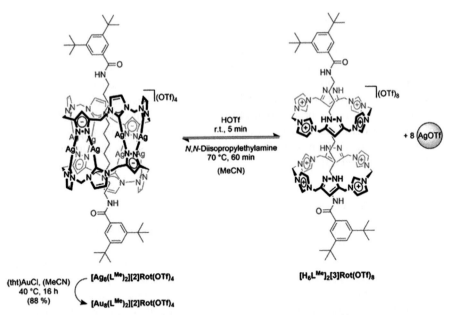

Figure 6.32 Reversible acid/base-assisted interconversion of [Ag$_8$(L^{Me})$_2$][2]Rot(OTf)$_4$ and (H$_6$L^{Me})$_2$[3]Rot(OTf)$_8$ as well as transmetallation to the isostructural complex [Au$_8$(L^{Me})$_2$] [2]Rot(OTf)$_4$ reprinted with permission from (Altman and Pöthig 2017).

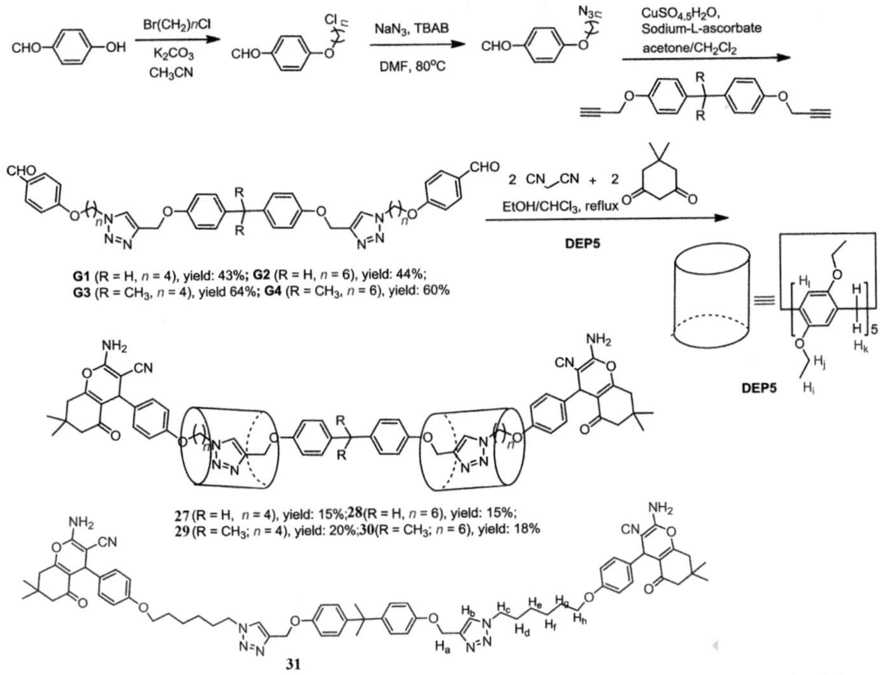

G1 (R = H, n = 4), yield: 43%; **G2** (R = H, n = 6), yield: 44%; **G3** (R = CH$_3$, n = 4), yield 64%; **G4** (R = CH$_3$, n = 6), yield: 60%

27 (R = H, n = 4), yield: 15%; **28** (R = H, n = 6), yield: 15%; **29** (R = CH$_3$, n = 4), yield: 20%; **30** (R = CH$_3$; n = 6), yield: 18%

Figure 6.33 A synthetic route to guests G1-G4 and [3]rotaxanes reprinted with permission from (Ye et al. 2020).

Ye and group (Ye et al. 2020) successfully synthesized and characterized the structure and self-assembly behaviors of four pillar[5]arene based [3]rotaxanes **(27–30)** involving two 1,4-diethoxypillar[5]arene **(DEP5)** ring and dumbbellshaped molecules **(G1–G4)** containing two triazole sites, one long bridge and two multicomponent stoppers, **32** (Fig. 6.33). They demonstrated the pH-responsive property of rotaxane **30** in water. DLS and TEM studies showed that the addition of HCl resulted in the protonation of the synthesized rotaxane due to the presence of amino groups and triazole units which resulted in an amphiphilic nature of the compound that can self-assemble into regular vesicles through hydrophilic-hydrophobic interactions.

pH-responsive [4]rotaxane

Tokunaga et al. (Tokunaga et al. 2016) synthesized and studied a macromolecule containing more than one [2]rotaxane units. The macromolecule under study was a branched [4]rotaxane featuring a 1,3,5-tris(4-aminophenyl)benzene core (TAPB) and containing three switching arms with both secondary ammonium cation and aniline binding sites for threaded crown ethers (three [2]rotaxane subunits). The triple reversible shuttling motion of the dibenzo[24]crown-8 macrocycle between the ammonium and aniline stations was driven by the addition of base and acid, respectively featuring a TAPB core (Fig. 6.34). ^{1}H NMR spectroscopy was used to study the translational isomerization of the [4]rotaxane and different absorption maxima and the oxidation potential for the [4]rotaxane were monitored by ultraviolet spectroscopy and electrochemical studies, respectively.

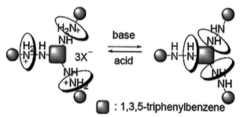

Figure 6.34   Schematic of the molecular shuttling of branched [4]rotaxane reprinted with permission from (Tokunaga et al. 2016).

pH-responsive [c2]daisy chain Rotaxanes

Topologically, they consist of the double thread of two symmetric macrocycles that are covalently linked to an axle bearing two switchable stations and a bulky stopper to avoid unthreading.

Molecular muscles, which exhibit expansion and contraction on the nanoscale, are another typical mode of motion by mimicking biology. Rotaxane architectures, especially supramolecular daisy chains, (Ashton et al. 1998b; Romuald et al. 2010) have been a promising candidate for the design and synthesis of artificial molecular muscles (Goujon et al. 2019; Tao et al. 2019). A daisy chain is a special case of MIMs created by interlocked monomers that consist of both a linear thread (guest) and a ring (host). In these systems, shuttling of the macrocycle along the thread results in a relative motion that either contracts or expands the entire molecule under the external stimuli. [c2]Daisy chains, with the double-threaded architecture of the rotaxane dimer, are an ideal module for the construction of artificial molecular muscles by means of antiparallel sliding among its two interlocked rings. A large proportion of [c2]daisy chains are based on crown ethers and dialkylammonium ions, where contracting and stretching motions are readily actuated by acid/ base. Stoddart and co-workers (Wu et al. 2008) for the first time reported the pH-controllable [c2]daisy chain, in which the two DB24C8 rings switch between two different recognition sites, dibenzylammonium and bipyridinium.

Goujon et al. (Goujon et al. 2017) have explored a branched covalent polymer involving pH-sensitive bistable [c2] daisy chain rotaxanes as monomers (Fig. 6.35) and established that in the chemically cross-linked network topology, the pH-actuation of the mechanical bond at the molecular level translates to large and fully reversible macroscopic contraction and expansion of the swollen chemical gel by ~50% of its volume.

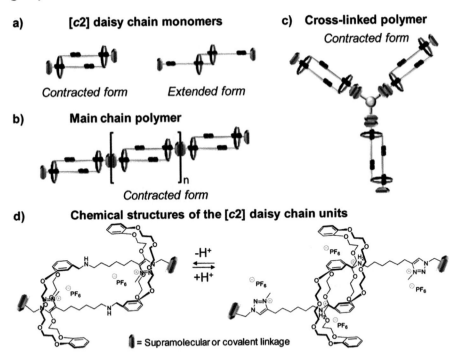

Figure 6.35 (a) Schematic representation of contracted and extended [c2]daisy chain monomers. (b) Schematic representation of the main chain polymer based on [c2]daisy chain monomers. (c) Schematic representation of a cross-linked polymer network based on [c2]daisy chain monomers. (d) Detailed chemical structure of the [c2]daisy chain unit used in the present study reprinted with permission from (Goujon et al. 2017).

Wolf et al. (Wolf et al. 2019) designed and synthesized two unsymmetrical pH-switchable [c2]daisy chain rotaxanes, which contain two different electroactive units as stoppers (an electron donor triarylamine, and an electron acceptor perylene bisimide unit, Fig. 6.36). Using a combination of 1D and 2D NMR along with cyclic voltammetry, they demonstrated that the pH actuation of the mechanical bond can be used to modulate the electrochemical properties of the bistable [c2]daisy chain rotaxanes when switching between the contracted and extended forms (Fig. 6.36) which may be important for the fabrication of complex electrochemically active molecular machines.

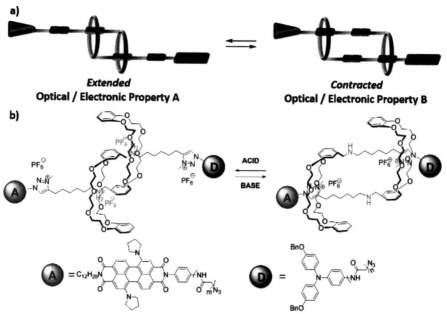

Figure 6.36   (a) General representation of the mechanical actuation of an unsymmetric [c2]daisy chain rotaxane incorporating electron donor (blue) and electron acceptor (red) stoppers; (b) Molecular structures of the pH-switchable [c2]daisy chain rotaxanes reprinted with permission from (Wolf et al. 2019).

pH-responsive Rotaxane Based Polymers/Gel/Rotaxane Immobilized into the Framework

Shukla et al. (Shukla et al. 2017) studied supra molecular interactions of orthogonal H-bonded mechanically interlocked system, the axle polymer **P1** and polyrotaxane **P2** retaining a conjugated backbone were synthesized by Suzuki coupling of fluorine-based monomer **F** with phenanthroimidazole (PIZ)-based axle monomer **A** and rotaxane monomer **R**, respectively (Fig. 6.37). The polyrotaxane **P2** containing H-bonded cavities showed sensitive acid-base molecular switching capability as reversible on-off-on fluorescence by the electronic energy transfer between the polymer backbone and topologically constrained cavities (via the molecular motion along with the axle unit). The morphological conversion of H-bonded polyrotaxane units was seen to be converted from a bloated form into flake form on protonation. The group further proposed that responsive MIPs

(Mechanically Interlocked Polyrotaxanes) strategy with controllable molecular switching capability towards acid-base would create a new direction for optical and electronic applications.

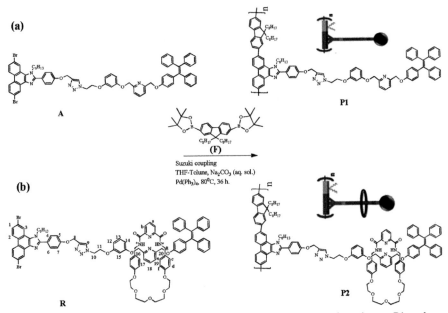

Figure 6.37 The general synthetic procedure of side-chain axle polymer P1 and polyrotaxane P2. (Shukla et al. 2017).

Conventional thermo- and pH-responsive microgels have been formed from the random copolymerization of acrylamide derivatives and charged have an intrinsic unsolved problem that is the individual mode of response usually interfering with each other. For instance, carboxyl groups are very strongly charged for the gel network to undergo a volume transition to acrylamide-based polymers. Moreover, in the presence of oppositely charged molecules, uncontrollable deswelling occurs due to charge screening (Kureha et al. 2016). To solve this problem Kureha et al. (Kureha et al. 2017) developed temperature and pH-sensitive mechanically cross-linked microgels where rotaxane cross linkers (RCs) were introduced during the precipitation polymerization, which minimized the disintegration of the mechanically cross-linked structure (Fig. 6.38a). The controllable aggregation/disaggregation states of γ-cyclodextrin (γ-CD) in the RC were used to tune the swelling capacity of the RC network, which led to pH-responsive swelling/deswelling behavior. The aggregation/disaggregation of CDs in RC occurred due to the formation of hydrogen bonds in acidic or neutral solutions, whereby the hydrogen bonds weakened in an aqueous alkali solution. The mechanical cross-links were obtained on treatment of PEG macromonomers with γ-CD, resulted in the formation of mechanical cross-links (Fig. 6.38b, milky solution). To minimize the disintegration of RC network and the aggregation of the microgels, a modified precipitation polymerization was used (Fig. 6.38c), in which the concentration of

N-isopropyl methacrylamide (NIPMAm) monomer and RC was fixed at 150 mM and reaction temperature was increased from low to high in order to suppress the disintegration of the RC, which occurs due to thermal motion in the early polymerization stages. It was further proposed that such RC microgel systems should apply to a variety of vinyl monomers, and may thus lead to the development of new design guidelines for microgels with applications that require responses to changes of pH value that should ideally not interfere with any other modes of response.

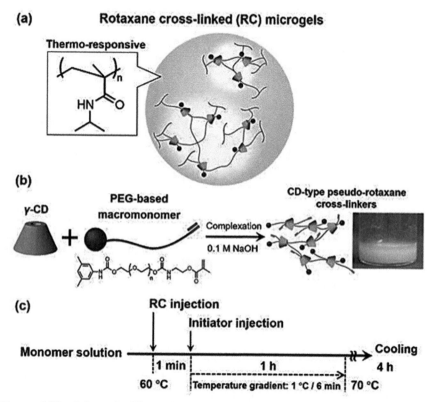

Figure 6.38 Schematic illustration of (a) rotaxane cross-linked (RC) microgels, (b) a cyclodextrin-based pseudo-rotaxane cross-linker, and (c) the procedure for the modified precipitation polymerization for RC microgels reprinted with permission from (Kureha et al. 2017).

Gao et al. (Gao et al. 2016) synthesized a pH-driven molecular shuttle, immobilized into the framework of the Periodic Mesoporous Organosilica (PMOs) that possessed enough free space to accommodate a mechanical motion of β-cyclodextrins (β-CDs). The structure of molecular shuttle was composed of β-CDs, a symmetrical molecular thread composed of a biphenyl unit, two ureido and propyl groups, and was end-trapped mechanically by two siloxane stoppers. The β-CD was observed to shuttle back and forth between the biphenyl unit and propyl spacer by pH stimuli in the rigid framework, accompanying the change of fluorescence emission of biphenyl units (Fig. 6.39). It was observed that the release rate of the guest molecules inside the mesopore was much faster in an acidic medium (pH 4) than in neutral pH. The rotaxane inserted PMOs not only helped to improve the water-dispersibility and biocompatibility of the materials but also decreased the toxicity of the nanocarriers with the presence of β-CDs. The rotaxane hybrid PMOs materials with such unique properties were proposed to be promising for the development of molecular devices in biosensor and drug delivery applications (due to their ability to encapsulate a payload of therapeutic compounds).

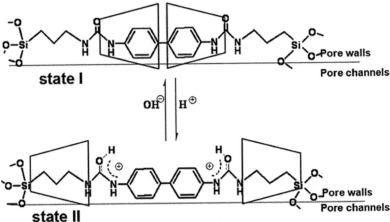

Figure 6.39 A pH-driven molecular shuttle was immobilized into the framework of the PMOs in which the β-CDs could shuttle mechanically reprinted with permission from (Gao et al. 2016).

Interconversion of Pseudorotaxane and Rotaxanes on the Addition of Acid/Base

The concept of electrostatic stoppers in mechanically interlocked molecules was reported by Luna-Ixmatlahua et al. (Luna-Ixmatlahua et al. 2019) wherein they displayed stoppers as chemically sensitive end groups on a linear axle molecule that allowed the conversion of a pseudo-rotaxane species into a rotaxane complex by a change in the medium acidity. Phosphono functional group was used as chemically sensitive end groups which can switch from neutral (phosphonic acid) to a dianionic state (phosphonate) by controlling the acidity of the medium. The pseudo-rotaxane structure consisted of an axle that comprises of a viologen fragment as a positively charged recognition motif, flanked by chemically responsive phosphono end groups and a sulfonated dibenzo-24-membered crown ether, $[DSDB24C8]^{2-}$, as a ring (Fig. 6.40) where the presence of the highly acidic sulfonate groups on the macrocycle assures that the ring would remain anionic under the experimental conditions, regardless of pH. Under acidic conditions, the axle featured electrostatically neutral phosphonic acid end groups, $[PEV-H]^{2+}$, allowing for unrestricted ring threading/dethreading, that is, a pseudo-rotaxane host-guest complex. In acidic medium, bright-yellow color solution (λ_{max}=380 nm) indicated the formation of a charge-transfer complex between the electron-rich crown ether and the electron-poor viologen. There was one viologen unit per macrocycle; this 1:1 stoichiometry was confirmed by high-resolution mass spectrometry. On addition of base, end groups transformed into negatively charged phosphonate units, [PEV]$^{2-}$) which resulted in rendering the dethreading energy barrier insurmountable for the anionic crown ether host, thereby giving rise to an interlocked species, a [2]rotaxane. The reason for this kind of behavior was a fast proton-transfer-reaction between a base and the pseudo-rotaxane complex which induced end groups deprotonation (phosphonate) before ring extrusion could occur, resulted in the formation of a rotaxane complex with electrostatic stoppers.

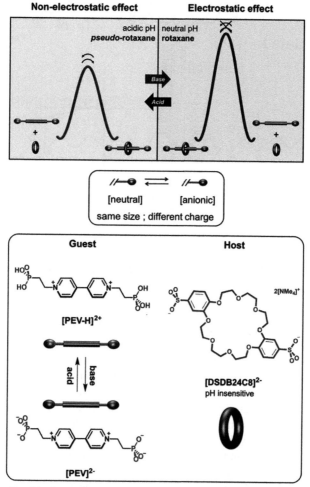

Figure 6.40 Graphical representation of the interconversion between a pseudorotaxane and a rotaxane species by means of a chemical stimulus affecting the electrostatic charge on the axle end groups (top). Proposed guest and host compounds (bottom) with permission from reprinted (Luna-Ixmatlahua et al. 2019).

By using resonance-stabilized imine bonds with enhanced hydrolytic stability (Blanco-Gómez et al. 2019), Neira et al. (Neira et al. 2019) reported the self-assembly of [2]rotaxane $3a^{4+}$; in aqueous media at pD 4 by the threading into CB[7] of an appropriate axle ended with acyl hydrazine groups and its concurrent capping with two molecules of a triphenylphosphonium aldehyde as stoppers (threading–capping strategy). The dynamism of the rotaxane was demonstrated (Fig. 6.41) under acidic conditions and was reported to get diminished by solvent swapping, pH modulation, or by the removal of the carbonyl groups on the axle (i.e., by using axle $32b^{2+}$ instead of $32a^{2+}$ on the kinetically controlled synthesis of the analogue $33b^{4+}$). These results open the door for the design of a reliable general strategy for the self-assembly of new rotaxane architectures of adjustable dynamism in aqueous media by using classic imine chemistry.

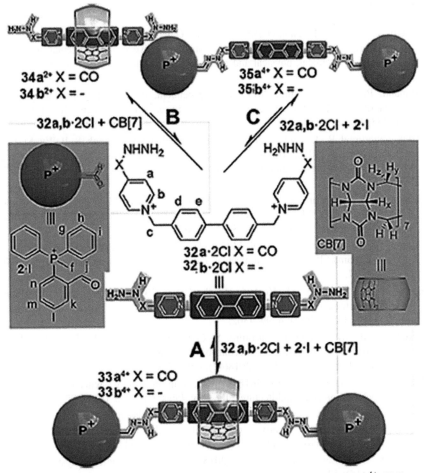

Figure 6.41 (a) Designed Self-Assembly Strategy for the [2]Rotaxanes $3a,b^{4+}$ (b) Formed by Orthogonal Hydrophobic/Electrostatic-Guided Slipping of Axles $1a,b^{2+}$ into CB[7] (c) and the Imine Capping of the corresponding Pseudorotaxanes reprinted with permission from (Neira et al. 2019)

CONCLUSION

In this chapter, the pH-controlled motions that involve non-covalent interactions, including hydrophobic effects, hydrogen bonding, electrostatic interactions, and others between the subcomponents of switchable assembly are usually sensitive to pH variation, which makes them one of the most powerful and commonly used means of controlling molecular motions. By adopting various host-guest recognition units, different types of motions including association/dissociation mobility, shuttling of the wheel along the axle and rotation in pseudorotaxanes can be efficiently controlled by acid-base chemistry.

REFERENCES

Alberti, S., G.J. Soler-Illia and O. Azzaroni. 2015. Gated supramolecular chemistry in hybrid mesoporous silica nanoarchitectures: controlled delivery and molecular transport in response to chemical, physical and biological stimuli. Chem. Commun. 51(28): 6050–6075.

Altmann, P.J. and A. Pöthig. 2017. A pH-dependent, mechanically interlocked switch: organometallic[2]rotaxane vs. organic[3]rotaxane. Angew. Chem. Int. Ed. 56(49): 15733–15736.

Arumugaperumal, R., P. Venkatesan, T. Shukla, P. Raghunath, R. Singh, S.-P. Wu, M.-C. Lin and H.-C. Lin. 2018. Multi-stimuli-responsive high contrast fluorescence molecular controls with a far-red emitting BODIPY-based [2]rotaxane. Sens. Actuators, B. 270: 382–395.

Ashton, P.R., R. Ballardini, V. Balzani, I. Baxter, A. Credi, M.C. Fyfe, M.T. Gandolfi, M. Gómez-López, M.-V. Martínez-Díaz and A. Piersanti. 1998a. Acid–base controllable molecular shuttles. J. Am. Chem. Soc. 120(46): 11932–11942.

Ashton, P.R., I. Baxter, S.J. Cantrill, M.C. Fyfe, P.T. Glink, J.F. Stoddart, A.J. White and D.J. Williams. 1998b. Supramolecular daisy chains. Angew. Chem. Int. Ed. 37(9): 1294–1297.

Baroncini, M., S. Silvi, M. Venturi and A. Credi. 2012. Photoactivated directionally controlled transit of a non-symmetric molecular axle through a macrocycle. Angew. Chem. Int. Ed. 124(17): 4299–4302.

Baroncini, M., M. Canton, L. Casimiro, S. Corra, J. Groppi, M. La Rosa, S. Silvi and A. Credi. 2018. Photoactive molecular-based devices, machines and materials: recent advances. Eur. J. Inorg. Chem. 2018(42): 4589–4603.

Bazargan, G. and K. Sohlberg. 2018. Advances in modelling switchable mechanically interlocked molecular architectures. Int. Rev. Phys. Chem. 37(1): 1–82.

Belowich, M.E., C. Valente, R.A. Smaldone, D.C. Friedman, J. Thiel, L. Cronin and J.F. Stoddart. 2012. Positive cooperativity in the template-directed synthesis of monodisperse macromolecules. J. Am. Chem. Soc. 134(11): 5243–5261.

Blanco, V., D.A. Leigh and V. Marcos. 2015. Artificial switchable catalysts. Chem. Soc. Rev. 44(15): 5341–5370.

Blanco-Gómez, A., I. Neira, J.L. Barriada, M. Melle-Franco, C. Peinador and M.D. García. 2019. Thinking outside the "Blue Box": from molecular to supramolecular pH-responsiveness. Chem. Sci. 10(46): 10680–10686.

Blanco-Gómez, A., P. Cortón, L. Barravecchia, I. Neira, E. Pazos, C. Peinador and M.D. García. 2020. Controlled binding of organic guests by stimuli-responsive macrocycles. Chem. Soc. Rev. 49: 3834–3862.

Bleger, D. and S. Hecht. 2015. Visible-light-activated molecular switches. Angew. Chem. Int. Ed. 54(39): 11338–11349.

Braunschweig, A.B., C.M. Ronconi, J.Y. Han, F. Aricó, S.J. Cantrill, J.F. Stoddart, S.I. Khan, A.J. White and D.J. Williams. 2006. Pseudorotaxanes and rotaxanes formed by viologen derivatives. Eur. J. Org. Chem. 2006(8): 1857–1866.

Browne, W.R. and B.L. Feringa. 2010. Light and redox switchable molecular components for molecular electronics. Chimia 64(6): 398–403.

Bruns, C.J., M. Frasconi, J. Iehl, K.J. Hartlieb, S.T. Schneebeli, C. Cheng, S.I. Stupp and J.F. Stoddart. 2014a. Redox switchable daisy chain rotaxanes driven by radical–radical interactions. J. Am. Chem. Soc. 136(12): 4714–4723.

Bruns, C.J., J. Li, M. Frasconi, S.T. Schneebeli, J. Iehl, H.P. Jacquot de Rouville, S.I. Stupp, G.A. Voth and J.F. Stoddart 2014b. An electrochemically and thermally switchable donor–acceptor [c2] daisy chain rotaxane. Angew. Chem. Int. Ed. 126(7): 1984–1989.

Bruns, C.J. and J.F. Stoddart. 2014. Rotaxane-based molecular muscles. ACC. Chem. Res. 47(7): 2186–2199.

Bruns, C.J. and J.F. Stoddart. 2016. The Nature of the Mechanical Bond: From Molecules to Machines. John Wiley & Sons.

Cao, D., Y. Kou, J. Liang, Z. Chen, L. Wang and H. Meier. 2009. A facile and efficient preparation of pillararenes and a pillarquinone. Angew. Chem. Int. Ed. 48(51): 9721–9723.

Cao, Y., X.-Y. Hu, Y. Li, X. Zou, S. Xiong, C. Lin, Y.-Z. Shen and L. Wang. 2014. Multistimuli-responsive supramolecular vesicles based on water-soluble pillar[6]arene and SAINT complexation for controllable drug release. J. Am. Chem. Soc. 136(30): 10762–10769.

Chen, M., S. Han, L. Jiang, S. Zhou, F. Jiang, Z. Xu, J. Liang and S. Zhang. 2010. New switchable [2]pseudorotaxanes formed by pyridine N-oxide derivatives with diamide-based macrocycles. Chem. Commun. 46(22): 3932–3934.

Chen, Q., J. Sun, P. Li, I. Hod, P.Z. Moghadam, Z.S. Kean, R.Q. Snurr, J.T. Hupp, O.K. Farha and J.F. Stoddart. 2016. A redox-active bistable molecular switch mounted inside a metal–organic framework. J. Am. Chem. Soc. 138(43): 14242–14245.

Corra, S., C. de Vet, J. Groppi, M. La Rosa, S. Silvi, M. Baroncini and A. Credi. 2019. Chemical on/off switching of mechanically planar chirality and chiral anion recognition in a [2]rotaxane molecular shuttle. J. Am. Chem. Soc. 141(23): 9129–9133.

Coutrot, F., E. Busseron and J.-L. Montero. 2008a. A very efficient synthesis of a mannosyl orthoester [2]rotaxane and mannosidic [2]rotaxanes. Org. Lett. 10(5): 753–756.

Coutrot, F., C. Romuald and E. Busseron. 2008b. A new pH-switchable dimannosyl [c2] daisy chain molecular machine. Org. Lett. 10(17): 3741–3744.

Coudert, F.-X. 2015. Responsive metal–organic frameworks and framework materials: under pressure, taking the heat, in the spotlight, with friends. Chem. Mater. 27(6): 1905–1916.

Cragg, P.J. and K. Sharma. 2012. Pillar[5]arenes: fascinating cyclophanes with a bright future. Chem. Soc. Rev. 41(2): 597–607.

Credi, A., V. Balzani, S.J. Langford and J.F. Stoddart. 1997. Logic operations at the molecular level. An XOR gate based on a molecular machine. J. Am. Chem. Soc. 119(11): 2679–2681.

Crini, G.J. 2014. Review: a history of cyclodextrins. Chem. Rev. 114(21): 10940–10975.

Croft, A.P. and R.A. Bartsch. 1983. Synthesis of chemically modified cyclodextrins. Tetrahedron. 39(9): 1417–1474.

Crowley, J.D., K.D. Hänni, D.A. Leigh and A.M.Z. Slawin. 2010. Diels–Alder active-template synthesis of rotaxanes and metal-ion-switchable molecular shuttles. J. Am. Chem. Soc. 132(14): 5309–5314.

Dietrich, B., J. Lehn and J. Sauvage. 1969a. Diaza-polyoxa-macrocycles et macrobicycles. Tetrahedron Lett. 10(34): 2885–2888.

Dietrich, B., J. Lehn and J. Sauvage. 1969b. Les cryptates. Tetrahedron Lett. 10(34): 2889–2892.

Ding, Z.-J., H.-Y. Zhang, L.-H. Wang, F. Ding and Y. Liu. 2011. A heterowheel [3]pseudorotaxane by integrating β-cyclodextrin and cucurbit[8]uril inclusion complexes. Org. Lett. 13(5): 856–859.

Ding, C., Y. Liu, T. Wang and J. Fu. 2016. Triple-stimuli-responsive nanocontainers assembled by water-soluble pillar[5]arene-based pseudorotaxanes for controlled release. J. Mater. Chem. B. 4(16): 2819–2827.

Diniz, A.M., N. Basílio, H. Cruz, F. Pina and A.J. Parola. 2015. Spatiotemporal control over the co-conformational switching in pH-responsive flavylium-based multistate pseudorotaxanes. Faraday Discuss. 185: 361–379.

Dragulescu-Andrasi, A., S.-R. Kothapalli, G.A. Tikhomirov, J. Rao and S.S. Gambhir. 2013. Activatable oligomerizable imaging agents for photoacoustic imaging of furin-like activity in living subjects. J. Am. Chem. Soc. 135(30): 11015–11022.

Duan, Q., W. Xia, X. Hu, M. Ni, J. Jiang, C. Lin, Y. Pan and L. Wang. 2012. Novel [2]pseudorotaxanes constructed by self-assembly of bis-urea-functionalized pillar[5] arene and linear alkyl dicarboxylates. Chem. Commun. 48(68): 8532–8534.

Farahani, N., K. Zhu and S.J.J.C. Loeb. 2016. Rigid, bistable molecular shuttles combining T-shaped benzimidazolium and Y-shaped imidazolium recognition sites. ChemPhysChem. 17(12): 1875–1880.

Feringa, B.L. and W.R. Browne. 2011. Molecular Switches. John Wiley & Sons.

Fihey, A., A. Perrier, W.R. Browne and D. Jacquemin. 2015. Multiphotochromic molecular systems. Chem. Soc. Rev. 44(11): 3719–3759.

Franchi, P., C. Poderi, E. Mezzina, C. Biagini, S. Di Stefano and M. Lucarini. 2019. 2-Cyano-2-phenylpropanoic acid triggers the back and forth motions of an acid–base-operated paramagnetic molecular switch. J. Org. Chem. 84(14): 9364–9368.

Frisch, H. and P.J.M.R.C. Besenius. 2015. pH-Switchable self-assembled materials. Macromol. Rapid Commun. 36(4): 346–363.

Fujita, H., T. Ooya, M. Kurisawa, H. Mori, M. Terano and N. Yui. 1996. Thermally switchable polyrotaxane as a model of stimuli-responsive supramolecules for nano-scale devices. Macromol. Rapid Commun. 17(8): 509–515.

Gao, M., S. Han, Y. Hu, J.J. Dynes, X. Liu and D. Wang. 2016. A pH-driven molecular shuttle based on rotaxane-bridged periodic mesoporous organosilicas with responsive release of guests. RSC Adv. 6(33): 27922–27932.

Garaudée, S., S. Silvi, M. Venturi, A. Credi, A.H. Flood and J.F. Stoddart. 2005. Shuttling dynamics in an acid–base-switchable [2]rotaxane. ChemPhysChem. 6(10): 2145–2152.

Gibson, H.W., H. Wang, C. Slebodnick, J. Merola, W.S. Kassel and A.L. Rheingold. 2007. Isomeric 2, 6-pyridino-cryptands based on dibenzo-24-crown-8. J. Org. Chem. 72(9): 3381–3393.

Gil, E.S. and S.M. Hudson. 2004. Stimuli-reponsive polymers and their bioconjugates. Prog. Polym. Sci. 29(12): 1173–1222.

Gokel, G.W., W.M. Leevy and M.E. Weber. 2004. Crown ethers: sensors for ions and molecular scaffolds for materials and biological models. Chem. Rev. 104(5): 2723–2750.

Gong, H.-Y., B.M. Rambo, E. Karnas, V.M. Lynch, K.M. Keller and J.L. Sessler. 2011. Environmentally responsive threading, dethreading, and fixation of anion-induced pseudorotaxanes. J. Am. Chem. Soc. 133(5): 1526–1533.

Goujon, A., T. Lang, G. Mariani, E. Moulin, G. Fuks, J. Raya, E. Buhler and N. Giuseppone. 2017. Bistable [c2] daisy chain rotaxanes as reversible muscle-like actuators in mechanically active gels. J. Am. Chem. Soc. 139(42): 14825–14828.

Goujon, A., E. Moulin, G. Fuks and N. Giuseppone. 2019. [c2] Daisy chain rotaxanes as molecular muscles. CCS Chem. 1(1): 83–96.

Guan, Y., M. Ni, X. Hu, T. Xiao, S. Xiong, C. Lin and L. Wang. 2012. Pillar[5]arene-based polymeric architectures constructed by orthogonal supramolecular interactions. Chem. Commun. 48(68): 8529–8531.

Guang-Huan, W., Z. Bin, Y. Yang and L. Shijun. 2015. Acid/base-controllable molecular machines and molecular switches. Chin. J. Org. Chem. 35(2): 309–324.

Guha, S., G.K. Shaw, T.M. Mitcham, R.R. Bouchard and B.D. Smith. 2016. Croconaine rotaxane for acid activated photothermal heating and ratiometric photoacoustic imaging of acidic pH. Chem. Commun. 52(1): 120–123.

Han, T. and C.-F. Chen. 2006. A triptycene-based bis (crown ether) host: complexation with both paraquat derivatives and dibenzylammonium salts. Org. Lett. 8(6): 1069–1072.

Han, C., F. Ma, Z. Zhang, B. Xia, Y. Yu and F. Huang. 2010. DIBPillar[n]arenes (n = 5, 6): syntheses, X-ray crystal structures, and complexation with n-Octyltriethyl ammonium hexafluorophosphate. Org. Lett. 12(19): 4360–4363.

Han, C., G. Yu, B. Zheng and F. Huang. 2012. Complexation between pillar[5]arenes and a secondary ammonium salt. Org. Lett. 14(7): 1712–1715.

Harada, A., A. Hashidzume, H. Yamaguchi and Y. Takashima. 2009. Polymeric rotaxanes. Chem. Rev. 109(11): 5974–6023.

He, Z., Q. Gu, J. Ma, T. Cheng, J. Ou, H. Wang and J. Sun. 2018. The pH-controlled [2] or [3]pseudorotaxanes based on stilbene dye SD⊂CB [7]. Supramol. Chem. 30(11): 955–959.

Holler, M., T. Stoerkler, A. Louis, F. Fischer and J.F. Nierengarten. 2019. Mechanochemical solvent-free conditions for the synthesis of pillar[5]arene-containing [2]rotaxanes. Eur. J. Org. Chem. 2019(21): 3401–3405.

Hu, X.-B., L. Chen, W. Si, Y. Yu and J.-L. Hou. 2011. Pillar[5]arene decaamine: synthesis, encapsulation of very long linear diacids and formation of ion pair-stopped [2]rotaxanes. Chem. Commun. 47(16): 4694–4696.

Hu, X.-Y., X. Liu, W. Zhang, S. Qin, C. Yao, Y. Li, D. Cao, L. Peng and L. Wang. 2016. Controllable construction of biocompatible supramolecular micelles and vesicles by water-soluble phosphate pillar[5,6]arenes for selective anti-cancer drug delivery. Chem. Mater. 28(11): 3778–3788.

Huang, F., J.W. Jones, C. Slebodnick and H.W. Gibson. 2003. Ion pairing in fast-exchange host−guest systems: concentration dependence of apparent association constants for complexes of neutral hosts and divalent guest salts with monovalent counterions. J. Am. Chem. Soc. 125(47): 14458–14464.

Hwang, I., W.S. Jeon, H.J. Kim, D. Kim, H. Kim, N. Selvapalam, N. Fujita, S. Shinkai and K.J.A.C. Kim. 2007. Cucurbit[7]uril: a simple macrocyclic, pH-triggered hydrogelator exhibiting guest-induced stimuli-responsive behavior. Angew. Chem. Int. Ed. 119(1-2): 214–217.

Im Jun, S., J.W. Lee, S. Sakamoto, K. Yamaguchi and K. Kim. 2000. Rotaxane-based molecular switch with fluorescence signaling. Tetrahedron Lett. 41(4): 471–475.

Isaacs, L. 2014. Stimuli responsive systems constructed using cucurbit[n]uril-type molecular containers. Acc. Chem. Res. 47(7): 2052–2062.

Jena, N.K. and N.A. Murugan. 2013. Solvent-dependent conformational states of a [2]rotaxane-based molecular machine: a molecular dynamics perspective. J. Phys. Chem. C. 117(47): 25059–25068.

Jeong, K.-S., K.-J. Chang and Y.-J. An. 2003. A pseudorotaxane-based molecular machine controlled by light and thermal stimuli. Chem. Commun. (12): 1450–1451.

Ji, X., M. Zhang, X. Yan, J. Li and F. Huang. 2013. Synthesis of a water-soluble bis (m-phenylene)-32-crown-10-based cryptand and its pH-responsive binding to a paraquat derivative. Chem. Commun. 49(12): 1178–1180.

Jiménez, M.C., C. Dietrich-Buchecker and J.P. Sauvage. 2000. Towards synthetic molecular muscles: contraction and stretching of a linear rotaxane dimer. Angew. Chem. Int. Ed. 39(18): 3284–3287.

Karimi, M., A. Ghasemi, P.S. Zangabad, R. Rahighi, S.M.M. Basri, H. Mirshekari, M. Amiri, Z.S. Pishabad, A. Aslani and M. Bozorgomid. 2016. Smart micro/nanoparticles in stimulus-responsive drug/gene delivery systems. Chem. Soc. Rev. 45(5): 1457–1501.

Kay, E.R., D.A. Leigh and F. Zerbetto. 2007. Synthetic molecular motors and mechanical machines. Angew. Chem. Int. Ed. 46(1-2): 72–191.

Kay, E.R. and D.A. Leigh. 2015. Rise of the molecular machines. Angew. Chem. Int. Ed. 54(35): 10080–10088.

Keaveney, C.M. and D.A. Leigh. 2004. Shuttling through anion recognition. Angew. Chem. Int. Ed. 116(10): 1242–1244.

Khan, A.R., P. Forgo, K.J. Stine and V.T. D'Souza. 1998. Methods for selective modifications of cyclodextrins. Chem. Rev. 98(5): 1977–1996.

Kimura, M., T. Mizuno, M. Ueda, S. Miyagawa, T. Kawasaki and Y. Tokunaga. 2017. Four-state molecular shuttling of [2]rotaxanes in response to acid/base and alkali-metal cation stimuli. Chem. Asian J. 12(12): 1381–1390.

Klein, H.A. and P.D. Beer. 2019. Iodide discrimination by tetra-iodotriazole halogen bonding interlocked hosts. Chem. Asian J. 25(12): 3125–3130.

Ko, Y.H., E. Kim, I. Hwang and K. Kim. 2007. Supramolecular assemblies built with host-stabilized charge-transfer interactions. Chem. Commun. (13): 1305–1315.

Kureha, T., T. Shibamoto, S. Matsui, T. Sato and D. Suzuki. 2016. Investigation of changes in the microscopic structure of anionic poly (N-isopropylacrylamide-co-acrylic acid) microgels in the presence of cationic organic dyes toward precisely controlled uptake/release of low-molecular-weight chemical compound. Langmuir. 32(18): 4575–4585.

Kureha, T., D. Aoki, S. Hiroshige, K. Iijima, D. Aoki, T. Takata and D. Suzuki. 2017. Decoupled thermo- and pH-responsive hydrogel microspheres cross-linked by rotaxane networks. Angew. Chem. Int. Ed. 56(48): 15393–15396.

Kyba, E.P., R.C. Helgeson, K. Madan, G.W. Gokel, T.L. Tarnowski, S.S. Moore and D.J. Cram. 1977. Host-guest complexation. 1. Concept and illustration. J. Am. Chem. Soc. 99(8): 2564–2571.

Lee, J.W., S. Samal, N. Selvapalam, H.-J. Kim and K. Kim. 2003. Cucurbituril homologues and derivatives: new opportunities in supramolecular chemistry. Acc. Chem. Res. 36(8): 621–630.

Lehn, J.-M. 2002. Toward self-organization and complex matter. Science. 295(5564): 2400–2403.

Lehn, J.-M. 2007. From supramolecular chemistry towards constitutional dynamic chemistry and adaptive chemistry. Chem. Soc. Rev. 36(2): 151–160.

Lehn, J.-M. 2009. Towards complex matter: supramolecular chemistry and self-organization. Eur. Rev. (2): 263–280.

Lehn, J.-M. 2013. Perspectives in chemistry—steps towards complex matter. Angew. Chem. Int. Ed. 52(10): 2836–2850.

Lehn, J.-M. 2015. Perspectives in chemistry—aspects of adaptive chemistry and materials. Angew. Chem. Int. Ed. 54(11): 3276–3289.

Lehn, J.-M. 2017. Supramolecular chemistry: Where from? where to? Chem. Soc. Rev. 46(9): 2378–2379.

Leung, K.C.-F., C.-P. Chak, C.-M. Lo, W.-Y. Wong, S. Xuan and C.H. Cheng. 2009. Focus Reviews. Chem. Asian J. 4: 364–381.

Li, C., L. Zhao, J. Li, X. Ding, S. Chen, Q. Zhang, Y. Yu and X. Jia. 2010. Self-assembly of [2]pseudorotaxanes based on pillar[5]arene and bis (imidazolium) cations. Chem. Commun. 46(47): 9016–9018.

Li, C., X. Shu, J. Li, S. Chen, K. Han, M. Xu, B. Hu, Y. Yu and X. Jia. 2011. Complexation of 1,4-Bis(pyridinium) butanes by negatively charged carboxylatopillar[5]arene. J. Org. Chem. 76(20): 8458–8465.

Li, S., G.H. Weng, B. Zheng, X. Yan, J. Wu, W. Lin, H.X. Chen and X.C. Zhang. 2012. Complexation of paraquat and diazapyrenium derivatives with dipyrido[30]crown-10. Eur. J. Org. Chem. 2012(33): 6570–6575.

Li, F.-z., L. Mei, K.-q. Hu, S.-w. An, S. Wu, N. Liu, Z.-f. Chai and W.-q. Shi. 2019. Uranyl compounds involving a weakly bonded pseudorotaxane linker: combined effect of pH and competing ligands on uranyl coordination and speciation. Inorg. Chem. 58(5): 3271–3282.

Li, C., H. Li, J. Guo, L. Li, X. Xi and Y. Yu. 2020a. Biocompatible supramolecular pseudorotaxane hydrogels for controllable release of doxorubicin in ovarian cancer SKOV-3 cells. RSC Adv. 10(2): 689–697.

Li, J., W. Jia, G. Ma, X. Zhang, S. An, T. Wang and S. Shi. 2020b. Construction of pH sensitive smart glutathione peroxidase (GPx) mimics based on pH responsive pseudorotaxanes. Org. Biomol. Chem. 18(16): 3125–3134.

Lin, R.-L., R. Li, H. Shi, K. Zhang, D. Meng, W.-Q. Sun, K. Chen and J.-X. Liu. 2020. Symmetrical-tetramethyl-cucurbit[6]uril-driven movement of cucurbit[7]uril gives rise to heterowheel[4]pseudorotaxanes. J. Org. Chem. 85(5): 3568–3575.

Liu, Y., S. Saha, S.A. Vignon, A.H. Flood and J.F. Stoddart. 2005. Template-directed syntheses of configurable and reconfigurable molecular switches. Synth. 2005(19): 3437–3445.

Liu, Z., S.K.M. Nalluri and J.F. Stoddart. 2017. Surveying macrocyclic chemistry: from flexible crown ethers to rigid cyclophanes. Chem. Soc. Rev. 46(9): 2459–2478.

Loeb, S. and J. Wisner. 1998a. 1,2-Bis(4, 4′-dipyridinium) ethane: a versatile dication for the formation of [2]rotaxanes with dibenzo-24-crown-8 ether. Chem. Commun. (24): 2757–2758.

Loeb, S.J. and J.A. Wisner. 1998b. A new motif for the self-assembly of [2]pseudorotaxanes; 1, 2-Bis(pyridinium) ethane axles and [24] crown-8 ether wheels. Angew. Chem. Int. Ed. 37(20): 2838–2840.

Luna-Ixmatlahua, R.A., A. Carrasco-Ruiz, R. Cervantes, A. Vela and J. Tiburcio. 2019. An anionic ring locked into an anionic axle: a metastable rotaxane with chemically activated electrostatic stoppers. Chem. Eur. J. 25(62): 14042–14047.

Lutz, J.-F., J.-M. Lehn, E. Meijer and K. Matyjaszewski. 2016. From precision polymers to complex materials and systems. Nat. Rev. Mater. 1(5): 1–14.

Ma, Y., X. Ji, F. Xiang, X. Chi, C. Han, J. He, Z. Abliz, W. Chen and F. Huang. 2011. A cationic water-soluble pillar[5]arene: synthesis and host–guest complexation with sodium 1-octanesulfonate. Chem. Commun. 47(45): 12340–12342.

Mandl, C.P. and B. König. 2004. Chemistry in motion—unidirectional rotating molecular motors. Angew. Chem. Int. Ed. Engl. 43(13): 1622–1624.

Marti-Centelles, V., M.D. Pandey, M.I. Burguete and S.V. Luis. 2015. Macrocyclization reactions: the importance of conformational, configurational, and template-induced preorganization. Chem. Rev. 115(16): 8736–8834.

Martínez-Díaz, M.V., N. Spencer and J.F. Stoddart. 1997. The self-assembly of a switchable [2]rotaxane. Angew. Chem. Int. Ed. Engl. 36(17): 1904–1907.

Mateo-Alonso, A., G. Fioravanti, M. Marcaccio, F. Paolucci, D.C. Jagesar, A.M. Brouwer and M. Prato. 2006. Reverse shuttling in a fullerene-stopped rotaxane. Org. Lett. 8(22): 5173–5176.

Mendes, P.M. 2008. Stimuli-responsive surfaces for bio-applications. Chem. Soc. Rev. 37(11): 2512–2529.

Mercer, D.J. and S.J. Loeb. 2011. Complexes of a [2]rotaxane ligand with terminal terpyridine groups. Dalton Trans. 40(24): 6385–6387.

Mock, W.L. and J. Pierpont. 1990. A cucurbituril-based molecular switch. J. Chem. Soc., Chem. Commun. 21: 1509–1511.

Mura, S., J. Nicolas and P. Couvreur. 2013. Stimuli-responsive nanocarriers for drug delivery. Nat. Mater. 12(11): 991–1003.

Murray, J., K. Kim, T. Ogoshi, W. Yao and B.C. Gibb. 2017. The aqueous supramolecular chemistry of cucurbit[n]urils, pillar[n]arenes and deep-cavity cavitands. Chem. Soc. Rev. 46(9): 2479–2496.

Nam, J., N. Won, H. Jin, H. Chung and S. Kim. 2009. pH-induced aggregation of gold nanoparticles for photothermal cancer therapy. J. Am. Chem. Soc. 131(38): 13639–13645.

Neira, I., A. Blanco-Gómez, J.M. Quintela, C. Peinador and M.D. García. 2019. Adjusting the dynamism of covalent imine chemistry in the aqueous synthesis of cucurbit[7]uril-based [2]rotaxanes. Org. Lett. 21(22): 8976–8980.

Nierengarten, I., S. Guerra, M. Holler, J.-F. Nierengarten and R. Deschenaux. 2012. Building liquid crystals from the 5-fold symmetrical pillar[5]arene core. Chem. Commun. 48(65): 8072–8074.

Noureddine, A., L. Lichon, M. Maynadier, M. Garcia, M. Gary-Bobo, J.I. Zink, X. Cattoën and M.W.C. Man. 2015. Controlled multiple functionalization of mesoporous silica nanoparticles: homogeneous implementation of pairs of functionalities communicating through energy or proton transfers. Nanoscale. 7(26): 11444–11452.

Ogoshi, T., S. Kanai, S. Fujinami, T.-a. Yamagishi and Y. Nakamoto. 2008. Para-bridged symmetrical pillar[5]arenes: their Lewis acid catalyzed synthesis and host–guest property. J. Am. Chem. Soc. 130(15): 5022–5023.

Ogoshi, T., M. Hashizume, T.-a. Yamagishi and Y. Nakamoto. 2010. Synthesis, conformational and host–guest properties of water-soluble pillar[5]arene. Chem. Commun. 46(21): 3708–3710.

Ogoshi, T. 2012. Synthesis of novel pillar-shaped cavitands "Pillar[5]arenes" and their application for supramolecular materials. J. Incl. Phenom. Macrocycl. Chem. 72(3): 247–262.

Ogoshi, T., R. Shiga and T.-a. Yamagishi. 2012a. Reversibly tunable lower critical solution temperature utilizing host–guest complexation of pillar[5]arene with triethylene oxide substituents. J. Am. Chem. Soc. 134(10): 4577–4580.

Ogoshi, T., H. Kayama, D. Yamafuji, T. Aoki and T.-a. Yamagishi. 2012b. Supramolecular polymers with alternating pillar[5]arene and pillar[6]arene units from a highly selective multiple host–guest complexation system and monofunctionalized pillar[6]arene. Chem. Sci. 3(11): 3221–3226.

Ogoshi, T., D. Yamafuji, T. Aoki and T.-a. Yamagishi. 2012c. Thermally responsive shuttling behavior of a pillar[6]arene-based [2]rotaxane. Chem. Commun. 48(54): 6842–6844.

Ogoshi, T., T.-a. Yamagishi and Y. Nakamoto. 2016. Pillar-shaped macrocyclic hosts pillar[n]arenes: new key players for supramolecular chemistry. Chem. Rev. 116(14): 7937–8002.

Qu, D.-H., Q.-C. Wang, Q.-W. Zhang, X. Ma and H. Tian. 2015. Photoresponsive host–guest functional systems. Chem. Rev. 115(15): 7543–7588.

Ragazzon, G., M. Baroncini, S. Silvi, M. Venturi and A. Credi. 2015. Light-powered, artificial molecular pumps: a minimalistic approach. Beilstein J. Nanotechnol. 6(1): 2096–2104.

Rambo, B.M., H.-Y. Gong, M. Oh and J.L. Sessler. 2012. The "Texas-sized" molecular box: A versatile building block for the construction of anion-directed mechanically interlocked structures. Acc. Chem. Res. 45(8): 1390–1401.

Riss-Yaw, B., P. Waelès and F. Coutrot. 2016. Reverse anomeric effect in large-amplitude pyridinium amide-containing mannosyl [2]rotaxane molecular shuttles. ChemPhysChem. 17(12): 1860–1869.

Romuald, C., E. Busseron and F. Coutrot. 2010. Very contracted to extended co-conformations with or without oscillations in two- and three-station [c2] daisy chains. J. Org. Chem. 75(19): 6516–6531.

Safarowsky, O., B. Windisch, A. Mohry and F. Vögtle. 2000. Nomenclature for catenanes, rotaxanes, molecular knots, and assemblies derived from these structural elements. J. Prakt. Chem. 342(5): 437–444.

Saha, S., K.F. Leung, T.D. Nguyen, J.F. Stoddart and J.I.J.A.F.M. Zink. 2007. Nanovalves. Adv. Funct. Mater. 17(5): 685–693.

Sanchez, C., P. Belleville, M. Popall and L. Nicole. 2011. Applications of advanced hybrid organic–inorganic nanomaterials: from laboratory to market. Chem. Soc. Rev. 40(2): 696–753.

Schwarz, J.A., C.I. Contescu and K. Putyera. 2004. Dekker Encyclopedia of Nanoscience and Nanotechnology. CRC Press.

Seco, A., A. M. Diniz, J. Sarrato, H. Mourão, H. Cruz, A.J. Parola and N. Basílio. 2020. A pseudorotaxane formed from a cucurbit[7]uril wheel and a bioinspired molecular axle with pH, light and redox-responsive properties. Pure Appl. Chem. 92(2): 301–313.

Share, A.I., K. Parimal and A.H. Flood. 2010. Bilability is defined when one electron is used to switch between concerted and stepwise pathways in Cu(I)-based bistable [2/3] pseudorotaxanes. J. Am. Chem. Soc. 132(5): 1665–1675.

Shi, H., K. Zhang, R.L. Lin, W.Q. Sun, X.F. Chu, X.H. Liu and J.-X. Liu. 2019. pH-Controlled multiple interconversion between cucurbit[7]uril-based molecular shuttle, [3]pseudorotaxane and [2]pseudorotaxane. Asian J. Org. Chem. 8(3): 339–343.

Shinkai, S. and B. Feringa. 2001. Molecular Switches. Wiley-VCH Verlag GmbH: Weinheim, Germany.

Shinohara, A., C. Pan, L. Wang and H. Shinmori. 2020. Acid–base controllable singlet oxygen generation in supramolecular porphyrin–gold nanoparticle composites tethered by rotaxane linkers. J. Porphyr. Phthalocyanines. 24(01n03): 171–180.

Shukla, T., R. Arumugaperumal, P. Raghunath, M.-C. Lin, C.-M. Lin and H.-C. Lin. 2017. Novel supramolecular conjugated polyrotaxane as an acid-base controllable optical molecular switch. Sens. Actuators B Chem. 243: 84–95.

Sinn, S. and F. Biedermann. 2018. Chemical sensors based on cucurbit[n]uril macrocycles. Isr. J. Chem. 58(3-4): 357–412.

Sollogoub, M. 2013. Site-selective heterofunctionalization of cyclodextrins: discovery, development, and use in catalysis. Synlett. 24(20): 2629–2640.

Song, N. and Y.-W. Yang. 2015. Molecular and supramolecular switches on mesoporous silica nanoparticles. Chem. Soc. Rev. 44(11): 3474–3504.

Stoddart, J.F. 2009. The chemistry of the mechanical bond. Chem. Soc. Rev. 38(6): 1802–1820.

Strutt, N.L., R.S. Forgan, J.M. Spruell, Y.Y. Botros and J.F. Stoddart. 2011. Monofunctionalized pillar[5]arene as a host for alkanediamines. J. Am. Chem. Soc. 133(15): 5668–5671.

Strutt, N.L., D. Fairen-Jimenez, J. Iehl, M.B. Lalonde, R.Q. Snurr, O.K. Farha, J.T. Hupp and J.F. Stoddart. 2012. Incorporation of an A1/A2-difunctionalized pillar[5]arene into a metal–organic framework. J. Am. Chem. Soc. 134(42): 17436–17439.

Strutt, N.L., H. Zhang, S.T. Schneebeli and J.F. Stoddart. 2014. Functionalizing pillar[n] arenes. Acc. Chem. Res. 47(8): 2631–2642.

Stuart, M.A.C., W.T. Huck, J. Genzer, M. Müller, C. Ober, M. Stamm, G.B. Sukhorukov, I. Szleifer, V.V. Tsukruk and M. Urban. 2010. Emerging applications of stimuli-responsive polymer materials. Nat. Mater. 9(2): 101–113.

Sun, H.L., H.Y. Zhang, Z. Dai, X. Han and Y. Liu. 2017. Insights into the difference between rotaxane and pseudorotaxane. Asian J. Chem. 12(2): 265–270.

Tao, R., Q. Zhang, S. Rao, X. Zheng, M. Li and D. Qu. 2019. Supramolecular gelator based on a [c2] daisy chain rotaxane: efficient gel-solution transition by ring-sliding motion. Sci. China Chem. 62(2): 245–250.

Thibeault, D. and J.-F. Morin. 2010. Recent advances in the synthesis of ammonium-based rotaxanes. Molecules. 15(5): 3709–3730.

Tokunaga, Y., M. Kimura, M. Ueda, S. Miyagawa, T. Kawasaki and K. Hisada. 2016. Base–acid-induced translational isomerism in a branched [4]rotaxane. Tetrahedron Lett. 57(10): 1120–1123.

Torres, J.M.G.T., E. Nichols, J.F. MacGregor and T. Hoare. 2014. Designing multi-responsive polymers using latent variable methods. Polymer 55(2): 505–516.

Ueda, M., S. Terazawa, Y. Deguchi, M. Kimura, N. Matsubara, S. Miyagawa, T. Kawasaki and Y. Tokunaga. 2016. Five-state molecular shuttling of a pair of [2]rotaxanes: distinct outputs in response to acid and base stimuli. Asian J. Chem. 11(16): 2291–2300.

Wang, X.Z., X.Q. Li, X.B. Shao, X. Zhao, P. Deng, X.K. Jiang, Z.T. Li and Y.Q. Chen. 2003. Selective rearrangements of quadruply hydrogen-bonded dimer driven by donor–acceptor interaction. Asian J. Chem. 9(12): 2904–2913.

Wang, Y., G. Ping and C. Li. 2016. Efficient complexation between pillar[5]arenes and neutral guests: from host–guest chemistry to functional materials. Chem. Commun. 52(64): 9858–9872.

Wenz, G., B.-H. Han and A. Müller. 2006. Cyclodextrin rotaxanes and polyrotaxanes. Chem. Rev. 106(3): 782–817.

Wolf, A., J.J. Cid, E. Moulin, F. Niess, G. Du, A. Goujon, E. Busseron, A. Ruff, S. Ludwigs and N. Giuseppone. 2019. Unsymmetric bistable [c2] daisy chain rotaxanes which combine two types of electroactive stoppers. Eur. J. Org. Chem. 2019(21): 3421–3432.

Wu, J., K.C.F. Leung, D. Benítez, J.Y. Han, S.J. Cantrill, L. Fang and J.F. Stoddart. 2008. An acid–base-controllable [c2] daisy chain. Angew. Chem. Int. Ed. 47(39): 7470–7474.

Wu, W., S. Song, X. Cui, T. Sun, J.-X. Zhang and X.-L. Ni. 2018. pH-Switched fluorescent pseudorotaxane assembly of cucurbit[7]uril with bispyridinium ethylene derivatives. Chin. Chem. Lett. 29(1): 95–98.

Wu, D., F. Pan, G.-C. Fan, Z. Zhu, L. Gao, Y. Tao and Y. Kong. 2019. Efficient enantiorecognition of amino acids under a stimuli-responsive system: synthesis, characterization and application of electroactive rotaxane. Analyst. 144(21): 6415–6421.

Xia, D., L. Wang, X. Lv, J. Chao, X. Wei and P. Wang. 2018. Dual-responsive [2]pseudorotaxane on the basis of a pH-sensitive pillar[5]arene and its application in the fabrication of metallosupramolecular polypseudorotaxane. Macromolecules. 51(7): 2716–2722.

Xiao, T., L. Qi, W. Zhong, C. Lin, R. Wang and L. Wang. 2019. Stimuli-responsive nanocarriers constructed from pillar[n]arene-based supra-amphiphiles. Mater. Chem. Front. 3(10): 1973–1993.

Xu, T., J. Li, J. Cao, W. Gao, L. Li and B. He. 2017. The effect of α-cyclodextrin on poly (pseudo) rotaxane nanoparticles self-assembled by protoporphyrin modified poly (ethylene glycol) for anticancer drug delivery. Carbohydr. Polym. 174: 789–797.

Xue, Z. and M.F. Mayer. 2010. Actuator prototype: capture and release of a self-entangled [1]rotaxane. J. Am. Chem. Soc. 132(10): 3274–3276.

Xue, M., Y. Yang, X. Chi, Z. Zhang and F. Huang. 2012. Pillararenes, a new class of macrocycles for supramolecular chemistry. Acc. Chem. Res. 45(8): 1294–1308.

Xue, M., Y. Yang, X. Chi, X. Yan and F. Huang. 2015. Development of pseudorotaxanes and rotaxanes: from synthesis to stimuli-responsive motions to applications. Chem. Rev. 115(15): 7398–7501.

Yakimova, L., D. Shurpik, L. Gilmanova, A. Makhmutova, A. Rakhimbekova and I. Stoikov. 2016. Highly selective binding of methyl orange dye by cationic water-soluble pillar[5] arenes. Org. Biomol. Chem. 14(18): 4233–4238.

Yamauchi, A., Y. Sakashita, K. Hirose, T. Hayashita and I. Suzuki. 2006. Pseudorotaxane-type fluorescent receptor exhibiting unique response to saccharides. Chem. Commun. (41): 4312–4314.

Yang, Y.-W., Y.-L. Sun and N. Song. 2014. Switchable host–guest systems on surfaces. Acc. Chem. Res. 47(7): 1950–1960.

Yang, K., S. Chao, F. Zhang, Y. Pei and Z. Pei. 2019. Recent advances in the development of rotaxanes and pseudorotaxanes based on pillar[n]arenes: from construction to application. Chem. Commun. 55(88): 13198–13210.

Yao, Y., M. Xue, J. Chen, M. Zhang and F. Huang. 2012. An amphiphilic pillar[5]arene: synthesis, controllable self-assembly in water, and application in calcein release and TNT adsorption. J. Am. Chem. Soc. 134(38): 15712–15715.

Ye, J., R. Zhang, W. Yang, Y. Han, H. Guo, J. Xie, C. Yan and Y. Yao. 2020. Pillar[5] arene-based [3]rotaxanes: convenient construction via multicomponent reaction and pH responsive self-assembly in water. Chin. Chem. Lett. 31(6): 1550–1553.

Yu, G., C. Han, Z. Zhang, J. Chen, X. Yan, B. Zheng, S. Liu and F. Huang. 2012a. Pillar[6] arene-based photoresponsive host–guest complexation. J. Am. Chem. Soc. 134(20): 8711–8717.

Yu, G., M. Xue, Z. Zhang, J. Li, C. Han and F. Huang. 2012b. A water-soluble pillar[6] arene: synthesis, host–guest chemistry, and its application in dispersion of multiwalled carbon nanotubes in water. J. Am. Chem. Soc. 134(32): 13248–13251.

Yu, G., Z. Zhang, C. Han, M. Xue, Q. Zhou and F. Huang. 2012c. A non-symmetric pillar[5] arene-based selective anion receptor for fluoride. Chem. Commun. 48(24): 2958–2960.

Yu, G., X. Zhou, Z. Zhang, C. Han, Z. Mao, C. Gao and F. Huang. 2012d. Pillar[6]arene/paraquat molecular recognition in water: high binding strength, pH-responsiveness, and application in controllable self-assembly, controlled release, and treatment of paraquat poisoning. J. Am. Chem. Soc 134(47): 19489–19497.

Yu, S., N.D. McClenaghan and J.-L. Pozzo. 2019. Photochromic rotaxanes and pseudorotaxanes. Photochem. Photobiol. Sci. 18(9): 2102–2111.

Zhang, H., Q. Wang, M. Liu, X. Ma and H. Tian. 2009. Switchable V-type [2]pseudorotaxanes. Org. Lett. 11(15): 3234–3237.

Zhang, H., N.L. Strutt, R.S. Stoll, H. Li, Z. Zhua and J.F. Stoddart. 2011. Dynamic clicked surfaces based on functionalised pillar[5]arene. Chem. Commun. 47(41): 11420–11422.

Zhang, Z., Y. Luo, J. Chen, S. Dong, Y. Yu, Z. Ma and F. Huang. 2011. Formation of linear supramolecular polymers that is driven by C–H⋯π interactions in solution and in the solid state. Angew. Chem. Int. Ed. 50(6): 1397–1401.

Zhang, Z., Y. Luo, B. Xia, C. Han, Y. Yu, X. Chen and F. Huang. 2011. Four constitutional isomers of BMpillar[5]arene: synthesis, crystal structures and complexation with n-octyltrimethyl ammonium hexafluorophosphate. Chem. Commun. 47(8): 2417–2419.

Zhang, Z., G. Yu, C. Han, J. Liu, X. Ding, Y. Yu and F. Huang. 2011. Formation of a cyclic dimer containing two mirror image monomers in the solid state controlled by van der Waals forces. Org. Lett. 13(18): 4818–4821.

Zhang, Z.-J., H.-Y. Zhang, L. Chen and Y. Liu. 2011. Interconversion between [5]pseudorotaxane and [3]pseudorotaxane by pasting/detaching two axle molecules. J. Org. Chem. 76(20): 8270–8276.

Zhang, Y., C.Y. Ang, M. Li, S.Y. Tan, Q. Qu, Z. Luo and Y. Zhao. 2015. Polymer-coated hollow mesoporous silica nanoparticles for triple-responsive drug delivery. ACS Appl. Mater. Interfaces. 7(32): 18179–18187.

Zhang, X., F. Zhang, F. Zhu, X. Zhang, D. Tian, R.P. Johnson and H. Li. 2019. Bioinspired γ-cyclodextrin pseudorotaxane assembly nanochannel for selective amino acid transport. ACS Appl. Bio Mater. 2(8): 3607–3612.

Zhao, Q., Y. Chen, B. Sun, C. Qian, M. Cheng, J. Jiang, C. Lin and L. Wang. 2019. Pillar[5] arene based pseudo[1]rotaxane operating as acid/base-controllable two state molecular shuttle. Eur. J. Org. Chem. 2019(21): 3396–3400.

Zhou, H.-Y., Y. Han, Q. Shi and C.-F. Chen. 2019. A triply operable molecular switch: anion-, acid/base- and solvent-responsive [2]rotaxane. Eur. J. Org. Chem. 2019(21), 3406–3411.

Zhou, H.-Y., Y. Han and C.-F. Chen. 2020. pH-Controlled motions in mechanically interlocked molecules. Mater. Chem. Front. 4(1): 12–28.

Zhu, K., L. Wu, X. Yan, B. Zheng, M. Zhang and F. Huang. 2010. Anion-assisted complexation of paraquat by cryptands based on bis(m-phenylene)-[32] crown-10. Chem. Asian J. 16(20): 6088–6098.

Zhu, K., V.N. Vukotic, N. Noujeim and S.J. Loeb. 2012. Bis(benzimidazolium) axles and crown ether wheels: a versatile templating pair for the formation of [2]rotaxane molecular shuttles. Chem. Sci. 3(11): 3265–3271.

Zhu, K., V.N. Vukotic and S.J. Loeb. 2016. Acid-base switchable [2]- and [3]rotaxane molecular shuttles with benzimidazolium and bis(pyridinium) recognition sites. Chem. Asian J. 11(22): 3258–3266.

Zhuang, J., M.R. Gordon, J. Ventura, L. Li and S. Thayumanavan. 2013. Multi-stimuli responsive macromolecules and their assemblies. Chem. Soc. Rev. 42(17): 7421–7435.

Zubillaga, A., P. Ferreira, A. Parola, S. Gago and N. Basílio. 2018. pH-Gated photoresponsive shuttling in a water-soluble pseudorotaxane. Chem. Commun. 54(22): 2743–2746.

Chapter 7

Macrocyclic Receptors for Biomolecules and Biochemical Sensing

INTRODUCTION

Recognition and sensing are the fundamental processes in nature. One has five sensing organs, i.e., the nose, tongue, ears, eyes and skin. Among these, ear, eyes and skin are perceived to be physical sensors as they detect physical sensations of sound, light and heat, etc., respectively. In contrast, the nose and tongue are chemical sensors where the sense of smell and taste is perceived through chemoreceptors by complicated mechanism. Although the nose or tongue as sense organs cannot quantify the concentration of a particular chemical that is being sensed the biorecognition property of these can provide a guiding principle in the development of artificial sensing systems for the identification and measurement of biomolecules. Since detection of biologically important molecules specifically biomolecules are very important in catalysis, drug delivery, surface immobilization and sensing a small section of cavitands, namely, cyclodextrins, cyclophanes, cucurbiturils, and molecular tweezers, which display specific, well-characterized recognition properties toward biomolecules (nucleic acid, carbohydrates, proteins and lipids) are described here.

INTRODUCTION TO STRUCTURE AND PROPERTIES: NUCLEOTIDES, NUCLEOSIDES AND NUCLEIC ACID

Nucleosides and Nucleotides

A nucleotide is the basic monomeric unit of Deoxyribose Nucleic Acid (DNA) and Ribose Nucleic Acid (RNA) that controls one's hereditary characteristics.

It also serves as an energy carrier such as adenosine triphosphate (ATP), Uridine 5′-triphosphate (UTP), Guanosine-5′-triphosphate (GTP) and acts as co-factors for enzymatic reactions, such as, the oxidized form of Nicotinamide Adenine Dinucleotide (NAD⁺), Flavin Adenine Dinucleotide (FAD), functions as a chemical messenger in cellular-signaling pathways like cyclic adenosine monophosphate (cAMP), cyclic guanosine monophosphate (cGMP), guanosine 3′, 5′-bispyrophosphate (ppGpp) and participates in phosphorylation chemistry. A nucleotide consists of a phosphate residue attached to a pentose sugar and a nucleobase as shown in Fig. 7.1. ATP is the power currency for cellular functions where energy is stored in phosphate groups attached by covalent bonds. It also acts as a co-transmitter in puringeric neurotransmission in both sympathetic and parasympathetic nerve purinergic neurotransmission in both the sympathetic and parasympathetic nerves. Other nucleotide triphosphates (GTP, UTP, CTP; Fig. 7.2) differ from each other due to the presence of various bases attached to sugar. They have similar chemical properties but perform different functions (Zala et al. 2017). Bioenergetically, GTP is similar to ATP and it is used in protein synthesis and signal transduction through G-protein and tubulin polymerization (Carvalho et al. 2015). UTP and CTP are used in polysaccharide and phospholipid synthesis, respectively. Redox nucleotide pairs (Fig. 7.3) nicotinamide adenine dinucleotide (NADH) and its oxidized form (NAD⁺) act as a coenzyme for many dehydrogenases. It is involved in mitochondrial function, energy metabolism, oxidative stress, calcium homeostasis, aging and apoptosis (Heikal 2010). Thus mitochondrial anomalies can be correlated with the redox pair (NAD⁺/NADH). Similarly, flavin adenine mononucleotide (FMN) and FAD are co-enzymes.

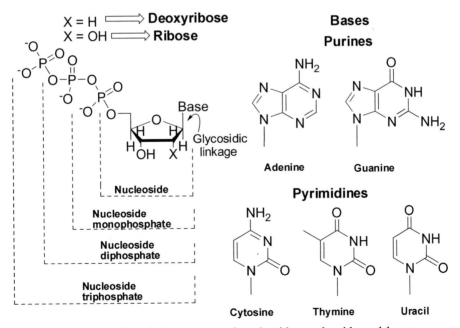

Figure 7.1 Chemical structure of nucleotide, nucleoside and bases.

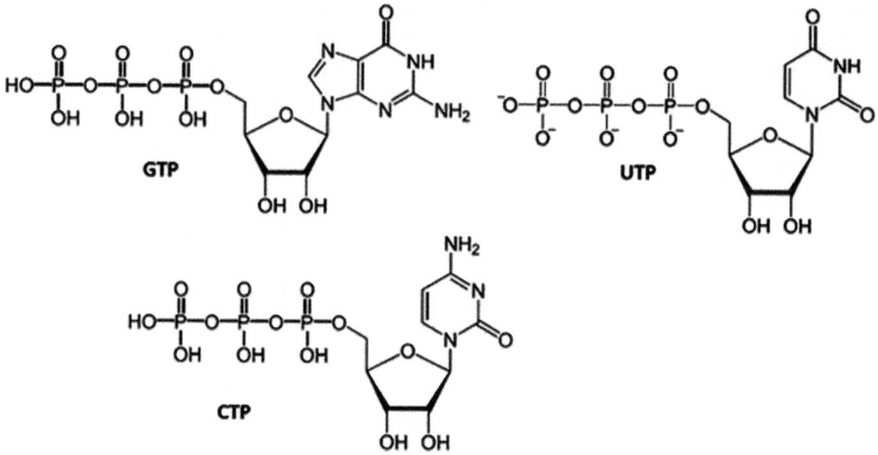

Figure 7.2   Structure of nucleoside triphosphates.

Dinucleotides (FAD) are co-enzymes present in flavoproteins, associated with respiration in eukaryotic cells. These are auto fluorescent in nature which have diagnostic value as a natural biomarker to study various pathophysiological conditions of cells (Benson et al. 1979).

Cyclic nucleotides, adenosine-3′,5′-monophosphate (cAMP) and cyclic guanosine-3′,5′-monophosphate (cGMP) are important 'second messengers' involved as modulators of physiological processes, such as regulating neuronal, glandular, cardiovascular, immune mechanism, nervous system, cell growth and differentiation. The nucleotides relay signals received at receptors on the cell surface—such as the arrival of protein hormones, growth factors, etc., to target molecules in the cytosol and/or nucleus. But in addition to their job as relay molecules, second messengers serve to greatly amplify the strength of the signal. These small cyclic nucleotides as second messengers can bind to cyclic nucleotide-gated ion channels and target proteins like protein kinases (protein kinase A and G) (Lucas et al. 2000; Kaupp and Seifert 2002). Modulation of intracellular cAMP/cGMP concentrations occurs by activation or inhibition of adenylyl/guanylyl cyclases, the cAMP and cGMP synthesizing enzymes. To inhibit signaling, both second messengers are degraded by different phosphodiesterases (PDEs) with more or less specificity for either cAMP or cGMP (Beavo and Brunton 2002; Conti and Beavo 2007; Oeckl and Ferger 2012). Owing to the wide distribution of this second messenger system, an imbalance in its homeostatic regulation leads to a variety of pathological states, and the system is targeted for the treatment of several diseases such as cancer, cardiovascular, neurodegenerative and psychiatric disorders (Reffelmann and Kloner 2009; Reneerkens et al. 2009; Savai et al. 2010). Guanosine 3′,5′-bispyrophosphate (ppGpp) is another important nucleotide found in bacteria. It is usually produced during amino acid starvation which further slows down protein synthesis. Hence cAMP and cGMP measurement could serve as valuable biomarkers to indicate normal biological and pathogenic processes as well as pharmacological responses to a therapeutic intervention.

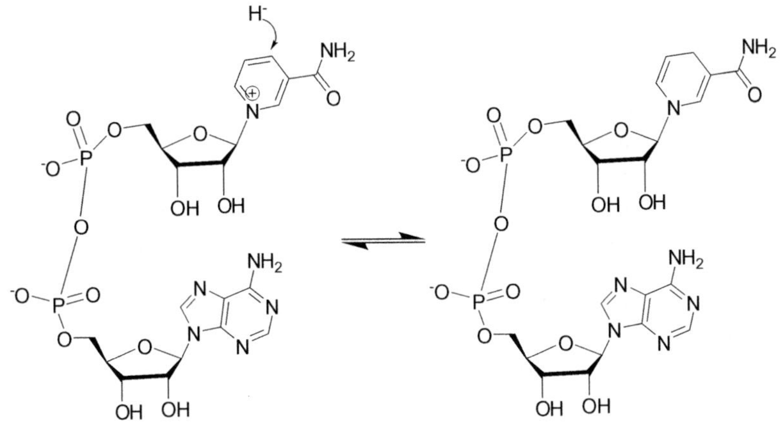

Figure 7.3 Redox nucleotide pairs.

Synthetic nucleosides analogues are pharmaceutically important molecules that resemble naturally occurring nucleosides. These are largely used as anti-viral agents (Acyclovir), a drug for cancer and rheumatologic diseases (azathioprine, allopurinol) and even bacterial infections (trimethoprim). The chemical modification of three-component of a nucleoside, i.e., purine or pyrimidine bases, five-membered sugar (ribose/deoxyribose) and hydroxymethyl group as a polar group are the key steps for the synthesis of these pharmaceutically active compounds. The interaction of nucleoside analogs with different viral polymerases (DNA polymerases/reverse transcriptase/RNA polymerases) follows different inhibition pathways for viral replication. Acyclovir is an antiviral agent that has been used against herpes simplex and varicella-zoster virus infections.

Despite their promising action, the toxicity causes severe side effects to human health. Macrocyclic compounds having potential binding ability towards these synthetic guests and can be used for drug sensing and their screening application (Ranganathan et al. 2019).

Oligonucleotides and Nucleic Acids

There are two main types of nucleic acids: Deoxyribose Nucleic Acid (DNA) and ribose nucleic acid (RNA) found in living beings. It consists of chains of nucleotide units containing pentose sugars, phosphate units and nucleobases as the polymeric backbone. RNA is single-stranded and DNA is double-stranded with specific nucleobase pairing, usually known as the Watson-Crick model which is the structural basis of the B-DNA (Fig. 7.4). The two anti-parallel complementary strands of DNA are coiled around each other forming an α-helix structure which is the typical conformation of DNA. The crucial difference between DNA and RNA lies in the pentose sugar-containing –H or –OH group at 2′ position respectively as shown in Fig. 7.1. The supercoiling of DNA-helix occurs in a directional manner forming tertiary structures with minor and major grooves. These grooves are essential for supramolecular interactions of nucleic acids with proteins or small molecules leading to mixed adducts stabilized by shape fitting and non-covalent bonds.

Double-stranded nucleic acid strands can adopt a variety of conformations. Small DNA fragments can fold to form a hair-pin structure. The stabilized forms of hairpins are associated with the disease. G-quadruplex is another unique secondary structure of DNA which are formed by four guanine bases associated through Hoogsteen hydrogen bonding to form a square planar structure called a guanine tetrad (G-tetrad or G-quartet), two or more guanine tetrads (G-tracts, continuous runs of guanine) can stack on top of each other to form a G-quadruplex. These structures bind to a variety of proteins and form complexes, which play essential roles in G-quadruplex-mediated regulation processes. Many of these G-quadruplex binding proteins and/or their complexes with G-quadruplexes are potential drug targets.

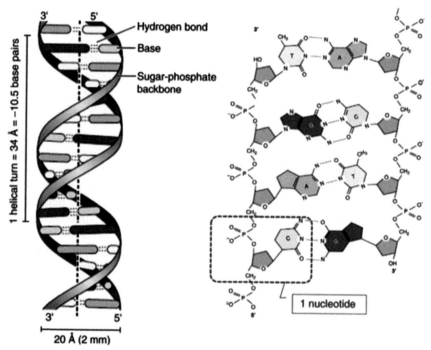

Figure 7.4 Schematic representation of the DNA double helix structure displaying nucleotide structure and hydrogen binding reprinted with permission from (Saddow 2016).

Single-stranded RNA undergoes folding like proteins and form secondary and tertiary structures. RNA mainly adopts A-form helical structure. The important function of RNA in a living system is to convert the genetic code stored in DNA into polypeptides by the transcription and translation process. RNA molecules play numerous other roles in both normal cellular processes and disease state.

An oligonucleotide with different base sequences can be used as receptors for the recognition of a complementary strand by a process called hybridization which can be exploited to detect DNA defects. Recognition and sensing of nucleic acid also help to understand and mimic biological processes, such as molecular recognition, protein interactions or DNA intercalation, host-pathogen interaction, etc. It can be exploited for the development of drug candidates, diagnostic agents or Drug Delivery Systems (DDS).

MACROCYCLIC RECEPTORS FOR NUCLEOTIDES, NUCLEOSIDES AND NUCLEIC ACID

Initial attempts were made to design the synthetic receptors for the recognition of nucleobases. The DNA double helix structure suggests that hydrogen bonding between adenine and thymine (or uracil) derivatives and between guanine and cytosine derivatives are highly selective. Earlier studies on purine and pyrimidine bonding forces in non-aqueous media (Kyogoku et al. 1967) confirmed the importance of hydrogen bonding interactions and aromatic–aromatic stacking between adjacent sets of base pairs in the structure of the double-stranded nucleic acids. Thus initial attempts were made for nucleobase recognition focused on the utilization of hydrogen bonding forces. Hamilton and his group synthesized various receptors for binding studies of thymine (Hamilton and Van Engen 1987), guanine (Hamilton and Pant 1988), cytosine (Jeong and Rebek 1988) and adenine derivatives (Goswami et al. 1989). These host-guest complexation studies were done with the NMR method. It was confirmed that receptors for adenine derivative bind through multi-site interaction by hydrogen-bonding and $\pi–\pi$ stacking interactions. To achieve a high affinity toward nucleobase- adenine, they modified the macrocyclic host molecules by incorporating the dipyridylethane group, which are suitable for oriented hydrogen bonding and aromatic stacking interactions. Thus strong and selective complexes with complementary substrates are formed as shown in Fig. 7.5. Adenine derivative **2** act as a guest molecule contains four pyrimidine nitrogen which are an accessible site for hydrogen bonding and aromatic pyridyl moiety present in the host **1a** & **1b** can act as aromatic π-stacking component for hydrophobic interactions. The receptor **1b** and butyl derivative of adenine **2** base shows strong binding **3** as shown in Fig.7.5 with association constant (K_a) of 3200 M^{-1} in organic medium, much higher than natural base-pairing association constant.

In earlier reports, the majority of nucleobase targeting studies are performed in non-aqueous media which makes them ineligible for application in sensing and measurement of biomolecules in physiological medium. Model studies of macrocyclic polyammonium compounds with inorganic phosphate, AMP, ADP and ATP as substrate at physiological pH (Kimura et al. 1982) paved the way for the synthesis of organic receptors targeting nucleotide in an aqueous system. Cyclophane based lipophilic receptors **4a** & **4b** synthesized by Schneider and Dhaenens (Schneider et al. 1992; Dhaenens et al. 1993) tend to form inclusion complexes in buffer with nucleoside, nucleotide and analogs by electrostatic interaction as shown in Fig. 7.6. Van der Waals contribution showed selectivity towards the specific nucleotides.

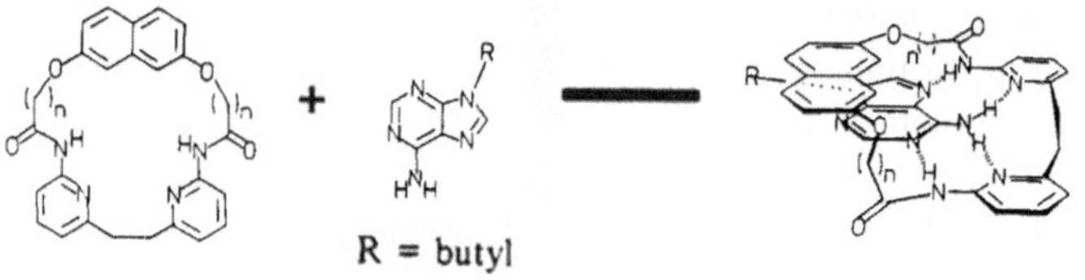

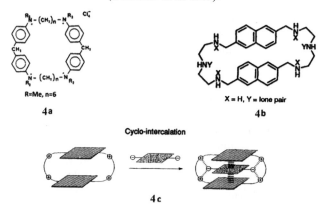

R = butyl

1a : n = 2
1b : n = 3

2

3

Figure 7.5 Synthetic receptors (**1a** &**1b**) for adenine derivatives (**2**). Adapted from (Goswami et al. 1989).

R=Me, n=6
4a

X = H, Y = lone pair
4b

Cyclo-intercalation

4c

Figure 7.6 Water-soluble synthetic receptors for nucleotides (**4a** & **4b**) and schematic representation of the cyclo-intercalation process for the binding of an anionic substrate by a cyclo-bis-intercaland (**4c**). Adapted with permission from (Schneider et al. 1992, Dhaenens et al. 1993).

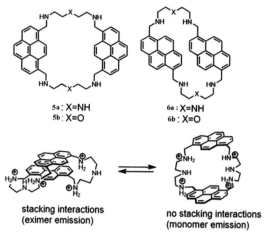

5a : X=NH
5b : X=O

6a : X=NH
6b : X=O

stacking interactions
(eximer emission)

no stacking interactions
(monomer emission)

Figure 7.7 Structure of receptors **5a**, **5b**, **6a**, **6b** and conformational equilibrium of receptor **5a** in aqueous solution, which describes the presence of the excimer and monomer emission in fluorescence spectra. Reprinted with permission from (Agafontsev et al. 2020).

Complexation and sensing characteristics of macrocyclic molecules can be applied for the measurement of nucleotide concentration in an aqueous sample by the further development of fluorescent features in the cyclophane derivatives. Pyrene-based fluorescent cylophanes (**5a, 5b, 6a** and **6b**) are reported for the ratiometric detection of nucleotide concentration in a buffered aqueous medium (Agafontsev et al. 2020). The fluorescence measurement of all receptors (Fig. 7.7) showed that pyrene-based receptor molecules display an ensembly of monomeric fluorescence emission peaks (380 nm) and an additional band (called excimer) at ~480 nm in an aqueous medium. Bellows-type sensing was observed which involves binding of nucleobases to the cyclophane structure through intercalation between two pyrene moieties present in the receptor molecules (Fig. 7.8). The intercalation process breaks the characteristic $\pi-\pi$ stacking interaction of pyrene units thereby quenching excimer emission. The highest selectivity was observed for ATP in a 50 mm MES buffer (pH 6.2) and a 50-fold change in the monomer-excimer emission ratio was measured from fluorescence studies. Pyrophosphate or other nucleotides such as GTP showed no obvious enhancement in monomer emission. As a result, a large fluorescence intensity ratio for ATP (I_{380}/I_{480}) enabled monitoring of the enzymatic reaction of creatine kinase could be achieved.

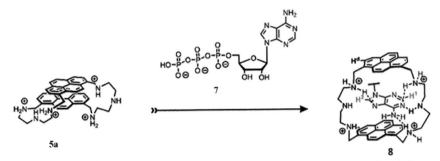

Figure 7.8 Complexation of ATP (**7**) with pyrene-cyclophane derivative (**5a**). Reprinted with permission from (Agafontsev et al. 2020).

Amphiphilic design of cyclodextrins (CDs) with a hydrophobic cavity and polar rim at the openings of the CD structure can also be tailored to create specific cone-structure selective towards amphiphilic nucleotides. Modification of β-cyclodextrin (β-CD) with methylamines at the lower rim were studied by Schneider group (Eliseev and Schneider 1994) in the aqueous phase. It has been found that substitution of the primary hydroxyl group at the cyclodextrin exterior by aminomethyl groups can form polytopic receptors (**9a, b**) selective towards nucleotides (Fig. 7.9). At pH 7, methylamine moieties are protonated which can interact with an oppositely charged moiety of the guest molecule with electrostatic forces. The electrostatic host-guest interaction forms thermodynamically stable complexes. CD derivatives containing two aminomethyl groups could differentiate pentose sugar (ribose or de-ribose) and identify phosphate group position (3′ Or 5′). A study of nuclear Overhauser effects shows that β-CD encapsulates only the sugar part leaving a nucleobase outside the cavity. Therefore, the specific binding mode of the host could play a decisive role in determining the overall selectivity for a nucleotide. Darcy and coworkers modified the structure of cyclodextrins synthesized by Schneider's via attaching a hydroxyethylamine moiety on the smaller opening (**9c**) and evaluated its ability for ATP recognition (Schwinté et al. 1998). This simple modification of β-CD produced a host with a remarkably increased binding affinity towards ATP, ($K_a = 6 \times 10^9$ M^{-1}), which is 2000-fold higher than that of **9b**. The protonated hydroxyethylamino side chains at the smaller opening not only served as a complementary binding site for phosphates but also formed an extended hydrophobic cavity which partially excluded water molecules and reinforced the electrostatic interaction between ammonium and phosphate.

It is now possible to detect unlabeled nucleotide present in a genome to a single molecular level by using β-cyclodextrin derivatives as an adapter's molecule in nanopore-based DNA sequencing method (Clarke et al. 2009). In the DNA sequencing method for continuous, label-free nucleoside monophosphate detection, the nucleic acid is first treated with exonuclease enzyme to produce individual nucleotide molecules present in DNA and subsequently, these monomers are driven through engineered α-hemolysin (αHL) channels acting as a filter for nucleotides. Biological channel protein (Ying et al. 2013), αHL is modified with β-cyclodextrin derivatives sensitive towards nucleotides which endorse detection of specific nucleotides. The conductivity of this channel is governed by the binding of β-cyclodextrin derivatives with the nucleotide. Thus, native or modified β-CD, which mediates channel blocking by the analyte/nucleotide bases reduces the conductivity property of the biological pore (Fig. 7.10) and it acts as a basis to identify and quantify the analyte-nucleotide (Banerjee et al. 2010).

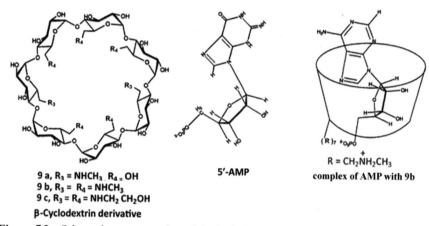

9 a, R_3 = NHCH$_3$ R_4 = OH
9 b, R_3 = R_4 = NHCH$_3$
9 c, R_3 = R_4 = NHCH$_2$ CH$_2$OH
β-Cyclodextrin derivative

5'-AMP

$(R)_7$ $^{+}$O$_2$PO
$^{+}$
R = CH$_2$NH$_2$CH$_3$
complex of AMP with 9b

Figure 7.9   Schematic representation of the inclusion complex formation by cyclodextrin derivative (**9b**) with AMP. Reprinted with permission from (Eliseev and Schneider 1994).

The amphiphilic character of β-CDs has been also utilized for the construction of an electrochemical sensor platform for the detection of DNA fragments with specific nucleotide sequences. Diao et al, (Jiang et al. 2017) reported a novel strategy for detection of DNA and miRNA by using a glassy carbon electrode modified with nitrogen-doped reduced graphene oxide/β-cyclodextrin polymer (NRGO/β-CDP) nanocomposites. The target detection is based on the combination of the cyclic cleavage reaction of the Mg^{2+}-dependent deoxyribozyme, also known as DNAzyme and the host-guest inclusion between a ferrocene-labeled DNA hairpin probe (H-1) molecule and the modified glassy carbon electrode (Fig. 7.11). The steps of sensing target DNA involves the interaction of S-1 in the loop region with the specific DNA sequence which led to the opening of the hair-pin structure of S-1. Subsequently the changes in the S-1 structure generate an active DNAzyme responsible for cleavage of the H-1 motif in the presence of Mg^{+2}. Thus the ferrocene-labeled H-1 probe undergoes cleavage into two single-stranded oligonucleotides. Finally, β-CD cavity with fixed dimension present in the nanocomposite layer captures the ferrocene moiety at the glassy carbon electrode surface resulting in increased current as the output signal. In the absence of the target DNA sequences, the hybridization reaction of the stem domain of subunit DNA (S-1) inhibits the formation of Mg^{2+}-dependent DNAzyme. Consequently, the uncleaved hairpin probe (H-1) is incidentally recognized by the β-CDP on the electrode because of dimension matching of the two closely associated ferrocenes, leading to a weak current response.

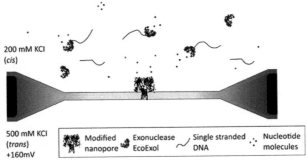

Figure 7.10 Experimental set-up for the detection of nucleotides cleaved from ssDNA by exonuclease. Nucleotides liberated by the enzyme are detected by the nanopore modified by β-CDs. (reprinted with permission from Clarke et al. 2009).

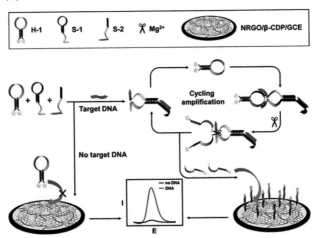

Figure 7.11 Illustration of electrochemical DNA sensor based on host-guest interaction of β-CDs and Mg^{2+} assistant DNA re-cycling. Reprinted with permission from (Jiang et al. 2017).

The proposed sensing system was also used for the detection of miRNA. The strategical application of a fabricated electrochemical transducer system decorated with β-CDP can be used for DNA fragment analysis of human samples. The controllable input responsive function demonstrated by synthetic host-guest binding and DNA hairpin assembly in aqueous media can be exploited as a drug delivery agent, modulable catalystsand biosensors.

The conjugation of hosts with oligonucleotides is still in its youth. But the continuous progress in the molecular recognition of nucleic acid has the potential to develop real-world applications.

INTRODUCTION TO STRUCTURE AND PROPERTIES OF SUGARS/CARBOHYDRATES

Carbohydrates are the most important and abundant substrate in biology. They are the building blocks for a more complex molecule like genetic material, important for cellular functions like energy generation and act as building materials in plants (cellulose). Carbohydrates are defined as polyhydroxy hydrocarbon, mainly divided into four groups: monosaccharides, disaccharides, oligosaccharides and polysaccharides. Three-dimensional structures of carbohydrates represent the spatial arrangements of individual sugar residues. Most commonly occurring mammalian complex carbohydrates consist of sugar residues that exist in the pyranose ring form, the most stable and rigid conformation of which are the chair forms. When two sugar residues are joined together covalently in a glycosidic linkage, the monomer units are free to rotate around the glycosidic oxygen atom between the two rings, and the resulting disaccharide can therefore assume a number of different conformations corresponding to the rotations about these two bonds. The most common monosaccharide is glucose which can exist in a six-membered (pyranose) or five-membered (furanose) ring structure is shown in Fig. 7.12 (Wikipedia). Simple monosaccharides are usually neutral, hydrophilic and possess a geometrically well-defined structure with peripheral hydroxyl groups. Some hydroxyl groups in monosaccharides can be modified and converted into other polar functionalities like amines, acylated amines, sulfates or carboxylates. Figure 7.13 shows the most abundant monosaccharide units, which are used as building blocks to create mammalian oligosaccharide structures (Peters 2014). Inside the cell, monosaccharide units polymerize in a step-wise manner to form oligosaccharides containing 3–10 monomer units. The cellular process of glycosylation (Cummings 2009) in the cells synthesizes glycolipids and glycoproteins important as cell-surface receptors, cell-adhesion molecules, immunoglobulins and tumor antigens (Elbein 1991). Oligosaccharides bound to lipid molecules are known as glycolipids important for cell recognition. Molecular recognition of sugars play an important role in different biological events, occurring in living organisms such as immune response (Lehninger 1982; Engström et al. 2005), viral and bacterial infection (Williams and Davies 2001; Olofsson and Bergström 2005), drug activity (Okamoto et al. 1997), metastasis of tumor cells (Gorelik et al. 2001), stem cell differentiation and cell-cell adhesion. Perhaps the most well-known saccharide is β-D-Glucose — a biomarker for diabetes that could be detected using synthetic receptors. The ligand-receptor sensing could be exploited for blood glucose monitoring (Sun and James 2015) and delivery of glucose-responsive insulin (Bakh et al. 2017). Thus molecular recognition of different carbohydrates by macrocycle could be used for the development of synthetic antibodies for cancer cells, anti-infectives, anti-inflammatories and many diagnostic applications (Solís et al. 2015; Francesconi and Roelens 2019; Tommasone et al. 2019).

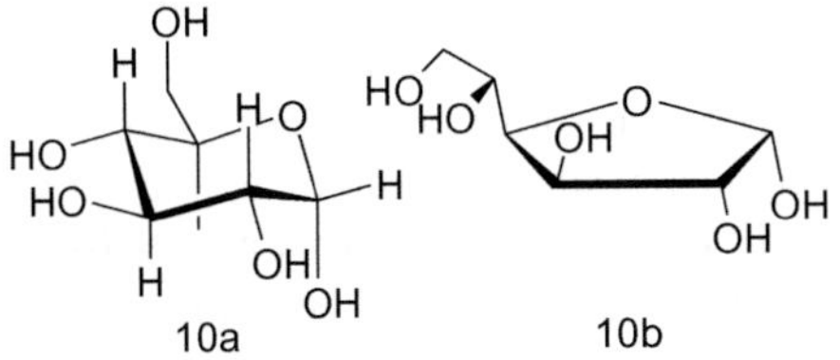

Figure 7.12 Monosaccharides: a. α-D Glucopyranose; b. α-D Glucofuranose.

As the name suggests (carbohydrates), the major functional groups are –OH similar to water molecules (HOH) and –O– which bind strongly to water molecules. The –CH– chain present in the structure forms the hydrophobic faces which are the center for non-polar hydrophobic interactions. Thus the strong binding of host-guest necessitates complementarity with both polar and non-polar features in the synthetic receptor and the carbohydrate. Additionally, the multiple stereogenic centers in the sugar molecules lead to a large number of stereo-isomeric forms. As a result, sensing by synthetic receptors depends on molecular recognition of the stereospecific structure of sugars.

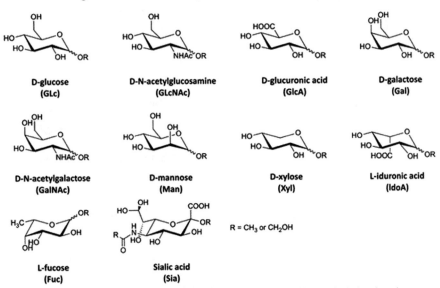

Figure 7.13 Structures of mammalian monosaccharide units and their abundance, as evaluated from a statistical analysis of a glycan databank of mammalian oligosaccharide structures. Reprinted with permission from (Peters 2014).

MACROCYCLIC RECEPTORS FOR CARBOHYDRATES

Early attempts were made to study the binding properties of organic receptors with carbohydrates in organic solvents. Carbohydrates guest molecules soluble in chloroform were used as a substrate (Kikuchi et al. 1992; Francesconi and Roelens 2019) where weak non-covalent interactions between host-guest complexes were observed as shown in Fig. 7.14. The binding events have been investigated by ^{1}H NMR spectroscopy, Circular Dichroism (CD), Isothermal Titration Calorimetry (ITC) and absorption spectroscopy. This information can be integrated effectively to ascertain recognition between substrate and host. Resorcinarene (Fig. 7.14a) with calixarene-like structure showed complexation to a lesser hydrophilic saccharides-alkyl glucoside, (Fig. 7.14c) with a selectivity towards β-anomer with low binding constant. But the saccharide binding could be improved by increasing the electron density available to the aromatic system through either structural modifications (Yanagihara and Aoyama 1994) or deprotonation (Kobayashi et al. 1992). These studies demonstrated that high affinities could be achieved with highly polar groups present in an appropriate medium. Cyclophane framework (Fig. 7.14b) synthesized by the Davis group (Bonar-Law et al. 1990) form complexes with α-glucoside (Fig. 7.14d) showing large binding constant ($K_a = 1700$ M^{-1}). The selective binding of monosaccharides depends on the stereo-spatial arrangement of –OH and the nature of substituent which affects the lipophilicity of monosaccharides. A guest with higher lipophilicity binds strongly with the host in organic solvents.

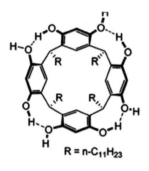

14 a. Resorcinarene

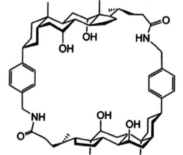

14 b. Cyclophanes framework

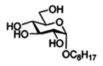

14 c. β-alky glucoside   **14 d. a-alkyl glucoside**

Figure 7.14 Biomimetic carbohydrate receptors (a, b) designed to operate in organic solvents, with standard substrates c and d. Adapted with permission from (Davis 2020).

Initially the general strategy applied for molecular recognition of carbohydrates using macrocyclic receptors was based on the use of hydrogen bonding in an aqueous medium. Aoyama et al. (Yanagihara and Aoyama 1994) developed synthetic polyhydroxy cyclotetramers as shown in Fig. 7.15. Their structure is similar to calixarenes showing a cavity-like structure for carbohydrate recognition. The binding studies demonstrated that unmodified saccharides show less affinity for water-soluble polyhydroxy cyclotetramers derivative (Fig. 7.15a). ^{1}H NMR data suggest that the polyhydroxy aromatic cavity of receptor molecule encapsulates the hydrophobic moiety of sugar forming the stable complex via CH–π interactions. Thus hydrophobic sugar-like fucose binds with the three macrocycles (7.15, 11a, b, c) with K_a = 2, 6, and 8 M^{-1} respectively.

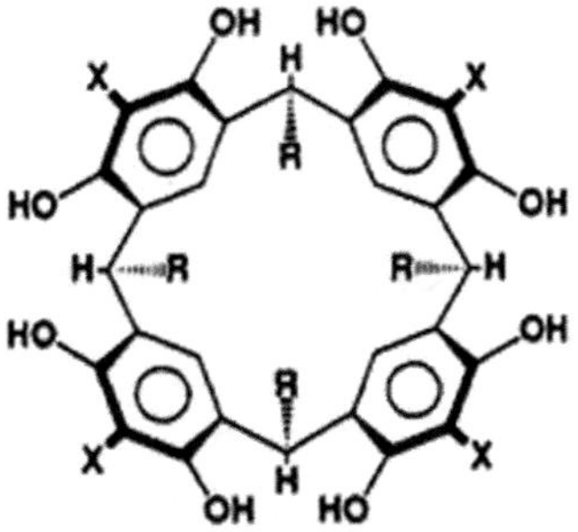

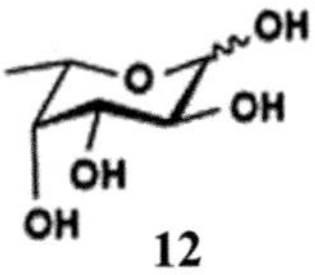

11a: R = (CH$_2$)$_2$ SO$_3$Na; X   H

11b: R = (CH$_2$)$_2$ SO$_3$Na; X   CH$_3$

11c: R = (CH$_2$)$_2$ SO$_3$Na; X   OH

Figure 7.15   Water-soluble resorcinarenes (a-c) and its substrate Fucose (Reprinted with permission from (Kobayashi et al. 1992).

Davis and coworkers (Klein et al. 2005; Ferrand et al. 2009) developed macropolycyclic cage molecules as efficient receptors for carbohydrates which completely encapsulate the molecules forming stable complexes. The macrocycles (Fig. 7.16) form complexes with monosaccharides containing an equatorial hydroxyl group or substituents (β-glucopyranose and β-glucosides). The molecular architecture contains a cage structure with a hydrophobic aromatic ring at the roof and floor of the cage supported by polar side columns. The receptor design with hydrophobic and polar patches could allow interaction with the substrate using CH–π interactions and hydrogen bonding. It could bind to O-linked β-N-acetylglucosamine (GlcNAc β-OMe) with an association constant (K_a) 630 M^{-1} in water – a metabolic link between diabetes and cancer (Slawson et al. 2010). The binding affinity of water-soluble sugars like glucose, galactose, and mannose was also measured and good selectivity for galactose and mannose were observed. For the development of the assay method for water-soluble saccharides, the so-called 'Temple' receptor's design was further modified with anthracene component for making it fluorescent, showing a good binding affinity for D-Glucose (~55 M^{-1}) and a selectivity of 50:1 over D-galactose or D-mannose. The fluorescent receptors show a prospect for the development of fluorescent sugar sensors.

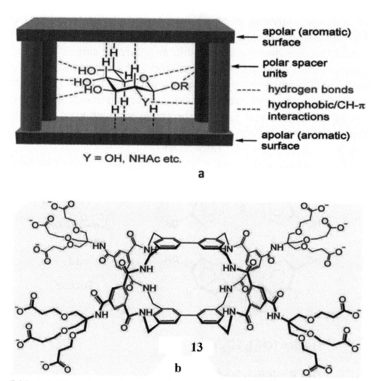

Figure 7.16 (a) The "temple" approach to receptors for all-equatorial carbohydrates. (b) The first water-soluble temple. Reprinted with permission from (Davis 2020).

INTRODUCTION TO STRUCTURE AND PROPERTIES OF AMINO ACID, PEPTIDE AND PROTEINS

Amino acids are the building blocks of peptides and protein which play a key role in various metabolic pathways as an intermediate, a neuro-transmitter and regulate cellular functions, involved in cell signaling process, catalysis and communication. In a living system, 20 different L-α-amino acids are commonly used for protein synthesis. These are versatile molecules carrying a wider range of chemical functionality. They differ from each other in the side chain or R group attached to an asymmetric carbon atom. In addition to this, D-α-amino acids, β-amino acids, and selenocysteine also play important roles in our bodies. A list of 20 different amino acids is shown in Fig. 7.17. Glutathione (GSH), cysteine (Cys), and homocysteine (Hcy) are necessary for regulating redox homeostasis, detoxification and metabolism. Charged amino acids like lysine and arginine are essential for various biological phenomena, as they can provide both short-range and long-range interactions for protein folding, helix aggregation (Lew et al. 2003), membrane protein anchoring (Liu et al. 2005) the sensing of membrane potentials (Jiang et al. 2003). They act as a precursor for hormone synthesis (Wu 2009). Deficiency of amino acids and their derivatives causes several diseases including slow growth, liver damage, edema and hematopoiesis decrease (Refsum et al. 1998). Different amino acids are covalently linked through amide linkage (–NH–C=O) to form peptides. They have an oligo amide backbone with an ammonium group at the N-terminus and carboxylate group at the C-terminus of the peptide chain. Peptide chains containing large amounts of amino acid residues are capable of spontaneously folding to form well-defined three-dimensional structures, dictated by the sequence of amino acids which permit highly specific chemical interactions with other molecules. Recent studies show that peptides from different sources such as dairy products, plants, animals and seafood have a wide range of bioactivities, e.g., antimicrobial, immune-modulatory, antihypertensive activities and others.

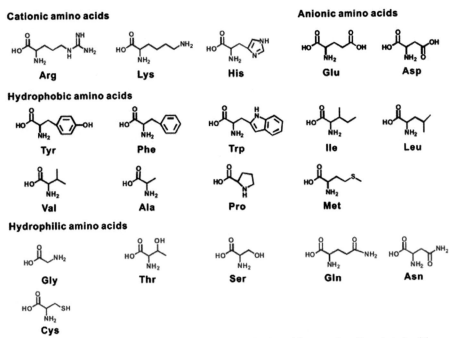

Figure 7.17   Structure of common amino acids found in proteins. Reprinted with permission from (Wang et al. 2016).

Protein hydrolysates and peptides from natural resources can be used as 'functional foods' and 'nutraceuticals' based on their bioactivity or as technological components (Hajfathalian et al. 2018). In the body, proteins are broken down into peptides on digestion by endogenous enzymes in the gastrointestinal system. Recognition of peptides by macromolecular receptors is ubiquitous in nature. These are important targets for artificial receptors which could give insight into their recognition process, screening of peptide mixtures and their application in the nature of receptor response. The development of receptors for selective recognition of these biomolecules is desired for many technological, industrial, biomedical and environmental applications (Fredericks and Hamilton 1996).

MACROCYCLIC RECEPTORS FOR AMINO ACIDS, PEPTIDES, AND PROTEINS

There are several challenges in designing macrocyclic receptors for sensing amino acids as they exist as zwitter ions within a biologically relevant pH range. They are highly solvated in an aqueous medium and form folded or self-assembled structures. Initially, the receptors were designed primarily on the basis of H-bonding interactions in organic solvents. Cyclodextrins are very successful receptors for forming stable inclusion complexes with different organic compounds in water (Szejtli 1998). Their hydrophobic cavity of fixed size can accommodate hydrophobic side-chains of amino acids. Unmodified CDs have a very low affinity towards zwitterionic amino acids but modification with carboxylic and ammonium functional groups increases their binding efficiency. Corradini et al (Corradini et al. 1994) studied the mechanism of chiral recognition of unmodified aromatic amino acids (phenylalanine, tyrosine and tryptophan) by using copper (II) complex of histamine modified cyclodextrin **14** [6-deoxy-6-N-histamine-β-cyclodextrin=CuCDhm)] as shown in Fig. 7.18. The quantitative data obtained through potentiometry, calorimetry and spectroscopic studies suggests that enantio-selective discrimination can be achieved by the ternary complex formation with aromatic amino acids in aqueous solution. Enthalpy favors the formation of ternary complexes with D-amino acid as stability constant for D-enantiomers are high. Thus, stereo-selectivity of copper (II) complex of histamine functionalized β-cyclodextrin could be used for HPLC-based separation of chiral aromatic amino acids. Later the same group modified the CD derivatives by introducing a specific metal-binding site and a dansyl fluorophore–6-deoxy-6-N-(N^{α}-[(5-dimethylamino-1-naphthalenesulfonyl)aminoethyl] phenylalanylamino-β-cyclodextrin (Pagliari et al. 2004). The conformational study of the CD derivatives by circular dichroism, two-dimensional NMR spectroscopy and time-resolved fluorescence spectroscopy confirmed that the introduction of a dansyl fluorophore linked through a phenyl (Phe) bridge (D or L) shows different behavior. Cyclodextrin derivative with L-phenylalanine moieties at the side chain showed predominant conformation in which the dansyl group is self-included in the cyclodextrin cavity, but the dansyl group is outside the CD cavity when D-phenylalanine is present in the structure. Thus the corresponding Cu(II) complex can form diastereomeric ternary complex with specific enantiomer of an amino acid in water. The self-inclusion of fluorophore in the cyclodextrin cavity strongly increases the chiral discrimination ability of the copper(II) complex. This suggested that the enantio-selective fluorescent receptors can be used for fast screening application of water-soluble amino acid.

Recently, Jurczack and co-workers (Stepniak et al. 2017) modified β-CDs (**15**) by incorporating urea groups in the side chain and studied their binding properties with hydrophobic L-amino acids guests (alanine, phenylalanine, tryptophan) in a phosphate buffer. A high-affinity constant (K_a) was observed for tryptophan at pH 8.0. The presence of hydrophilic urea moieties **15** (Fig. 7.19) in the β-CD side chain contributes to the hydrogen bond donor resulting in a strongly bonded complex with high association constants. Interestingly, the presence of isopropyl functionality allows the formation of 'active aggregates' which allow the urea functionality to establish interactions isolated from the bulk solvent. Based on this design, the same group studied a series of compounds and found binding affinities for non-polar amino acids from 2300 M^{-1} for alanine to 54,800 M^{-1} for tryptophan. Importantly carboxylates lacking the amino group did not remain bound. The binding strength was shown to be highly pH-dependent with good affinities at pH 8 but almost no binding at pH 6.

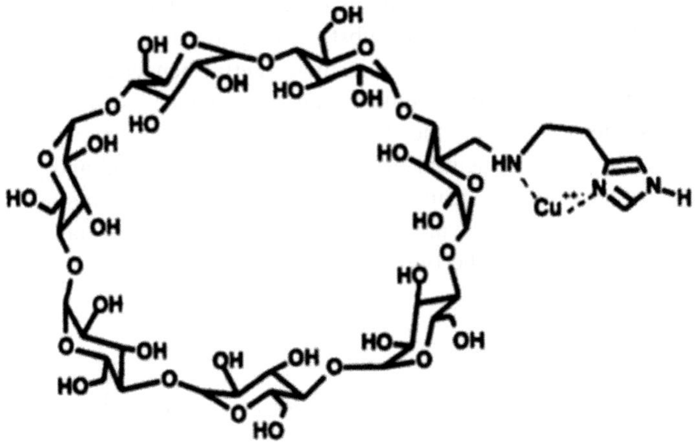

14

Figure 7.18 Cyclodextrin-based receptor 14 for enantio-selective discrimination of chiral D-, L-tryptophan amino acid.

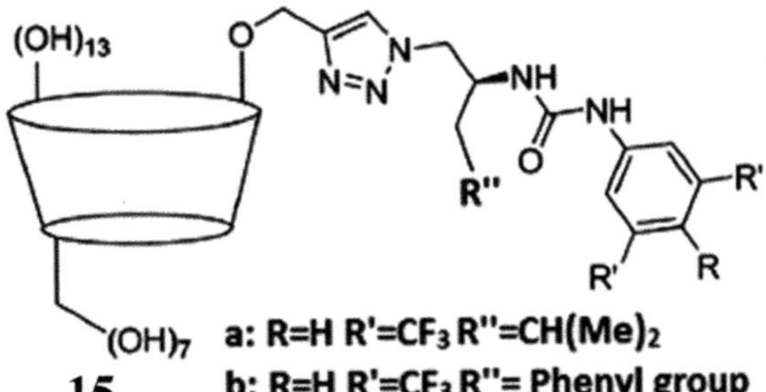

a: R=H R'=CF$_3$ R''=CH(Me)$_2$

b: R=H R'=CF$_3$ R''= Phenyl group

15

Figure 7.19 Cyclodextrin derivative 15 as a receptor for Tryptophan amino acid Adapted with permission from (Stepniak et al. 2017).

Tweezer molecules are supramolecular structures possessing a central, parallel, torus-shaped cavity with a surrounding belt of convergent aromatic rings that can bind to a range of cationic and neutral guests, including several types of biomolecules (Klärner and Schrader 2013). Schrader and co-workers (Fokkens et al. 2005) synthesized a molecular tweezer (Fig. 7.20) with aromatic torus-shaped cavity 16 adorned with two peripheral anionic phosphonate groups 16e for recognition of lysine and arginine (Fig. 7.20). They tested the inclusion complex formation by using NMR and ITC titration methods for free amino acid as well as arginine-based tripeptide RGD, responsible for many cell-surface recognition events. Peptide containing N-terminal lysine or the lysine moiety present in the middle of the chain showed inclusion complex formation efficiently in buffered aqueous solution, irrespective of the amino acid's position within the peptide. The binding properties of these complexes were justified on the basis of a combination of van der Waals interactions and substantial electrostatic contributions for stable inclusion complex formation. Later, in their studies, the tweezer molecule was modified (Dutt et al. 2013) with substituents-phosphate, methanephosphonate, sulfate or OCH_2-carboxylate groups in their central benzene bridge (Fig. 7.20 b, c, e, and f). The complexation studies using fluorescence, NMR and ITC titration experiments showed that all the new tweezer molecules are water-soluble and form stable complexes at neutral pH with various amino acid and peptide guests containing either lysine or arginine moieties. The complexation study in different solvents suggested that the hydrophobic effect in water is the major force driving the guest molecule inside the tweezer.

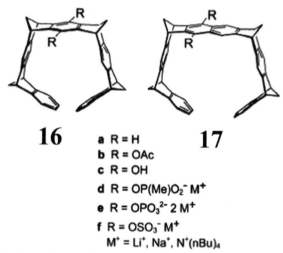

16   **a** R = H   **17**

b R = OAc

c R = OH

d R = OP(Me)O_2^- M^+

e R = OPO_3^{2-} 2 M^+

f R = OSO_3^- M^+

M^+ = Li^+, Na^+, $N^+(nBu)_4$

Figure 7.20   Structure of molecular tweezers Adapted from (Klärner and Schrader 2013).

Tweezers are also known to modulate functions of protein-containing lysine residues and protein aggregation behavior causing many diseases. The tweezers **16** (Fig. 7.20) can bind the alkane chains of aliphatic guest molecules inside their cavity by CH–π interactions, while tweezers **17** (Fig. 7.20) can bind aromatic guests via both CH–π and π–π interactions. There are seven different types of 14-3-3 proteins present in human beings which act as adaptor proteins and are involved in many cellular processes. Protein-protein interactions of 14-3-3 proteins regulate the activity of kinase C-Raf and regulate the tumor suppressor protein p53. Among them 14-3-3σ contains 17 lysine residues. Molecular tweezers **16e** (Fig. 7.20) with negatively charged hydrogen phosphate groups can target these lysine residues and inhibits binding between the 14-3-3 protein (Bier et al. 2013) and two of the partner proteins—a phosphorylated (C-Raf) protein and an unphosphorylated one (ExoS)—in a concentration-dependent manner. Thus macrocyclic structural features can be used as a tool for regulating protein-protein interactions by interference mechanism.

The lysine selective complexation behavior of tweezer molecule **16e** (Fig. 7.20) can be applied for the aggregation of proteins. G. Bitan and group (Sinha et al. 2011) reported that Aβ protein associated with Alzheimer's disease contains misfolded and aggregated amino acid moieties at the surface which can be recognized by the tweezer molecule. In fact NMR and mass spectrometric studies showed that the complex formation at the protein surface reduces the rate of protein aggregation and β-sheet formation. The reported strategy can be applied to redirect the aggregation of misfolded proteins into nontoxic species.

INTRODUCTION TO STRUCTURE AND FUNCTION OF BIOLOGICAL LIPIDS

Lipids are an important class of biomolecules, which contain chemically diverse groups with the characteristic feature of insolubility in water. It plays a crucial role in membrane architecture/compartment formation, intracellular traffic, signaling, hormone regulation, inflammation, energy storage and metabolism (Van Meer et al. 2008; Fahy et al. 2011). The structure of different lipids found in a living system is complex and diversity arises due to variable chain length, a multitude of oxidative, reductive, substitutional and ring-forming biochemical transformations as well as a modification with sugar residues and other functional groups of different biosynthetic origin. These diverse groups of lipid molecules can be organized into major groups as fatty acyls, glycerolipids, glycerophospholipids, sphingolipids, saccharolipids, polyketides (derived from the condensation of ketoacyl subunits); sterol lipids, and prenol lipids (derived from the condensation of isoprene subunits) (Fahy et al. 2011). Major classes of lipids found in biological systems and their role in a living system is shown in Fig. 7.21.

Polar lipids such as glycerophospholipids and sterols consisting of polar heads and nonpolar hydrocarbon chain are major components of the cell membrane. Biophysical properties of plasma membrane such as surface curvature and internal fluidity can be modulated by their lipid compositions. The presence of sterols, such as cholesterol or ergosterol, greatly affects the membrane packing and structural integrity (Dufourc 2008). Interaction of lipid molecules with membrane proteins regulates G-protein and GPCRs activity. Sterols are crucial membrane components in the formation of local liquid-ordered membrane phases called 'lipid rafts', which also contain transmembrane proteins and are involved in biological signaling and trafficking processes (Simons and Sampaio 2011). Essential Fatty Acids (EFA) are indispensable for one's health as they serve as dietary precursors for the formation of prostanoids and other eicosanoids important for health and modulation of disease conditions. Poly-Unsaturated Fatty Acids (PUFA) like Omega-3 (ω-3) fatty acids is an essential requirement for one's health which cannot be synthesized by the human physiological system and must therefore be obtained through a nutritional diet (Uauy 1999). The endoplasmic reticulum is the bulk supplier of lipids to other organelles in the eukaryotic cell. The transport and metabolism of lipid is governed by various protein receptors that selectively associate with lipids and transfer them to specific locations with the help of supramolecular transporting assemblies (Holthuis and Menon 2014).

New approaches based on detailed molecular-recognition insights could allow the development of selective artificial receptors for the detection of lipid biomarkers, they can act as solubilizing agents for hydrophobic lipids such as terpenoids/steroids needed for pharmaceutical applications. The discrimination of similar lipid molecule with different biological function in a complex cellular milieu is a daunting task which requires a profound understanding of their molecular recognition principle. Here synthetic receptors for steroid (cholesterol) and polar lipids are described giving insight into the principles governing the molecular recognition of these molecules.

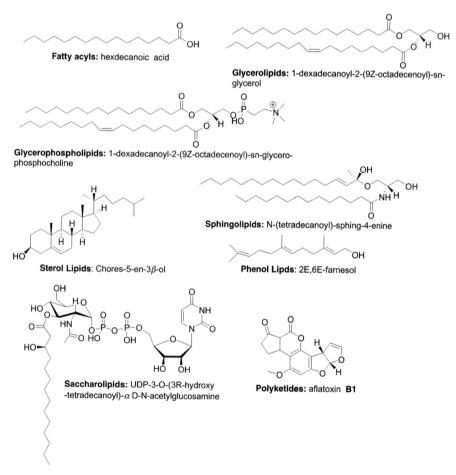

Figure 7.21 Examples of lipid classes: Representative structures from each of the 8 LIPID MAPS lipid categories. Reprinted with permission from (Fahy et al. 2011).

MACROCYCLIC RECEPTORS FOR MOLECULAR RECOGNITION OF LIPIDS

Most molecular recognition studies were initially conducted on steroids. Steroids are oxidative derivatives of sterols containing four fused rings, three with six carbons and one with five rings. The steroid nucleus is almost planar and is relatively rigid; the fused rings do not allow rotation about C–C bonds. Breslow and co-workers (Breslow and Zhang 1996) synthesized cyclodextrin dimers linked to each other by different spacer units which can bind to hydrophobic molecules like sterols. Complexation studies between dimeric cyclodextrin derivative **18** and cholesterol (Fig. 7.22) were studied using the solubility technique. The binding constant was found to be 200–300 times greater than monomeric β-cyclodextrins. The strong complexation could be observed due to the ability of the two cyclodextrin rings to align correctly around the cholesterol molecule for cooperative binding in between. This strategy of complexation can be used to increase the solubility of steroids, which are sparingly soluble in water and used as pharmaceutical compounds. Various other container type molecules like calixarenes, cyclophanes and cucurbituril have also shown complexation behavior with cholesterol and other hydrophobic guest used in pharmaceutical applications. An outline of complexation strategy with different steroid molecules is reviewed (Hishiya et al. 1999; Schneider et al. 2013).

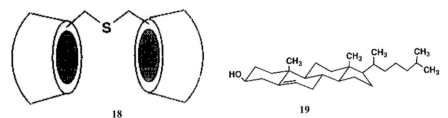

Figure 7.22 Structure of cyclodextrin dimer **18** and host molecule for cholesterol **19**.

There is a wide array of synthetic receptors for the phosphate, carboxylate and ammonium groups that are present within the polar head groups of lipids. Very few ligands designed to capture the aliphatic chain portion of the polar lipid are reported. Glass and co-workers (Avetta et al. 2012) designed the fluorescent molecular tubes sensitive towards the hydrophobic chain of lipids. These hydrocarbon chains of the polar lipid molecules lack a distinct functional group and vary only in the size and shape of their hydrophobic surface. The naphthalene-based molecular tube design was based on the shape-selective hydrophobic interactions with straight-chain alkanes in water. These fluorescent receptors **20** and **21** are formed from two naphthol dimers that are linked by either an amide bond or a trans olefin (Fig. 7.23). The molecular tubes with amide linkage could exist in two conformations, an 'open' conformation in which both amide bonds are trans, and a 'closed' conformation where one amide switches to the cis-rotamer, leading to the collapsed structure. But **21** can exist as a rigid open structure as it contains a trans-olefin linker and its rigid hydrophobic architecture can bind lipids

~100 more strongly than **20**. The construction of fluorescent sensors for bioactive lipids is a useful tool for studying various biophysical processes.

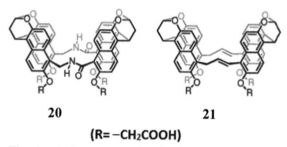

20　　　　　　**21**

(R= −CH₂COOH)

Figure 7.23　The chemical structures of molecular tubes **20** and **21**. Adapted with permission from (Avetta et al. 2012).

Mosca et al. (Mosca et al. 2015) studied the binding capacity of deep cavitands (Fig. 7.24) with lipophilic cavity and hydrophilic surfaces. These cavitands were insoluble in water but in presence of DMSO, formed a stable structure through non-covalent intra-molecular interactions. The upper rim of the cavity is laced with secondary amino groups which creates a seam of hydrogen bonds at the gate and the adjacent aromatic panel active as stacking interactions which provided sufficient stabilization of the vase (such as kite) to maintain a receptive, deep pocket. These attractive interactions maintain an open binding pocket even in absence of hydrophobic guests. Cavitand showed a binding affinity for a range of physiologically important fatty acids, specifically *ω*-fatty acids. The cavity of the cavitand remains open and forms an inclusion complex with different fatty acids. The tailor-made synthetic receptors could be further designed to a sensor molecule for quantitative measurements of *ω*-fatty acids.

Many pharmacological objectives such as detection, solubilization in aqueous systems, enhanced transport and targeted delivery can be achieved in the future by the development of macrocyclic ligand as host molecules for various lipids as guests.

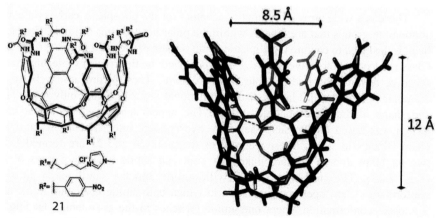

Figure 7.24　Chemical structure of cavitand receptor **22** and its modeled vase conformation (right). Adapted with permission from (Mosca et al. 2015).

CONCLUSION AND OUTLOOK

In this chapter, the complexation of four major classes of biomolecules has been described which encompasses the binding properties of macrocyclic molecules with water-soluble structures. Remarkable advances have been achieved over the last decade. Intensive effort in this direction will produce new macrocyclic receptors and probes with high translational impact in healthcare and biomedical research.

REFERENCES

Agafontsev, A.M., T.A. Shumilova, A.S. Oshchepkov, F. Hampel and E.A. Kataev. 2020. Ratiometric detection of ATP by fluorescent cyclophanes with bellows-type sensing mechanism. Chem. Eur. J. 26(44): 9991.

Avetta, C.T., B.J. Shorthill, C. Ren and T.E. Glass. 2012. Molecular tubes for lipid sensing: tube conformations control analyte selectivity and fluorescent response. J. Org. Chem. 77(2): 851–857.

Bakh, N.A., A.B. Cortinas, M.A. Weiss, R.S. Langer, D.G. Anderson, Z. Gu, S. Dutta and M.S. Strano. 2017. Glucose-responsive insulin by molecular and physical design. Nat. Chem. 9(10): 937–944.

Banerjee, A., E. Mikhailova, S. Cheley, L.-Q. Gu, M. Montoya, Y. Nagaoka, E. Gouaux and H. Bayley. 2010. Molecular bases of cyclodextrin adapter interactions with engineered protein nanopores. Proc. Natl. Acad. Sci. U.S.A. 107(18): 8165–8170.

Beavo, J.A. and L.L. Brunton. 2002. Cyclic nucleotide research—still expanding after half a century. Nat. Rev. Mol. Cell Biol. 3(9): 710–717.

Benson, R., R. Meyer, M. Zaruba and G. McKhann. 1979. Cellular autofluorescence—is it due to flavins? J. Histochem. Cytochem. 27(1): 44–48.

Bier, D., R. Rose, K. Bravo-Rodriguez, M. Bartel, J.M. Ramirez-Anguita, S. Dutt, C. Wilch, F.G. Klärner, E.S. Garcia and T. Schrader. 2013. Molecular tweezers modulate 14-3-3 protein–protein interactions. Nat. Chem. 5(3): 234–239.

Bonar-Law, R.P., A.P. Davis and B.A. Murray. 1990. Artificial receptors for carbohydrate derivatives. Angew. Chem. Int. Ed. 29(12): 1407–1408.

Breslow, R. and B. Zhang. 1996. Cholesterol recognition and binding by cyclodextrin dimers. J. Am. Chem. Soc. 118(35): 8495–8496.

Carvalho, A.T., K. Szeler, K. Vavitsas, J. Åqvist and S.C. Kamerlin. 2015. Modeling the mechanisms of biological GTP hydrolysis. Arch. Biochem. Biophys. 582: 80–90.

Clarke, J., H.-C. Wu, L. Jayasinghe, A. Patel, S. Reid and H. Bayley. 2009. Continuous base identification for single-molecule nanopore DNA sequencing. Nat. Nanotechnol. 4(4): 265–270.

Conti, M. and J. Beavo. 2007. Biochemistry and physiology of cyclic nucleotide phosphodiesterases: essential components in cyclic nucleotide signaling. Annu. Rev. Biochem. 76: 481–511.

Corradini, R., A. Dossena, G. Impellizzeri, G. Maccarrone, R. Marchelli, E. Rizzarelli, G. Sartor and G. Vecchio. 1994. Chiral recognition and separation of amino acids by means of a copper (II) complex of histamine monofunctionalized. beta.-cyclodextrin. J. Am. Chem. Soc. 116(22): 10267–10274.

Cummings, R.D. 2009. The repertoire of glycan determinants in the human glycome. Mol. Biosyst. 5(10): 1087–1104.

Davis, A.P. 2020. Biomimetic carbohydrate recognition. Chem. Soc. Rev. 49(9): 2531–2545.

Dhaenens, M., J.-M. Lehn and J.-P. Vigneron. 1993. Molecular recognition of nucleosides, nucleotides and anionic planar substrates by a water-soluble bis-intercaland-type receptor molecule. J. Chem. Soc., Perkin Trans. 2(7): 1379–1381.

Dufourc, E. 2008. Sterols and membrane dynamics. J. Chem. Biol. 1(1-4): 63–77.

Dutt, S., C. Wilch, T. Gersthagen, P. Talbiersky, K. Bravo-Rodriguez, M. Hanni, E.S. Garcia, C. Ochsenfeld, F.G. Klärner and T. Schrader. 2013. Molecular tweezers with varying anions: a comparative study. J. Org. Chem. 78(13): 6721–6734.

Elbein, A.D. 1991. The role of N-linked oligosaccharides in glycoprotein function. Trends Biotechnol. 9(1): 346–352.

Eliseev, A.V. and H.-J. Schneider. 1994. Molecular recognition of nucleotides, nucleosides, and sugars by aminocyclodextrins. J. Am. Chem. Soc. 116(14): 6081–6088.

Engström, H.A., P.O. Andersson and S. Ohlson. 2005. Analysis of the specificity and thermodynamics of the interaction between low affinity antibodies and carbohydrate antigens using fluorescence spectroscopy. J. Immunol. Methods 297(1-2): 203–211.

Fahy, E., D. Cotter, M. Sud and S. Subramaniam 2011. Lipid classification, structures and tools. Biochim. Biophys. Acta, Mol. Cell. Biol. Lipids. 1811(11): 637–647.

Ferrand, Y., E. Klein, N.P. Barwell, M.P. Crump, J. Jiménez-Barbero, C. Vicent, G.-J. Boons, S. Ingale and A.P. Davis. 2009. A synthetic lectin for O-linked β-N-acetylglucosamine. Angew. Chem. 121(10): 1775-1779.

Fokkens, M., T. Schrader and F.-G. Klärner. 2005. A molecular tweezer for lysine and arginine. J. Am. Chem. Soc. 127(41): 14415–14421.

Francesconi, O. and S. Roelens. 2019. Biomimetic carbohydrate-binding agents (CBAs): Binding affinities and biological activities. ChemBioChem 20(11): 1329–1346.

Fredericks, J. and A. Hamilton. 1996. Hydrogen bonding control of molecular self-assembly: Recent advances in design, synthesis, and analysis. pp. 565-594. In: J.P. Sauvage and M.W. Hosseini (eds). Comprehensive Supramolecular Chemistry, Vol. 9. Pergamon Press: Newyork.

Gorelik, E., U. Galili and A. Raz. 2001. On the role of cell surface carbohydrates and their binding proteins (lectins) in tumor metastasis. Cancer Metastasis Rev. 20(3): 245–277.

Goswami, S., D. Van Engen and A.D. Hamilton. 1989. Nucleotide base recognition: a macrocyclic receptor for adenine employing hydrogen bonding and aromatic stacking interactions. J. Am. Chem. Soc. 111(9): 3425–3426.

Hajfathalian, M., S. Ghelichi, P.J. García-Moreno, A.-D. Moltke Sørensen and C. Jacobsen. 2018. Peptides: production, bioactivity, functionality, and applications. Crit. Rev. Food Sci. Nutr. 58(18): 3097–3129.

Hamilton, A.D. and N. Pant. 1988. Nucleotide base recognition: ditopic binding of guanine to a macrocyclic receptor containing naphthyridine and naphthalene units. J. Chem. Soc., Chem. Commun. (12): 765–766.

Hamilton, A.D. and D. Van Engen. 1987. Induced fit in synthetic receptors: nucleotide base recognition by a molecular hinge. J. Am. Chem. Soc. 109(16): 5035–5036.

Heikal, A.A. 2010. Intracellular coenzymes as natural biomarkers for metabolic activities and mitochondrial anomalies. Biomark Med. 4(2): 241–263.

Hishiya, T., M. Shibata, M. Kakazu, H. Asanuma and M. Komiyama. 1999. Molecularly imprinted cyclodextrins as selective receptors for steroids. Macromolecules. 32(7): 2265–2269.

Holthuis, J.C. and A.K. Menon. 2014. Lipid landscapes and pipelines in membrane homeostasis. Nature. 510(7503): 48–57.

Jeong, K. and J. Rebek. 1988. Molecular recognition: hydrogen bonding and aromatic stacking converge to bind cytosine derivatives. J. Am. Chem. Soc. 110(10): 3327–3328.

Jiang, J., X. Lin and G. Diao. 2017. Smart combination of cyclodextrin polymer host–guest recognition and Mg2+-assistant cyclic cleavage reaction for sensitive electrochemical assay of nucleic acids. ACS Appl. Mater. Interfaces. 9(42): 36688–36694.

Jiang, Y., A. Lee, J. Chen, V. Ruta, M. Cadene, B.T. Chait and R. MacKinnon. 2003. X-ray structure of a voltage-dependent K+ channel. Nature 423(6935): 33–41.

Kaupp, U.B. and R. Seifert. 2002. Cyclic nucleotide-gated ion channels. Physiol. Rev. 82(3): 769–824.

Kikuchi, Y., Y. Tanaka, S. Sutarto, K. Kobayashi, H. Toi and Y. Aoyama. 1992. Highly cooperative binding of alkyl glucopyranosides to the resorcinol cyclic tetramer due to intracomplex guest-guest hydrogen-bonding: solvophobicity/solvophilicity control by an alkyl group of the geometry, stoichiometry, stereoselectivity, and cooperativity. J. Am. Chem. Soc. 114(26): 10302–10306.

Kimura, E., M. Kodama and T. Yatsunami. 1982. Macromonocyclic polyamines as biological polyanion complexons. 2. Ion-pair association with phosphate and nucleotides. J. Am. Chem. Soc. 104(11): 3182–3187.

Klärner, F.-G. and T. Schrader. 2013. Aromatic interactions by molecular tweezers and clips in chemical and biological systems. Acc. Chem. Res. 46(4): 967–978.

Klein, E., M.P. Crump and A.P. Davis. 2005. Carbohydrate recognition in water by a tricyclic polyamide receptor. Angew. Chem. Int. Ed. 44(2): 298–302.

Kobayashi, K., Y. Asakawa, Y. Kato and Y. Aoyama. 1992. Complexation of hydrophobic sugars and nucleosides in water with tetrasulfonate derivatives of resorcinol cyclic tetramer having a polyhydroxy aromatic cavity: importance of guest-host CH-. pi. interaction. J. Am. Chem. Soc 114(26): 10307–10313.

Kyogoku, Y., R. Lord and A. Rich. 1967. The effect of substituents on the hydrogen bonding of adenine and uracil derivatives. Proc. Natl. Acad. Sci. U.S.A. 57(2): 250.

Lehninger, A.L. 1982. Principles of Biochemistry, Part II. Worth Pubs Inc., New York, USA.

Lew, S., G.A. Caputo and E. London. 2003. The effect of interactions involving ionizable residues flanking membrane-inserted hydrophobic helices upon helix-helix interaction. Biochem. 42(36): 10833–10842.

Liu, A., N. Wenzel and X. Qi. 2005. Role of lysine residues in membrane anchoring of saposin C. Arch. Biochem. Biophys. 443(1–2): 101–112.

Lucas, K.A., G.M. Pitari, S. Kazerounian, I. Ruiz-Stewart, J. Park, S. Schulz, K.P. Chepenik and S.A.Waldman. 2000. Guanylyl cyclases and signaling by cyclic GMP. Pharmacol. Rev. 52(3): 375–414.

Mosca, S., D. Ajami and J. Rebek. 2015. Recognition and sequestration of ω-fatty acids by a cavitand receptor. Proc. Natl. Acad. Sci. 112(36): 11181–11186.

Oeckl, P. and B. Ferger. 2012. Simultaneous LC–MS/MS analysis of the biomarkers cAMP and cGMP in plasma, CSF and brain tissue. J. Neurosci. Methods. 203(2): 338–343.

Okamoto, K., T. Oki, Y. Igarashi, M. Tsurudome, M. Nishio, M. Kawano, H. Komada and Y. Ito. 1997. Enhancement of human parainfluenza virus-induced cell fusion by pradimicin, a low molecular weight mannose-binding antibiotic. Med. Microbiol. Immunol. 186(2): 101–108.

Olofsson, S. and T. Bergström. 2005. Glycoconjugate glycans as viral receptors. Ann. Med. 37(3): 154–172.

Pagliari, S., R. Corradini, G. Galaverna, S. Sforza, A. Dossena, M. Montalti, L. Prodi and R. Marchelli. 2004. Enantioselective fluorescence sensing of amino acids by modified cyclodextrins: role of the cavity and sensing mechanism. Chem. Eur. J. 10(11): 2749–2758.

Peters, J.A. 2014. Interactions between boric acid derivatives and saccharides in aqueous media: structures and stabilities of resulting esters. Coord. Chem. Rev. 268: 1–22.

Ranganathan, P., B. Mutharani, S.-M. Chen and P. Sireesha. 2019. Biocompatible chitosan-pectin polyelectrolyte complex for simultaneous electrochemical determination of metronidazole and metribuzin. Carbohydr. Polym. 214: 317–327.

Reffelmann, T. and R.A. Kloner. 2009. Phosphodiesterase 5 inhibitors: are they cardioprotective? Cardiovasc. Res. 83(2): 204–212.

Refsum, H., P. Ueland, O. Nygård and S. Vollset. 1998. Homocysteine and cardiovascular disease. Annu. Rev. Med. 49(1): 31–62.

Reneerkens, O.A., K. Rutten, H.W. Steinbusch, A. Blokland and J. Prickaerts. 2009. Selective phosphodiesterase inhibitors: a promising target for cognition enhancement. Psychopharmacol. (Ber.) 202(1): 419–443.

Saddow, S.E. 2016. Silicon Carbide Biotechnology: A Biocompatible Semiconductor for Advanced Biomedical Devices and Applications. Elsevier.

Savai, R., S.S. Pullamsetti, G.-A. Banat, N. Weissmann, H.A. Ghofrani, F. Grimminger and R.T. Schermuly. 2010. Targeting cancer with phosphodiesterase inhibitors. Expert Opin. Invest. Drugs. 19(1): 117–131.

Schneider, H.J., T. Blatter, B. Palm, U. Pfingstag, V. Ruediger and I. Theis. 1992. Complexation of nucleosides, nucleotides, and analogs in an azoniacyclophane. Van der Waals and electrostatic binding increments and NMR shielding effects. J. Am. Chem. Soc. 114(20): 7704–7708.

Schneider, H.-J., P. Agrawal and A.K. Yatsimirsky. 2013. Supramolecular complexations of natural products. Chem. Soc. Rev. 42(16): 6777–6800.

Schwinté, P., R. Darcy and F. O'Keeffe. 1998. Ditopic binding of nucleotides by heptakis (6-hydroxyethylamino-6-deoxy)-β-cyclodextrin. J. Chem. Soc., Perkin Trans. 2(4): 805–808.

Simons, K. and J.L. Sampaio. 2011. Membrane organization and lipid rafts. Cold Spring Harbor Perspect. Biol. 3(10): a004697.

Sinha, S., D.H. Lopes, Z. Du, E.S. Pang, A. Shanmugam, A. Lomakin, P. Talbiersky, A. Tennstaedt, K. McDaniel, R. Bakshi, P-Y. Kuo, M. Ehrmann, G.B. Benedek, J.A. Loo, F-G. Klärner, T. Schrader, C. Wang and G. Bitan. 2011. Lysine-specific molecular tweezers are broad-spectrum inhibitors of assembly and toxicity of amyloid proteins. J. Am. Chem. Soc. 133(42): 16958–16969.

Slawson, C., R. Copeland and G.W. Hart. 2010. O-GlcNAc signaling: a metabolic link between diabetes and cancer? Trends Biochem. Sci. 35(10): 547–555.

Solís, D., N.V. Bovin, A.P. Davis, J. Jiménez-Barbero, A. Romero, R. Roy, K. Smetana Jr. and H.-J. Gabius. 2015. A guide into glycosciences: how chemistry, biochemistry and biology cooperate to crack the sugar code. Biochim. Biophys. Acta. 1850(1): 186–235.

Stepniak, P., B. Lainer, K. Chmurski and J. Jurczak. 2017. The effect of urea moiety in amino acid binding by β-cyclodextrin derivatives: a 1000-fold increase in efficacy comparing to native β-cyclodextrin. Carbohydr. Polym. 164: 233–241.

Sun, X. and T.D. James. 2015. Glucose sensing in supramolecular chemistry. Chem. Rev. 115(15): 8001–8037.

Szejtli, J. 1998. Introduction and general overview of cyclodextrin chemistry. Chem. Rev. 98(5): 1743–1754.

Tommasone, S., F. Allabush, Y.K. Tagger, J. Norman, M. Köpf, J.H. Tucker and P.M. Mendes. 2019. The challenges of glycan recognition with natural and artificial receptors. Chem. Soc. Rev. 48(22): 5488–5505.

Uauy, R. 1999. Nutrition throughout the life cycle. Eur. J. Clin. Nutr. 53(3): S8.

Van Meer, G., D.R. Voelker and G.W. Feigenson. 2008. Membrane lipids: where they are and how they behave. Nat. Rev. Mol. Cell Biol. 9(2): 112–124.

Wang, F., K. Hu and Y. Cheng. 2016. Structure–activity relationship of dendrimers engineered with twenty common amino acids in gene delivery. Acta. Biomater. 29: 94–102.

Williams, S.J. and G.J. Davies. 2001. Protein–carbohydrate interactions: learning lessons from nature. Trends Biotechnol. 19(9): 356–362.

Yanagihara, R. and Y. Aoyama. 1994. Enhanced sugar-binding ability of deprotonated calix[4]resorcarene in water: balance of CH-π interaction and hydrophobic effect. Tetrahedron Lett. 35(52): 9725–9728.

Ying, Y.L., J. Zhang, R. Gao and Y.T. Long. 2013. Nanopore-based sequencing and detection of nucleic acids. Angew. Chem. Int. Ed. 52(50): 13154–13161.

Zala, D., U. Schlattner, T. Desvignes, J. Bobe, A. Roux, P. Chavrier and M. Boissan. 2017. The advantage of channeling nucleotides for very processive functions. F1000Research 6: 724.

Chapter 8

Macrocyclic Receptors for Biologically Relevant Metal Ions

INTRODUCTION

Metals and their complexes are ubiquitous in life processes. They are essential for human health as they serve both structural and functional roles including the maintenance of cellular functions involved in a wide range of biological activities. Alkali and alkaline earth metals are involved in many biological processes such as maintenance of membrane charge balance and electrical conductivity (Da Silva and Williams 2001). First-row transition elements such as iron, copper, zinc, manganese, cobalt, chromium, molybdenum, vanadium and nickel are required in trace amounts. These are essential inorganic nutrients as their deficiency results in the impairment of a physiological function (Underwood and Mertz 1987). Numerous proteins incorporate metal ions in their structure. These metal centers can act as strong Lewis acids that can activate coordinated ligands for reactivity, for example, a water molecule coordinated to a Zn(II) center becomes a potent nucleophile for amide bond hydrolysis of a protein substrate (Bertini et al. 2007). In a large number of biological molecules, transition metals showing variable valency are present as enzyme co-factors and involved in different redox reactions. Electron transfer units such as cytochromes, iron-sulfur clusters and blue copper proteins shuttle electrons to other proteins that require redox chemistry for their function, while many other redox proteins catalyze multi-electron oxidation/reduction reactions directly on a substrate.

There are many metal ions toxic in our body (Cd^{2+}, Hg^{2+}, Pb^{2+}) (Da Silva and Williams 2001) and others are used as a therapeutic and diagnostic agent at an appropriate concentration (Haas and Franz 2009). Properties of metal ions in biological systems can be modified and fine-tuned through their coordination environment. Owing to the so-called 'macrocyclic effect', metal complexes can attain higher kinetic inertness and thermodynamic stability (Gloe 2005). Metal complexes can also impart additional functionality not found naturally. A large number of metal-complexes have been employed for the development of sensors, diagnostic agents, therapeutic agents, etc. The most striking feature is in visualization, where the photophysical, magnetic and radioactive properties of metals make possible studies based on luminescence, magnetic resonance imaging modalities. Here the focus will be on some of the important aspects of the synthetic macrocycle complexes with metal applicable in biomedical science.

METALLOCYCLES AS AN ANTICANCER DRUG

Anticancer drug development by the coordination-driven self-assembly has been exploited in a number of synthetic strategies resulting in numerous diverse compounds applicable for targeting different tumor sites in cancer cells. Platinum-based drugs are the first successful therapeutic agent used for the treatment of a wide range of cancers. After the success story of cisplatin, the Food and Drug Administration of USA approved cisplatin and similar platinum complexes such as carboplatin and oxaplatin, etc. (Fig. 8.1) for clinical use (Alame and Huq 2016). These organometallic drug-molecules target nuclear DNA and inhibit the DNA replication process. These drug molecules are DNA intercalators and the metal centers present in the organometallic complexes directly coordinates to a DNA base present in the duplex strand forming Pt-DNA adducts (Rabik and Dolan 2007). But their wide-spread application is limited as it causes cellular toxicity to cancerous as well as healthy cells. Besides, platinum (Pt) metal induces drug resistance in the human body (Brabec and Kasparkova 2005). Thus, new organometallic complexes with other metals and different designs have been explored for targeting nuclear DNA and many other targeting sites (multi-targeting drugs) present in the tumor cells. It is assumed that the new compound with a different structure will show improved interaction with nuclear DNA which may lead to a different spectrum of biological activity with better performance.

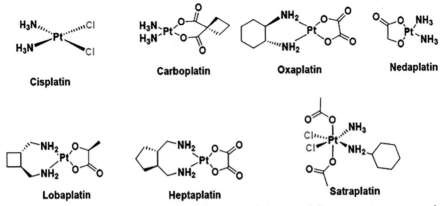

Figure 8.1 Chemical structure of platinum-based drug used for cancer treatment and clinical trial studies. Adapted from (Ellahioui et al. 2017).

Tocher et al. reported the ruthenium (Ru)-metronidazole complex as an anti-cancer agent (Dale et al. 1992). In the Ru (III)-based complex (**1**), the antibiotic molecule-metronidazole acts as a ligand present in the complexation sphere-forming small drug molecule [Ru(η^6-C$_6$H$_6$) (metronidazole)Cl$_2$] (Fig. 8.2). The complex shows higher cytotoxicity in comparison to the normal antibiotic.

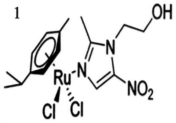

Figure 8.2 The first metal—arene compound was evaluated for anticancer activity. The ligand in red is the antibiotic agent metronidazole. Reprinted with permission of ACS (Cook et al. 2013).

Later Therrien's group assembled Ru-based (Therrien et al. 2008) trigonal prismatic cage molecules for testing its efficacy against A2780 human ovarian cancer cells. The cage molecule was synthesized by using trispyridyl-1,3,5-triazine as a donor and 1,4-benzoquinonato-based arene–Ru as an acceptor in the presence of AgO$_3$SCF$_3$ (AgOTf) as shown in Fig. 8.3. The trigonal prismatic cage could encapsulate [M(acac)$_2$] (M = Pt^{2+}, Pd^{2+}) complexes. The half-minimum inhibitory concentration (IC$_{50}$-value) values were determined for different cage structures which showed a correlation with their solubility and stability characteristics in water. It was observed that Pt^{2+} and Pd^{2+} complexes [(acac)$_2$M] (M = Pd or Pt) are sparingly soluble in water and do not show any inhibitory activity towards cancer cells while water-soluble Ru-cages encapsulating [(acac)$_2$M] (M = Pd or Pt) were more effective (IC$_{50}$ = 12 and 1 µM respectively) than the empty cage (IC$_{50}$ = 23 µM). Further, the same group studied the activities and stabilities of Ru-cage structure in a biological media containing different biomolecules, including amino acids, ascorbic acid and glutathione by using electrospray ionization mass spectrometry (ESI–MS) and Nuclear Magnetic Resonance (NMR) spectroscopy. Some basic amino acids including arginine, histidine and lysine disassemble the cage structure whereas methionine had no effect. Catalytic oxidations of ascorbic acid and glutathione by the metallocages were suggested to underlie their cytotoxicity (Paul et al. 2012).

Therrien et al. (Mattsson et al. 2009) and the Navarro group's (Linares et al. 2009) further modified their strategy to develop [2 + 2] cage structures with Ru(II) metal ions. New donor molecules having rigid structure, i.e., pyrazine, dipyridyl or 1,2-bis-(4-pyridyl)ethylene were used with 1,4-benzoquinonato based Ru-clips as an acceptor for the development of self-assembled and rectangular metallomacrocycles. The flexibility of Ru-cage structures inherited from the presence of 1,2-bis-(4-pyridyl)ethylene was correlated to synergic enhancement of anti-tumor activity in comparison to bipyridine-containing coordination cages (Mattsson et al. 2009). Navarro et al. (Linares et al. 2009) separately investigated the anticancer and other biological activities of Ru-based [2 + 2] self-assembled metallomacrocycles synthesized using rigid donor molecules-dipyridyl and 4,7-phenanthroline and hoxonato-based Ru-clips as an acceptor molecule. These new Ru-cage structures exhibited non-covalent bonding interactions to nuclear DNA and the nature of induced resistance in the cell was significantly different than observed with cisplatin. They emphasized that hoxonato-based macrocyclic systems are less cytotoxic to cancer cells and might be cell-type specific. Recently, Therrien's group evaluated a series of metallacycles (Therrien 2012) including octanuclear *p*-cymene Ru tetragonal metallocages prepared using tetrapyridyl porphyrin panels; many of them showed IC$_{50}$ values in the nanomolar range (Gupta et al. 2016). It was suggested that apoptosis is induced by these complexes and that interaction with DNA can be in part responsible for the high cytotoxicity.

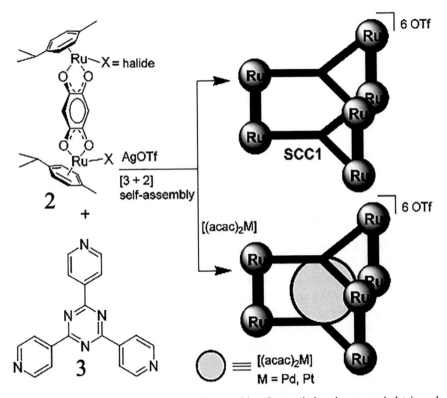

Figure 8.3 Coordination-driven self-assembly of occupied and unoccupied trigonal prismatic cages. Reprinted with permission of ACS (Cook et al. 2013).

Chi and coworkers developed Ru-based polynuclear nanocages using 1,3,5-tris(pyridin-4-ylethynyl)benzene as donor ligands (L1) and four different acceptors (**4A** – **7A**) as shown in Fig. 8.4 (Vajpayee et al. 2011).

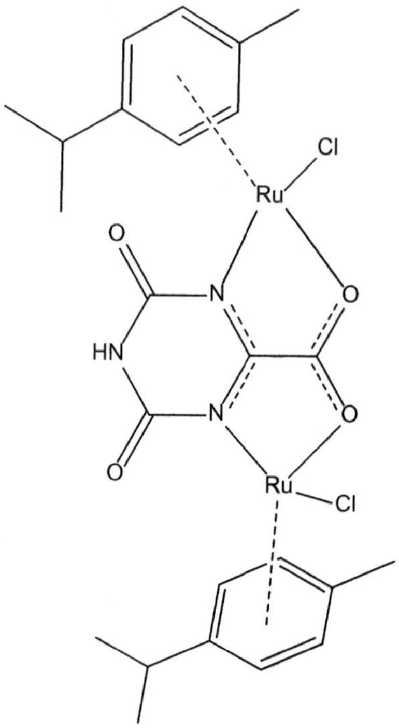

Figure 8.4 Self-assembled arene-ruthenium nanocages. (Adapted with permission of RSC (Vajpayee et al. 2011).

The 3D prismatic cage structures synthesized by Ru-based acceptors (**4A to 8A**) were confirmed by NMR spectroscopy and proliferative activity was tested with five cell lines: SK-hep-1 (liver), HCT-15 (colon), HeLa (cervix), A-549 (lung), and MDA-MB-231 (breast). Compound C (Fig. 8.4) showed comparable cytotoxicity with respect to cisplatin drug molecules. The IC_{50} values obtained showed that tetracenquinonato (**7A**) and bezenquinonato (**5A**) spacer derived prismatic-cages are inactive against all cell lines, whereas oxalate and naphthoquinonato based cages showed cytotoxicity comparable to cisplatin. It was suggested that the presence of an extended π-conjugated system activates inhibition of tumor cell proliferation through interference into regulatory pathways of the cell cycle by apoptosis. These anti-cancer activities demonstrated that cytotoxicity does not depend on the acceptor clip or donor size but the right donor/acceptor combination. The results are useful for understanding drug action mechanisms and how cavity dimensions and π-conjugation influence selectivity. Based on these observations, they further developed a series of rectangular complexes for targeting cancer cells (Figs. 8.5 and 8.6) using the same arene–Ru acceptors (**4A–8A**, shown in Fig. 8.4) and series of di-pyridyl donors (Vajpayee et al. 2011, Mishra et al. 2012, Vajpayee et al. 2012, Mishra et al. 2014).

The heterometallic hexanuclear metal rectangles were also synthesized to test the effect of enhanced functionalities in the form of mixed-metal, heterometallic scaffolds. The rectangle metallacages were characterized by multinuclear NMR,

UV-visible and fluorescence studies. Besides anti-tumor activities were determined against various human cancer cell-lines that suggested the naphthaquinonato acceptor **6**-based metallomacrocycles were most promising showing IC_{50} values similar to doxorubicin and much smaller than cisplatin. The large metalla-rectangles containing **3**, **13** to **18** ligands with oxalate- and benzoquinone-bridged arene–Ru complexes did not show significant antitumor activity. The Pt-containing donor **5** in combination with the naphthaquinonato acceptor **6** showed the highest activity among the metallomacrocycles studied. Recently, the same group reported the triazole-based Ru(II) metallo-macrocycles (Singh et al. 2019). Triazole is an important heterocyclic structure that can help in the addition of various functional groups to the macrocyclic structure. It also displays a wide range of catalytic and medicinal use. The self-assembling of the metalla-cage structure is shown in Fig. 8.7.

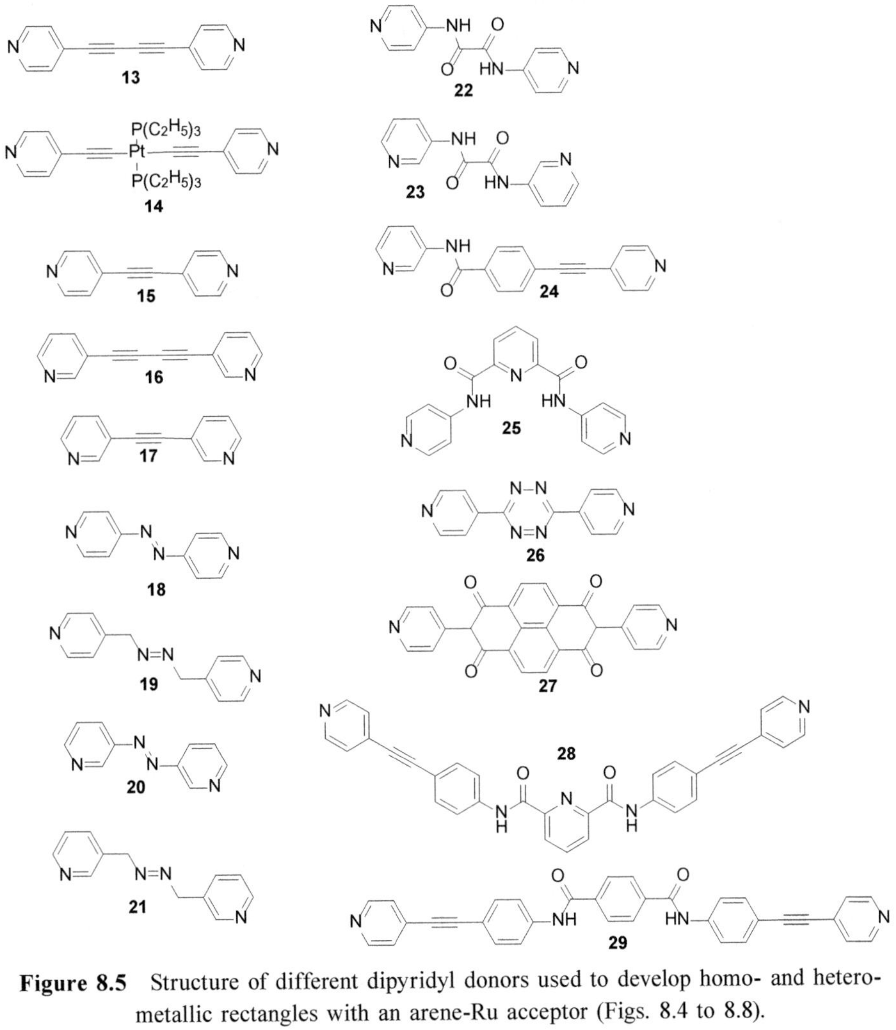

Figure 8.5 Structure of different dipyridyl donors used to develop homo- and hetero-metallic rectangles with an arene-Ru acceptor (Figs. 8.4 to 8.8).

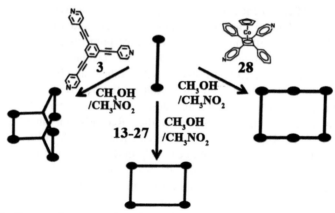

Figure 8.6 Systematic presentation of coordination-driven self-assemblies using acceptor (**4–8**, shown in Fig. 8.4) and ligands (**13–29** as shown in Fig. 8.5). (Reprinted with permission of ACS (Singh et al. 2019).

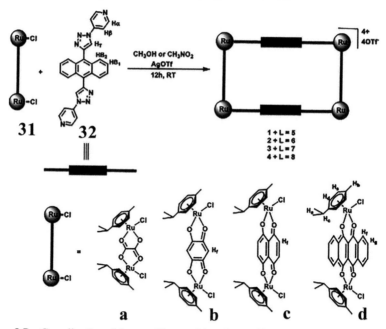

Figure 8.7 Coordination-driven self-assembly of metallomacrocycles using triazole-based Ligand (**32**) and Ru-based acceptors (**31a-d**). Reprinted with permission of ACS (Singh et al. 2019).

The new cage molecules containing napthoquinato moiety as Ru-based acceptors inhibited the growth of A549, AGS, HCT-15 and HepG2 more effectively than doxorubicin, which suggests its potential use as the basis of anticancer drug development.

A large number of ruthenium-based macrocyclic complexes have been studied. But their fate in a living system is not yet established. However, two of the most promising Ru complexes, NAMI-A and KP1019/1339, are now undergoing clinical trials (Fig. 8.8) (Alessio and Messori 2019). Although these complexes have Ru(III) centers, they are reduced to Ru(II) within the body (Hartinger et al. 2006).

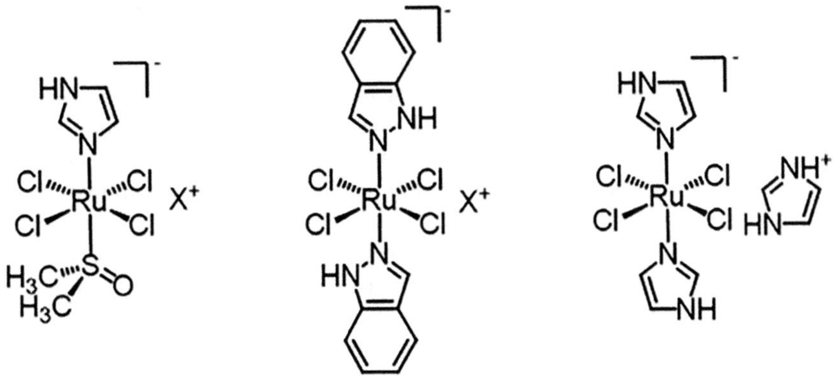

X⁺ = Na⁺   **NAMI**   X⁺ = Na⁺   **KP1339**   **KP418**

X⁺ = ImH⁺ **NAMI-A**   X⁺ = IndH⁺ **KP1019**

Figure 8.8   Schematic structures of promising Ruthenium-based anticancer drug molecules: NAMI-A, KP1019/KP1339 and KP418.

A large number of self-assembled metallomacrocycles incorporating various other metal ions, such as Ru (Singh et al. 2014), Pd (Schmidt et al. 2014), Pt (Aliprandi et al. 2015), Os (Pigeon et al. 2005), or Ir (You and Nam 2012; Gupta et al. 2014), have been studied for cytotoxicity against different cancer cell lines. A wide variety of preclinical and clinical studies with anticancer metallodrugs have been also reported containing various metal ions such as gallium, titanium, gold, cobalt, ruthenium and tin (Ellahioui et al. 2017) which targets multiple other sites and their biological chemistry is also different form cis-platin. A perfect molecule with clinically acceptable anti-cancer activity is yet to be synthesized.

METALLO-MACROCYCLE AS MAGNETIC RESONANCE IMAGING (MRI) CONTRAST AGENT

The biological applications of metal coordinated macrocycles are not limited to anticancer drug development, a particularly exciting emerging application is Magnetic Resonance Imaging (MRI). The opaque biological structures can be visualized as three-dimensional images with a relatively high spatial resolution ($\sim$10 μM) in a noninvasive procedure using MRI. The MR images are obtained from the nuclear magnetic resonance of water protons present in the biological samples. The contrast in the MR image depends on the local concentration of water and on the longitudinal (T_1) and transverse (T_2) relaxation times of its protons (Caravan et al. 1999). Inversion in the magnetic vector of water protons from their preferential alignment with the external magnetic field can be induced by a radiofrequency pulse. And the realignment of proton spin with the external magnetic field in a specific time interval (T_1) can be significantly reduced if the spins are in contact with a local paramagnetic center, thereby brightening the image.

Coordination complexes containing paramagnetic metal centers that leave open coordination sites for water molecules to access the inner-sphere of metal ions (particularly Gd^{3+} with 7 unpaired electrons in 4f orbitals, but also high-spin Fe^{3+} and Mn^{2+} with unpaired electrons in 5d orbitals) act as a Contrast Agent (CA) that enhance MR images via a T_1 mechanism. Coordination interactions are important in transmitting the effect of the paramagnetic ion to the bulk solvent. In recent times, the approval of gadolinium (III) ion-based contrast agent [Gd (DTPA) $(H_2O)]^{2-}$ has revolutionized the imaging modality for a clinical study of physiological events. These metal complexes are routinely used for imaging lesions in the central nervous system, breast and abdomen, angiographic imaging, cardiac imaging and articular imaging, among numerous additional applications (Wahsner et al. 2019). A large number of different types of CAs are reported in literature. All existing Gadolinium-Based Contrasting Agents (GBCAs) that are used in the clinic are based on octadentate polyaminocarboxylate ligands. These Gd (III) based chelates are nine-coordinate complexes in which a ligand occupies eight binding sites at the metal center and the ninth coordination site is occupied by a solvent water molecule (Fig. 8.9). The ideal geometry for the 9-coordinated complex $[ML_9]$ is Tricapped Trigonal Prism (TTP) and capped square antiprism (CSAP) (Caravan et al. 1999). Two types of gadolinium complexes are widely used to enhance the contrast of images in MR imaging procedures: (i) the macrocyclic molecules where Gd^{3+} is caged in the pre-organized cavity of the ligand and (ii) the linear molecules. Gadolinium chelates containing linear and macrocyclic ligands differ in their thermodynamic stability constants and kinetic stabilities. Thermodynamic stability of gadolinium complexes relates to the Gd^{3+} releases at an equilibrium under specific environments while kinetic stability indicates the rate of Gd^{3+} release. Recently, the U.S. Food and Drug Administration (FDA) approved a new molecule– Clariscan, a macrocyclic, ionic, gadolinium-based, contrast agent (Clough et al. 2019), and the structure of other approved CAs are shown in Fig. 8.9.

The MRI contrast agent [Gd (OH_2) (DTPA)$]^{2-}$ (Magnevist®) and Gd (DOTA)⁻ (Dotarem®) were the first complexes approved for clinical practice and used as gold standards. The dianionic [Gd $(OH_2)(DTPA)]^{2-}$ contains acyclic ligand DTPA, an amino polycarboxylic acid (IUPAC name: 2-[Bis[2-[bis(carboxymethyl)amino] ethyl] amino]acetic acid). It forms a stable octa-coordinated chelate as distorted TTP and the ninth coordination site was occupied by a water molecule (Caravan et al. 1999). Gd (DOTA)⁻ (Dotarem®) also forms a very stable lanthanide chelates with macrocyclic ring structure of DOTA (1,4,7,10-tetraazacyclododecane-1,4,7,10-tetraacetic acid). This ligand consists of the macrocycle cyclen, which is N-functionalized with four acetic acid pendant arms and is octadentate with four nitrogen and four oxygen donor atoms as shown in Fig. 8.9. The solid-state X-ray structure of Gd (DOTA))⁻ indicates that the Gd^{3+} is situated in the center of a CASP cage, with the water molecule in an axial position. These complexes are hydrophilic and in solution, they exist as stereoisomeric interchangeable molecules with a

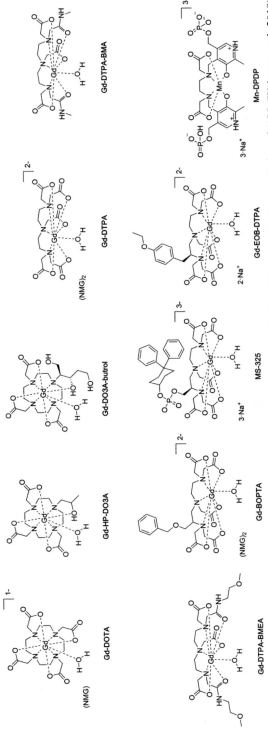

Figure 8.9 Commercially approved T_1 contrast agents containing linear and cyclic ligand (Reprinted with permission of ACS (Wahsner et al. 2019).

different structure (square antiprism = SA and twisted square antiprism = TSA) as shown in Fig. 8.10. High complex stability is required for good contrast images and stability constant for [Gd(OH$_2$)(DOTA)]$^-$ ranges from 24.0 to 27.0 as reported in the literature (Clough et al. 2019). The minor isomer (TSA) exhibits faster water exchange than the major isomer (SA), suggesting that CA should adopt the TSA structure.

The GdDO3A is another ligand with a higher image contrasting nature and substantial stability. The ligand structure contains a 12-membered cyclen macrocycle similar to the DOTA ligand with an acetate group as arms. The ligand design opens up coordination of two water molecules to metal ions which helps in improving image contrasting nature (Aime et al. 1998). But the clinical application of CAs with acyclic ligands later showed Nephrogenic Systemic Fibrosis (NSF) and kidney failure in patients (Grobner 2006). It was also observed that gadolinium has an ionic size similar to Ca^{2+} which enables them to interfere with calcium-mediated signaling pathways (Darnall and Birnbaum 1973, Spencer et al. 1998). The release of the metal ion depends on the thermodynamic stability (stability constant, log K) and kinetic stability (complex dissociation rate $-t_{1/2}$) parameter of the coordinate complexes (Sherry et al. 2009, Ramalho et al. 2016). Frenzel et al. (Frenzel et al. 2008) measured the dissociation rate ($t_{1/2}$) of the gadolinium complex under the physiological condition in serum solutions (pH 7.4, 37°C) and determined the inertness of all approved Gd-based CAs. The results showed that the Gd (III) release order was non-ionic linear > ionic linear > Macrocycles. Thus new design features were needed which can improve the leaching behaviors of Gd (III) from its complexes. Recently the GD-based CA designs have been mostly oriented on modifications of the structural motifs of DOTA, DTPA, and DO3A. A large number of different ligands were synthesized with 12-membered cyclen macrocycle ligands bearing functionalities different from carboxylic acids in the pendant arms, such as amides, alcohols, phosphonic or phosphinic acids and phenols, their lanthanide (III) complexes were studied which are reviewed in literature (Caravan et al. 1999, Bottrill et al. 2006, Wahsner et al. 2018, Clough et al. 2019). The efficacy of these CAs were measured as the ability of its 1 mM solution to increase the longitudinal relaxation rate (r_1) ($= 1/T_1$) of water protons which is known as relaxivity and labeled as r_1.

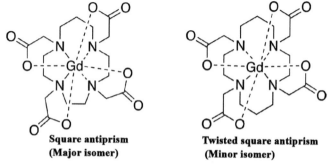

Square antiprism
(Major isomer)

Twisted square antiprism
(Minor isomer)

Figure 8.10 The minor and major isomers of DOTA-based MRI contrast agents (Reprinted with permission of ACS (Wahsner et al. 2018).

From a chemical point of view, water ligand directly attached to the Gd(III) in the inner complex cation sphere undergoes an exchange with the nearby bulk water molecules and transmits the relaxation effect to the bulk. The Twisted-Square Antiprismatic (TSAP) isomer of Gd(III)-DOTA shows faster water exchange reaction due to steric crowding at the water-binding site. It was reported that modification of pendant arms in the DOTA ligands by substitution on the α-position of the pendant arms slows or eliminates the arm rotation process (Di Bari et al. 2000; Dunand et al. 2001) while substitution on the cyclen backbone restricts the ring inversion motion (Woods et al. 2003; Woods et al. 2004). Substitution at both positions effectively locks the macrocycle into a single conformation (Avecilla et al. 2003; Rudovský et al. 2005). In general, macrocyclic chelates such as Gd-DOTA or Gd-HP-DO3A or their derivatives are more stable than linear molecules. These scaffolds could be further used to develop safer contrast agents.

There are other high spin cations—namely high spin Mn(II), high spin Mn(III), high spin Fe(III) and Eu(II) that can serve as effective relaxation agents in MRI imaging (Lauffer 1987). They can form thermodynamically stable and kinetically inert complexes with macrocyclic ligands (Drahoš et al. 2012). Manganese is a biogenic metal, needed as an essential micronutrient in our body. Mn(II) ion with d^5 configuration is a very effective relaxation agent as it prefers a high-spin (S = 5/2) electronic configuration and is characteristically long T_1. In contrast to Gd(III) complexes, aqueous complexes of Mn(II) typically have coordination numbers (CN) of 6 or 7 (Drahoš et al. 2012). Various Mn(II) complexes have been investigated with linear and macrocyclic ligands. The better stability and high inertness of macrocyclic metal complexes have led to the synthesis of Mn(II) complexes containing ligands with different donor atoms and cavity sizes (Fig. 8.11). Derivatives of triazacyclononane (NOTA) and diaza-oxacyclononane (ENOTA) containing 9-membered ring structure were reported as a ligand for Mn(II) complexes which showed low relaxivities as compared to Gd-complexes. The low relaxation time exclusively indicates the presence of other-sphere contribution, thus confirming non-hydration of the complex (Geraldes et al. 1986). Mn-complexes, [Mn (DOTA)]$^{2-}$ and [Mn (DO3A)]$^-$ with cyclen derivatives as ligands were also synthesized like Gd-complexes (Geraldes et al, 1986). The proton relaxation behavior was found to be similar to [Mn (NOTA)]$^-$. Porphyrin and 1,2-phenylenediamine derived bis-amidate ligands form high spin complexes with Mn(III) as Mn-porphyrin and Mn-PDA respectively (Koenig et al. 1987, Barandov et al. 2016). These complexes (Fig. 8.12) demonstrated better performance as MRI agents (Koenig et al. 1987). The relaxivity of Mn (III)-porphyrins has been described as 'anomalously' considering greatly the S = 2 spin state. It is believed that elongation of the singly occupied dz^2 orbital which lies across the Mn(III)-OH$_2$ bond axis and further development of the asymmetric structure of Mn(III)-porphyrin reduces the distance between the Mn(III) spin density and the water ^{1}H nuclei. These Mn-complexes are considered as Gd-free contrast agents.

High spin complexes of Fe(III) have been evaluated as MRI contrast agents (Lauffer et al. 1985; Jenkins et al. 1991). They are considered hepatobiliary-specific contrast agents. The Eu(II) ion is isoelectronic with Gd(III) and complexes of Eu(III) have been demonstrated to be potent relaxation agents. But the oxidation state of Eu(III) is unstable as compared to the diamagnetic Eu(III) which can be stabilized by different cavitands and macrocyclic structures (Burai et al. 2002).

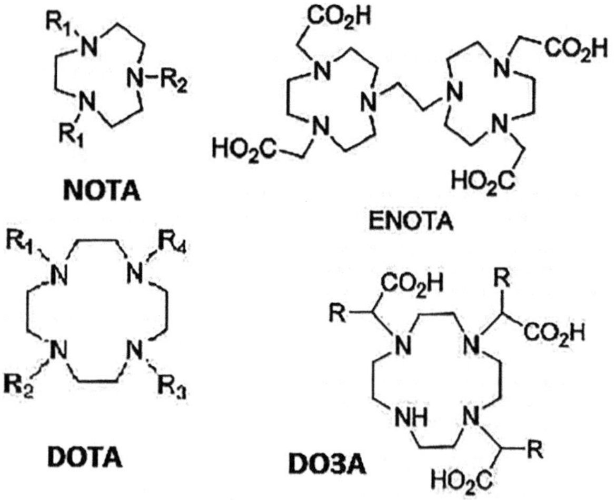

NOTA

ENOTA

DOTA

DO3A

Figure 8.11 Structure of macrocyclic ligand used to study Mn(II) complexes as MRI agent. Printed by permission of John Wiley and Sons (Drahoš et al. 2012).

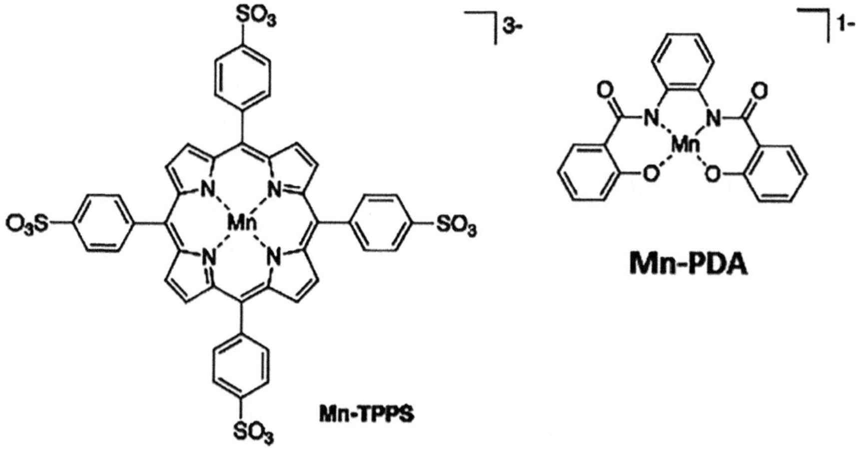

Mn-TPPS

Mn-PDA

Figure 8.12 Structure of Mn(III) complexes discussed in the chapter. Printed with permission of ACS (Barandov et al. 2016).

The Eu(II)(aq) ion has one of the fastest water-exchange rates measured (Caravan et al. 1999; Seibig et al. 2000). Animal model studies showed that Eu(III) complexes with cavitands rapidly oxidized in the bloodstream, providing no positive contrast enhancement 3 minutes after intravenous injection in mice (Ekanger et al. 2016). Thus the development of more stable complexes with better-modified structures is required for intravenous injection.

Macrocyclic ligand-based contrast agents can also be used to study and detect biological processes and diseases. Stimulus responsive MRI contrast agent responds to changes in pH, O_2 partial pressure, small molecules like glutathione, proteins, enzymatic activity and specific metal ions (Hingorani et al. 2015).

METALLOMACROCYCLES AS SYNTHETIC ENZYME

Nature performs many important biological functions by using metal-ion macrocyclic systems such as the conversion of light into chemical energy using chlorophyll, oxygen transport by hemoglobin, etc. These thermodynamically stable and kinetically inert macrocyclic complexes are attractive scaffolds for assembling metal complexes for a multitude of applications as their structure and stability do not impede the biological role of these species. The metal ion-ligating architecture present within these biological systems can be broadly divided into two classes: (1) the metal ion coordination sphere is 'saturated' or remains unchanged during its function and (2) ligand exchange in the metal-ligand complexation sphere is critical to function. Mimicking the structural and control aspects of these macrocyclic complexes has provided new directions to design synthetic architectures that can mimic some of their properties, such as providing binding sites and hydrophobic pockets where substrate selection and transformation can occur. Here the focus is on the design and construction of copper-containing metalloenzyme that mimics what use of the principles of supramolecular chemistry and catalysis.

Rebek pioneered the field of enzyme mimics. He developed cavitands for host-guest catalysis, investigated autocatalysis in self-replicators (Tjivikua et al. 1990) and allosteric catalysts (Rebek 1984). Initially small supramolecular catalysts were synthesized which combined a binding site linked to a reactive center. Likewise, catalysts acting as templates and possessing two or more binding sites bringing together the two reactants were also investigated (Wolfe et al. 1988; Tecilla et al. 1990).

The catalytic reaction can be envisioned as intramolecular for the host–guest-catalyst complex. Spiccia et al. have reported (Joshi et al. 2015) the design and synthesis of 'artificial nucleases' with the possibility of producing novel therapeutics that not only bind to targeted DNA/RNA sequences, but also cleave the phosphate ester bonds present within the sugar-phosphate backbones, rendering them inactive. Copper complexes of ligands derived from the small, macrocycle, 1,3,5-tris(1,4,7-triazacyclonon-1-ylmethyl) benzene, (tacn; Fig. 8.13a), which is capable of binding three metal centers in close proximity were synthesized. The tacn ligand was used to self-assemble $\{[Cu_3tacn(\mu\text{-}OH) (\mu_3\text{-}HPO_4) (H_2O)] [PF_6]_3 \cdot 3H_2O\}_n$. The X-ray crystallographic studies and ^{31}P NMR analysis confirmed that the compound is polymeric with tri-nuclear metal sites in which three copper(II) centers are linked by an HPO_4^{2-} phosphate bridge and two of the coppers are further linked by a hydroxo bridge (Fig. 8.13b). Each copper center in the complex exists in tetragonally distorted square-pyramidal geometry.

The polymeric structure contains a hydroxo bridge between two metal centers indicating similarity to trinuclear Zn(II)-containing active sites of phospholipase C. Additionally the magnetic susceptibility studies and electron spin resonance measurement suggested that the self-assembled complex features metal—metal separation and metal—metal interaction by hydroxo bridged and alkoxo-bridged multinuclear structure which resembles the active sites of ascorbate oxidase and laccase (Fig. 8.13b) (Spiccia et al. 1997).

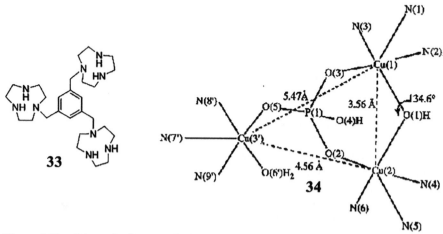

Figure 8.13   Schematic diagram of tri-nuclear Cu(II) complex (**34**) formed by tacn-based ligands **33** used to develop metallobiosite models (Adapted with permission of ACS (Joshi et al. 2015).

They further modified the tacn ligand structure as N-alkyl tacn (Me$_3$tacn = 1,4,7-trimethyl-1,4,7-triazacyclononane) and synthesized their Cu(II) complexes (Cu(II)-Me$_3$tacn)(Desbouis et al. 2012). These complexes were used to study their ability to cleave phosphate-ester bond in BNPP [bis(p-nitrophenyl) phosphate] and NPP [nitrophenylphosphate] as a model substrate. The important criteria to act as a catalyst for phosphor-ester bond cleavage are the presence of a coordination site at the metal center for the coordination of the phosphate ester and a water ligand cis to the coordinated phosphate ester, and formation of hydroxide (nucleophile) as attacking reagent for phosphate ester bond (Hendry and Sargeson 1989). The kinetic measurement using the spectrophotometric method revealed that the rate of bond cleavage in BNPP is about 100-times faster than NPP. A comparison of catalytic activity of similar complexes shows a lower rate than complex containing methylated tacn ligands. The mechanism of hydrolysis is shown in Fig. 8.7 which is postulated to involve deprotonation of [Cu (Me$_3$tacn) (OH$_2$)$_2$]$^{2+}$ to form [Cu (Me$_3$tacn) (OH$_2$) (OH)]$^+$, substrate coordination and rate-determining phosphate ester cleavage (Fig. 8.14). The catalytic activity of the mononuclear complex was much lower than bi-nuclear and subsequently trinuclear complexes confirming metal-ion cooperativity as observed in enzyme active sites. The copper complexes were applicable to DNA and RNA phosphate ester hydrolysis (Belousoff et al. 2008).

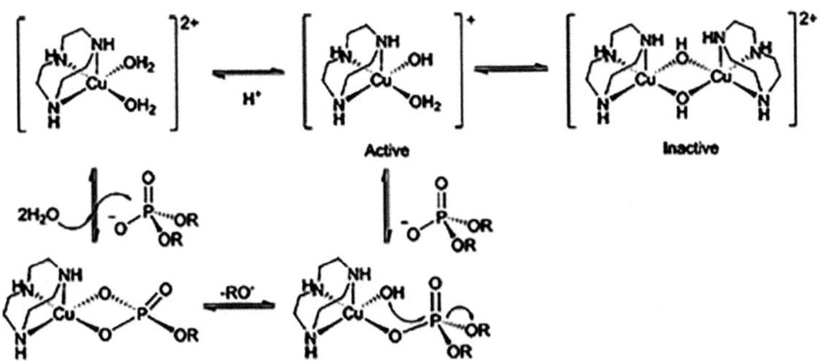

Figure 8.14 Proposed Mechanism for Phosphate Diester Hydrolysis by Cu(II)−tacn Complexes (Me$_3$tacn)(Reprinted with permission of ACS (Joshi et al. 2015).

CONCLUSION AND OUTLOOK

Some selected examples of the possibilities offered by the combination of macrocyclic–metal ion complexes have been described. The design and synthesis of macrocyclic complexes with metal ions are attractive due to their diverse applications in catalysis, sensing new drug molecule and other fields. Significant advances have been made in different fields reviewed in this chapter. With the current effort devoted to the development of the structure-activity relationship, it will be fascinating to see how much more chemists and their collaborators can achieve in the near future.

REFERENCES

Aime, S., M. Botta, S.G. Crich, G. Giovenzana, R. Pagliarin, M. Sisti and E. Terreno. 1998. NMR relaxometric studies of Gd(III) complexes with heptadentate macrocyclic ligands. Magn. Reson. Chem. 36(S1): S200–S208.

Alam, M.N. and F. Huq. 2016. Comprehensive review on tumour active palladium compounds and structure–activity relationships. Coord. Chem. Rev. 316: 36–67.

Alessio, E. and L. Messori. 2019. NAMI-A and KP1019/1339, two iconic ruthenium anticancer drug candidates face-to-face: a case story in medicinal inorganic chemistry. Molecules. 24(10): 1995.

Aliprandi, A., D. Genovese, M. Mauro and L. De Cola. 2015. Recent advances in phosphorescent Pt(II) complexes featuring metallophilic interactions: properties and applications. Chem. Lett. 44(9): 1152–1169.

Avecilla, F., J.A. Peters and C.F. Geraldes. 2003. X-ray crystal structure of a sodium salt of [Gd(DOTP)]$^{5-}$: Implications for its second-sphere relaxivity and the 23Na NMR hyperfine shift effects of [Tm (DOTP)]$^{5-}$. Eur. J. Inorg. Chem. 2003(23): 4179–4186.

Barandov, A., B.B. Bartelle, B.A. Gonzalez, W.L. White, S.J. Lippard and A. Jasanoff. 2016. Membrane-permeable Mn(III) complexes for molecular magnetic resonance imaging of intracellular targets. J. Am. Chem. Soc. 138(17): 5483–5486.

Belousoff, M.J., B. Graham and L. Spiccia. 2008. Copper(II) Complexes of N-methylated derivatives of ortho- and meta-Xylyl-Bridged Bis(1,4,7-triazacyclononane) ligands:

synthesis, X-ray structure and reactivity as artificial nucleases. Eur. J. Inorg. Chem. 2008(26): 4133–4139.

Bertini, I., G. Bertini, H. Gray, H.B. Gray, E. Stiefel, J.S. Valentine and E.I. Stiefel. 2007. Biological Inorganic Chemistry: Structure and Reactivity. Sausalito, CA, University Science Books.

Bottrill, M., L. Kwok and N.J. Long. 2006. Lanthanides in magnetic resonance imaging. Chem. Soc. Rev. 35(6): 557–571.

Brabec, V. and J. Kasparkova. 2005. Modifications of DNA by platinum complexes: relation to resistance of tumors to platinum antitumor drugs. Drug Resist. Updat. 8(3): 131–146.

Burai, L., R. Scopelliti and É. Tóth. 2002. EuII-cryptate with optimal water exchange and electronic relaxation: a synthon for potential pO_2 responsive macromolecular MRI contrast agents. Chem. Commun.(20): 2366–2367.

Caravan, P., J.J. Ellison, T.J. McMurry and R.B. Lauffer. 1999. Gadolinium(III) chelates as MRI contrast agents: structure, dynamics, and applications. Chem. Rev. 99(9): 2293–2352.

Caravan, P., É. Tóth, A. Rockenbauer and A.E. Merbach. 1999. Nuclear and Electronic Relaxation of Eu2+ (aq): An Extremely Labile Aqua Ion1. J. Am. Chem. Soc. 121(44): 10403–10409.

Clough, T.J., L. Jiang, K.-L. Wong and N.J. Long. 2019. Ligand design strategies to increase stability of gadolinium-based magnetic resonance imaging contrast agents. Nat. Commun. 10(1): 1–14.

Cook, T.R., V. Vajpayee, M.H. Lee, P.J. Stang and K.-W. Chi. 2013. Biomedical and biochemical applications of self-assembled metallacycles and metallacages. Acc. Chem. Res. 46(11): 2464–2474.

Da Silva, J.F. and R.J.P. Williams. 2001. The biological chemistry of the elements: the inorganic chemistry of life. Oxford, Oxford University Press.

Dale, L., J. Tocher, T. Dyson, D. Edwards and D. Tocher. 1992. Studies on DNA damage and induction of SOS repair by novel multifunctional bioreducible compounds. II. A metronidazole adduct of a ruthenium-arene compound. Anticancer Drug Des. 7(1): 3–14.

Darnall, D.W. and E.R. Birnbaum. 1973. Lanthanide ions activate α-amylase. Biochemistry 12(18): 3489–3491.

Desbouis, D., I.P. Troitsky, M.J. Belousoff, L. Spiccia and B. Graham. 2012. Copper(II), zinc(II) and nickel(II) complexes as nuclease mimetics. Coord. Chem. Rev. 256(11–12): 897–937.

Di Bari, L., G. Pintacuda and P. Salvadori. 2000. Solution equilibria in YbDOTMA, a chiral analogue of one of the most successful contrast agents for MRI, GdDOTA. Eur. J. Inorg. Chem. 2000(1): 75–82.

Drahoš, B., I. Lukeš and É. Tóth. 2012. Manganese (II) complexes as potential contrast agents for MRI. Eur. J. Inorg. Chem. 2012(12): 1975–1986.

Dunand, F.A., R.S. Dickins, D. Parker and A.E. Merbach. 2001. Towards rational design of fast water-exchanging Gd (dota-like) contrast agents? Importance of the M/m ratio. Chem. Eur. J. 7(23): 5160–5167.

Ekanger, L.A., L.A. Polin, Y. Shen, E.M. Haacke and M.J. Allen. 2016. Evaluation of EuII-based positive contrast enhancement after intravenous, intraperitoneal, and subcutaneous injections. Contrast Media Mol. Imaging. 11(4): 299–303.

Ellahioui, Y., S. Prashar and S. Gomez-Ruiz. 2017. Anticancer applications and recent investigations of metallodrugs based on gallium, tin and titanium. Inorganics. 5(1): 4.

Frenzel, T., P. Lengsfeld, H. Schirmer, J. Hütter and H.-J. Weinmann. 2008. Stability of gadolinium-based magnetic resonance imaging contrast agents in human serum at 37 C. Invest. Radiol. 43(12): 817–828.

Geraldes, C., A. Sherry, R. Brown III and S. Koenig. 1986. Magnetic field dependence of solvent proton relaxation rates induced by Gd^{3+} and Mn^{2+} complexes of various polyaza macrocyclic ligands: implications for NMR imaging. Magn. Reson. Med. 3(2): 242–250.

Gloe, K. 2005. Macrocyclic Chemistry. The Netherlands, Springer.

Grobner, T. 2006. Gadolinium–a specific trigger for the development of nephrogenic fibrosing dermopathy and nephrogenic systemic fibrosis? Nephrol. Dial. Transplant. 21(4): 1104–1108.

Gupta, G., J.M. Kumar, A. Garci, N. Nagesh and B. Therrien. 2014. Exploiting natural products to build metalla-assemblies: The anticancer activity of embelin-derived Rh(III) and Ir(III) metalla-rectangles. Molecules. 19(5): 6031–6046.

Gupta, G., G.S. Oggu, N. Nagesh, K.K. Bokara and B. Therrien. 2016. Anticancer activity of large metalla-assemblies built from half-sandwich complexes. CrystEngComm. 18(26): 4952–4957.

Haas, K.L. and K.J. Franz. 2009. Application of metal coordination chemistry to explore and manipulate cell biology. Chem. Rev. 109(10): 4921–4960.

Hartinger, C.G., S. Zorbas-Seifried, M.A. Jakupec, B. Kynast, H. Zorbas and B.K. Keppler. 2006. From bench to bedside–preclinical and early clinical development of the anticancer agent indazolium *trans*-[tetrachlorobis (1H-indazole) ruthenate(III)] (KP1019 or FFC14A). J. Inorg. Biochem. 100(5–6): 891–904.

Hendry, P. and A.M. Sargeson. 1989. Metal ion promoted phosphate ester hydrolysis. Intramolecular attack of coordinated hydroxide ion. J. Am. Chem. Soc. 111(7): 2521–2527.

Hingorani, D.V., A.S. Bernstein and M.D. Pagel. 2015. A review of responsive MRI contrast agents: 2005–2014. Contrast Media Mol. Imaging. 10(4): 245–265.

Jenkins, B.G., E. Armstrong and R.B. Lauffer. 1991. Site-specific water proton relaxation enhancement of iron(III) chelates noncovalently bound to human serum albumin. Magn. Reson. Med. 17(1): 164–178.

Joshi, T., B. Graham and L. Spiccia. 2015. Macrocyclic metal complexes for metalloenzyme mimicry and sensor development. Acc. Chem. Res. 48(8): 2366–2379.

Koenig, S.H., R.D. Brown III and M. Spiller. 1987. The anomalous relaxivity of Mn^{3+} ($TPPS_4$). Magn. Reson. Med. 4(3): 252–260.

Lauffer, R.B. 1987. Paramagnetic metal complexes as water proton relaxation agents for NMR imaging: theory and design. Chem. Rev. 87(5): 901–927.

Lauffer, R.B., W.L. Greif, D.D. Stark, A.C. Vincent, S. Saini, V.J. Wedeen and T.J. Brady. 1985. Iron-EHPG as an hepatobiliary MR contrast agent: initial imaging and biodistribution studies. J. Comput. Assist. Tomogr. 9(3): 431–438.

Linares, F., M.A. Galindo, S. Galli, M.A. Romero, J.A. Navarro and E. Barea. 2009. Tetranuclear coordination assemblies based on half-sandwich ruthenium(II) complexes: noncovalent binding to DNA and cytotoxicity. Inorg. Chem. 48(15): 7413–7420.

Mattsson, J., P. Govindaswamy, A.K. Renfrew, P.J. Dyson, P. Stepnicka, G. Süss-Fink and B. Therrien. 2009. Synthesis, molecular structure, and anticancer activity of cationic arene ruthenium metallarectangles. Organometallics 28(15): 4350–4357.

Mishra, A., Y.J. Jeong, J.H. Jo, S.C. Kang, M.S. Lah and K.W. Chi. 2014. Anticancer Potency studies of coordination driven self-assembled arene–ru-based metalla-bowls. ChemBioChem. 15(5): 695–700.

Mishra, A., H. Jung, J.W. Park, H.K. Kim, H. Kim, P.J. Stang and K.-W. Chi. 2012. Anticancer activity of self-assembled molecular rectangles via arene–ruthenium acceptors and a new unsymmetrical amide ligand. Organometallics. 31(9): 3519–3526.

Paul, L.E., B. Therrien and J. Furrer. 2012. Investigation of the reactivity between a ruthenium hexacationic prism and biological ligands. Inorg. Chem. 51(2): 1057–1067.

Pigeon, P., S. Top, A. Vessières, M. Huché, E.A. Hillard, E. Salomon and G. Jaouen. 2005. Selective estrogen receptor modulators in the ruthenocene series. Synthesis and biological behavior. J. Med. Chem. 48(8): 2814–2821.

Rabik, C.A. and M.E. Dolan. 2007. Molecular mechanisms of resistance and toxicity associated with platinating agents. Cancer Treat. Rev. 33(1): 9–23.

Ramalho, J., R. Semelka, M. Ramalho, R. Nunes, M. AlObaidy and M. Castillo. 2016. Gadolinium-based contrast agent accumulation and toxicity: an update. Am. J. Neuroradiol. 37(7): 1192–1198.

Rebek, J. 1984. Binding forces, equilibria and rates: new models for enzymic catalysis. Acc. Chem. Res. 17(7): 258–264.

Rudovský, J., P. Cígler, J. Kotek, P. Hermann, P. Vojtíšek, I. Lukeš, J.A. Peters, L. Vander Elst and R.N. Muller. 2005. Lanthanide(III) complexes of a mono (methylphosphonate) analogue of H4dota: the influence of protonation of the phosphonate moiety on the TSAP/SAP isomer ratio and the water exchange rate. Chem. Eur. J. 11(8): 2373–2384.

Schmidt, A., A. Casini and F.E. Kühn. 2014. Self-assembled M2L4 coordination cages: synthesis and potential applications. Coord. Chem. Rev. 275: 19–36.

Seibig, S., É. Tóth and A.E. Merbach. 2000. Unexpected differences in the dynamics and in the nuclear and electronic relaxation properties of the isoelectronic $[Eu^{II}(DTPA)(H_2O)]^{3-}$ and $[Gd^{III}(DTPA)(H_2O)]^{2-}$ complexes (DTPA = diethylenetriamine pentaacetate). J. Am. Chem. Soc. 122(24): 5822–5830.

Sherry, A.D., P. Caravan and R.E. Lenkinski. 2009. Primer on gadolinium chemistry. Magn. Reson. Imaging 30(6): 1240–1248.

Singh, A.K., D.S. Pandey, Q. Xu and P. Braunstein. 2014. Recent advances in supramolecular and biological aspects of arene ruthenium(II) complexes. Coord. Chem. Rev. 270: 31–56.

Singh, J., D.W. Park, D.H. Kim, N. Singh, S.C. Kang and K.-W. Chi. 2019. Coordination-driven self-assembly of triazole-based apoptosis-inducible metallomacrocycles. ACS Omega 4(6): 10810–10817.

Spencer, A., S. Wilson and E. Harpur. 1998. Gadolinium chloride toxicity in the mouse. Hum. Exp. Toxicol. 17(11): 633–637.

Spiccia, L., B. Graham, M.T. Hearn, G. Lazarev, B. Moubaraki, K.S. Murray and E.R. Tiekink. 1997. Towards synthetic models for trinuclear copper active sites of ascorbate oxidase and laccase: self-assembly, crystal structure and magnetic properties of the copper(II) complexes of 1,3,5-tris(1,4,7-triazacyclonon-1-ylmethyl) benzene. J. Chem. Soc., Dalton Trans. (21): 4089–4098.

Tecilla, P., S.K. Chang and A.D. Hamilton. 1990. Transition-state stabilization and molecular recognition: acceleration of phosphoryl-transfer reactions by an artificial receptor. J. Am. Chem. Soc. 112(26): 9586–9590.

Therrien, B., G. Süss-Fink, P. Govindaswamy, A.K. Renfrew and P.J. Dyson. 2008. The "Complex-in-a-Complex" cations [(acac) 2M⊂Ru6 (p-iPrC6H4Me)6(tpt)2(dhbq)3]6+: a trojan horse for cancer cells. Angew. Chem. Int. Ed. Engl. 120(20): 3833–3836.

Therrien, B. 2012. Drug delivery by water-soluble organometallic cages. Top. Curr. Chem. 319: 35–56.

Tjivikua, T., P. Ballester and J. Rebek Jr. 1990. Self-replicating system. J. Am. Chem. Soc. 112(3): 1249–1250.

Underwood, E.J. and W. Mertz. 1987. Introduction. pp. 1–19. *In*: W. Mertz (ed.). Trace Elements in Human and Animal Nutrition, 5th Ed. San Diego, Academic Press.

Vajpayee, V., Y.J. Yang, S.C. Kang, H. Kim, I.S. Kim, M. Wang, P.J. Stang and K.-W. Chi. 2011. Hexanuclear self-assembled arene-ruthenium nano-prismatic cages: potential anticancer agents. Chem. Commun. 47(18): 5184–5186.

Vajpayee, V., Y.H. Song, Y.J. Jung, S.C. Kang, H. Kim, I.S. Kim, M. Wang, T.R. Cook, P.J. Stang and K.-W. Chi. 2012. Coordination-driven self-assembly of ruthenium-based molecular-rectangles: synthesis, characterization, photo-physical and anticancer potency studies. Dalton Trans. 41(10): 3046–3052.

Wahsner, J., E.M. Gale, A. Rodríguez-Rodríguez and P. Caravan. 2018. Chemistry of MRI contrast agents: current challenges and new frontiers. Chem. Rev. 119(2): 957–1057.

Wolfe, J., D. Nemeth, A. Costero and J. Rebek. 1988. Convergent functional groups: catalysis of hemiacetal cleavage in a synthetic molecular cleft. J. Am. Chem. Soc. 110(3): 983–984.

Woods, M., Z. Kovacs, S. Zhang and A.D. Sherry. 2003. Towards the rational design of magnetic resonance imaging contrast agents: isolation of the two coordination isomers of lanthanide DOTA-type complexes. Angew. Chem. 115(47): 6069–6072.

Woods, M., Z. Kovacs, R. Kiraly, E. Brücher, S. Zhang and A.D. Sherry. 2004. Solution dynamics and stability of lanthanide(III)(S)-2-(p-nitrobenzyl) DOTA complexes. Inorg. Chem. 43(9): 2845–2851.

You, Y. and W. Nam. 2012. Photofunctional triplet excited states of cyclometalated Ir(III) complexes: beyond electroluminescence. Chem. Soc. Rev. 41(21): 7061–7084.

Chapter 9

Molecular Machines based on Macrocyclic Receptors: Switches and Motors

INTRODUCTION

A *molecular machine*, sometimes also known as a nanomachine, is that part of a molecule or multicomponent assembly which produces quasi-mechanical movements or specific mechanical movements as outputs on application of a defined energy or stimuli such as pH change, redox process, light pulse, etc., or of a chemical gradient as input (Ballardini et al. 2001; Sauvage 2003; Balzani et al. 2006a, b; Balzani et al. 2008; Credi et al. 2014; Abendroth et al. 2015).

In nature, macromolecular machines often perform essential functions of life such as DNA replication and ATP synthesis and many protein-based motors use chemical energy generated by hydrolysis of ATP (Vale and Milligan 2000; Schliwa and Woehlke 2003; Vale 2003; Amos 2008) to perform mechanical work (move the microtubule filaments) (Hoyt et al. 1997; Lodish et al. 2000; Alberts et al. 2002; Kassem 2017).

Terminology

Molecular motor, a subset of molecular machines, can utilize energy from different sources to generate unidirectional mechanical motion of its sub-molecular components and can do work repetitively and progressively on a system. On the other hand, a *molecular switch* is a molecular device, which can display reversible stimuli-responsive switching between two or more different states differing in relative positions of its sub-molecular components but undoes any mechanical effect caused when it returns to its original position and hence it cannot be used to do work repetitively and progressively. To distinguish between the two terms, one needs to understand that although both the terms are used to indicate molecular systems whose coordinated bond movements such as conformation, configuration, translation or circum-rotational motion, etc., can be controlled using external stimuli but in the case of a motor, these controlled movements can be used to perform work because the path through which the component returns to their initial position is different from the path it adopts to leave the initial state (Kay et al. 2007; Erbas-Cakmak et al. 2015).

Macrocyclic receptors have been intensively investigated for their immense potential as machines: both switches as well motors (Kassem et al. 2017; Blanco-Gómez et al. 2020) Although integration of synthetic molecular machines into hierarchical architectures in order to generate a useful momentum/work is still a challenge (Lancia et al. 2019), significant advancements have been made in the field. This chapter focuses on the developments in the field of use of synthetic macrocyclic receptors as molecular machines or their use as sub-molecular components of these devices during the past few decades with an emphasis on progress made during the last three decades and also the focus is on their structures, features and characteristics.

Here we first cover some of the most popular families of organic macrocyclic hosts with emphasis on their structure-binding relationships with different guests to try to understand their those features which make them inherently stimuli-responsive architectures. Complexes based on supramolecular macrocyclic architecture display interesting operational properties that are useful for device development (Erbas-Cakmak et al. 2015). This part also includes a brief discussion about the ease of their functionalization which is important to make them stimuli-responsive molecular devices, in case they are not inherently stimuli-responsive in their native state and often becomes a bottleneck for the synthetic approachability of new switchable architectures or molecular machine. This description is important as often these macrocycles serve as the starting point in the design and construction of the artificial molecular machines based on the macrocyclic assemblies.

Macrocycles Hosts: Implements of the Trade

Cyclodextrins (CDs)

It is perhaps the most widely utilized family of the naturally occurring, bucket-shaped macrocyclic hosts (Sollogoub 2013) formed by joining of glucose subunits via α-1,4-glycosidic bonds and with a history of more than a century since their discovery by Villiers (Crini 2014). CD derivatives which include commercially available cyclic oligosaccharides **5** (α-form), **6** (β-form) or **7** (γ-form) (Fig. 9.1) α-D-glucopyranoside subunits, made by enzymatic conversion from starch, are mostly non-toxic and water-soluble. Functionalization is easy (Croft and Bartsch 1983; Khan et al. 1998) as hydroxyl groups are on the larger primary face and hydroxymethyl functions are on the secondary face, whereas the inner surface is hydrophobically lined with the ether-like anomeric oxygen atoms and the C3-H and C5-H hydrogen atoms making it an amphipathic tori. No wonder, a large number of CD derivatives have been reported with diverse binding abilities which can complex with a variety of organic guests in its inner cavity in aqueous media. Owing to the different cavity diameters (α-: 0.49 nm; β-: 0.62 nm; γ-: 0.8 nm) linear alkanes and other monocyclic aromatic guests can be accommodated by a α-CD host, more bulky hydrocarbons such as adamantane or polyaromatic compounds such as naphthalene and anthracene derivatives can be incorporated in a β-CD and even larger guests like two aromatic guests can be accommodated in the cavity of γ-CD or a fullerene molecule can be sandwiched.

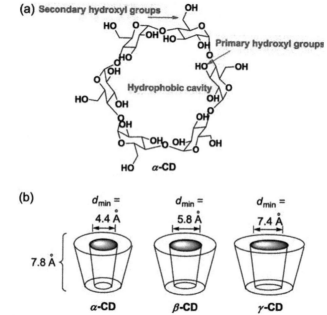

Figure 9.1 (a) Chemical structure of α-CD, and (b) cartoon representations of α-, β-, γ-CD. Reprinted with permission from (Xue et al. 2015).

Coronands and Cryptands

Coronands are macrocyclic polydentate oligomers, usually uncharged with three or more coordinating ring atoms, often oxygen or nitrogen, that are or may become suitably close for easy formation of chelate complexes with metal ions or other cationic species. They are also known as crown compounds and their chelate complexes are called coronates. Their invention, in 1967 by Pedersen, (Pedersen and Frensdorff 1972) is viewed as one of the most important discoveries of the 20th century. It is proclaimed to be the birth of supramolecular chemistry. Since it is the first man-made receptor capable of structure-specific metal ion recognition. [m]Crown-n ethers (m = total number of atoms, n = number of oxygen atoms) is a subclass of coronands oligomers containing ethylene oxide as a repeating unit with just oxygen as coordinating atom. Although more commonly used for coordinating inorganic cations (Gokel et al. 2004) to solubilize them in organic solvents exploiting the hydrophobic nature of the exterior of the host, crown ethers can also act as receptors in organic media of ammonium salts, and even appropriate neutral molecules, based on multitopic hydrogen bonding interactions (Izatt et al. 1992). Supramolecular chemistry of coronands derivatives has attracted the attention of the researchers because of their easy structural modification by incorporating nitrogen, (Krakowiak et al. 1989) phosphorous (Caminade and Majoral 1994) or sulfur (Cooper 1988) in place of oxygen.

Cryptands, the term coined by Lehn, meaning hidden, are called so because they wrap around and hide the cation. These are molecular entities (Dietrich et al. 1969; Lehn 1978a, b) that are three-dimensional, polycyclic equivalents of crown ethers, but have comparatively better complexation abilities containing three or more binding sites in each cycle that are held together by covalent bonds and possessing nitrogen (sometimes phosphorus or sulfur) as the bridge groups.

Coronands and cryptands such as aryl containing coronands and their cryptand like equivalents can form complex with organic guests. Interaction between simple aryl-containing coronands and organic dications were initially studied by Stoddart et al. (Stoddart 1988) for instance, bis(p-phenylene)-34-crown-10 ether's ability to bind with organic cations such as herbicide paraquat (N,N′-dimethyl-4,4′-bipyridinium cation, MV) and its derivatives (viologens, MVs) was studied in acetone (Allwood et al. 1987). Later, aryl-containing crown ethers, such as bis(m-phenylene)-32-crown-10 ether or dibenzo-24-crown-8 ether derivatives, have been extensively studied for making the host-guest complexes with MVs or ammonium salts (Allwood et al. 1987). These have also been used as binding motifs within mechanically interlocked molecules based on $\pi-\pi$ interactions (Barin et al. 2012). Zhang et al. demonstrated the modification of aryl-containing coronands to produce cryptand-like structures which renders superior hosts (K_a values in the 105–106 range for 1:1 complex) with MVs (Zhang et al. 2014).

Cyclic Arenes and Analogous Deep Cavitands

Calix[n]arene: is a macrocyclic oligomer of n phenolic units interlinked by methylene bridges at their 2, 6 positions. Its genesis was in the laboratories of Adolf von Baeyer (Baeyer 1872) over a century ago, it was made by Alois Zinke (Zinke and Ziegler 1941) and later popularized (Gutsche 1983, 1989). Owing to the presence of intramolecular hydrogen bonds between the phenolic moieties, these macrocycles adopt unique calix-shaped (Greek vase known as calix crater), structures. The upper rim of CX[n]s is wider and is of a hydrophobic nature due to the methylene bridges, whereas the lower rim is hydrophilic because of the phenolic oxygen atoms. By adjusting the conditions of the reaction between phenolic derivatives and formaldehyde, CX[n] analogues (n = 4, 6, 8) may be prepared in reasonably good yields. However, corresponding odd-numbered homologues of calix[n]arenes CX[n]s (n = 5, 7, 9) and other large homologues, can only be obtained in comparatively lower yields. The basis of host-guest interactions with organic substrates is usually $\pi-\pi$ stacking, cation–π, ion-dipole and hydrogen-bonding interactions. Functionalization of calix[x]arenes has been widely studied and reported, with changes made on the methylene bridges, on the rims and/or the meta-position of the phenolic rings (Gutsche and Lin 1986; Böhmer 1995). Although CX[n]s are able to form host-guest complexes with aromatic cationic species in organic media due to the electron-donating nature of the phenolic units (Shinkai 1993) Shinkai et al. reported the first water-soluble CX[n]s (Shinkai et al. 1984) by introducing sulfonate moieties on their structures that were able to form host-guest complexes not only with cationic molecules but also with neutral organic species because of the hydrophobic effect (Guo et al. 2008; Guo and Liu 2014).

A class of container-shaped macrocycles, stiff and constrained by structure and made by modification of the upper rim of calix[n]arenes, roughly belongs to cavitands. These are very versatile receptors with rich host-guest chemistry owing to their highly preorganized structures (Biros and Rebek 2007).

Calix[4]pyrroles (Gale et al. 1996) are pyrrole-based. These are hetero-analogues of calix[n]arene, that are constructed from four pyrrole rings connected by 2- and 5-positions (α-positions) or mesolike positions by sp^3 hybridized carbon atoms. Although the first few macromolecules of the family are attributed to Baeyer (Gale et al. 1996), this family of macrocycles was promoted by Sessler et al. mainly for the recognition of inorganic anions (Saha et al. 2015) and ion pairs, (Kim and Sessler 2014). Ease of their of functionalization (Sessler et al. 1998) resulted in a number of analogues that have the ability to act as receptors for organic moieties (Sessler et al. 1998; Allen et al. 1996; Bähring et al. 2019).

Resorcin[n]arenes are another subclass of cyclic arenes developed by Cram (Moran et al. 1982) containing n arene rings prepared by the acid-catalyzed condensation reaction between resorcinol and aliphatic or aromatic aldehydes. Resorcin[4]arenes are the most widely studied resorcin[n]arenes which also have two different rims in terms of structure and a very similar host-guest chemistry like calix[4]arenes. However, Resorcin[4]arenes have eight hydroxyl groups on the upper rim that can participate in hydrogen bonding interactions and can be exploited for further functionalization. Also depending on the aldehyde starting material, the lower rim may possess four affixed groups, which determine its properties and are usually preferred to give it optimal solubility (Jain and Kanaiya 2011). It is not surprising to find RA[4]s complexes with a variety of guest molecules in organic media (Schneider and Schneider 1994).

Pillar[n]arenes are a relatively new class of cyclic arenes containing hydroquinone units linked by methylene bridges at para-positions. The first pillararene was reported in 2008 (Ogoshi et al. 2008) made by condensation of hydroquinone and paraformaldehyde. Pillar[n]arenes (n = 5–15) features a symmetrical pillar-like architecture with two identical cavity gates that are comparatively conformationally rigid and easy to functionalize (Strutt et al. 2014). Their derivatives are appropriately soluble in aqueous or organic solvents, where they display a very rich host-guest chemistry (Xue et al. 2012; Ogoshi et al. 2016). Due to the presence of hydroquinone rings as units, native Pillar[n]arenes exhibit Electron Donor (ED) properties and ionophoric (a chemical species that reversibly binds ions) rims, which make these macrocycles appropriate hosts for electron-poor organic dications species, such as MVs in organic media. Additionally as demonstrated by the complexation ability of Pillar[5]arenes, these macrocycles can form an association complex with a wide range of neutral organic guests such as aliphatic nitriles, alcohols, esters, aldehydes and ketones or haloalkanes, in addition to other guests exploiting dipole-dipole, π–π, CH–π, and hydrogen bonding interactions (Wang et al. 2016).

Cucurbiturils

Cucurbit[n]urils or CB[n]s (n = 5–11) contain n glycoluril units joined together by 2n methylene bridges generating highly symmetric hollow macrocyclic moieties usually with inner hydrophobic barrel-shaped cavities, due to the absence of any functional groups or lone pairs, accessible through two entrances ringed by carbonyl groups at the top and bottom of the internal cavity (Lee et al. 2003).

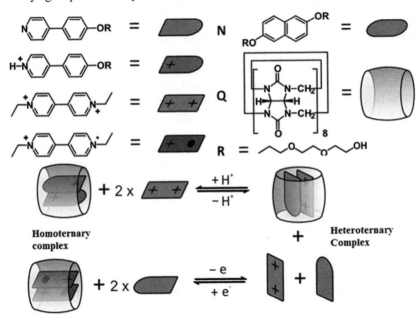

Figure 9.2 1:2 Homoternary and 1:1:1 heteroternary (bottom) CB[n]-based inclusion complexes and implementation of the latter types into an orthogonal pH/redox CB[8]-based supramolecular switch. Adapted with permission by (Schoder and Schalley 2017).

These are named so because of close resemblance of their shape to that of a pumpkin which belongs to the Cucurbitaceae family. They were first produced by acidic condensation of glycoluril and excess formaldehyde in 1905 by Behrend et al. (Behrend et al. 1905) to obtain an insoluble compound; its molecular structure was, however, assigned by Mock et al. in 1981 (Freeman et al. 1981). Due to the above mentioned structural features and electronic characteristics which is unparallel to other families of macrocyclic cavitands, CB[n]s display rich host-guest chemistry for guest molecules with complementary size, shape and charge/polarity that has been comprehensively reviewed (Barrow et al. 2015). During the last three decades, the supramolecular chemistry of CB[n]s has developed quickly. Cucurbit[n]urils in general, have the ability to form various binary aggregates with both neutral as well as cationic species including organic ones. While having hydrophobic cavity of CB[n]s with cavity height of 9.1 Å, CB[n]s is able to encapsulate neutral molecules through hydrophobic interactions, a combination of ion-dipole interactions and hydrogen-bonding interactions with carbonyl gateways aiding the encapsulation of positively charged or cationic species. Optimization of host-guest packing coefficient is also often the basis for recognizing substrates. As the cavity dimensions (volume) of CB[n]s heightens from 82 Å^3 for CB[5] to 870 Å^3 for CB[10] with the increasing number of methylene-bridged glycoluril units, recognition properties become different. CB[5] has the smallest cavity size (Buschmann et al. 2001) and has the ability to encapsulate various alkali, alkaline earth and ammonium cations. CB[6] reveals binding with a variety of alkylammonium, alkyldiammonium ions in addition to bipyridinium and imidazolium derivatives. However, even larger but hydrophobic cavity dimensions of the CB[n]s enables it to form complexes of the type homoternary 1:2 (HG$_2$) and heteroternary 1:1:1 (HGG′, G is electron donor/ G′ is electron acceptor). In the latter case, increased charge transfer interactions established between two complementary guests are the basis of formation (Ko et al. 2007). Due to its moderate water solubility, cavity size comparable to that of β-cyclodextrin and large enough to encapsulate molecules such as ferrocene and adamantylamine derivatives, CB[7] has attracted the attention of many chemists. CB[7] with pKa = 2.2, reversibly forms hydrogels in acidic aqueous media (Hwang et al. 2007) which is a less explored aspect of the cavitand family. These can be regarded as intrinsic pH-based Molecular Switches (MSs) as per their stimuli-responsiveness However, no other CB[n] derivatives acting like a MSs have been reported to date which may be due to their non-trivial functionalization (Kim et al. 2007). Although, the main aim of this chapter is the study of controlled binding of organic guests on application of stimulus on hosts (macrocyclic) that may be useful in the generation of molecular machines, this is not the only case in which the inclusion and release of a guest can be achieved on stimulation of a host-guest system. CB[8] displays the formation of unusual 1:1:1 heteroternary complex and hence a considerably large number of CB[8]-based supramolecular switches have been reported, by using appropriate aromatic Electron Donors (EDs) or aromatic Electron Acceptors (EAs) as stimuli-responsive guests (Pazos et al. 2019). For example, Schalley et al. combined the characteristics of an MV as redox-responsive EA, and a pH-sensitive ED, i.e., phenylpyridine derivative to achieve the dual-responsiveness (with both redox potentials and pH) within a CB[8]:MV: ED heteroternary complex (Fig. 9.2) (Schoder and Schalley 2017).

Blue Box and its Derivatives (Exboxes)

This is a group of rigid, electron-deficient, pyridinium-based polycationic cyclophanes which was primarily developed by Stoddart et al. with a aim to construct topological structures and molecular machines (Dale et al. 2016).

The first member of this rectangular macrocycle group is known as little Blue Box. It is one of the most extensively investigated building blocks in the construction of Mechanically Interlocked Molecules (MIM) and by structure it, tetracationic cyclobis(paraquat-*p*-phenylene) (CBPQT^{4+}) (Odell et al. 1988) which was also invented by J. Fraser Stoddart, it is composed of two π electron-deficient 4,4'-bipyridinium (BIPY^{2+}) units, connected, in a cyclic manner, by two *p*-xylylene spacers and is reported to and form host-guest aggregates with suitable electron-rich aromatics in both organic media (Anelli et al. 1992) and water (Bria et al. 2007). Blue box not only serves as a model for the development of other stimuli-responsive hosts, but also as a continuous inspiration in the field of (supramolecular) chemistry and molecular machines. The extended version of the blue box is known as Ex^{n}Box^{4+} (n = 1–3), where n is the number of *p*-phenylene linkers between the pyridinium units. ExBox^{4+} was reported to have its applications as an extractor (Barnes et al. 2013) of Polyaromatic Aromatic

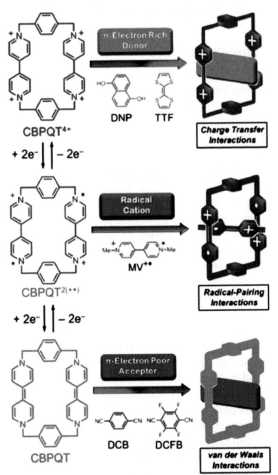

Figure 9.3 Types of inclusion complexes formed by the molecular host 'blue box' depending on its reduction state. Reprinted with permission from (Frasconi et al. 2015).

Hydrocarbons (PAHs) due to its high binding affinities towards PAH and a catalyst for the bowl-to-bowl inversion (Juríček et al. 2014) of corannulene. Other analogs with different modifications have also been reported depending on the nature of the short/large sides of the macrocycle including, (Dale et al. 2016) cage-like derivatives (Dale et al. 2014; Liu et al. 2020). Due to the presence of two Methyl Viologen (MV) units in the macrocyclic ring, the 'blue box'(CBPQT^{4+}) has two accessible redox states [$E_{1/2}$(CBPQT^{4+} – CBPQT$^{2(+.)}$) = 328 mV, $E_{1/2}$ (CBPQT$^{2(+.)}$ – CBPQT) = 753 mV], a feature that renders it a very different behavior as a host depending on its reduction state (Frasconi et al. 2015). As a result, the macrocycle may form inclusion complexes of the following three different types (Fig. 9.3):

(i) ED $\subset$ CBPQT^{4+} (where ED is an aromatic electron donor guest) Host-guest affinity by π-donor/π-acceptor interactions,

(ii) MV$^{+.}$ $\subset$ CBPQT$^{2(+.)}$, Host-guest affinity by radical pairing, and

(iiii) EA $\subset$ CBPQT (where EA is aromatic electron acceptors) host-guest affinity by van der Waals interactions (Fig. 9.3).

Stimuli-Responsive Molecular Switches

Molecular Switches (MSs) are molecular systems that exhibit reversible stimuli-responsive switching between two or more different states based on bistability, are being widely studied in the solution phase. Molecular switches that are able to exhibit good control, rapid responsiveness as well as reversibility are well suited for the design of molecular devices used in drug delivery, information or sensing functions.

Application of stimulus normally cause a change in the configuration (or conformation) and molecular properties of MS. Although the return of a molecular-level switch to its original position undoes any mechanical effect it has on an external system (energy from the switching stimulus is usually dissipated to heat) the alternation in observed properties of molecular switches induced a strong interest among researchers around the world. Materials based on these switching systems displaying a dynamic change in chemical or physical properties can be developed into molecular machines mimicking the action of an enzyme. They also have the potential application in areas like optoelectronic devices, storage, and smart materials.

Molecular switches can be controlled through an application of external stimuli. Inputs that are capable of a reversible extrinsic control of the molecular switching system include pH (pH: S = +H$^+$ and S$_o$ = –H$^+$), electrochemical/redox, (S = +e$^-$ and S$_o$ = –e$^-$) photochemical (S = hv and S$_o$ = hv_0/D) metal ions and anions change. These effects also include Effectors, E (by the interaction of protons, electrons or photons or metal ions as E with specific binding moieties of a supramolecule) that may produce a conformational change of the host due to its binding in the so-called 'allosteric site', resulting in activation or deactivation of the association of another substrate at a different primary binding site.

Macrocyclic compounds (discussed above) based host-guest systems, are the conventional focus of the supramolecular chemistry, encircling around fine-tuning of affinities particularly non-covalent ones to achieve selectivity of the host towards a guest. The contemporary emphasis from the past few decades is also on the external/internal control of the inherently dynamic behavior and development of stimuli-responsive host-guest systems including switching systems based on macrocyclic systems and their practical applications (Blanco-Gómez et al. 2020). Controlled motion in these supramolecular systems is usually achieved based on the assembly of cooperative non-covalent interactions which are individually weak, but collectively can generate strong interactions.

The construction of molecular machines using supramolecular macrocyclic systems is a challenging task, in which competing processes such as association and dissociation of sub-components involved in motion generation from and to the bulk must be restricted (Siegel 1996; Amendola et al. 2001) as these unbound elements can either interfere with the integrity and/or with the process of motion generation except when such a binding process is an indispensable part of the motion generation. If host and guest interactions are exploited to assemble a machine, then the control on their exchange with hosts or guests from the bulk is extremely essential which is even true for most of the biological functional assemblies to maintain the integrity and kinetic stability of the machine. For example, mRNA and the ribosome, at the protein synthesis site, form a kinetically stable complex where the exchange of ribosome subunit with the bulk is restricted, and thus the enzyme is processively able to produce large proteins If this was not the case, rapid exchange of the ribosome subunits from the complex or individual subunits of the ribosome during protein synthesis would result in the detachment of the enzyme from the substrate mRNA and would release a partially synthesized peptide, which would not be of any use (Kaempfer 1968). However, in certain cases, disassembly or the exchange process may be exploited intentionally and purposefully to provide communication with external species.

In a typical macrocyclic supramolecular switching system, under tight thermodynamic control, three different types of molecular mobility processes e.g. *interactional, motional and constitutional* (Fig. 9.4) can be envisioned (Fig. 9.4).

(i) Change(s) in shape, conformation or molecular motion due to isomerization of double bonds or other functional groups formation or disruption of intramolecular hydrogen bonding without change in the (internal or external) constitution, in a given macrocyclic moiety, depicts ***motional mobility*** processes,

(ii) Reversible self-unification of a number of (sub)-components (alike or different) to a unit (reorganization, deconstruction and reconstruction) constitute ***constitutional mobility*** processes and

(iii) The ability to form or break non-covalent bonds such as formation or disruption of intramolecular hydrogen bonding, $\pi-\pi$ interactions, CH-π interactions, etc., demonstrates **reactional/interactional mobility processes** (Lehn 2002a, b, 2013, 2015; Lutz et al. 2016).

Tremendous efforts have been devoted to the development of stimuli-responsive host-guest systems as well as their practical applications (Gil and Hudson 2004; Kay et al. 2007; Mendes 2008; Mura et al. 2013; Zhuang et al. 2013; Blanco et al. 2015; Coudert 2015). However, in most of them, the control over the dynamic behavior has been achieved through the use of host-guest molecular switches as the guest part of such assemblies (Kaifer and Go'mez-Kaifer 1999; Feringa 2001; Leung et al. 2009; Browne and Feringa 2010a, b; Bleger and Hecht 2015; Fihey et al. 2015; Frisch and Besenius 2015). On the contrary, stimuli-responsive macrocyclic hosts based molecular switches are significantly less in number because of the inherent synthetic difficulty to realize the target.

A) Interactional [G⊂H⇌H + G]:
B) Motional [H⇌H']:

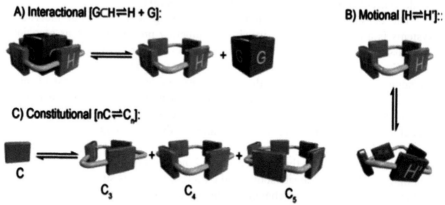

C) Constitutional [nC⇌C$_n$]:

Figure 9.4 Schematic representation of (A) Interactional (B) Motional and (C) Constitutional dynamism for a given macrocyclic Host H. Reprinted with permission from (Blanco-Gómez 2020).

Mechanically Interlocked Molecular Structures

Mechanically Interlocked Molecular Architectures (MIMAs) are supramolecular assemblies that are made up of molecular components that are mechanically linked. Rotaxanes and pseudorotaxanes (Fig. 9.5) are important members, from the aspect of this chapter, that is broad, topologically complex class of supramolecular assemblies.

Figure 9.5 Cartoon representation of rotaxane (left) and pseudorotaxane structure (right).

Rotaxane

A *rotaxane* is a mechanically interlocked molecular architecture consisting of a 'dumbbell-shaped molecule' as a 'shaft which is threaded through a 'macrocycle' (Fig. 9.6). However, *if a rotaxane* can dissociate into its two components without breaking the covalent bonds since its shaft molecule lacks the bulky groups at the ends which would lock the surrounding cycle on its position is called a *pseudorotaxane* (Fig. 9.5), however technically pseudorotaxanes are not part of

MIMA. In MIMAs, subunits typically interact through non-covalent interactions such as π–π stacking, hydrogen bonding and dispersion interactions (Bazargan and Sohlberg 2018). Depending on how many different components have been combined by non-covalent interactions to assemble the MIM architecture is called [n]rotaxane or [n]pseudorotaxane, where n in the square bracket represents the number of each component.

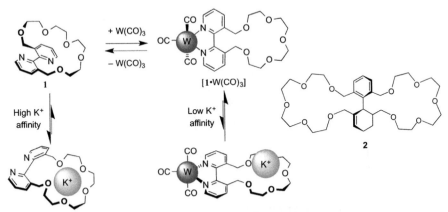

Figure 9.6 A negative heterotopic allosteric receptor **1** (Rebek Jr et al. 1979) reprinted with permission from (Kay and Leigh 2015) and positive allosteric effect allosteric receptor **2** (right) adapted with permission from (Rebek Jr. et al. 1985).

The study of MIMA is understood to have begun in 1960 (Wasserman 1960; Sauvage and Gaspard 2011). At first this was reported as an example of a donor/acceptor-based. Pseudorotaxane was reported in 1987 (Allwood et al. 1987; Ashton et al. 1987).

Due to the presence of the mechanical bond in these MIM structures, relative degrees of freedom of the interlocked components is strictly restricted in most directions, permitting motion of large amplitude in only a few allowed modes (Kassem et al. 2017) and also MIM components cannot be exchanged with those present in bulk without breaking the covalent bonds. For example, in a rotaxane, one or more mechanically interlocked macrocycles are restricted to move only in one or two directions, namely, random movement of the ring back and forth along the thread ('shuttling') (Ashton et al. 1989; Anelli et al. 1991) and non-directional rotation around the thread ('pirouetting') and is mechanically not allowed from dethreading, due to the presence of bulky 'stoppers' usually at the end of the molecular thread.

During the past three decades, a good range of different types of mechanically interlocked molecular (MIM) systems has been synthesized. Although MIM structures are not the only way forward to building molecular machines, they provided the first feasible synthetic molecular architecture to address, study and exploit well-defined large-amplitude molecular-level motions (Balzani et al. 2000a, b; Barbara 2001; Kinbara and Aida 2005; Kay et al. 2007; Balzani et al. 2008; Browne and Feringa 2010a, b). It is an attractive subject in ongoing research as the environmental responsiveness of these systems makes them desirable candidates for incorporation into molecular machines. MIMs have also been applied as drug delivery agents and as optical bio-imaging agents. Molecular motors and switches are examples of applications of these molecules in materials research. The chemical applications of these molecules range from catalysts and sensors to polymers. But bringing these structures from the laboratory to the real world or finding a use for these molecules is still a difficult task.

Pseudorotaxane

Since there are no bulky groups at the ends acting as stoppers at the terminals of the thread, dissociation of the pseudorotaxane complex to free molecular components is an easy process in the solution and it is always equilibrated with the constituents in an unbounded state. Different switches based on pseudorotaxanes have also been reported in the last three decades. Shuttling of the macrocyclic wheel along the bistable thread/axle is a type of mobility for pseudorotaxanes apart from a thread in and dethreading mobility (Erbas-Cakmak et al. 2015; Xue et al. 2015).

MOLECULAR SWITCHES

Conformational Control using Host-guest Interaction

Host-guest binding often brings or is useful in bringing conformational changes which range from small, localized bond deformations to long-range structural changes and therefore bring functional changes in biological systems. These can be of great importance in the design of practical molecular machines. The structural changes at distal sites are often exploited to alter the binding through allosteric regulation. There are several examples of stimuli-induced changes in conformation that had been used to control molecular recognition properties and these relatively large molecular level motions were produced on guest binding.

Important examples with the use of metal ions as guests for generating such changes, discussed above, include the use of negative allosteric interaction with tungsten as guest influencing (reducing) binding affinity for potassium ion at the second site of receptor **1** (Rebek Jr et al. 1979; Rebek Jr and Wattley 1980; Rebek Jr 1980) and the positive allosteric effect of mercury reducing the conformational freedom of the distal site in receptor **2** in a manner which is favorable to binding (Rebek Jr et al. 1985) (Fig. 9.6).

Few other, similar types of synthetic switching systems based on allosteric receptors, have also been reported (Van Veggel et al. 1994; Linton and Hamilton 1997; Robertson and Shinkai 2000, (Shinkai et al. 1981; Leighton and Sanders 1984; Abraham et al. 1985; Chong et al. 2001; Shinkai et al. 2001; Kovbasyuk and Krämer 2004; Degenhardt et al. 2005; Hunter and Anderson 2009; Kuwabara et al. 2009; Kennedy et al. 2013; Mendez-Arroyo et al. 2014). Allosterically regulated pseudo[1]rotaxane comprising of a cryptand **3** where dethreading of the electron-deficient bipyridinium such as paraquat takes place with allosteric regulation of the crown cavity with Na⁺ salts have been reported (Muraoka et al. 2014) and the controlled release of the pesticide carbaryl, containing a naphthalene group has also been reported by using the calix[4]arene-crown ether chimera **4** as the receptor possessing two sites with distinct binding preferences carbaryl moiety and sodium ions (Luo et al. 2015) (Fig. 9.7).

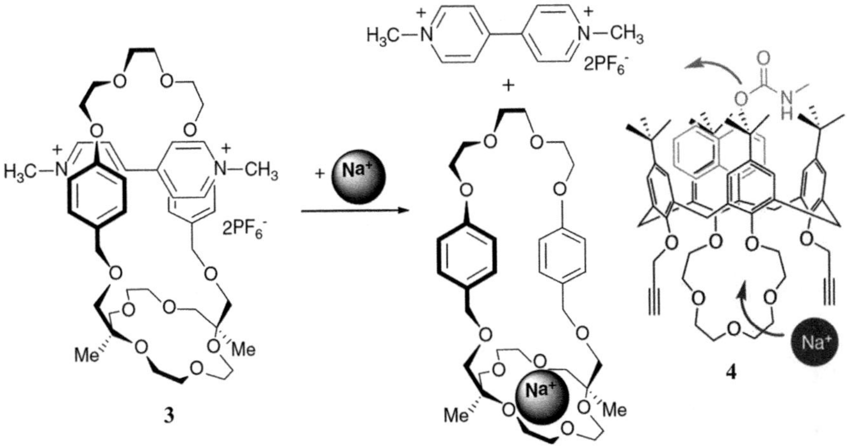

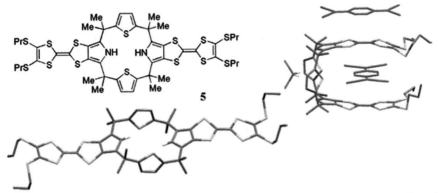

Figure 9.7 Schematic representation of dethreading molecular motion in pseudo[1] rotaxane comprising of a cryptand **3** (left) reprinted with permission from (Muraoka et al. 2014) and in receptor **4** (right) reprinted with permission from (Blanco-Gómez et al. 2020).

Few similar yet interesting, allosteric (chimeric) host have been reported by (Haino et al. 1998).

A large-amplitude structural alteration was observed in **5** while a 'sandwich' form is adopted to maximize binding with organic guest TCNQ leaving otherwise more stable extended structure (Fig. 9.8) (Poulsen et al. 2007).

Figure 9.8 Representation of the extended and 'sandwich' form of receptor **5** adapted with permission from (Poulsen et al. 2007).

Interconverting vase-kite mixture of resorcin[4]arenes based cavitand **6** (Fig. 9.9), having large conformational ariability (Moran et al. 1982, 1991; Timmerman et al. 1996) is converted exclusively to the kite form on protonation or by resorting to metal coordination (Skinner et al. 2001; Azov et al. 2003, 2004; Frei et al. 2004; Knipe et al. 2015) or by incorporating redox-active centers by modifying the design (Pochorovski et al. 2012, 2014). Stabilization of the vase form has also been achieved by substitution of the cavitand with amide groups or by guest complexation (Tucker et al. 1989; Ikeda and Shinkai 1997; Rudkevich et al. 1997, 1998; Ma et al. 1999; Haino et al. 2000; Tucci et al. 2000; Amrhein et al. 2002a, b; Far et al. 2002).

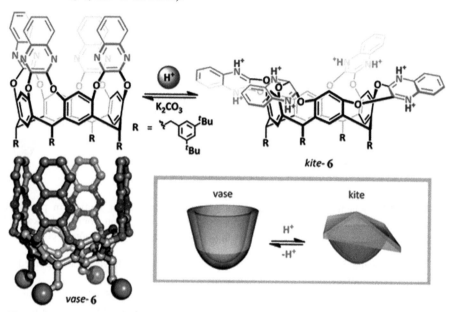

Figure 9.9 Schematic representation of acid-mediated interconversion between a vase and kite cryptands **6** (R-groups are represented as orange spheres). Reprinted with permission from (Knipe et al. 2015).

A series of molecular switches based on macrocyclic amides adopting a chair conformation and undergoing an unfolding process on anion addition stimuli has been reported (Szumna and Jurczak 2001; Chmielewski and Jurczak 2004; Chmielewski et al. 2004).

Using Capping on the Macrocyclic Hosts

pH-Driven Threading and Dethreading (Pseudorotaxane)

Gibson et al. (Huang et al. 2005) reported the development of three armed pH-responsive bis(m-phenylene) crown ether cryptand **7** (Fig. 9.10) where the host is able to bind with the paraquat dication in its neutral state to form pseudorotaxane and dethreading takes place on protonation of the pyridyl nitrogen atom by changing the solution pH owing to electrostatic repulsions between the charged binding site and the dication.

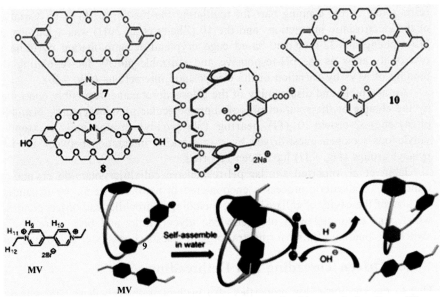

Figure 9.10 (Top) Structures of **7, 8, 9, 10** reprinted with permission from (Huang et al. 2005; Zhang et al. 2010; Ji et al. 2013) (Zhang et al. 2011) respectively and cartoon representation of the pH-controlled complexation between MV (Bottom right) and **9** reprinted with permission from (Ji et al. 2013).

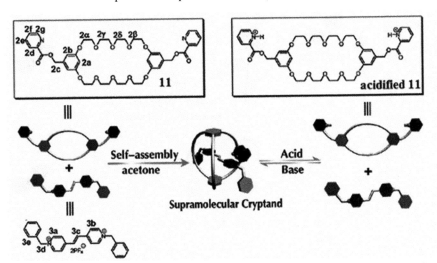

Figure 9.11 Supramolecular cryptand **11** used in this study and cartoon representations of an acid-base controllable [2]pseudorotaxane formation based on it (Yan et al. 2011).

Few other analogous pH-responsive supramolecular 32-crown-10 based cryptand switches (Fig. 9.10) were also reported during the last decade. Water soluble **8** (Zhang et al. 2010) and **9** (Ji et al. 2013) are three-armed cryptands containing bis(1,2,3-phenylene) and bis(1,3,5-phenylene) respectively and having ionizable moieties pyridine and phenylene with two anionic carboxylate groups, respectively as the capping part for regulating the binding of MVs by variation of the electrostatic interactions and the **10** (Zhang et al. 2011) was a four-armed bis(m-phenylene)-32-crown-10-based cage cryptand accommodating pyridine in two of its arms as the pH-responsive and ionizable moiety for regulating the binding of MVs by variation of the electrostatic interactions (Fig. 9.10).

An assembly and disassembly of the [2]pseudorotaxane, reversibly controlled by the change of the solution pH, of supramolecular pseudocryptand, bis(meta-phenylene)-32-crown-10 (**11**) bearing two pyridyl groups and π-extended vinylogous viologen guest-driven by π–π stacking interaction between the two pyridyl groups (Fig. 9.11) have been reported.

Jabin et al. reported similar pH-responsive calix[6]cryptamide cryptands, indicative of allosteric processes encountered in natural systems, by expanding the size of the cavity of calix[6]arenes, including hydrophilic calix[6]cryptamides which was decorated with oligo(ethylene glycol) units and both were having switchable binding affinity for neutral organic guests (Lascaux et al. 2010, 2016).

Redox Driven Threading and Dethreading

Due to the specific redox properties of a bisthiotetrathiafulvalene unit which is attached in the third, the bridging arm of the bis(meta-phenylene)-32-crown-10 cryptand **12**, a redox responsive switchable cryptand architecture has been realized (Wang et al. 2014a) (Fig. 9.12).

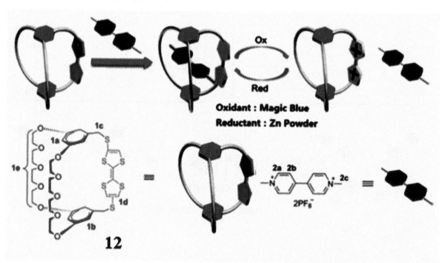

Figure 9.12   Chemical structure of **12** (bottom left) and its schematic representation of threading and dethreading (top) reprinted from (Wang et al. 2014a).

Light-driven Switching

An earlier paradigmatic example of the capping strategy was reported by Ueno et al. (Ueno et al. 1979) where the use of azobenzene moiety as the capping moiety of β-CD converted the resulting moiety **13** into a photo-responsive one that can adopt a different structure due to the switching of azobenzene moiety between the (E)- and (Z)-isomers. Significantly, the photo-isomerization also altered the association affinity, measured in terms of association constant K_a, of the host towards selected organic guests. K_a was of lower value for isomer (E) isomer than those for (Z) isomer. Furthermore, an initial example of the potential application of such capped supramolecular switch was also reported (Ueno et al. 1981) where CD-mediated hydrolysis of acetate esters of *p*-nitrophenol, a case of switchable catalysis, by **13** [(E) ↔ (Z)] was studied (Fig. 9.13).

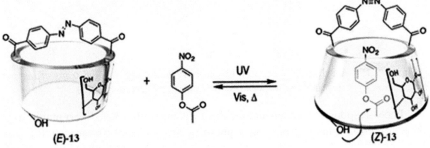

Figure 9.13   Application of the photoresponsive β-CD-based cryptand **13** for the controlled hydrolysis of esters reprinted with permission from (Blanco-Gómez et al. 2020).

A photoswitch based on 1,10-diaza-18-crown-6 ether cryptand containing a bridged photo-responsive azobenzene has been reported by Shinkai et. al. Here also, a photo stimuli driven E–Z isomerization of azobenzene was used for altering the cation-binding properties of crown ethers based host **14** for controlling the process of molecular recognition of these cryptands and also to suppress the thermal relaxation of Z isomer, so formed by cations like K^+ and NH_4^+ group (Shinkai et al. 1979, 1980, 1987; Kay and Leigh 2015) (Fig. 9.14).

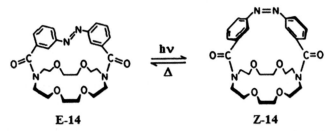

Figure 9.14   Photoswitchable binding of a crown ether-based cryptand, E-14 selective for small alkali cations and Z-14 selective for large alkali cations reprinted with permission from (Shinkai et al. 1980).

A photo switchable cylindrical receptor **15** containing two hexa protonated azocrown macrocycles connected with two *trans*-azobenzene units was reported by Bencini et al. (Bencini et al. 2001) (Fig. 9.15) whose fully protonated form bind the $[Co(CN)_6]^{3-}$ apparently by Cl^- exchange with a reasonable binding affinity of 5000 M^{-1} in water. The molecule could be readily photoisomerized to the *cis–cis* state through the *cis–trans* state. Although synthesized for testing anion-switching operation, it could not be evaluated. The authors believed that the photoisomerized *cis*-isomer would have might have a stronger affinity towards this larger anion. (Lee and Flood 2013).

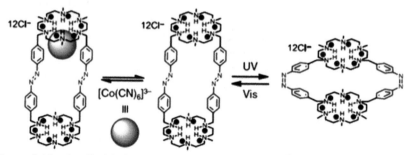

Figure 9.15 A cylindrical receptor 15 can bind $[Co(CN)_6]^{3-}$ in the *trans–trans*-state reprinted with permission from (Lee and Flood 2013).

Photo-responsive azobenzene-bridged dibenzo-30-crown-10 cryptand **16** is a three-armed cryptand displaying a photo or heat-induced isomerization possibly between E and Z isomers was reported (Liu et al. 2010). It exhibits ON-OFF binding affinity with 2,7-diazapyrenium (DAP) derivative **17**, a fluorescence active molecule, only Z isomer of the host binds with the DAP derivatives making the host-guest systems detectable with fluorescence, which makes the study of the system convenient (Fig. 9.16).

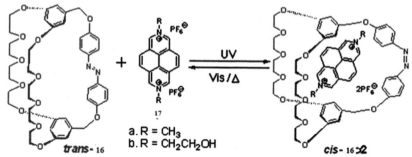

Figure 9.16 Chemical structure of **16** schematic representations of its ON-OFF binding ability with **17** reprinted with permission from (Liu et al. 2010).

A similar photoswitchable three-armed bis(*m*-phenylene)-32-crown-10 based cryptand photoswitch was also developed and reported (Xu et al. 2014).

Using a capping/bridging of a photo switchable stilbene derivative as a design control in the calix[4]arene derivative **18** (Rojanathanes et al. 2005), a switchable control over its cone and pinched cone conformations have been reported. The resultant calix[4]arene derivative is no longer able to form a complex with alkali cations but acts as a partial supramolecular switch for small organic molecules like acetonitrile or nitromethane. A more compact Z-form of the molecule (meta-isomer) is a highly selective receptor for these small neutral electron-deficient organic molecules (Fig. 9.17).

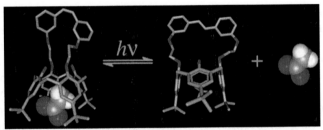

Figure 9.17   Schematic representation of switchable control of **18** over of its cone and pinched cone conformations in order to bind nitromethane reprinted with permission from (Rojanathanes et al. 2005).

Molecular Tweezers

Relatively sophisticated devices such as the molecular tweezers can be by constructedly exploiting the switching macrocyclic systems (Shinkai et al. 1981; Zimmerman 1993; Hardouin–Lerouge et al. 2011; Leblond and Petitjean 2011).

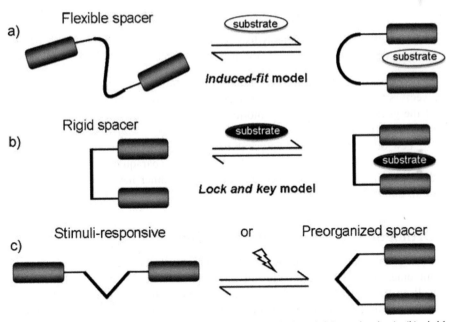

Figure 9.18   Representation of (a) flexible tweezers ('induced-fit' mechanism), (b) rigid tweezers ('lock and key' model) and (c) stimuli-responsive and pre-organized spacers (Leblond and Petitjean 2011).

Molecular switches can be designed based on responsive tweezers architectures (Fig. 9.18), with either motion-induced and regulated by application of an external stimulus including the binding of guest molecules (Leblond and Petitjean 2011).

One of the initial instances of a macrocyclic molecular switch **19** that was a 'molecular tweezer' was based on the photo responsive *cis-trans* isomerization of azobenzene with flanking crown ether macrocycles on both the sides of azobenzene (Fig. 9.19) (Shinkai et al. 1981, 1985). The more stable open *trans* form has the usual selective binding ability to bind smaller cations like Na$^+$, but the otherwise less stable *cis* form can be stabilized by larger alkali metal ions like Rb$^+$ by forming a sandwich complex causing the photo-isomerized molecule to thermally relax to the *trans* isomer much more slowly. Therefore photoisomerization-dependent changes altered the affinity of the receptor towards the guests and the kinetics of the photoisomerization was also altered by the guest binding.

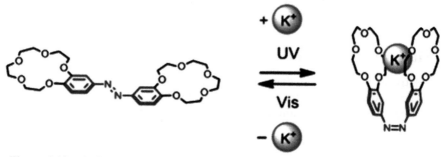

Figure 9.19 Archetypal butterfly-like azobenzene-fused switch **19** binds and releases potassium ions reprinted with permission from (Lee and Flood 2013).

Sandwiching of guests (both organic and metal ions) by macrocyclic molecular tweezers having two π-systems has also been explored where binding and release of the guest is modified by the motion of the receptor components controlled by stimulus (Tulyakova et al. 2010; Ulrich et al. 2010).

Few other instances of triptycene-based containing macrocyclic crown ethers molecular tweezers and acting as host for MVs (paraquat derivatives) in both dynamic (induced fit type) lock and key type manner have been reported (Cao et al. 2009; Chen 2011).

Orthogonal Control over Switching

Orthogonal control on the switching which manifests itself in a wide array of outcomes due to the application of distinct stimuli differing in both nature as well as the order of the stimuli is an improvement over the mono-functional switching and also offer an advantage over it. To achieve the orthogonal control over the switching in a supramolecular macrocyclic entity and to realize different outcomes in such a type of host-guest architecture, a cucurbit[8]uril (CB[8]) host which has the ability to form a heteroternary complex accommodating both a redox responsive guest, a methylviologen (MV^{2+}) and photo-responsive guest, a *trans*-azobenzene derivative have been used (Tian et al. 2012). It is well known that the *trans* form of the azobenzene can be isomerized to its *cis*-form using photo-irradiation and also an MV^{2+} which is redox-active and can be reduced to its smaller form, MV$^{+\bullet}$ in a reversible manner by one-electron reduction (Fig. 9.20).

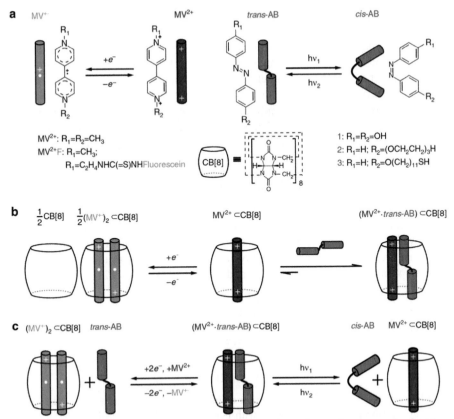

Figure 9.20 Schematic representation of the design, formation and orthogonal switching of a heteroternary complex of CB[8]. (a) The redox-induced reversible transition between MV^{2+} and $MV^{+\cdot}$ and light-driven photoisomerization of azobenzene derivatives. (b) Stepwise formation of Cucurbit[8]uril (CB[8])-mediated heteroternary complex with MV^{2+} as the first guest and an azobenzene derivative as the second guest, as well as the one-electron reduction of $MV^{2+}\subset CB[8]$. (c) The redox-driven reversible transition between a 'close' heteroternary complex and a 'close' homoternary complex with the *trans*-isomer, as well as the light-driven reversible transition between a 'close' heteroternary complex and an 'open' inclusion complex with the *cis*-isomer reprinted with permission from (Tian et al. 2012).

The CB[8] complex was found to respond to orthogonal stimuli in a controlled and reversible manner creating a multifunctional switch between a 'closed' heteroternary complex state accommodating *cis*-azobenzene and MV^{2+} as guests which are a redox-driven, a 'closed' homoternary complex state encapsulating two units of $MV^{+\cdot}$ as guests and a photo-driven 'open' state inclusion complex with the *cis*- isomer of the azobenzene. This means that the heteroternary complex of CB[8] can be orthogonally switched to two distinct complexed states. On one side, application of two-electron reduction electrochemical stimuli in the presence of $MV^{+\cdot}$ results in the formation of homoternary complex by removal of the *cis*-azobenzene guest from the heteroternary complex, and on the other hand, on applying the photo-stimulus, capable of isomerization of encapsulated *trans*-azobenzene to its *cis*-isomer, results in the removal of again the azobenzene molecule, but due to steric repulsion. Accordingly, these three distinct, orthogonal stimuli-dependent types of complexation state offer better control over molecular dynamic behavior. The (1,1,1) heteroternary complex was prepared from a CB[8] moiety in a two-step process, the homoternery CB[8] complex (with two $MV^{+\bullet}$) mediated formation of Cucurbit[8]uril complex with MV^{2+} was achieved in the first step by resorting to one-electron oxidation process and complexation with an azobenzene derivative, as the second guest in the second step.

Intramolecular Translocation of Ions with Stimuli

The initial example of the reversible intramolecular translocation of the alkaline earth cations such Ca^{2+}, Sr^{2+}, Ba^{2+} inside a ditopic receptor, between the two equivalent binding compartments, comprising of two 12-membered N_2O_2 crowns **20** that are linked together by diethylene oxide spacers with the help of nitrogen atoms was reported in 1974 (Lehn and Stubbs 1974) (Fig. 9.21).

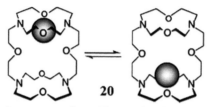

Figure 9.21   Schematic representation of intramolecular translocation of the alkaline earth cation in **20** reprinted with permission by Lehn and Stubbs 1974.

A molecular syringe' mimic **21**, comprising of calix[4]arene in a 1,3-alternate conformation, possessing a nitrogen-containing macrocyclic crown ether cap at one of the sides and two ethoxyethoxy groups at the other was reported (Ikeda et al. 1997). It can also be perceived as a chemical stimulus switchable molecular shuttle tunneling a silver metal ion between two accessible complexation sites through its π-basic tube/cavity owing to the difference in binding affinity for the metal ion. The calix[4]arene edge of the molecular architecture bearing a crown ether cap has a comparatively higher binding affinity for the Ag^+ ion than the other edge having a bis-(ethoxyethoxy) group as demonstrated by the relatively fast dissociation of the complex with Ag^+ from the bis(ethoxyethoxy) side while the nitrogen of the crown cap is protonated (Zheng et al. 2010) (Figs. 9.22).

An earlier example of dual translocation of two distinct metal cations on a ditopic ligand architecture based on a macrocyclic calix[6]arene with three imidazoles rings attached at the upper smaller rim and also three triazoles (Tria) rings appended at the upper wider rim was reported (Colasson et al. 2010).

A few examples of switchable intramolecular ion (anion) translocation between two binding sites on the application of the electrochemical stimuli to drive the intramolecular movement of the ion have also been reported (De Santis et al. 1997; Fabbrizzi et al. 1999; Amendola et al. 2011).

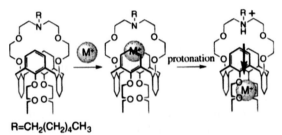

R=CH$_2$(CH$_2$)$_4$CH$_3$

Figure 9.22 Schematic representation of the metal tunneling through a π-basic tube of 21 reprinted with permission by (Ikeda et al. 1997).

Excited-State Intramolecular Proton Transfer (ESIPT)

A few macrocyclic molecules display distinct photo-physical properties in the case of Excited-State Intramolecular Proton Transfer (ESIPT) such as a variation in the pK$_a$ values of the molecule in the excited state which contributes to the swift intramolecular movement of the proton, i.e., dissociation from one part and re-association with another part acting as an acceptor. ESIPT has attracted the attention of chemists due to the applications in molecular probes, luminescent materials and molecular logic gates, etc., (Zhao et al. 2012). The ESIPT process along with significant relaxation in the geometry of the chromophore and additionally, the geometry of the emissive state, i.e., keto tautomer which is considerably different from the enol form which in turn may lead to the observation of large Stokes shift. It has been reported (Wu et al. 2004) that both macrocyclic compounds **22** and **23** differing in the aldehydic group can in principle access two conformations in solutions. However, A is the dominant conformer for **22** in acetonitrile, responsible for emission at 585 nm (ESIPT) whereas for **23**, dual emission is observed in the acetonitrile solution first at 355 nm, due to the conformer C and at 432 nm, due to the zwitterionic form of conformer A and the dominant species is again the intramolecular hydrogen-bonded conformer A (Fig. 9.23). On addition of sodium cations to the solution number, the emission at 585 nm decreases , due to reduction in ESIPT to nitrogen of the crown ether ring. However, with the addition of sodium cations to the solution of molecule **23**, the emission at 432 nm decreases due to the inhibition of the ESIPT of the conformer A and the emission band at 355 nm was enhances due to the increase of the conformer C (Fig. 9.23).

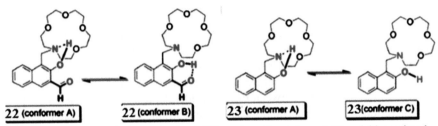

| 22 (conformer A) | 22 (conformer B) | 23 (conformer A) | 23(conformer C) |

Figure 9.23 Chemical structures of different conformation of **22** and **23** reprinted with permission from (Wu et al. 2004).

SELF-COMPLEXING PSEUDOROTAXANES

Self-complexing host-guest switching molecular system or a self-inclusion pseudorotaxane comprising of macrocyclic host covalently attached to the guest allows producing interesting topologies such as [1]rotaxane, daisy chains, cyclic or linear supramolecular oligomers/polymers through intermolecular host-guest recognition (Fig. 9.24), and a few interesting features in the field of molecular recognition (Blanco-Gómez et al. 2020).

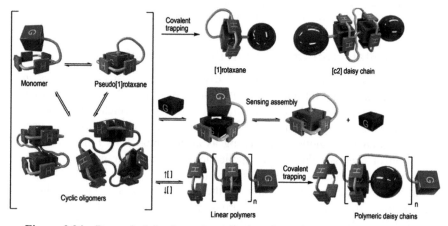

Figure 9.24 Dynamic behavior and topologies of covalently attached host-guest molecules reprinted with permission by (Blanco-Gómez et al. 2020)

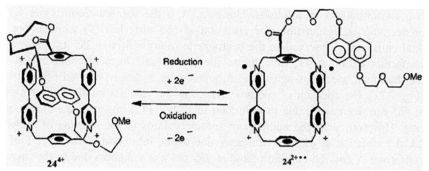

Figure 9.25 Electrochemically driven dethreading-rethreading of 24^{4+} reprinted by permission from (Ashton et al. 1997a).

An early representative example of stimuli regulated self-complexing scorpion **24** like system composed of the macrocycle, a blue box stable in the self-complexed form (a threaded form of [1]pseudorotaxane) till the application of trigger such as electrochemical or photochemical (electrochemical reduction in this case) decreases the interaction between the [1]pseudorotaxane subcomponents, resulting in alteration of its conformation to an unrestrained, unencapsulated form by the release of the intramolecular guest from the host cavity shown in Fig. 9.25 (Ashton et al. 1997a).

A self-complexing scorpion **25** comprising of the macrocycle, a 'blue box' displaces its intramolecular guest hydroquinone ring attached to it through a flexible arm from the cavity of the macrocycle in the presence of the dioxynaphthalene derivative **26**, an external intermolecular guest and forms a 1:1 complex with an external guest disrupting its self-complexation. The change is accompanied by the alteration in the UV/Vis absorbance and color of the solution— from red to purple (Ashton et al. 1997b) (Fig. 9.26).

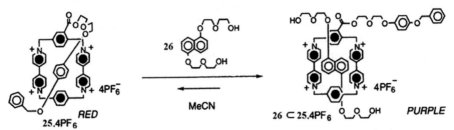

Figure 9.26   A self-complexing pseudorotaxane **25** acting as a chromophoric receptor for **26** reprinted with permission by (Ashton et al. 1997b).

Gated Regulation of the Binding Site of the Host in Self-inclusive Pseudorotaxanes

The above mentioned examples also exhibits that a stimuli-driven switching systems can be constructed by application of a trigger on either the self-complexing pendant guest or a macrocyclic host of self-complexing pseudorotaxane architecture to lock/unlock a gated dynamism for regulating the entry external guest into the host cavity.

Early examples of such photo-responsive switching systems with intramolecular guest-induced gating of the external guest have been reported in self-complexing β or γ-CD appended with photoisomerizable azobenzene (Ueno et al. 1983, 1990; Fukushima et al. 1991; Nakagama et al. 2001). However in these reports, both the (E)- and (Z)-isomers were found to form inclusion complexes with intramolecular guests since the structural alteration in the intramolecular guest on photoisomerization of azobenzene is not large enough to force the intramolecular out of the cavity of the macrocyclic host. Potential application of such *exo*-functionalized self-complexing β-cyclodextrin in the field of photo-responsive switchable supramolecular catalysis was also reported (Lee and Ueno 2001).

A remarkable illustration of a photo-stimuli-controlled switching in a self-complexing pseudo[1]rotaxane to realize the intramolecular guest-induced gating of the external guest has been reported by Coulson et al. (Coulston et al. 2006). In the report, the initial N-methylpropanamide prototype of the molecular machine **27** is described as a nano molecular pump. **27** comprises of β-cyclodextrin, an aryl substituent attached to the CD moiety through N-methylpropanamide connector and an external guest 1-adamantanol depicted respectively as a piston, as a cylinder and as a fuel of the nano-sized pump (Fig. 9.27). During the compression stroke of the molecular piston, the intramolecular complexation of the aryl substituent within the cavity of the cyclodextrin host takes place (in the aqueous solution) forcing the competitive guest, 1-adamantanol out of the cyclodextrin cavity. The output energy of the molecular recognition event is channelized to overcome the amide group's torsional barrier and to alter the ratio of the syn- and anti-isomers. During the decompression piston stroke, 1-adamantanol enters the cavity again. The work generated by the engine is harnessed to force the alkyl moiety out of the hydrophobic cavity of the cyclodextrin to water, by modifying a dihedral angle of the amide moiety. Initial N-methylpropanamide prototype is converted into photo-switchabale E- and Z-isomers **28** and **29** by bringing in a photoisomerizable double bond where the reversible on/off states are regulated by the isomerization of the N-methylcinnamide moiety (Fig. 9.27). However, (Z)-form of the pendant is unable to self-include within the cavity of β-CD preventing it from getting involved in the compression stroke.

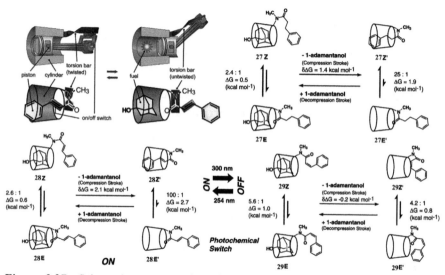

Figure 9.27 Schematic representation of a mechanical machine and its molecular counterpart (top left), operation of **27** as a nano molecular pump (top right) operation of **28** and **29** as a molecular machine having a photochemical on/off switch (bottom) reprinted with permission from (Coulston et al. 2006).

The pH-triggered switching between the formation of self-inclusion pseudorotaxane and the binding with external guest, i.e., intramolecular guest gated regulation has also been reported in the fluorophoric dansyl appended cyclodextrin derivatives differing from each other in terms of the spacer length between cyclodextrin and the dansyl moiety (Nelissen et al. 1997). Another very interesting report about the pH-responsive self-inclusion complexes of mono-benzimidazole functionalized β-cyclodextrins which were attached on the external surface of mesoporous silica and displaying gated regulation of the external guest was published by Fu et al. (Wang et al. 2014b). These were called Mechanized Silica Nanoparticles (MSNPs) by the authors. The MSNPs were found to be able to reversibly block the macrocycle's cavity on protonation/deprotonation owing to the responsive aromatic pendant. These MSNPs exhibited themselves as supramolecular nano-valves by pH-triggered reversible release of cargo such as *p*-coumaric acid and have the potential to be investigated for their performance for cellular targeted drug delivery (Fig. 9.28).

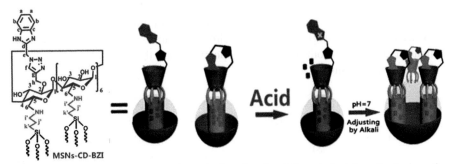

Figure 9.28 Structure of benzimidazole functionalized β-cyclodextrins attached on the external surface of mesoporous silica (MSNPs) and the cartoon representation of the working as pH-controlled drug delivery agent reprinted with permission by (Wang et al. 2014b).

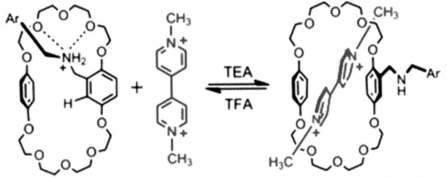

Figure 9.29 Schematic representation pH regulated gated switching of **30** bearing crown ether-based host with paraquat reprinted with permission of (Jiang et al. 2016).

The gated regulation switching of external guest paraquat in self-including pseudorotaxane **30** forming a sailboat-shaped self-complex; possessing bis(p-phenylene)-34-crown-10-based macrocycle as the host and (Fig. 9.29) bearing a dibenzylamine side arm was reported (Jiang et al. 2016) In this case, the pendant guest of the substituted macrocycle binds into the cavity of the crown macrocycle only when the amine is protonated.

A redox-responsive example of gated regulation of external guest in a self-inclusive pseudorotaxane based on ferrocene (Fc)-β-cyclodextrin conjugate **31** was reported (Casas-Solvas et al. 2009). It is based on the ferrocene-based responsive pendant on the wider rim of the Cyclodextrin. Different types of self-association behavior is shown by the reduced and oxidized forms of **31**. It is reported that redox controllable head-to-head homodimer Janus [2]pseudorotaxane, i.e., interpenetration of the pendant ferrocene units into the facing macrocycle cavities is in equilibrium with a monomeric form where the ferrocene moiety is intramolecularly self-included in one of the β-CD cavities. However, only one distinguishable form of the oxidized state of the conjugate is detectable in aqueous solution, corresponding to the ferrocene cation stood outside the cavity of the host. Values of binding constants for the complexes formed in an aqueous solution between the conjugates and a series of bile salts is also determined. The authors demonstrated the redox-sensing abilities of the synthesized conjugates toward bile salts through the use of sensitivity parameters and based on the guest-induced changes in both the half-wave potential and the current peak intensity of the electroactive moiety (Fig. 9.30).

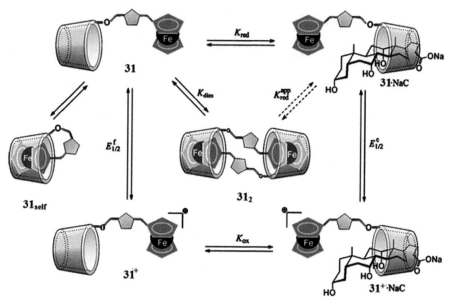

Figure 9.30 Chemical and electrochemical equilibria proposed for the interaction of **31** with bile salts. The structure of complex **31⁺·NaC** is tentative. Reprinted with permission from (Casas-Solvas et al. 2009).

Self-inclusive Pseudorotaxnes with an Externally Switchable Configuration of the Pendant for Gated Regulation of Binding Site

It is also possible to control the approach of the internal intramolecular guest attached to the linker part of the pendant towards the binding site of the host in the self-inclusive rotaxane by use of external stimulus resulting in linker-regulated gating mechanism rendering locked in/out states of the binding site which can regulate the binding of the external guest as well.

An initial exemplary illustration of this type of design displaying self-hosting behavior as discussed above is reported (Shinkai et al. 1985) in which a coronand (**32-H⁺**) having a pendant chain ending in ammonium alkyl [H_4N^+-(CH_2) *n* = 4,6,10] group which is attached to the benzo crown ether through phenyl diazo linkage resulting in the formation of diazo benzene on the side chain. These carefully designed crown ethers have been made so that intramolecular complexation of the ammonium group of the pendant group can only take place in to the *cis*-forms of the pendant formed by photoisomerization of *trans*-diazobenzne. The first-order rate constants for thermal *cis–trans* isomerization of *cis*-(32-H$^+$) were observed to be smaller by 1.6–2.2-fold than those having analogous free amines (Fig. 9.31).

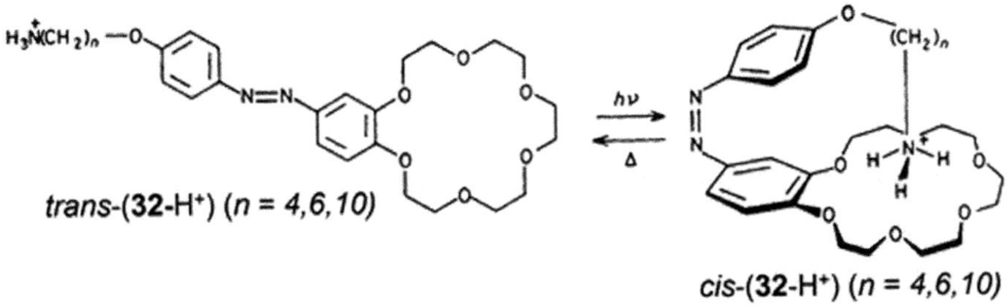

Figure 9.31 Schematic representation of photoresponsive linker-regulated gating mechanism in **32-H$^+$** reprinted with permission by (Shinkai et al. 1985).

Such self-inclusive pseudorotaxanes architecture possessing photo-isomerizable azobenzene linker can even be used for photo-regulating binding of an external guest as it is achieved in the case of resorcinarene based cavitand (Berryman et al. 2011a, b).

Allosteric Regulation of Linker Controlled Gated Regulation of the Binding Site

Gated regulation of the binding site of the host in the self-complexing pseudorotaxanes can also be achieved allosterically by meticulous design either by resorting to an 'allosteric linker' or by incorporating an 'allosteric internal guest' (Durola and Rebek 2010; Bähring et al. 2013).

SHUTTLING IN ROTAXANES

Rotaxane-based molecular shuttle contains two or more same/different potential macrocycle-binding sites ('stations') connected by a negotiable pathway on the molecular thread. The shuttle with two degenerate stations is called degenerate molecular shuttles (Fig. 9.5). The macrocyclic ring moves randomly between the two stations provided sufficient thermal energy is available to break the inter-component non-covalent interactions. However, on average the time spent by macrocycle is more on the station with the higher binding affinity co-conformation (a co-conformation referring to relative positions of the mechanically interlocked components concerning each other) (Kay and Leigh 2008). At a molecular-level, molecular shuttles based on rotaxane offer an ideal experimental basis for the realization of mechanisms conducted a few years ago by physicists (Sauvage 2003).

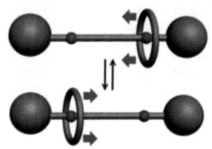

Figure 9.32 Cartoon representation of shuttling or translational motion in a [2]rotaxane.

Degenerate Molecular Shuttles with Two Stations

The first example of 'molecular shuttle' based on [2]rotaxane system with two degenerate stations (Fig. 9.33) was reported by Stoddart's group which is essentially considered as the beginning of the field of artificial molecular machinery (Anelli et al. 1991). A few other degenerate shuttles based on [2]rotaxanes were reported later (Ashton et al. 1992a, b; Lane et al. 1997)

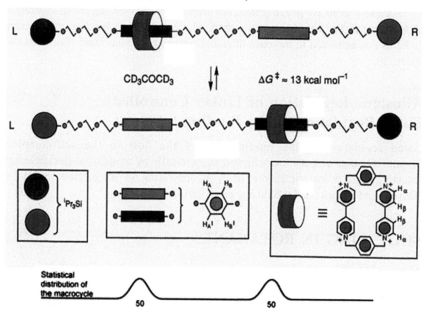

Figure 9.33 A representative example of a molecular shuttle in action and statistical distribution of macrocycle over the two stations. Reprinted with permission from (Anelli et al. 1991).

Stimuli-responsive Molecular Shuttles

The distribution of macrocyclic ring component among the available binding sites of a rotaxane is according to the binding affinity of the ring component towards each station and that both strengths of station–macrocycle interactions and available thermal energy play an important role in determining the rates of shuttling between the stations and the occupancy of the different stations in bistable rotaxanes and that the molecular motion in kinetically stable MIM architectures can be controlled by using multiple binding sites by varying affinities of the station for the macrocycle ring component under different conditions. These molecular devices have been used to perform a variety of tasks.

Light controlled shuttling in a single station rotaxanes is achieved in a [2]rotaxane, **33** by light-induced electron transfer from the electron-rich dioxyarene station to the 'blue box' macrocycle with a life time sufficient to permit secondary electron transfer to an appended ferrocene stoppers, thereby inducing shuttling of the macrocycle away from the station (Benniston et al. 1993) (Fig. 9.34).

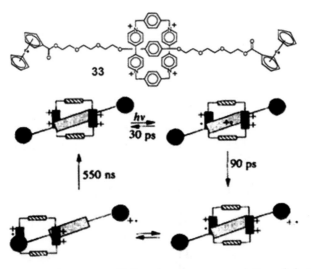

Figure 9.34 Chemical structure of **33** and cartoon representation of shuttling in **33**. Reprinted with permission from (Benniston et al. 1993).

In another approach, photoisomerizable azobenzene or stilbene-based station has been used for shuttling in the single station rotaxane (Murakami et al. 1997; Nakashima et al. 1998)

Stimuli-Responsive Molecular Shuttles with Two or More Stations

The ring component can be driven toward a new equilibrium state under the influence of available thermal energy by a perturbation of the binding affinities (which are in turn are dependent on binding energies) of any of the stations. The statistical distribution of the ring component between the stations changes in response to external triggers or stimuli such as chemical, electrochemical or photochemical stimuli to control the translational movement of one molecular component concerning the other which usually, amends the structure of one of the binding sites to alter the relative binding affinities of the stations for the macrocyclic ring components, switching the system out of co-conformational equilibrium (Pezzato et al. 2017) (Fig. 9.35).

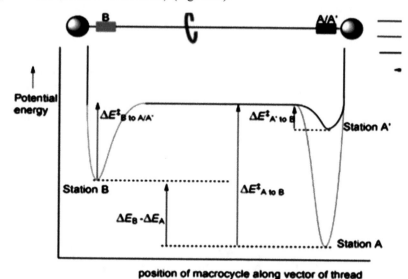

Figure 9.35   Potential energy diagram of a rotaxane-based bistable molecular shuttle reprinted with permission by (Kay et al. 2007).

Molecular Shuttles Driven by pH Change

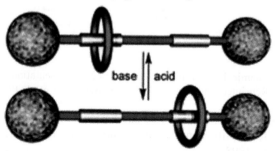

Figure 9.36   Cartoon representation of pH-driven molecular shuttle reprinted with permission from (Vella et al. 2007).

The pH driven molecular shuttling (Fig. 9.36) is covered in Chapter 6, pH triggered pseudorotaxanes and rotaxanes.

Redox Driven Molecular Shuttling

An illustrative example of a chemically controlled and redox driven shuttling process was reported by Stoddart et. al. (Tseng et al. 2003) in a rotaxane **34** consisting of redox-active, electron donor tetrathiafulvalene (TTF) and 1,5-dioxynaphthalene (DNP) as stations separated by a rigid terphenylene spacer in a dumbbellshaped component demonstrating a relative shuttling of the macrocyclic cyclobis(paraquat-p-phenylene) ($CBPQT^{4+}$) between the recognition units. A TTF binding site, a preferential site for tetracationic $CBPQT^{4+}$, with the addition of oxidant or electron acceptor $Fe(ClO_4)_3$ get oxidized and thus driving the tertacationic ring component towards the DNP station (Fig. 9.37) due to coloumbic repulsion.

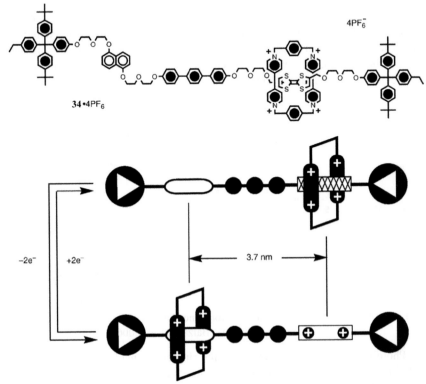

Figure 9.37 Chemical structure of rotaxane **34** (top) and cartoon representation of redox-driven shuttling in it (bottom) reprinted with permission from (Tseng et al. 2003).

Rotaxanes possessing electron donor-acceptor groups either at a binding site or in a macrocycle (Ashton et al. 1992a, b, c, d and e; Anelli et al. 1997) or involving the formation of transition metal complexes (Joosten et al. 2012) can be regulated by redox chemistry provided that the conditions, i.e., redox potentials of the interacting units are suitably chosen and that the redox products are stable on the shuttling time scale and that the rapid charge recombination may be prevented. Tetrathiafulvalene, viologen, benzidine, dioxynaphthalene and naphthalimide derivative have been extensively exploited as redox-active stations of the dumbbellshaped component and the blue box as a ring component.

Photodriven Molecular Shuttles

Photoinduced control on translational shuttling is significant owing to the simplicity with which it can be implemented. Additionally the overall process of the shuttling can become autonomous, if the restoration of the original state is spontaneous and does not require an additional stimulus, and thus the shuttling will take place till the light energy is supplied.

Use of continuous light may drive the system towards the steady-state distribution and hence should be avoided. If the kinetics of translation and photochemical processes are carefully considered then the photo-driven shuttling is usually an effective and efficient approach. However, it is important to note that if the photoinduced process goes by redox reaction, then either the excited states with a long lifetime are required or external reagents must be used to reduce or oxidize the species, otherwise rapid charge recombination may occur before the translocation of the macrocycle.

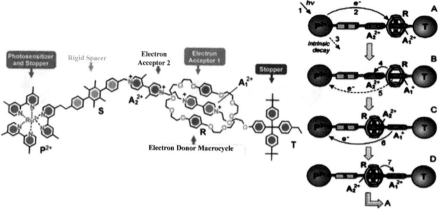

Figure 9.38   Structural formulas of rotaxane (left) exhibiting photoinduced ring displacement (right) (Balzani et al. 2006a).

Photoinduced ring displacement of the ring components between the viologen and dimethyl viologen stations which are placed along the dumbbellshaped component, involving the assistance of photosensitizer ruthenium trisbipyridine stopper to generate an autonomous linear motor powered by visible light of 532 nm was reported (Balzani et al. 2006a, b) where the whole process takes place in four strokes: destabilization of the stable conformation, displacement of the ring, electronic reset and nuclear reset. The interesting switching system was obtained by reduction of a viologen station by the electron generated from ruthenium trisbipyridine on photo-irradiation (Fig. 9.38). However, a relatively slow back electron transfer allowed the translocation of around 10% of the macrocycles to the dimethyl viologen station on each electron transfer, and continuous irradiation resulted in a 95:5 distribution of the macrocycle towards the dimethyl viologen station.

Ion-Induced Changes Driving Molecular Shuttles

Usually ions have the ability to interfere either with the existing hydrogen bonding, ion–dipole and dipole–dipole interactions present between the ring component and the station bringing the induced co-conformation change of the macrocycle on a station or can sterically disfavor the binding interaction prevalent, thus affecting its stability and the effect may be large enough to drive movement of the macrocycle to another station on the dumbbellshaped binding component.

For instance, metal ions can be chelated by the crown ether-based macrocycle which is acting as a ring component in a rotaxane, altering its affinity for the stations, and can even induce shuttling. One such example is reported in neutral bistable rotaxanes wheres shuttling of 1/5DNP38C10 crown ether-based macrocycles, incorporating π electron-rich 1,5-dioxynaphthalene ring systems, around their dumb-bell binding components from one of the π-electron-deficient naphthalimide station to other pyromellitic diimides binding site takes place due to the mediation of lithium cation in its presence presumably due to stronger ion–dipole interactions of the ring component with the later station (Vignon et al. 2004). However the addition of an excess of [12]crown-4 restored the original equilibrium distribution by seizing the lithium cation (Fig. 9.39).

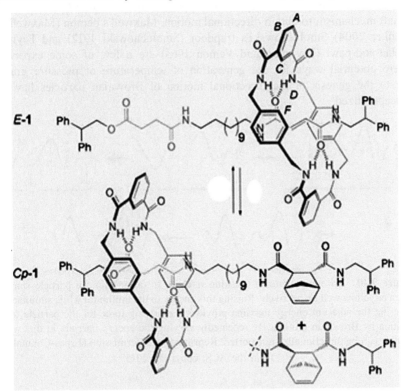

Figure 9.39 Shuttling through reversible covalent bond formation (ii) cyclopentadiene, d_6-DMSO, 80°C, 16 hours, 90%; (iii) 250°C, 10^{-2} Torr, 20 minutes, ~100%. Reprinted with permission from (Leigh and Pérez 2004).

Shuttling Driven by Reversible Covalent Modification

Use of dynamic covalent bonds created through the utilization of reversible chemical reactions can be of significance in the synthesis of supramolecular devices and machines. These chemical modifications act as a physical barrier to check random Brownian motion reverting the preferred translational motion during re-equilibration steps and permit the shuttles to do work by 'compartmentalization' the macrocycle on a station. A preferred binding site for the wheel component in the rotaxane architecture can be blocked by introduction of a new chemical group or by destabilization of the binding of a macrocycle by steric or electronic modification and thus the macrocycle can be chemically trapped on one station.

Block and unblocking of the better-binding fumaramide site in a rotaxane was realized by use of Diels–Alder and retro-Diels–Alder reactions respectively, thus controlling the shuttling between hydrogen-bonding station.

MOLECULAR MOTORS

Synthetic molecular motors depend on tapping random thermal fluctuations using a shaft mechanism to obtain directional motion. Maxwell's demon (Maxwell 1871; Shenker 2004) Smoluchowski's trapdoor (Smoluchowski 1912) and Feynman's ratchet-and-pawl (Feynman and Vemon 1963) are a few of some experiments where potential ways for the generation of temperature or pressure gradients due to the genesis of the directional motion of Brownian particles have been conceptualized.

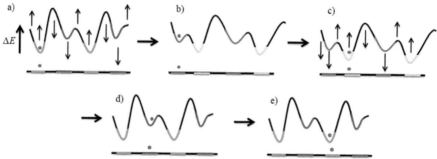

Figure 9.40 An energy ratchet (flashing ratchet). In (a) and (c) the particle starts in a green or yellow well respectively. Raising this energy to the minimum while simultaneously lowering the adjacent energy maxima provides the driving force for the particle to move position by Brownian motion. By repeatedly varying the energy barriers in this way, the particle can be directionally transported. Reprinted with permission (Erbas-Cakmak 2015) from the ACS, copyright 2015.

Directional Motion and Work under Brownian Motion

Regulating the directionality of movement and producing motion at the molecular level is the challenge in designing artificial or synthetic molecular motors. Molecular motors utilize random thermal fluctuations also known as Brownian motion to yield directed motion by employing ratchet mechanisms (Chatterjee et al. 2006; Astumian 2007; Kay et al. 2007; Erbas-Cakmak et al. 2015) and these motions are an important feature of any machine more sophisticated than a simple switch and falls into two distinct classes: energy ratchets (Chatterjee et al. 2006) and information ratchets (Serreli et al. 2007). Non-equilibrium statistical can be used to have a more accurate understanding of basic mechanisms and processes used by the current generation of molecular machines.

In its simple form, an energy ratchet consists of a series of periodic pairs of energy maxima and minima as depicted in Fig. 9.40. To start with the particle is in a green or yellow well (Figs. 9.40a or c respectively). By raising that energy minimum while simultaneously reducing the adjacent maxima and minima provides the thermodynamic push for the particle to move to a new position by exploiting Brownian motion (Figs. 9.40b and c or d and e). Similarly, by synchronizing the change in the relative heights of the maxima responsible for determining the direction in which transport proceeds with the simultaneous alteration in the relative depths of the minima providing the driving force for directional transport, the particle can be moved in a particular direction. An important point to note here is that the variation of the energy surface occurs irrespective of the particle's position.

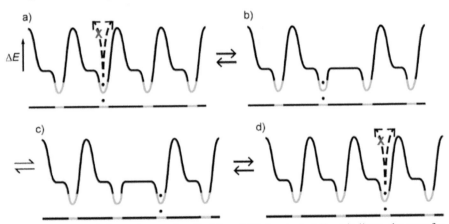

Figure 9.41 An information ratchet. In (a) and (d) the dashed lines indicate the transfer of information about the location of the particle. (b) The position of the particle selectively lowers the energy maxima to the right, but not to the left. (c) The particle moves by Brownian motion. Reprinted with permission from (Kassem et al. 2017).

However, in an information ratchet, the information of the position of the particle on the potential energy surface is important since directional transport is realized by selectively reducing the kinetic barriers to transport in front of the particle or by selectively raising barriers to transport behind the particle (Fig. 9.41).

Molecular Ratchet based on Rotaxanes/Pseudorotaxanes

In 2012 Credi described an energy ratchet in which photochemical and chemical stimuli responsive directional molecular motion of threading/dethreading of an asymmetric thread through a macrocycle was presented in psuedorotaxane architecture (Fig. 9.42). The thread comprises a bulky cyclopentane ring at one of the terminus to ensure that the threading takes place over the other terminus, photoisomerisable azobenzene acting as a 'gate' in the open state ('open' for E-isomer and closed for Z-isomer) for allowing the direction of threading motion and an ammonium binding site responsible for providing the driving force for directional threading motion of the crown ether-based macrocycle. Chemical stimuli in terms of potassium ions were used to achieve a dethreading motion and providing a thermodynamic push for it which takes place through the cyclopentane terminal due to the presence of a gate in a closed state which was realized by switching the azobenzene gate closed by application of photostimulus.

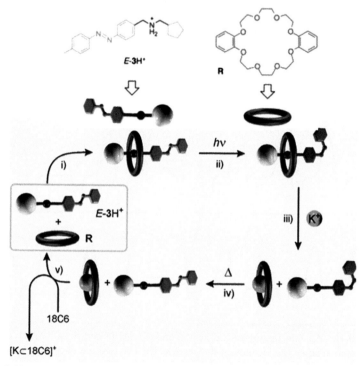

Figure 9.42   Structure formulas and cartoon representation of the examined axle and ring components Representation of the photochemically and chemically controlled transit of 3H+ through R adapted from (Baroncini et al. 2012).

CONCLUSION

In the present chapter regulated dynamic behavior of host-guest systems achieved by manipulation of well-known organic macrocyclic receptors and used in the construction of synthetic molecular machines have been described. Although substantial efforts have been made in the direction of producing unidirectional movements, the integration of these synthetic molecular machines into hierarchical architectures in order to generate a useful momentum/work is still a challenge, since there are fundamental differences between the macroscopic and molecular world and efforts to address these issues in designing functional molecular motors are being done and synthetic chemists lead the bottom-up journey.

REFERENCES

Abendroth, J.M., O.S. Bushuyev, P.S. Weiss and C. Barrett. 2015. Controlling motion at the nanoscale: rise of the molecular machines. ACS Nano 9(8): 7746–7768.

Abraham, R.J., P. Leighton and J.K. Sanders. 1985. Coordination chemistry and geometries of some 4, 4′-bipyridyl-capped porphyrins. Proton-and ligand-induced switching of conformations. J. Am. Chem. Soc. 107(12): 3472–3478.

Alberts, B., A. Johnson, J. Lewis, M. Raff, K. Roberts and P. Walter. 2002. Molecular Biology of the Cell, 4 Ed. Garland Science.

Allen, W.E., P.A. Gale, C.T. Brown, V.M. Lynch and J.L. Sessler. 1996. Binding of neutral substrates by calix[4]pyrroles. J. Am. Chem. Soc. 118(49): 12471–12472.

Allwood, B.L., N. Spencer, H. Shahriari-Zavareh, J.F. Stoddart and D.J. Williams. 1987. Complexation of paraquat by a bisparaphenylene-34-crown-10 derivative. J. Chem. Soc. Chem. Commun. (14): 1064–1066.

Amendola, V., L. Fabbrizzi, C. Mangano and P. Pallavicini. 2001. Molecular movements and translocations controlled by transition metals and signaled by light emission. pp. 79–115. *In*: [ed.] Molecular Machines and Motors, Springer.

Amendola, V., M. Bonizzoni and L. Fabbrizzi. 2011. Ion translocation within multisite receptors. pp. 361–398. *In*: B.L. Feringa [ed.]. Molecular Switches, Wiley Online Library.

Amos, L.A. 2008. Molecular motors: not quite like clockwork. Cell. Mol. Life Sci. 65(4): 509.

Amrhein, P., A. Shivanyuk, D.W. Johnson and J. Rebek. 2002a. Metal-switching and self-inclusion of functional cavitands. J. Am. Chem. Soc. 124(35): 10349–10358.

Amrhein, P., P.L. Wash, A. Shivanyuk and J. Rebek. 2002b. Metal ligation regulates conformational equilibria and binding properties of cavitands. Org. Lett. 4(3): 319–321.

Anelli, P.L., N. Spencer and J.F. Stoddart. 1991. A molecular shuttle. J. Am. Chem. Soc. 113(13): 5131–5133.

Anelli, P.L., P.R. Ashton, R. Ballardini, V. Balzani, M. Delgado, M.T. Gandolfi, T.T. Goodnow, A.E. Kaifer, D. Philp, M. Pietraszkiewicz, L. Prodi, M.V. Reddington, A.M.Z. Slawin, N. Spencer, J.F. Stoddart, S.C. Vicent and D.J. Williams. 1992. Molecular meccano. 1.[2] Rotaxanes and a [2] catenane made to order. J. Am. Chem. Soc. 114(1): 193–218.

Anelli, P.L., M. Asakawa, P.R. Ashton, R.A. Bissell, G. Clavier, R. Górski, A.E. Kaifer, S.J. Langford, G. Mattersteig, S. Menzer, D. Philp, A.M.Z. Slawin, N. Spencer, J.F. Stoddart, M.S. Tolley and D.J. Williams. 1997. Toward controllable molecular shuttles. Chem. Eur. J. 3(7): 1113–1135.

Ashton, P.R., A.M. Slawin, N. Spencer, J.F. Stoddart and D.J. Williams. 1987. Complex formation between bisparaphenylene-(3 n+ 4)-crown-n ethers and the paraquat and diquat dications. J. Chem. Soc., Chem. Commun. (14): 1066–1069.

Ashton, P.R.. T.T. Goodnow, A.E. Kaifer, M.V. Reddington, A.M.Z. Slawin, N. Spencer, J.F. Stoddart, C. Vicent and D.J. Williams. 1989. A [2]catenane made to order angew. Chem. Int. Ed. 28: 1396–1399.

Ashton, R. Ballardini, V. Balzani, S.E. Boyd, A. Credi, M.T. Gandolfi, M. Gómez-López, S. Iqbal, D. Philp, J.A. Preece, L. Prodi, H.G. Ricketts, J.F. Stoddart, M.S. Tolley, M. Venturi, M. Venturi, A.J.P. White and D.J. Williams. 1997a. Simple mechanical molecular and supramolecular machines: photochemical and electrochemical control of switching processes. Chem. Eur. J. 3(1): 152–170.

Ashton, P.R., M. Gómez-López, S. Iqbal, J.A. Preece and J.F. Stoddart. 1997b. A self-complexing macrocycle acting as a chromophoric receptor. Tetrahedron Lett. 38(20): 3635–3638.

Ashton, P.R., M.R. Johnston, J.F. Stoddart, M.S. Tolley and J.W. Wheeler. 1992a. The template-directed synthesis of porphyrin-stoppered [2]rotaxanes. J. Chem. Soc., Chem. Commun. (16): 1128–1131.

Ashton, P.R., D. Philp, N. Spencer and J.F. Stoddart. 1992b. A new design strategy for the self-assembly of molecular shuttles. J. Chem. Soc., Chem. Commun. (16): 1124–1128.

Ashton, P.R., R.A. Bissell, N. Spencer, J.F. Stoddart and M.S. Tolley. 1992c. Towards controllable molecular shuttles-1. Synlett. 1992(11): 914–918.

Ashton, P.R., R.A. Bissell, R. Górski, D. Philp, N. Spencer, J.F. Stoddart and M.S. Tolley. 1992d. Towards controllable molecular shuttles-2. Synlett 1992(11): 919–922.

Ashton, P.R., R.A. Bissell, N. Spencer, J.F. Stoddart and M.S. Tolley. 1992e. Towards controllable molecular shuttles-3. Synlett. 1992(11): 923–926.

Astumian, R.D. 2007. Design principles for Brownian molecular machines: how to swim in molasses and walk in a hurricane. Phys. Chem. Chem. Phys. 9(37): 5067–5083.

Azov, V.A., F. Diederich, Y. Lill and B. Hecht. 2003. Synthesis and conformational switching of partially and differentially bridged resorcin[4]arenes bearing fluorescent dye labels. preliminary communication. Helv. Chim. Acta 86(6): 2149–2155.

Azov, V.A., B. Jaun and F. Diederich. 2004. NMR Investigations into the vase-kite conformational switching of resorcin[4]arene cavitands. Helv. Chim. Acta 87(2): 449–462.

Baeyer, A. 1872. Ueber die verbindungen der aldehyde mit den phenolen und aromatischen kohlenwasserstoffen. Ber. Dtsch. Chem. Ges. 5(2): 1094–1100.

Bähring, S., G. Olsen, P.C. Stein, J. Kongsted and K.A. Nielsen. 2013. Coordination-driven switching of a preorganized and cooperative calix[4]pyrrole receptor. Chem. Eur. J. 19(8): 2768–2775.

Bähring, S., H.D. Root, J.L. Sessler and J.O. Jeppesen. 2019. Tetrathiafulvalene-calix[4] pyrrole: a versatile synthetic receptor for electron-deficient planar and spherical guests. Org. Biomol. Chem. 17(10): 2594–2613.

Ballardini, R., V. Balzani, A. Credi, M.T. Gandolfi and M. Venturi. 2001. Artificial molecular-level machines: which energy to make them work? Acc. Chem. Res. 34(6): 445–455.

Balzani, V., A. Credi, F.M. Raymo and J.F. Stoddart. 2000a. Artificial molecular machines. Angew. Chem. Int. Ed. 39(19): 3348–3391.

Balzani, V., A. Credi, F.M. Raymo and J.F. Stoddart. 2000b. Künstliche molekulare maschinen. Angew. Chem. 112(19): 3484–3530.

Balzani, V., M. Clemente-León, A. Credi, B. Ferrer, M. Venturi, A.H. Flood and J.F. Stoddart. 2006a. Autonomous artificial nanomotor powered by sunlight. Proc. Natl. Acad. Sci. U.S.A. 103(5): 1178–1183.

Balzani, V., M. Venturi and A. Credi. 2006b. Molecular Devices and Machines: A Journey Into The Nanoworld. John Wiley & Sons.

Balzani, V., A. Credi and M. Venturi. 2008. Molecular devices and machines: concepts and perspectives for the nanoworld. John Wiley & Sons.

Barbara, P.F. 2001. Molecular machines special issue. Acc. Chem. Res. 34(6): 409–409.

Barin, G., A. Coskun, M.M. Fouda and J.F. Stoddart. 2012. Mechanically interlocked molecules assembled by [pi]-[pi] recognition. ChemPlusChem. 77(3): 159.

Barnes, J.C., M. Juríček, N.L. Strutt, M. Frasconi, S. Sampath, M.A. Giesener, P.L. McGrier, C.J. Bruns, C.L. Stern, A.A. Sarjeant and J.F. Stoddart.. 2013. ExBox: a polycyclic aromatic hydrocarbon scavenger. J. Am. Chem. Soc. 135(1): 183–192.

Baroncini, M., S. Silvi, M. Venturi and A. Credi. 2012. Photoactivated directionally controlled transit of a non-symmetric molecular axle through a macrocycle. Angew. Chem. Int. Ed. 51(17): 4223–4226.

Barrow, S.J., S. Kasera, M.J. Rowland, J. del Barrio and O.A. Scherman. 2015. Cucurbituril-based molecular recognition. Chem. Rev. 115(22): 12320–12406.

Bazargan, G. and K. Sohlberg. 2018. Advances in modelling switchable mechanically interlocked molecular architectures. Int. Rev. Phys. Chem. 37(1): 1–82.

Behrend, R., E. Meyer and F. Rusche. 1905. über condensation-producte aus glycoluril und formaldehyd. Justus Liebigs Ann. Chem. 339(1): 1–37.

Bencini, A., A. Bianchi, C. Giorgi, E. Romagnoli, C. Lodeiro, A. Saint-Maurice, F. Pina and B. Valtancoli. 2001. Photochemical-and pH-switching properties of a new photoelastic ligand based upon azobenzene. Basicity and anion binding. Supramol. Chem. 13(2): 277–285.

Benniston, A.C. and A. Harriman. 1993. A light-induced molecular shuttle based on a [2]rotaxane-derived triad. Angew. Chem. Int. Ed. 32(10): 1459–1461.

Berryman, O.B., A.C. Sather, A. Lledó and J. Rebek Jr. 2011a. Switchable catalysis with a light-responsive cavitand. Angew. Chem. 123(40): 9572–9575.

Berryman, O.B., A.C. Sather and J. Rebek Jr. 2011b. A light controlled cavitand wall regulates guest binding. Chem. Commun. 47(2): 656–658.

Biros, S.M. and J. Rebek Jr. 2007. Structure and binding properties of water-soluble cavitands and capsules. Chem. Soc. Rev. 36(1): 93–104.

Blanco, V., D.A. Leigh and V. Marcos. 2015. Artificial switchable catalysts. Chem. Soc. Rev. 44(15): 5341–5370.

Blanco-Gómez, A., P. Cortón, L. Barravecchia, I. Neira, E. Pazos, C. Peinador and M.D. García. 2020. Controlled binding of organic guests by stimuli-responsive macrocycles. Chem. Soc. Rev. 49(12): 3834–3862.

Bleger, D. and S. Hecht. 2015. Visible-light-activated molecular switches. Angew. Chem. Int. Ed. 54(39): 11338–11349.

Böhmer, V. 1995. Calixarenes, macrocycles with (almost) unlimited possibilities. Angew. Chem. Int. Ed. Engl. 34(7): 713–745.

Bria, M., G. Cooke, A. Cooper, J.F. Garety, S.G. Hewage, M. Nutley, G. Rabani and P. Woisel. 2007. An investigation of the complexation properties of cyclobis (paraquat-p-phenylene) in water. Tetrahedron Lett. 48(2): 301–304.

Browne, W.R. and B.L. Feringa. 2010a. Making molecular machines work. pp. 79–89. *In*: [ed.] Nanoscience and Technology: A Collection of Reviews from Nature Journals. World Scientific.

Browne, W.R. and B.L. Feringa. 2010b. Light and redox switchable molecular components for molecular electronics. Chimia. 64(6): 398–403.

Buschmann, H.-J., E. Cleve, K. Jansen, A. Wego and E. Schollmeyer. 2001. Complex formation between cucurbit[n]urils and alkali, alkaline earth and ammonium ions in aqueous solution. J. Incl. Phenom. Macrocycl. Chem. 40(1–2): 117–120.

Caminade, A.-M. and J.P. Majoral. 1994. Synthesis of phosphorus-containing macrocycles and cryptands. Chem. Rev. 94(5): 1183–1213.

Cao, J., Y. Jiang, J.-M. Zhao and C.-F. Chen. 2009. A pentiptycene-based bis (crown ether) host: synthesis and its complexation with cyclobis (paraquat-p-phenylene). Chem. Commun. (15): 1987–1989.

Casas-Solvas, J.M., E. Ortiz-Salmerón, I. Fernández, L. García-Fuentes, F. Santoyo-González and A. Vargas-Berenguel. 2009. Ferrocene–β-cyclodextrin conjugates: synthesis, supramolecular behavior, and use as electrochemical sensors. Chem. Eur. J. 15(33): 8146–8162.

Chatterjee, M.N., E.R. Kay and D.A. Leigh. 2006. Beyond switches: ratcheting a particle energetically uphill with a compartmentalized molecular machine. J. Am. Chem. Soc. 128(12): 4058–4073.

Chen, C.-F. 2011. Novel triptycene-derived hosts: synthesis and their applications in supramolecular chemistry. Chem. Commun. 47(6): 1674–1688.

Chmielewski, M. and J. Jurczak. 2004. Size complementarity in anion recognition by neutral macrocyclic tetraamides. Tetrahedron Lett. 45(31): 6007–6010.

Chmielewski, M.J., A. Szumna and J. Jurczak. 2004. Anion induced conformational switch of a macrocyclic amide receptor. Tetrahedron Lett. 45(47): 8699–8703.

Chong, Y.S., M.D. Smith and K.D. Shimizu. 2001. A conformationally programmable ligand. J. Am. Chem. Soc. 123(30): 7463–7464.

Colasson, B., N.L. Poul, Y.L. Mest and O. Reinaud. 2010. Electrochemically triggered double translocation of two different metal ions with a ditopic calix[6]arene ligand. J. Am. Chem. Soc. 132(12): 4393–4398.

Cooper, S.R. 1988. Crown thio ether chemistry. Acc. Chem. Res. 21(4): 141–146.

Coudert, F.-X. 2015. Responsive metal–organic frameworks and framework materials: under pressure, taking the heat, in the spotlight, with friends. Chem. Mater 27(6): 1905–1916.

Coulston, R.J., H. Onagi, S.F. Lincoln and C.J. Easton. 2006. Harnessing the energy of molecular recognition in a nanomachine having a photochemical on/off switch. J. Am. Chem. Soc. 128(46): 14750–14751.

Credi, A., S. Silvi and M. Venturi. 2014. Molecular Machines and Motors Recent Advances and Perspectives Preface. Springer Int publishing Ag Gewerbestrasse 11, Cham, CH-6330, Switzerland.

Crini, G. 2014. A history of cyclodextrins. Chem. Rev. 114(21): 10940–10975.

Croft, A.P. and R.A. Bartsch. 1983. Synthesis of chemically modified cyclodextrins. Tetrahedron. 39(9): 1417–1474.

Dale, E.J., N.A. Vermeulen, A.A. Thomas, J.C. Barnes, M. Juricek, A.K. Blackburn, N.L. Strutt, A.A. Sarjeant, C.L. Stern, S.E. Denmark and J.F. Stoddart. 2014. ExCage. J. Am. Chem. Soc. 136(30): 10669–10682.

Dale, E.J., N.A. Vermeulen, M. Juricek, J.C. Barnes, R.M. Young, M.R. Wasielewski and J.F. Stoddart. 2016. Supramolecular explorations: exhibiting the extent of extended cationic cyclophanes. Acc. Chem. Res. 49(2): 262–273.

De Santis, G., L. Fabbrizzi, D. Iacopino, P. Pallavicini, A. Perotti and A. Poggi. 1997. Electrochemically switched anion translocation in a multicomponent coordination compound. Inorg. Chem. 36(5): 827–832.

Degenhardt, C.F., J.M. Lavin, M.D. Smith and K.D. Shimizu. 2005. Conformationally imprinted receptors: atropisomers with "write", "save", and "erase" recognition properties. Org. Lett. 7(19): 4079–4081.

Dietrich, B., J. Lehn and J. Sauvage. 1969. Les cryptates. Tetrahedron Lett. 10(34): 2889–2892.

Durola, F. and J. Rebek Jr. 2010. The ouroborand: a cavitand with a coordination-driven switching device. Angew. Chem. Int. Ed. 49(18): 3189–3191.

Erbas-Cakmak, S., D.A. Leigh, C.T. McTernan and A.L. Nussbaumer 2015. Artificial molecular machines. Chem. Rev. 115(18): 10081–10206.

Fabbrizzi, L., F. Gatti, P. Pallavicini and E. Zambarbieri. 1999. Redox-driven intramolecular anion translocation between transition metal centres. Chem. Eur. J. 5(2): 682–690.

Far, A.R., A. Shivanyuk and J. Rebek. 2002. Water-stabilized cavitands. J. Am. Chem. Soc. 124(12): 2854–2855.

Feringa, B.L. (ed.) 2001. Molecular Switches. Darmstadt, Wiley-VCH Verlag GmbH.

Feynman, R.P. and F.L. Vernon. 1963. The theory of a general quantum system interacting with a linear dissipative system. Ann. Phys. 24: 118–173.

Fihey, A., A. Perrier, W.R. Browne and D. Jacquemin. 2015. Multiphotochromic molecular systems. Chem. Soc. Rev. 44(11): 3719–3759.

Frasconi, M., I.R. Fernando, Y. Wu, Z. Liu, W.-G. Liu, S.M. Dyar et al. J.F. Stoddart. 2015. Redox control of the binding modes of an organic receptor. J. Am. Chem. Soc. 137(34): 11057–11068.

Fraserá-Stoddart, J. 1992. The template-directed synthesis of porphyrin-stoppered [2]rotaxanes. J. Chem. Soc., Chem. Commun. (16): 1128–1131.

Freeman, W., W. Mock and N. Shih. 1981. Cucurbituril. J. Am. Chem. Soc. 103(24): 7367–7368.

Frei, M., F. Marotti and F. Diederich. 2004. Zn II-induced conformational control of amphiphilic cavitands in langmuir monolayers. Chem. Commun. (12): 1362–1363.

Frisch, H. and P. Besenius. 2015. pH-Switchable self-assembled materials. Macromol. Rapid Commun. 36(4): 346–363.

Fukushima, M., T. Osa and A. Ueno. 1991. Photoswitchable multi-response sensor of azobenzene-modified γ-cyclodextrin for detecting organic compounds. Chem. Lett. 20(4): 709–712.

Gale, P.A., J.L. Sessler, V. Kral and V. Lynch. 1996. Calix[4]pyrroles: old yet new anion-binding agents. J. Am. Chem. Soc. 118(21): 5140–5141.

Gil, E.S. and S.M. Hudson 2004. Stimuli-reponsive polymers and their bioconjugates. Prog. Polym. Sci. 29(12): 1173–1222.

Gokel, G.W., W.M. Leevy and M.E. Weber. 2004. Crown ethers: sensors for ions and molecular scaffolds for materials and biological models. Chem. Rev. 104(5): 2723–2750.

Guo, D.-S., K. Wang and Y. Liu. 2008. Selective binding behaviors of p-sulfonatocalixarenes in aqueous solution. J. Incl. Phenom. Macrocycl. Chem. 62(1-2): 1–21.

Guo, D.-S. and Y. Liu. 2014. Supramolecular chemistry of p-sulfonatocalix[n]arenes and its biological applications. Acc. Chem. Res. 47(7): 1925–1934.

Gutsche, C.D. 1983. Calixarenes. Acc. Chem. Res. 16(5): 161–170.

Gutsche, C.D. and L.-G. Lin. 1986. Calixarenes 12: the synthesis of functionalized calixarenes. Tetrahedron. 42(6): 1633–1640.

Gutsche, C.D. 1989. Calixarenes. Royal Society of Chemistry, Cambridge.

Haino, T., Y. Katsutani, H. Akii and Y. Fukazawa. 1998. Allosteric receptor based on monodeoxycalix [4] arene crown ether. Tetrahedron Lett. 39(44): 8133–8136.

Haino, T., D.M. Rudkevich, A. Shivanyuk, K. Rissanen and J. Rebek, Julius. 2000. Induced-fit molecular recognition with water-soluble cavitands. Chem. Eur. J. 6(20): 3797–3805.

Hardouin–Lerouge, M., P. Hudhomme and M. Sallé. 2011. Molecular clips and tweezers hosting neutral guests. Chem. Soc. Rev. 40(1): 30–43.

Hoyt, M.A., A.A. Hyman and M. Bähler. 1997. Motor proteins of the eukaryotic cytoskeleton. Proc. Natl. Acad. Sci. U.S.A. 94(24): 12747–12748.

Huang, F., K.A. Switek and H.W. Gibson. 2005. pH-controlled assembly and disassembly of a cryptand/paraquat [2]pseudorotaxane. Chem. Commun. (29): 3655–3657.

Hunter, C.A. and H.L. Anderson. 2009. What is cooperativity? Angew. Chem. Int. Ed. 48(41): 7488–7499.

Hwang, I., W.S. Jeon, H.J. Kim, D. Kim, H. Kim, N. Selvapalam, N. Fujita, S. Shinkai and K. Kim 2007. Cucurbit[7]uril: a simple macrocyclic, pH-triggered hydrogelator exhibiting guest-induced stimuli-responsive behavior. Angew. Chem. 119(1-2): 214–217.

Ikeda, A. and S. Shinkai. 1997. Novel cavity design using calix[n]arene skeletons: toward molecular recognition and metal binding. Chem. Rev. 97(5): 1713–1734.

Ikeda, A., T. Tsudera and S. Shinkai. 1997. Molecular Design of a "Molecular Syringe" mimic for metal cations using a 1, 3-alternate calix[4]arene cavity. J. Org. Chem. 62(11): 3568–3574.

Izatt, R.M., J.S. Bradshaw, K. Pawlak, R.L. Bruening and B.J. Tarbet. 1992. Thermodynamic and kinetic data for macrocycle interaction with neutral molecules. Chem. Rev. 92(6): 1261–1354.

Jain, V. and P. Kanaiya. 2011. Chemistry of calix[4]resorcinarenes. Russ. Chem. Rev. 80(1): 75.

Ji, X., M. Zhang, X. Yan, J. Li and F. Huang. 2013. Synthesis of a water-soluble bis (m-phenylene)-32-crown-10-based cryptand and its pH-responsive binding to a paraquat derivative. Chem. Commun. 49(12): 1178–1180.

Jiang, F., M. Chen, J. Liang, Z. Gao, M. Tang, Z. Xu, B. Peng, S. Zhu and L. Jiang. 2016. Sailboat-shaped self-complexes that function as controllable rotary switches. Eur. J. Org. Chem. 2016(20): 3310–3315.

Joosten, A., Y. Trolez, J.-P. Collin, V. Heitz and J.-P. Sauvage. 2012. Copper (I)-assembled [3]rotaxane whose two rings act as flapping wings. J. Am. Chem. Soc. 134(3): 1802–1809.

Juríček, M., N.L. Strutt, J.C. Barnes, A.M. Butterfield, E.J. Dale, K.K. Baldridge, J.F. Stoddart and J.S. Siegel. 2014. Induced-fit catalysis of corannulene bowl-to-bowl inversion. Nat. Chem. 6(3): 222–228.

Kaempfer, R. 1968. Ribosomal subunit exchange during protein synthesis. Proc. Natl. Acad. Sci. U.S.A. 61(1): 106.

Kaifer, A. and M. Gómez-Kaifer. 1999. Supramolecular Electrochemistry. Weinheim, Germany: Wiley-VCH.

Kassem, S., T. van Leeuwen, A.S. Lubbe, M.R. Wilson, B.L. Feringa and D.A. Leigh. 2017. Artificial molecular motors. Chem. Soc. Rev. 46(9): 2592–2621.

Kay, E.R., D.A. Leigh and F. Zerbetto. 2007. Synthetic molecular motors and mechanical machines. Angew. Chem. Int. Ed. 46(1-2): 72–191.

Kay, E.R. and D.A. Leigh. 2008. Beyond switches: rotaxane-and catenane-based synthetic molecular motors. Pure Appl. Chem. 80(1): 17–29.

Kay, E.R. and D.A. Leigh. 2015. Rise of the molecular machines. Angew. Chem. Int. Ed. 54(35): 10080–10088.

Kennedy, R.D., C.W. Machan, C.M. McGuirk, M.S. Rosen, C.L. Stern, A.A. Sarjeant and C.A. Mirkin. 2013. General strategy for the synthesis of rigid weak-link approach platinum (II) complexes: tweezers, triple–layer complexes, and macrocycles. Inorg. Chem. 52(10): 5876–5888.

Khan, A.R., P. Forgo, K.J. Stine and V.T. D'Souza. 1998. Methods for selective modifications of cyclodextrins. Chem. Rev. 98(5): 1977–1996.

Kim, K., N. Selvapalam, Y.H. Ko, K.M. Park, D. Kim and J. Kim. 2007. Functionalized cucurbiturils and their applications. Chem. Soc. Rev. 36: 267–279.

Kim, S.K. and J.L. Sessler. 2014. Calix[4]pyrrole-based ion pair receptors. Acc. Chem. Res. 47(8): 2525–2536.

Kinbara, K. and T. Aida. 2005. Toward intelligent molecular machines: directed motions of biological and artificial molecules and assemblies. Chem. Rev. 105(4): 1377–1400.

Knipe, P.C., S. Thompson and A.D. Hamilton. 2015. Ion-mediated conformational switches. Chem. Sci. 6(3): 1630–1639.

Ko, Y.H., E. Kim, I. Hwang and K. Kim. 2007. Supramolecular assemblies built with host-stabilized charge-transfer interactions. Chem. Commun. (13): 1305–1315.

Kovbasyuk, L. and R. Krämer. 2004. Allosteric supramolecular receptors and catalysts. Chem. Rev. 104(6): 3161–3188.

Krakowiak, K.E., J.S. Bradshaw and D.J. Zamecka-Krakowiak. 1989. Synthesis of aza-crown ethers. Chem. Rav. 89(4): 929–972.

Kuwabara, J., H.J. Yoon, C.A. Mirkin, A.G. DiPasquale and A.L. Rheingold. 2009. Pseudo-allosteric regulation of the anion binding affinity of a macrocyclic coordination complex. Chem. Commun. (30): 4557–4559.

Lancia, F., A. Ryabchun and N. Katsonis. 2019. Life-like motion driven by artificial molecular machines. Nat. Rev. Chem. 3(9): 536–551.

Lane, A.S., D.A. Leigh and A. Murphy. 1997. Peptide-based molecular shuttles. J. Am. Chem. Soc. 119(45): 11092–11093.

Lascaux, A., S. Le Gac, J. Wouters, M. Luhmer and I. Jabin. 2010. An allosteric heteroditopic receptor for neutral guests and contact ion pairs with a remarkable selectivity for ammonium fluoride salts. Org. Biomol. Chem. 8(20): 4607–4616.

Lascaux, A., G. Leener and L. Fusaro 2016. F. Topic, K. Rissanen, M. Luhmer and I. Jabin. Org. Biomol. Chem 14: 738–746.

Leblond, J. and A. Petitjean. 2011. Molecular tweezers: concepts and applications. ChemPhysChem. 12(6): 1043–1051.

Lee, J. 2003. S, Samal, N. Selvapalam, H.-J. Kim, and K. Kim. Acc. Chem. Res. 36: 621.

Lee, W.S. and A. Ueno. 2001. Photocontrol of the catalytic activity of a β-cyclodextrin bearing azobenzene and histidine moieties as a pendant group. Macromol. Rapid Commun. 22(6): 448–450.

Lee, S. and A.H. Flood. 2013. Photoresponsive receptors for binding and releasing anions. J. Phys. Org. Chem. 26(2): 79–86.

Lehn, J. and M. Stubbs. 1974. Cryptates. XIII. Intramolecular cation exchange in [3]cryptates of alkaline earth cations. J. Am. Chem. Soc. 96(12): 4011–4012.

Lehn, J.M. 1978a. Cryptates: inclusion complexes of macropolycyclic receptor molecules Pure Appl. Chem. 50(9–10): 871–892.

Lehn, J.M. 1978b. Cryptates: the chemistry of macropolycyclic inclusion complexes. Acc. Chem. Res. 11(2): 49–57.

Lehn, J.-M. 2002a. Toward complex matter: supramolecular chemistry and self-organization. Proc. Natl. Acad. Sci. U.S.A. 99(8): 4763–4768.

Lehn, J.-M. 2002b. Toward self-organization and complex matter. Science. 295(5564): 2400–2403.

Lehn, J.M. 2013. Perspectives in chemistry–steps towards complex matter. Angew. Chem. Int. Ed. Engl. 52(10): 2836–2850.

Lehn, J.M. 2015. Perspectives in chemistry–aspects of adaptive chemistry and materials. Angew. Chem. Int. Ed. Engl. 54(11): 3276–3289.

Leigh, D.A. and E.M. Pérez. 2004. Shuttling through reversible covalent chemistry. Chem. Commun. (20): 2262–2263.

Leighton, P. and J.K. Sanders. 1984. A molecular switch for control of conformation: strained intramolecular co-ordination in 4, 4′-bipyridyl-capped zinc porphyrins. J. Chem. Soc., Chem. Commun. (13): 854–856.

Leung, K. C.-F., C.-P. Chak, C.-M. Lo, W.-Y. Wong, S. Xuan and C.H.K. Cheng. 2009. pH-controllable supramolecular systems. Chem. Asian J. 4(3): 364–381.

Linton, B. and A.D.J.C.r. Hamilton. 1997. Formation of artificial receptors by metal-templated self-assembly. Chem. Rev. 97(5): 1669–1680.

Liu, M., X. Yan, M. Hu, X. Chen, M. Zhang, B. Zheng, X. Hu, S. Shao and F. Huang. 2010. Photoresponsive host-guest systems based on a new azobenzene-containing cryptand. Org. Lett. 12(11): 2558–2561.

Liu, W., S. Bobbala, C.L. Stern, J.E. Hornick, Y. Liu, A.E. Enciso et al. J.F. Stoddart 2020. XCage: a tricyclic octacationic receptor for perylene diimide with picomolar affinity in water. J. Am. Chem. Soc. 142(6): 3165–3173.

Lodish, H., A. Berk, S.L. Zipursky, P. Matsudaira, D. Baltimore and J. Darnell. 2000. Molecular Cell Biology. W.H. Freeman, New York.

Luo, L., X. Zhang, N. Feng, D. Tian, H. Deng and H. Li. 2015. Cation-induced pesticide binding and release by a functionalized calix[4]arene molecular host. Sci. Rep. 5: 8982.

Lutz, J.-F., J.-M. Lehn, E. Meijer and K. Matyjaszewski. 2016. From precision polymers to complex materials and systems. Nat. Rev. Mater. 1(5): 1–14.

Ma, S., D.M. Rudkevich and J. Rebek, Julius. 1999. Supramolecular isomerism in caviplexes. Angew. Chem. Int. Ed. 38(17): 2600–2602.

Maxwell, J.C. 1871. Theory of Heat. Longmans, London: UK.

Mendes, P.M. 2008. Stimuli-responsive surfaces for bio-applications. Chem. Soc. Rev. 37(11): 2512–2529.

Mendez-Arroyo, J., J. Barroso-Flores, A.M. Lifschitz, A.A. Sarjeant, C.L. Stern and C.A. Mirkin. 2014. A multi-state, allosterically-regulated molecular receptor with switchable selectivity. J. Am. Chem. Soc. 136(29): 10340–10348.

Moran, J.R., J.L. Ericson, E. Dalcanale, J.A. Bryant, C.B. Knobler and D.J. Cram. 1991. Vases and kites as cavitands. J. Am. Chem. Soc. 113(15): 5707–5714.

Moran, J.R., S. Karbach and D.J. Cram. 1982. Cavitands: synthetic molecular vessels. J. Am. Chem. Soc. 104(21): 5826–5828.

Mura, S., J. Nicolas and P. Couvreur. 2013. Stimuli-responsive nanocarriers for drug delivery. Nat. Mater. 12(11): 991–1003.

Murakami, H., A. Kawabuchi, K. Kotoo, M. Kunitake and N. Nakashima. 1997. A light-driven molecular shuttle based on a rotaxane. J. Am. Chem. Soc. 119(32): 7605–7606.

Muraoka, M., M. Ohta, Y. Mizutani, M. Takezawa, A. Matsumoto and Y. Nakatsuji. 2014. Formation of a pseudorotaxane, capable of sensing cations via dethreading molecular motion, from a cryptand and bipyridinium salts. J. Inclusion Phenom. Macrocyclic Chem. 78(1–4): 137–144.

Nakashima, N., A. Kawabuchi and H. Murakami. 1998. Design and synthesis of cyclodextrin-based rotaxanes and polyrotaxanes. J. Inclusion Phenom. Mol. Recognit. Chem. 32(2–3): 363–373.

Nakagama, T., K. Hirasawa, K. Uchiyama and T. Hobo. 2001. Photo-responsive retention behavior of azobenzene-modified cyclodextrin stationary phase in micro-HPLC. Anal. Sci. 17(1): 119–124.

Nelissen, H., F. Venema, R. Uittenbogaard, M. Feiters and R. Nolte. 1997. J. Chem. Soc. Perkin Trans. 2. J. Chem. Soc. Perkin Trans. 2.

Odell, B., M.V. Reddington, A.M. Slawin, N. Spencer, J.F. Stoddart and D.J. Williams. 1988. Cyclobis (paraquat-p-phenylene). A tetracationic multipurpose receptor. Angew. Chem. Int. Ed. Engl. 27(11): 1547–1550.

Ogoshi, T., S. Kanai, S. Fujinami, T.-a. Yamagishi and Y. Nakamoto. 2008. Para-Bridged symmetrical pillar[5]arenes: their Lewis acid catalyzed synthesis and host-guest property. J. Am. Chem. Soc. 130(15): 5022–5023.

Ogoshi, T., T.-a. Yamagishi and Y. Nakamoto. 2016. Pillar-shaped macrocyclic hosts pillar[n]arenes: new key players for supramolecular chemistry. Chem. Rev. 116(14): 7937–8002.

Pazos, E., P. Novo, C. Peinador, A.E. Kaifer and M.D. García. 2019. Cucurbit[8]uril (CB [8])-based supramolecular switches. Angew. Chem. Int. Ed. Engl. 58(2): 403–416.

Pedersen, C.J. and H. Frensdorff. 1972. Macrocyclic polyethers and their complexes. Angew. Chem. Int. Ed. Engl. 11(1): 16–25.

Pezzato, C., C. Cheng, J.F. Stoddart and R.D. Astumian. 2017. Mastering the non-equilibrium assembly and operation of molecular machines. Chem. Soc. Rev. 46(18): 5491–5507.

Pochorovski, I., M.-O. Ebert, J.-P. Gisselbrecht, C. Boudon, W.B. Schweizer and F. Diederich. 2012. Redox-switchable resorcin[4]arene cavitands: molecular grippers. J. Am. Chem. Soc. 134(36): 14702–14705.

Pochorovski, I., J. Milić, D.A. Kolarski, C. Gropp, W.B. Schweizer and F.O. Diederich. 2014. Evaluation of hydrogen-bond acceptors for redox-switchable resorcin[4]arene cavitands. J. Am. Chem. Soc. 136(10): 3852–3858.

Poulsen, T., K.A. Nielsen, A.D. Bond and J.O. Jeppesen. 2007. Bis(tetrathiafulvalene)-calix [2]pyrrole[2]-thiophene and its complexation with TCNQ. Org. Lett. 9(26): 5485–5488.

Rebek Jr, J., J. Trend, R. Wattley and S. Chakravorti. 1979. Allosteric effects in organic chemistry. Site-specific binding. J. Am. Chem. Soc. 101(15): 4333–4337.

Rebek Jr, J. and R. Wattley. 1980. Allosteric effects. Remote control of ion transport selectivity. J. Am. Chem. Soc. 102(14): 4853–4854.

Rebek Jr, J. 1984. Binding forces, equilibria and rates: new models for enzymic catalysis. Acc. Chem. Res. 17(7): 258–264.

Rebek Jr, J., T. Costello, L. Marshall, R. Wattley, R.C. Gadwood and K. Onan. 1985. Allosteric effects in organic chemistry: binding cooperativity in a model for subunit interactions. J. Am. Chem. Soc. 107(25): 7481–7487.

Robertson, A. and S. Shinkai. 2000. Cooperative binding in selective sensors, catalysts and actuators. Coord. Chem. Rev. 205(1): 157–199.

Rojanathanes, R., T. Tuntulani, W. Bhanthumnavin and M. Sukwattanasinitt. 2005. Stilbene-bridged tert-butylcalix[4]arene as photoswitchable molecular receptors. Organic lett. 7(16): 3401–3404.

Rudkevich, D.M., G. Hilmersson and J. Rebek. 1997. Intramolecular hydrogen bonding controls the exchange rates of guests in a cavitand. J. Am. Chem. Soc. 119(41): 9911–9912.

Rudkevich, D.M., G. Hilmersson and J. Rebek 1998. Self-folding cavitands. J. Am. Chem. Soc. 120(47): 12216–12225.

Saha, I., J.T. Lee and C.H. Lee. 2015. Recent advancements in calix[4]pyrrole-based anion-receptor chemistry. Eur. J. Org. Chem. 2015(18): 3859–3885.

Sauvage, J.-P. 2003. Molecular Machines and Motors. Springer.

Sauvage, J.-P. and P. Gaspard. 2011. From Non-covalent Assemblies to Molecular Machines. John Wiley & Sons.

Schliwa, M. and G. Woehlke. 2003. Molecular motors. Nature. 422(6933): 759–765.

Schneider, H.-J. and U. Schneider. 1994. The host-guest chemistry of resorcinarenes. J. Incl. Phenom. Macrocycl. Chem. 19(1–4): 67–83.

Schoder, S. and C.A. Schalley. 2017. Orthogonal switching of self-sorting processes in a stimuli-responsive library of cucurbit[8]uril complexes. Chem. Commun. 53(69): 9546–9549.

Serreli, V., C.-F. Lee, E.R. Kay and D.A. Leigh. 2007. A molecular information ratchet. Nature. 445(7127): 523–527.

Sessler, J.L., P. Anzenbacher, K. Jursikova, H. Miyaji, J.W. Genge, N.A. Tvermoes, W.E. Allen, J.A. Shriver, P.A. Gale and V. Kral 1998. Functionalized calix[4]pyrroles. Pure Appl. Chem. 70(12): 2401–2408.

Shenker, O.R. 2004. Book review: Maxwell's Demon 2: Entropy, classical and quantum information, computing. Harvey Leff and Andrew Rex (eds), Pergamon. Stud. Hist. Philos. Sci. Part B: Stud. Hist. Philos. Mod. Phys. 35(3): 537–540.

Shinkai, S., T. Ogawa, T. Nakaji, Y. Kusano and O. Nanabe. 1979. Photocontrolled extraction ability of azobenzene-bridged azacrown ether. Tetrahedron Lett. 20(47): 4569–4572.

Shinkai, S., T. Nakaji, Y. Nishida, T. Ogawa and O. Manabe. 1980. Photoresponsive crown ethers. 1. *Cis-trans* isomerism of azobenzene as a tool to enforce conformational changes of crown ethers and polymers. J. Am. Chem. Soc. 102(18): 5860–5865.

Shinkai, S., T. Nakaji, T. Ogawa, K. Shigematsu and O. Manabe. 1981. Photoresponsive crown ethers. 2. Photocontrol of ion extraction and ion transport by a bis (crown ether) with a butterfly-like motion. J. Am. Chem. Soc. 103(1): 111–115.

Shinkai, S., S. Mori, T. Tsubaki, T. Sone and O. Manabe. 1984. New water-soluble host molecules derived from calix[6]arene. Tetrahedron Lett. 25(46): 5315–5318.

Shinkai, S., M. Ishihara, K. Ueda and O. Manabe. 1985. Photoresponsive crown ethers. Part 14. Photoregulated crown–metal complexation by competitive intramolecular tail (ammonium)-biting. J. Chem. Soc., Perkin Trans. 2(4): 511–518.

Shinkai, S. 1987. Switch-functionalized systems in biomimetic chemistry. Pure Appl. Chem. 59(3): 425–430.

Shinkai, S. 1993. Calixarenes-the third generation of supramolecules. Tetrahedron. 49(40): 8933–8968.

Shinkai, S., M. Ikeda, A. Sugasaki and M. Takeuchi. 2001. Positive allosteric systems designed on dynamic supramolecular scaffolds: toward switching and amplification of guest affinity and selectivity. Acc. Chem. Res. 34(6): 494–503.

Siegel, J.S. 1996. Supramolecular chemistry: concepts and perspectives. Science. 271(5251): 949–950.

Skinner, P.J., A.G. Cheetham, A. Beeby, V. Gramlich and F. Diederich. 2001. Conformational switching of resorcin[4]arene cavitands by protonation, preliminary communication. Helv. Chim. Acta 84(7): 2146–2153.

Smoluchowski, M. 1912. XIII. On opalescence of gases in the critical state. Lond. Edinb. Dubl. Phil. Mag. 23(133): 165–173.

Sollogoub, M.J.S. 2013. Site-selective heterofunctionalization of cyclodextrins: discovery, development, and use in catalysis. Synlett. 24(20): 2629–2640.

Stoddart, J.F. 1988. Conception and birth of new receptor chemistry from dibenzo-18-crown-6. Pure Appl. Chem. 60(4): 467–472.

Strutt, N.L., H. Zhang, S.T. Schneebeli and J.F. Stoddart. 2014. Functionalizing pillar[n] arenes. Acc. Chem. Res. 47(8): 2631–2642.

Szumna, A. and J. Jurczak 2001. A new macrocyclic polylactam-type neutral receptor for anions-structural aspects of anion recognition. Eur. J. Org. Chem. 2001(21): 4031–4039.

Tian, F., D. Jiao, F. Biedermann and O.A. Scherman 2012. Orthogonal switching of a single supramolecular complex. Nat. Commun. 3(1): 1–8.

Timmerman, P., W. Verboom and D.N. Reinhoudt. 1996. Resorcinarenes. Tetrahedron. 52(8): 2663–2704.

Tseng, H.R., S.A. Vignon and J.F. Stoddart 2003. Toward chemically controlled nanoscale molecular machinery. Angew. Chem. Int. Ed. 42(13): 1491–1495.

Tucci, F.C., D.M. Rudkevich and J. Rebek Jr. 2000. Velcrands with snaps and their conformational control. Chem. Eur. J. 6(6): 1007–1016.

Tucker, J.A., C.B. Knobler, K. Trueblood and D.J. Cram. 1989. Host-guest complexation. 49. Cavitands containing two binding cavities. J. Am. Chem. Soc. 111(10): 3688–3699.

Tulyakova, E.V., G. Vermeersch, E.N. Gulakova, O.A. Fedorova, Y.V. Fedorov, J.C. Micheau and S. Delbaere. 2010. Metal ions drive thermodynamics and photochemistry of the bis(styryl) macrocyclic tweezer. Chem. Eur. J. 16(19): 5661–5671.

Ueno, A., H. Yoshimura, R. Saka and T. Osa. 1979. Photocontrol of binding ability of capped cyclodextrin. J. Am. Chem. Soc. 101(10): 2779–2780.

Ueno, A., K. Takahashi and T. Osa. 1981. Photocontrol of catalytic activity of capped cyclodextrin. J. Chem. Soc., Chem. Commun. 1981(3): 94–96.

Ueno, A., Y. Tomita and T. Osa. 1983. Photoresponsive binding ability of azobenzene-appended γ-cyclodextrin. Tetrahedron Lett. 24(47): 5245–5248.

Ueno, A., M. Fukushima and T. Osa. 1990. Inclusion complexes and Z-E photoisomerization of β-cyclodextrin bearing an azobenzene pendant. J. Chem. Soc., Perkin Trans. 2(7): 1067–1072.

Ulrich, S., A. Petitjean and J.M. Lehn. 2010. Metallo-controlled dynamic molecular tweezers: design, synthesis, and self-assembly by metal-ion coordination. Eur. J. Inorg. Chem. 2010(13): 1913–1928.

Vale, R.D. and R.A. Milligan 2000. The way things move: looking under the hood of molecular motor proteins. Science. 288(5463): 88–95.

Vale, R.D. 2003. The molecular motor toolbox for intracellular transport. Cell. 112(4): 467–480.

Van Veggel, F.C., W. Verboom and D.N. Reinhoudt. 1994. Metallomacrocycles: supramolecular chemistry with hard and soft metal cations in action. Chem. Rev. 94(2): 279–299.

Vella, S.J., J. Tiburcio and S. Loeb. 2007. Optically sensed, molecular shuttles driven by acid–base chemistry. Chem. Commun. (45): 4752–4754.

Vignon, S.A., T. Jarrosson, T. Iijima, H.-R. Tseng, J.K. Sanders and J.F. Stoddart. 2004. Switchable neutral bistable rotaxanes. J. Am. Chem. Soc. 126(32): 9884–9885.

Wang, Q., M. Cheng, Y. Zhao, Z. Yang, J. Jiang, L. Wang and Y. Pan. 2014a. Redox-switchable host-guest systems based on a bisthiotetrathiafulvalene-bridged cryptand. Chem. Commun. 50(98): 15585–15588.

Wang, T., M. Wang, C. Ding and J. Fu. 2014b. Mono-benzimidazole functionalized β-cyclodextrins as supramolecular nanovalves for pH-triggered release of p-coumaric acid. Chem. Commun. 50(83): 12469–12472.

Wang, Y., G. Ping and C. Li. 2016. Efficient complexation between pillar[5]arenes and neutral guests: from host–guest chemistry to functional materials. Chem. Commun. 52(64): 9858–9872.

Wasserman, E. 1960. The preparation of interlocking rings: a catenane1. J. Am. Chem. Soc. 82(16): 4433–4434.

Wu, K.-C., Y.-S. Lin, Y.-S. Yeh, C.-Y. Chen, M.O. Ahmed, P.-T. Chou and Y.-S. Hon. 2004. Design and synthesis of intramolecular hydrogen bonding systems. Their application in metal cation sensing based on excited-state proton transfer reaction. Tetrahedron 60(51): 11861–11868.

Xu, J.-F., Y.-Z. Chen, L.-Z. Wu, C.-H. Tung and Q.-Z. Yang. 2014. Synthesis of a photoresponsive cryptand and its complexations with paraquat and 2, 7-diazapyrenium. Organic lett. 16(3): 684–687.

Xue, M., Y. Yang, X. Chi, Z. Zhang and F. Huang. 2012. Pillararenes, a new class of macrocycles for supramolecular chemistry. Acc. Chem. Res. 45(8): 1294–1308.

Xue, M., Y. Yang, X. Chi, X. Yan and F. Huang. 2015. Development of pseudorotaxanes and rotaxanes: from synthesis to stimuli-responsive motions to applications. Chem. Rev. 115(15): 7398–7501.

Yan, X., M. Zhang, P. Wei, B. Zheng, X. Chi, X. Ji and F. Huang. 2011. pH-responsive assembly and disassembly of a supramolecular cryptand-based pseudorotaxane driven by π–π stacking interaction. Chem. Commun. 47(35): 9840–9842.

Zhang, M., B. Zheng, B. Xia, K. Zhu, C. Wu and F. Huang. 2010. Synthesis of a bis(1, 2, 3-phenylene) cryptand and its dual-response binding to paraquat and diquat. Eur. J. Org. Chem. 2010(35): 6804–6809.

Zhang, M., B. Zheng and F. Huang. 2011. Synthesis of a four-armed cage molecule and its pH-controlled complexation with paraquat. Chem. Commun. 47(36): 10103–10105.

Zhang, M., X. Yan, F. Huang, Z. Niu and H.W. Gibson 2014. Stimuli-responsive host-guest systems based on the recognition of cryptands by organic guests. Acc. Chem. Res. 47(7): 1995–2005.

Zhao, J., S. Ji, Y. Chen, H. Guo and P. Yang. 2012. Excited state intramolecular proton transfer (ESIPT): from principal photophysics to the development of new chromophores and applications in fluorescent molecular probes and luminescent materials. Phys. Chem. Chem. Phys. 14(25): 8803–8817.

Zheng, X., X. Wang, K. Shen, N. Wang and Y. Peng. 2010. Molecular design of a "molecular syringe" mimic for metal cations using a 1, 3-alternate calix[4]arene cavity. J. Comput. Chem. 31(11): 2143–2156.

Zhuang, J., M.R. Gordon, J. Ventura, L. Li and S. Thayumanavan. 2013. Multi-stimuli responsive macromolecules and their assemblies. Chem. Soc. Rev. 42(17): 7421–7435.

Zimmerman, S.C. 1993. Rigid molecular tweezers as hosts for the complexation of neutral guests. pp. 71–102. *In*: [ed.] Supramolecular Chemistry I—Directed Synthesis and Molecular Recognition. Springer.

Zinke, A. and E. Ziegler. 1941. Zur kenntnis des härtungsprozesses von phenol-formaldehyd harzen, VII. Ber. Dtsch. Chem. Ges. B. 74: 1729–1736.

Chapter 10

Role of Macrocyclic Receptors in Surface Self-assembly

INTRODUCTION

Molecular self-assembly is a ubiquitous process in nature (Fang and Böhringer 2008). It may be defined as the process through which a molecular species spontaneously comes together to form a complex supramolecular species (Klajn et al. 2010). Through this process ordered aggregates forms spontaneously having intriguing functions and properties (Klajn et al. 2010). Amphiphilic molecules, which contain hydrophilic and hydrophobic moieties in their structure are usually important precursors in the self-assembling process (Lehn 1993, Lehn 2017). Examples of amphiphilic molecules include surfactants, peptides, lipids and substituted synthetic macrocyclic molecules. Supramolecular systems like calix[n] arene also possess a hydrophilic and a hydrophobic part in their molecular structures. The molecular self-assembly leads to a variety of structures like micelles (cylindrical or spherical), ribbons, vesicles, nanosheets, nanofibers, nanorods, nanoparticles and nanotubes, which is dependent on the external environment or application of external stimuli (Fig. 10.1). The application of external stimuli or changes in the external environment leads to a change in the relative volume fraction of hydrophobic and hydrophilic parts in the molecular unit. The external stimuli usually leads to a change in the nanoscale dimension of the molecular assembly where molecules link to each other through non-covalent interactions including electrostatic interaction, hydrogen-bonding and van der Waals type. The knowledge of different non-covalent interactions operating in macrocyclic systems is of fundamental importance to design an appropriate host-guest system.

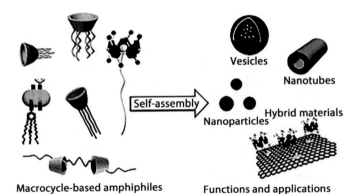

Figure 10.1 The formation of different type of structures during the association of macrocycles (Zhu et al. 2018).

Different types of interactions responsible for the self-assembly process in general can be classified into short range and long-range interactions. The short-range interactions mainly consist of coulombic and exchange type, which include covalent bonds and appear due to orbital overlap. Such interactions can be attractive or repulsive interactions. The long-range interactions usually include electrostatic, H-bond, $\pi\cdots\pi$, $C-H\cdots\pi$, and van der Waals type interactions. These interactions are usually proportional to r^{-m} (where r = internuclear distance and m is a positive integer) and mainly observed during the complex formation.

The non-covalent interactions and conformational changes that cause self-assembly can be influenced by external stimuli such as an analyte (anionic, cationic or uncharged molecular species), light, temperature, pH or redox potential. The removal of external stimuli may lead to the reversal of the assembly-disassembly process. The different interactions are briefly described including their effect on the assembly of macrocyclic systems.

The self-assembly/disassembly of the macrocyclic systems has enormous potential in the areas such as chemosensing, biosensing, bioimaging, stimuli-responsive materials, optoelectronic devices and functional self-assembly systems. The molecular self-assembly in macrocyclic receptors relies on careful molecular design considering the incorporation of specific functional groups, which induces the host-guest interactions.

Macrocyclic systems are structurally attractive as they display shape-persistent self-assembly. The molecular architecture of the macrocyclic systems provides a pre organized morphology that controls the arrangement observed in the self-assembly process. The confined cavity of a macrocyclic molecular system forms a unidirectional columnar assembly through self-stacking to generate a highly organized configuration like nanochannel tubules. The columnar assembly or macrocycles forms due to the sequential formation of intermolecular non-covalent interactions. However, such structures revert to the original configuration on exposure to an unfavorable external environment or stimuli. The reversible control of supramolecular assembling and the disassembling process can play a vital role in simulating the biological functions, such as the conversion of microtubules and tubulin dimers (Howard and Hyman 2003; Kueh and Mitchison 2009; Nogales and Zhang 2016). Different types of stimuli, such as light (Samanta and Klajn 2016; Shi et al. 2016b), pH (Amodio et al. 2016; Yang et al. 2016a), temperature (Izatt et al. 1983; Ruben et al. 2006), solvent composition (Mon et al. 2016) and specific guest can be used to control the assembling and disassembling process (Xiao et al. 2019). This approach was investigated intensively to design and develop interesting, well-defined molecular assemblies with diverse properties for applications in various areas of chemistry and biology (Ward and Raithby 2013). The non-covalent interactions between the host and guest are the main driving force for the reversible formation of supramolecular host assemblies (Chang et al. 2013). To achieve control of stimuli over the self-assembly process in the macrocyclic receptors, the molecular architecture of the receptors can be modified through chemical reactions. The optical control over the self-assembly process has a remarkable effect in controlling the size and pattern of the assembled structure. The intermolecular interactions can also be fine-tuned with appropriate guest species to achieve an appropriate particle size, pattern and optical signal. The J- and H-aggregates of organic dyes are also well-known examples of supramolecular assemblies with the optical response. The self-association of the macrocyclic amphiphilic system is a frequently observed event that occurs at the solid-liquid interface brought by non-covalent attractive forces. The aggregation may cause a hypsochromic shift (H for Hypsochromic effect) or a bathochromic shift (J for Jelly: The person who studied this effect). The shift in the bands can be explained on the basis of exciton theory (Kasha et al. 1965; Czikklely et al. 1970; Rösch et al. 2006). A face to face arrangement is usually observed in H-aggregates, while an edge to edge arrangement is observed in J-aggregates (Möbius 1995; Peyratout et al. 2001).

MACROCYCLIC AMPHIPHILIC STRUCTURES

Macrocyclic amphiphilic structures obtained by modification of easily available macrocyclic structures have drawn considerable interest during the last several decades due to their potential in the formation of tailor-made morphologies (Ravoo and Darcy 2000; Lee et al. 2005; Helttunen and Shahgaldian 2010; Yao et al. 2012; Jie et al. 2015). The intensive research in host-guest chemistry of macrocyclic compounds can provide knowledge of different interactions operating during the self-assembly process (Voskuhl and Ravoo 2009). The amphiphilic structures are generated from macrocyclic molecules, which can also be termed as 'surfactants' having analyte recognition sites. The host-guest interactions of macrocyclic self-assemblies can improve our understanding of various biological processes such as catalysis or transport through membranes. This chapter summarizes results obtained during past decades in the studies of self-assemblies of macrocyclic amphiphiles based on commonly encountered supramolecular macrocycles like calix[n]arenes, pillararenes, cucurbiturils, cyclodextrins and other macrocycles. Here the main focus is the molecular structure-dependent self-assembly behavior, stimuli response characteristic and their applications in different areas of daily life. Various techniques used for the characterization of self-assembled structures like Dynamic Light Scattering (DLS), Transmittance Electron Microscopy (TEM), Scanning Electron Microscopy (SEM), Atomic Force Microscopy (AFF), confocal laser scanning microscopy, etc., were also discussed. Next different types of macrocyclic amphiphiles are investigated for developing suitable morphologies will be described.

Macrocyclic Amphiphilic Structures based on Calix[n]arenes and Resorcinarenes

Calix[n]arenes are easily synthesizable and modifiable macrocycles having basket type structures (Gutsche 1998; Gutsche 2008; Chawla et al. 2010). Calix[n]arene skeleton can be modified through easy, high yield reactions at both upper and lower rims, which prompted investigators around the world to investigate their potential as a candidate to produce self-assembled architecture with varying morphologies (Helttunen and Shahgaldian 2010) discussed earlier. Regen et al. (Markowitz et al. 1989) stimulated this area by reporting the formation of vesicles based on *p-tert*-butylcalix[6]arene by injecting its solution in tetrahydrofuran into water. The DLS and TEM studies indicated the formation of self-assembled structures with size ranging from 500 nm to 1000 nm, which remained stable for a week. The successful formation of vesicles in case of *p-tert*-butylcalix[6]arene was followed by investigation on different types of calix[n]arene derivatives for self-assembling behavior. For instance, Meier et al. (Strobel et al. 2006) investigated the self-assembled structures of carboxylic acid or trimethylammonium (hydrophilic moiety) and alkyl group (hydrophobic moiety) substituted calix[4]arene (**1–3**). At low concentration (2.0 mg/mL), the calix[4]arene skeleton substituted with carboxylic acid and alkyl group displayed stable monolayers of vesicles in water, while ammonium and alkyl-substituted calix[4]arene display no aggregation behavior. At higher concentrations, all calixarene derivative displayed aggregation behavior to form rectangular lyotropic liquid crystals.

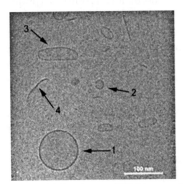

1: R_1 = COOH, R_2 = $C_{12}H_{25}$
2: R_1 = $CH_2N^+(CH_3)_3Cl^-$, R_2 = C_8H_{17}
3: R_1 = $CH_2N^+(CH_3)_3Cl^-$, R_2 = $C_{12}H_{25}$

Figure 10.2   A cryo-TEM image showing self-assembly **1** in 0.1 N NH₃(aq): (1) large size vesicle, (2) smaller size vesicle, (3) distorted vesicle, and (4) isolated lamella and rod-like micelle. Reproduced here with the permission of John Wiley and Sons (Strobel et al. 2006).

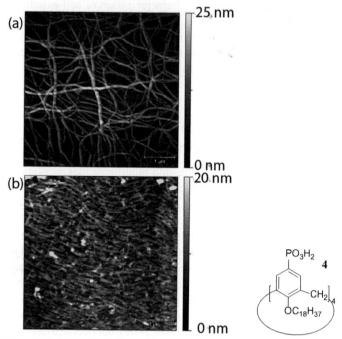

Figure 10.3   AFM of **4** showing nanofibers on mica (a), and graphite (b) Reproduced
here with the permission of RSC (Martin et al. 2011).

Raston et al. (Martin et al. 2011) deployed an upper rim substituted
phosphorylated and lower rim alkylated ($C_{18}H_{37}$) amphiphilic calix[4]arene using
toluene as a solvent to produce aggregated fibers chains on mica and graphite
surface, which were observed using Atomic Force Microscopy (AFM). The
computational calculations indicated the formation of the micellar structure of
the calix[4]arene derivative (**4**), where alkyl chains were occupying the inner
part of the composite material. The uptake of a fluorophore (7-nitro-4-(prop-2-
bynylamino benzofurazan) by nanofibers of **4** led to an increase in the diameter
of the nano-assemblies from 6 nm to 125 nm. The excess fluorophore was washed
with solvent (toluene) followed by the recording of the fluorescence spectrum,
which indicated a characteristic signal for the fluorophore incorporated into the
fiber. The dilution of a composite material solution followed by slow evaporation
regenerated the original 6 nm fibers.

Zheng et al. (Song et al. 2014) synthesized various amphiphilic calix[4]arene
derivatives (**5a-d**) by substituting the upper rim with hydrophilic carboxylic acid
groups and the lower rim with hydrophobic alkyl groups. The investigations
demonstrated that amphiphilic macrocycles can produce hydrogel by storing
their solution in an aqueous medium at room temperature for some time by the
dissolution-reassembly process. The study revealed that the hydrogel has been
produced through the conversion of unstable nanospheres into nanofibers in an
aqueous solution. This process has potential application in the preparation of
silica nanotube.

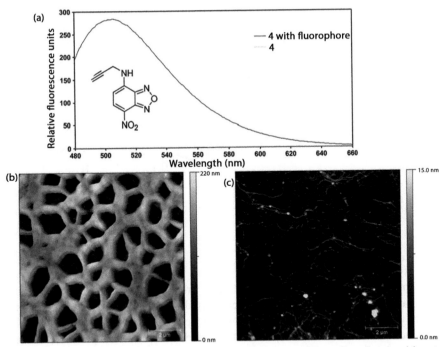

Figure 10.4 An emission spectrum of nano-fibers in toluene (a), after washing to remove excess fluorophore, (b) and after dilution that regenerated the original fibers (c). Reproduced here with the permission of RSC (Martin et al. 2011).

Similarly, Fang et al. (Yang et al. 2016b) reported three *p-tert*-butyl-calix[4] arene derivatives (**6–8**) bearing variable carboxyl groups at the lower rim and evaluated them for gel formation behavior in organoalkoxysilane. The study indicated that macrocycle possessing the highest number of carboxylic group (**6**) and can function as gelator and require a concentration of only 2%(w/v) in three different types of organosilanes out of 10 used in the study. At a concentration of 6% in PTMS, the macrocyclic receptor **6** displays thermo-reversible behavior and thixotropic property at room temperature. The gel displayed stability of up to 4 months at room temperature in a dry environment and was used successfully for film fabrication, injection molding and melting free deposition molding.

Jin et al. (Jin et al. 2005) prepared hybrid organic-inorganic materials by application of amphiphilic calix[4]arene substituted with sulfonate (upper rim) and alkyl groups (lower rim) over CdSe/ZnS quantum dots. The hybrid material has been used in the detection of the neurotransmitter acetylcholine (Ach). The trioctylphophine (TOPO) coated CDSe/ZnS water-insoluble quantum dots have been coated with amphiphilic calix[4]arene derivatives (**9a-c**). The quantum dots coated with **9a, b** displayed very weak fluorescence, while quantum dots coated with **9c** were fluorescent, stable and water-soluble. The emission of quantum dots coated with **9c** was quenched upon the addition of acetylcholine.

9a: R = H
9b: R= CH₃
9c: R = -CH₂(CH₂)₄CH₃

Prosperi et al. (Avvakumova et al. 2014) prepared an amphiphilic calix[4]arene derivative (**10**) having mannose units connected to the upper rim through a thiourea unit, which was used for coating gold nanoparticles by exploiting noncovalent interactions between calixarene backbone and dodecanthiol layer over the gold nanoparticles surface. The water-soluble nanoparticles decorated with calix[4]arene derivative displayed improved cancer cell targeting efficiency in comparison to the nanoparticles coated with non-macrocycle (**11**) coated nanoparticles.

Shahgaldian et al. (Shahgaldian et al. 2008) have used amphiphilic calix[4] arene derivative (**12**) as biomaterials and investigated their self-assembly in an air-water system. The study indicated that the cone conformer of the calix[4]arene derivative self-assemble to form stable Langmuir monolayers at the air-water interface in an orthogonal orientation regarding the interface.

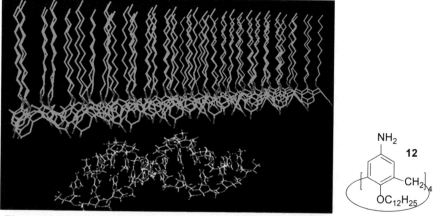

Figure 10.5 The self-assembly of **12** and its interaction with DNA. Reproduced here with the permission of ACS (Shahgaldian et al. 2008).

The addition of negatively charged DNA to the self-assembled system revealed interactions between anionic and cation surfaces, which caused the expansion of a monolayer along with phase transition to the liquid-expanded phase from the liquid-condensed phase. Besides, a decrease in the stability of the monolayers was also reported. The Photon Correlation Spectroscopy (PCS) and AFM indicated the formation of solid lipid nanoparticles of **12** having an average diameter of 190 nm (potential: +13.2 mV) with a PDI of 0.15 in water, which displayed electrostatic interaction with DNA. The study indicated the potential utility of such systems in gene therapy. Ungaro et al. (Sansone et al. 2006) have synthesized a series of upper rim guanidinium and lower rim alkyl group functionalized calix[n]arenes. The amphiphilic calix[4]arene derivatives have been used to investigate the influence of parameters such as conformation of the calix[4]arene ring, lipophilicity, and size on the self-assembly properties, interaction with DNA, condensation and cell transfection properties. Atomic force microscopy revealed that DNA binding to the cone conformer of the calix[4]arene derivative led to the formation of intramolecular DNA condensates, which have been characterized through DNA loops that emerged from the dense core. The study indicated that the addition of larger hexyl or octyl chains over the lower rim imparts cell transfection capabilities to the calix[n]arene derivatives. However, adducts having dominant inter-strand electrostatic interactions failed to promote cell transfection. Klymchenko et al. (Rodik et al. 2011) have synthesized calix[4] arene derivatives having cationic groups at the upper rim and alkyl group at lower the rim. The fluorescence correlation spectroscopy, DLS, gel electrophoresis and AFM studies indicated that **13a** formed micelles having a diameter of 3–4 nm, while **13b, c** formed micelles having a diameter of 6 nm.

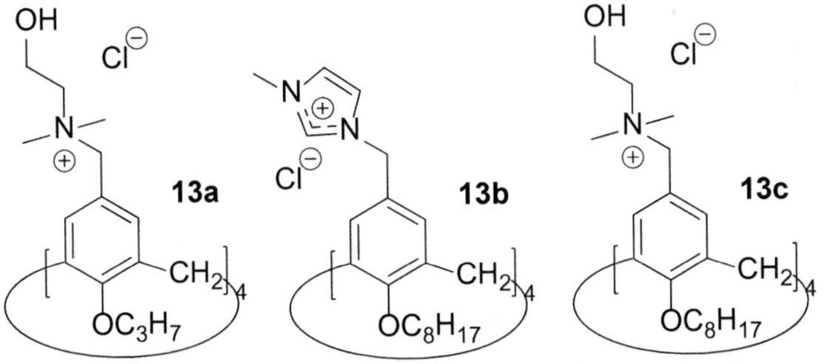

The macrocyclic calix[4]arene derivatives are an attractive candidate to induce the controlled assembly of DNA into nanoparticles owing to their unique one architecture and multiple positively charged groups. The study based on gel Electrophoresis and ethidium Bromide (EtBr) exclusion assay indicated that **13a** bind the DNA using mainly electrostatic interactions due to smaller alkyl group over it, while **13b, c** binds the DNA using both electrostatic and hydrophobic interactions in the micelles. The micelles of **13b, c** have successfully induced the condensation of DNA into nanoparticles with an average size of 65 and 55 nm, respectively. The exclusion studies of EtBr from calf thymus DNA have been taken as a model to study the extent of DNA condensation. The binding of CT-DNA to the calixarene produced a sharp decrease in the fluorescence intensity that reached a plateau at N/P ratio > 1, which confirmed the efficient binding of DNA to the calixarene derivatives. The DLS, FCS and AFM studies indicated that each **13b**/CT-DNA particle has a single CT-DNA molecular unit and nearly 700 **13b** micelles. The MTT (3-(4,5-dimethyl-2-thiazolyl)-2,5-diphenyltetrazolium bromide) assay and *in vitro* studies indicated low cytotoxicity for **13b, c**/CT-DNA nanoparticles along with transfection efficiency, which has been attributed to their high endocytosis efficiency and positive zeta potential (Rodik et al. 2011; Jie et al. 2015).

Macrocyclic Amphiphiles based on Calix[n]crowns

Calix[n]crown ethers are macrocyclic compounds produced from a combination of crown ether and calix[n]arene skeleton possessing superior properties in comparison to the parent macrocyclic compounds, which allow them to bind both anionic, cationic and neutral organic species. The formation of calix[n]crown ether requires the connection of a polyether bridge to the two hydroxyl groups of calix[4]arene skeleton, which limits the conformation mobility of the two aromatic rings that impart rigidity to the entire framework. The rigid structure of calix [n]crown ether allows the formation of stable self-assembled structures after the introduction of suitable moieties through chemical reactions to produce a balance of the hydrophobic and hydrophilic group.

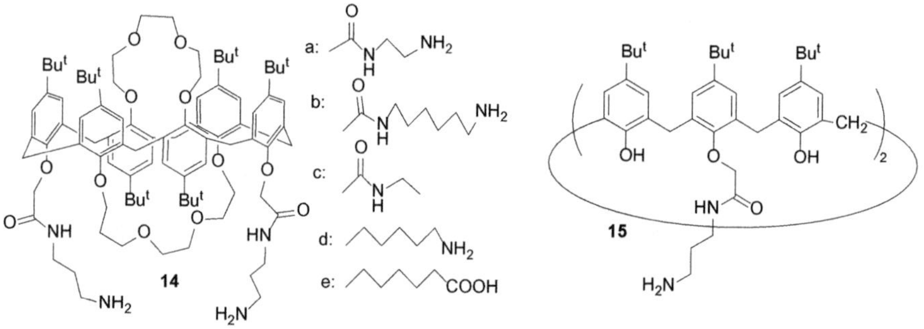

Jiang et al. (Guan et al. 2008) have designed amphiphilic calix[6]crown ether derivatives **14a** and **15** containing crown ether and the aminopropyl groups as the hydrophilic part and the tert-butyl groups as the hydrophobic groups. The self-assembly behavior of macrocycle **14a** was investigated in a water-ethanol mixture. At a ratio of 1:3 (water: ethanol), the SEM, DLS, AFM and TEM studies revealed that self-aggregation of **14a** into spherical vesicles showing an average diameter of 289 nm. At a ratio of 2:3 (water: ethanol), the SEM and TEM studies indicated the coexistence of both self-assembled vesicles and fibers, which suggest that in high polarity medium the vesicles gets converted into 1-D nanoaggregates. A further increase in the polarity of the medium, at 1:2 (water: ethanol) ratio, ultra-long fibrous aggregates with a length above 10 nm, and width in 100–200 nm were observed. No vesicles have been observed at this polarity. The TEM and

AFM studies indicated the presence of a tubular structure having a wall thickness of 20 nm. The studies indicated the formation of nanotubes on increasing the polarity of the solvent with EDX spectra indicating the presence of cesium ion within the nanotubes. To investigate the rigidity of the framework of **14a**, the results have been compared with a more flexible Structure **15**. The TEM studies indicated the formation of vesicles of **15** in water: ethanol (1:3) mixture. However, on increasing the water content the irregular and loose aggregates were observed, which suggested the utility of rigid structure in aggregation. Jiang et al. (Liang et al. 2011) further modified **14** (**14a-e**) with a variety of substituents and investigated the effect of structural factors on the formation of nanotubes and self-assembly. It was observed that the presence of an amide group allows **14b** to display a clear transition to nanotubes from a vesicular structure on increasing the water content. The macrocycle **14c** also displayed similar properties. For **14d, e**, a decrease in the size of the vesicles to 150 nm diameter, with no other changes on increasing the content of water was observed. Besides, Jiang et al. (Guan et al. 2009) also demonstrated the synthesis of metal-organic hybrid nanotubes under mild conditions using calix[6]crown ether as the template.

Macrocyclic Amphiphiles based on Resorcinarenes

Resorcinarenes are macrocyclic receptors obtained through the condensation resorcinol and aldehydes. The molecular skeleton of resorcinarene allows for their conversion into an amphiphilic structure through a simple chemical reaction. Interesting examples of amphiphilic macrocycles (**16–17**) for self-assembly were synthesized by Tanaka et al. (Tanaka et al. 1999). The intermolecular and intramolecular H-bonds between the OH group played a vital role in stabilizing the assembled vesicles. A particle size distribution between 30 to 240 nm was confirmed by the DLS measurement, where the population of the 53 nm particles was maximum.

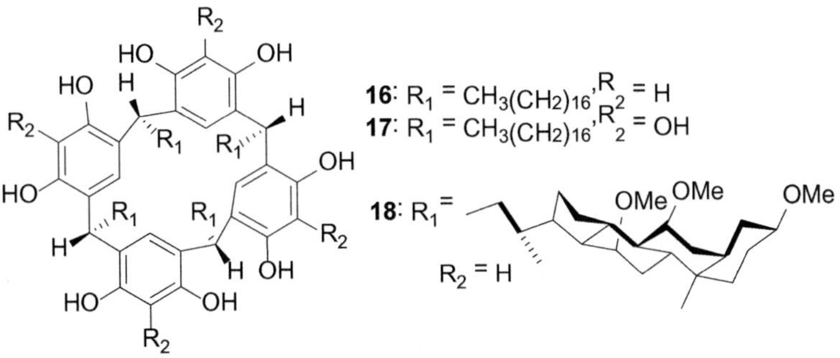

Kobuke et al. (Yoshino et al. 2001) reported another amphiphilic resorcinarenes substituted with cholic acid (**18**). The hydrophilic head and hydrophobic tail (cholic acid) containing resorcinarene assembled using tail to tail interactions to form a bilayer, which formed a transmembrane channel suitable to transport potassium ions. The macrocycle **18** can use the steroidal moiety to align itself to lipid bilayer to allow advantages of long-lasting ion channel in lipid bilayers.

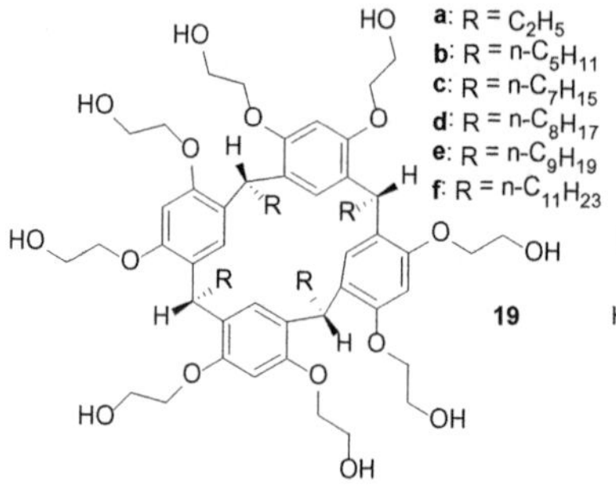

Zakharova et al. (Pashirova et al. 2014) have synthesized a series of amphiphilic resorcin[4]arenes with variable hydrophobic alkyl chain. The study observed that the surface properties and self-aggregation behavior were dependent on the structure and the nature of the solvent system. For instance, self-assembly was observed in a mixture of water and DMSO or DMF and not in the water-THF mixture. Aggregates of 300–400 nm have been observed in DMSO-water and DMF-water solvent system, which were formed through an open shell model. With an increase in hydrophobicity small micelle-like aggregates having 10–20 nm size were observed. The occurrence of the hydrophobic interior in the micellar structure were confirmed through the solubilization of dyes and drugs, which suggested their potential utility as a carrier of drug molecules. Zakharova et al. (Zakharova et al. 2009) further investigated the aggregation and potential catalytic application of amphiphilic resorcin[4]arene (**20**) and polyethyleneimine (PEI). Macrocycle **20** did not display any catalytic activity during the hydrolysis of *p*-nitrophenyl dimethyl phosphate. However, a tenfold improvement in catalytic effect was witnessed in a binary PEI-**20** system at a concentration below critical micelle concentration (cmc). The inclusive interactions within the cavity of macrocycle **20** are responsible for PEI-**20** self-assembly below the cmc value, which accelerated the rate of reaction. Shen et al. (Sun et al. 2008) reported the utility of amphiphilic resorcinarenes in the preparation of nanomaterials such as multiwalled microtubes in water. This study used various techniques to investigate the self-assemblies of **21**–**22**. Microtubules having a thickness of 300 nm, a diameter of several millimeters and lengths of several centimeters were successfully prepared in water.

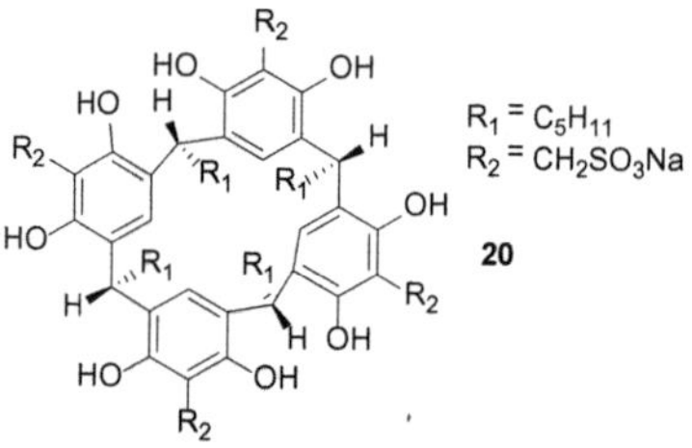

Shen et al. (Sun et al. 2010) decorated the microtubes of resorcinarenes (**21b**) with different nanoparticles like gold, silver, platinum and palladium nanoparticles. It was reported that the nanoparticles can be encapsulated into the microtubules during the resorcinarene self-assembling process. The process yielded unique organic-inorganic hybrid materials with narrow size distribution as observed in the case of gold nanoparticles described in the study as a model system. The process can be exploited for the separation of nanoparticles.

Resorcin[n]arenes have also been converted into amphiphilic biomaterials and evaluated for self-assembly (Aoyama et al. 2003; Hayashida et al. 2003; Nakai et al. 2003; Osaki et al. 2004). Ayoma et al. (Hayashida et al. 2003) reported amphiphilic resorcin[4]arenes (**23**) substituted with a hydrophobic alkyl group (undecyl) and hydrophilic oligosaccharides.

21a: $R_1 = $ n-$C_{11}H_{23}$, $R_2 = CH_2CH_2NH_2$
21b: $R_1 = $ n-C_5H_{11}, $R_2 = CH_2CH_2NH_2$
22: $R_1 = $ n-$C_{11}H_{23}$, $R_2 = NH_2$

23 $R_1 = (CH_2)_{10}CH_3$

$n = 2-7$

The studies based on DLS, Gel Permeation Chromatography (GPC) and TEM indicated micelles type nanoparticles having an average diameter of 3 nm in water. Agglutination of nanoparticles has been observed in the presence of Na_2HPO_4/NaH_2PO_4 buffer system (0.1 M, pH = 7.0). The oligosaccharide moieties in the macrocycle **23** act as a tab, while phosphate ions act as a glue, to increase the size of the nanoparticles to 60–100 nm. The agglutination process can be observed by using Surface Plasmon Resonance (SPR). Macrocycle **23n** displayed closely packed monolayers on the surface of the hydrophilized sensor chip, where oligosaccharide moieties were oriented towards the water.

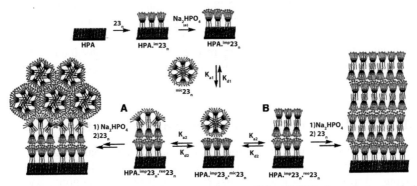

Figure 10.6 Immobilization of **23n** on a Sensor Chip of SPR through hydrophobic interactions followed by multilayer formation induced by the phosphate ions. Reproduced here with the permission of the ACS (Hayashida et al. 2003).

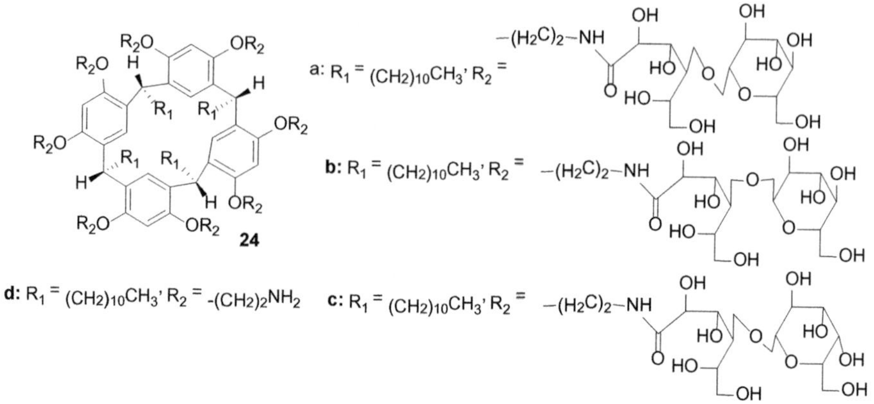

a: $R_1 = (CH_2)_{10}CH_3$, $R_2 =$

b: $R_1 = (CH_2)_{10}CH_3$, $R_2 = -(H_2C)_2-NH$

c: $R_1 = (CH_2)_{10}CH_3$, $R_2 = -(H_2C)_2-NH$

d: $R_1 = (CH_2)_{10}CH_3$, $R_2 = -(CH_2)_2NH_2$

Resorcinarenes having different types of sugar moieties as substituents displayed micellar aggregates termed as Glycocluster Nanoparticles (GNPs) in water. The stability of the glycocluster nanoparticles was attributed to lateral side-by-side H-bond interactions between the sugar moieties in water (Aoyama et al. 2003, Nakai et al. 2003). The complex formation between GNP and phosphate indicated the potential application of the resorcin[4]arenes based systems as a new type of DNA binder, especially gene carriers or vectors. Aoyama et al. (Nakai et al. 2003) have also reported examples of artificial glycoviral vectors created through the number and size-controlled gene (pCMVluc, 7040 bp) coatings using micellar GNPs (**24**). For example, in the case of GNPs obtained from self-aggregation of resorcin[4]arene derivatives (**24a**), the DLS and TEM studies indicated a compactly packed glycoviruses having a diameter of ~50 nm and a zeta potential of $\zeta \cong 0$ mV, which displayed sugar moiety dependent self-aggregation (α-Glc > β-Gal $\gg$ β-Glc) and a pinocytic form of endocytosis triggered transfect cell (HepG2 and Hela) cultures. Other resorcin[4]arene derivatives also delivered similar results with pCMVluc to produce glycoviruses having lesser aggregation. Aoyama et al. (Osaki et al. 2004) further used amphiphilic resorcinarene-saccharides macrocycles to coat the surface of lipophilic ^{TOPO}QDs, which produced novel QD-conjugated sugar balls having a diameter of 15 nm. The sugar balls have marked endosomes much more and much less efficiently in comparison to the micellar homo-aggregates of the amphiphiles having 5 nm diameter and virus-like GNP-DNA conjugates having 50 nm diameter, respectively. The results indicated that the control of the virus size is extremely important during the design of an artificial molecular delivery system.

Macrocyclic Amphiphiles based on Cyclodextrins

Cyclodextrins (CDs) are also macrocyclic oligosaccharides consisting of D-glucose (six: α, seven: β or eight: γ) units connected through α-1,4-glucosidic linkages having a toroidal shape (Szejtli 1998; Haider and Pikramenou 2005; Wang et al. 2009). The cyclodextrins have a highly polar exterior, where primary hydroxyl groups are located on the narrow side, while the secondary hydroxyl groups occupy the wider side. The inner cavities of cyclodextrins are non-polar, which makes them suitable for acting as a host system and constructing amphiphilic systems for self-aggregation (Li et al. 2007; Hu et al. 2014; Mura 2014). The presence of a non-polar and polar region in the molecular architecture of cyclodextrin led to reports of successful formation of monolayers on the air-water interface (Parrot-Lopez et al. 1992), micellar structures in water (Auzely-Velty et al. 2000) and nanoparticles for potential application in pharmaceutical chemistry (Gulik et al. 1998). Keeping in view the formation of self-assembled structures of cyclodextrins, Darcy et al. (Ravoo and Darcy 2000) prepared non-ionic derivatives of β-CD (**25a, b**) containing sulfur-alkyl chains on the secondary face and oligo(ethylene glycol) on the primary face. The macrocycles **25a, b** displayed bilayer vesicles in water having an average diameter of 170 nm.

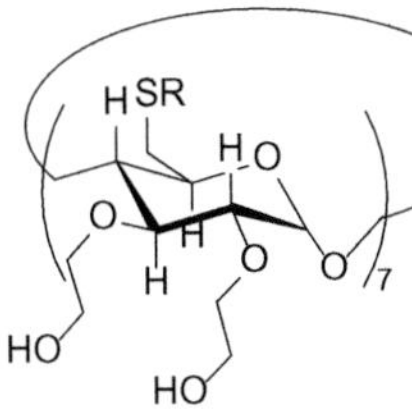

25a: R = $C_{12}H_{25}$
25b: R = $C_{16}H_{33}$

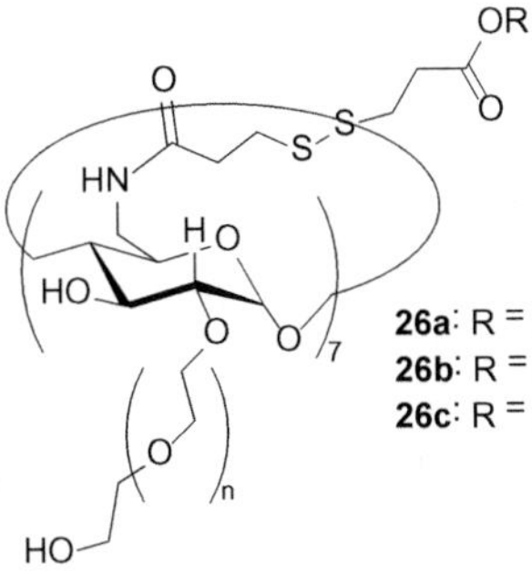

26a: R = C_8H_{17}
26b: R = $C_{12}H_{25}$
26c: R = $C_{16}H_{33}$

The successful formation of self-assembled structures with macrocycles **25a, b** led Darcy et al. to synthesize a series of substituted β-cyclodextrins. The cyclodextrin derivative containing a disulfide bridge was adjusted for hydrophobic/hydrophilic balance by the introduction of alkyl group chain on disulfide ester at the primary hydroxyl group side and oligo(ethylene glycol) moiety at the site of the secondary hydroxyl group. The vesicles formed by cyclodextrin derivatives (**26a, b, c**) in water can be disrupted through the addition of reducing agent dithiothreitol (DTT). The breaking of the sulfur bridge through reduction led to the release of model hydrophobic guest due to the destruction of nanoparticles.

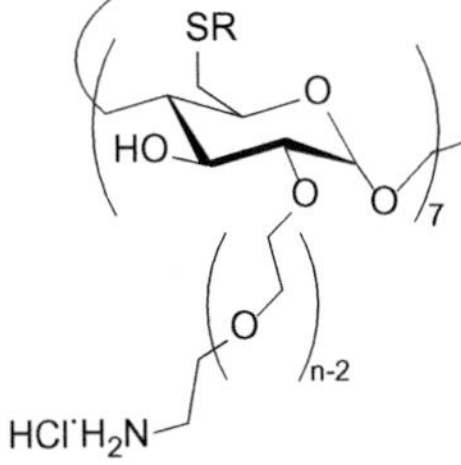

27a: R = C_6S_{13}
27b: R = $C_{16}H_{33}$

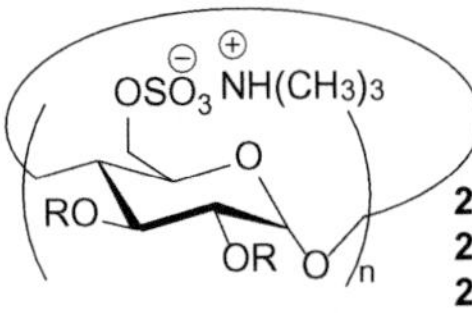

28a: n = 6
28b: n = 7
28c: n = 8

R = $CO(CH_2)_{14}CH_3$

Darcy et al. (Donohue et al. 2002) reported amphiphilic cationic β-CD derivatives (**27a, b**) having hydrophobic *n*-alkylthio chains on the primary hydroxyl side, while hydrophilic o-amino-oligo(ethylene glycol) moieties were placed on the secondary hydroxyl side. The bilayer vesicles possessing a diameter of 30–35 nm (**25b**) and nanoparticles having a diameter of ca. 120 nm (**27a**) were observed in water. The anionic cyclodextrins have been reported by Nishimura et al. (Sukegawa et al. 2002) by substituting cyclodextrins (α-, β- and γ-) with sulfonate groups (28a-c). The amphiphilic anionic cyclodextrins (28a-c) self-assembled to form monolayers at the air-water interface to display erythrocyte like liposomes.

To further investigate the self-assembling process in cyclodextrins, Darcy et al. (Falvey et al. 2005) designed a variety of non-ionic cyclodextrin derivatives having different cavity sizes (α-, β- and γ-) substituted with n-dodecyl or n-hexadecyl and oligo(ethylene glycol) moieties with a capability to bind guests (adamantyl and tert-butylbenzoyl derivatives) in water. The cyclodextrin derivatives displayed self-assembled host to form bilayer vesicles, which functioned as membranes to bind hydrophobic guests like the surface of cell membranes. The self-assembled systems displayed the potential to develop advanced carriers for biological applications. Ravoo et al. (Ravoo et al. 2003; Voskuhl et al. 2010b) also demonstrated enhanced host-guest interactions between the surface of the bilayer vesicles constructed from cyclodextrin derivative (31) and polymers. Similar results were observed between the surface of the bilayer vesicles obtained from other amphiphilic cyclodextrins and the molecular guests (Versluis et al. 2009; Voskuhl et al. 2010a, b; Vico et al. 2011). The interactions between the surface of the bilayer vesicles and the guest take place without any disruption to the structure of the vesicles and supported by orthogonal non-covalent interactions like hydrogen bonds, metal coordination and protein-carbohydrate interactions (Versluis et al. 2009; Voskuhl et al. 20100a, b; Vico et al. 2011). The orthogonal non-covalent interactions at the surface of the bilayer vesicles obtained from cyclodextrin derivatives and the guest produced vesicles, receptors clusters and vesicle-nanotube or nanofiber transitions (Lim et al. 2005). Ravoo et al. (Voskuhl et al. 2010b) further exploited the unilamellar vesicles of 31 decorated with maltose and lactose using host-guest interaction to demonstrate hierarchical self-assembly to form artificial glycocalix. To achieve the desired objective, two bi-functional conjugates (32–33) have been designed and synthesized through the coupling of adamantyl moiety to the lectins (concanavalin A (ConA)) (32) and peanut agglutinin (PNA) (33). Three carbohydrates 31–33 (β-cyclodextrin, maltose and lactose, respectively) were involved in highly specific orthogonal interactions. The macrocycle 31 acts as a receptor for adamantane, maltose behaves as a ligand for ConA, while lactose acts as a ligand for PNA. The addition of lectins was observed to induce the aggregation of vesicles, which involved no disruption of bilayer vesicles. However, it was observed that aggregation requires high carbohydrate density at the surface of the vesicles. The aggregation can be reversed on addition of competing binders, which can be either for binding **31** or lectins through non-covalent interactions.

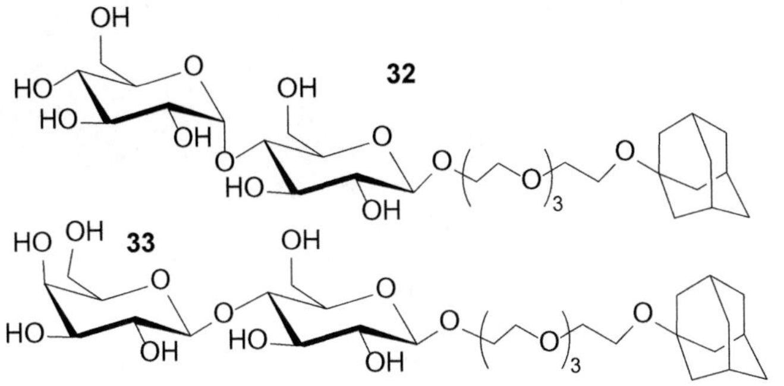

The studies have the potential to improve the understanding of molecular recognition at the cell surfaces and develop strategies for achieving dynamic self-assemblies using soft materials.

Self-assemblies of Macrocyclic Receptors based on Cucurbit[n]urils (CB[n])

Cucurbit[n]urils are an interesting family of macrocyclic receptors consisting of *n*-glycoluril units and a hydrophobic cavity to bind the non-polar guest (Isaacs et al. 2005; Kim et al. 2007; Ni et al. 2013; Assaf and Nau 2015). However, limited research has been published with amphiphilic macrocycles based on the modification of CB[n]. For instance, Kim et al. (Jon et al. 2003) reported a direct functionalization route, which allowed the synthesis of amphiphilic CB[n] derivatives. Following this route, Kim et al. (Lee et al. 2005) successfully synthesized an amphiphilic CB[6] derivative (**34**).

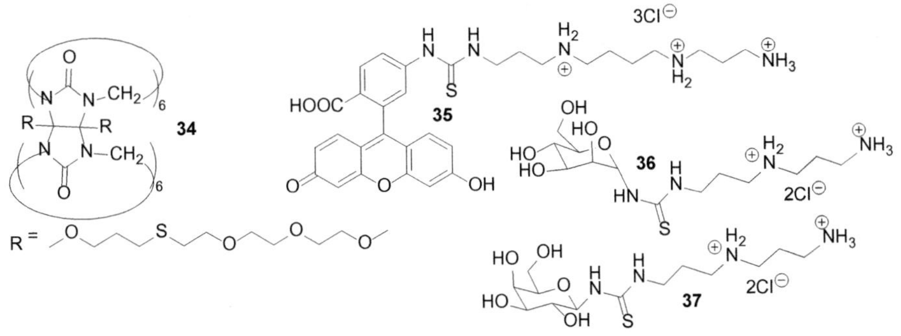

The TEM studies indicated the formation of self-assembled structures of **34** in the form of spherical bilayer vesicles of 30–1000 nm diameter. The high-resolution TEM images indicated the formation of hollow spheres of 170 ± 50 nm diameter and 6 ± 1 nm thickness (Fig. 10.7a).

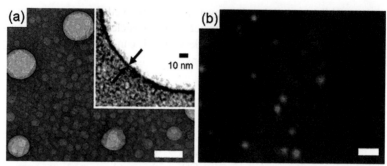

Figure 10.7 (a) High-resolution TEM image showing the formation of the vesicle of macrocycle **34** (0.4 mM) obtained by extrusion through a 200 nm pores membrane (scale bar = 200 nm). The small grains visible in the background are due to the staining agent uranyl acetate. The membrane thickness is shown in the inset with arrows (scale bar = 10 nm). (b) Confocal microscope image of the vesicles of the macrocycle **34** (0.4 mM), the surface was decorated with **35** (scale bar = 2 μm). Reproduced here with the permission of ACS (Lee et al. 2005).

Light scattering studies with a sample of **34** having monodisperse vesicles revealed that the radius of gyration (R_g = 54.7 nm) and hydrodynamic radius (R_h = 50.9 nm) were almost similar showing $\rho(R_g/R_h)$ = 1.06, which suggested the formation of vesicles (Chécot et al. 2002). The formation of a stable host-guest complex between CB[6] and polyamines (K > 10^6 M^{-1}) was exploited to modify the surface of the vesicles through non-covalent interactions between the accessible CB[6] surface in the vesicles and the polyamine derivatives (**35–37**). The binding of **34** with (fluorescein isothiocyanate)-spermine conjugate was exploited for the surface modification of vesicles with a green fluorescence tag, which was purified by size exclusion chromatography to obtain surface-modified vesicles. Similarly, the same method was used to decorate the vesicles with thiourea linked mannose-spermidine conjugate (**36**), which displayed specific interaction towards concanavalin A (ConA) by forming aggregated structures, a lectin that displays specificity towards mannose. Similarly, aggregation of vesicles was not observed either with **34** or vesicles decorated with galactose-spermidine conjugate (**37**) in presence of ConA. Kim et al. (Park et al. 2010) have synthesized a variety of amphiphilic reduction sensitive and biocompatible systems based on cucurbit[6]urils (**38–41**) having disulfide linkages to investigate them for the self-assembly process.

The TEM studies indicated the formation of vesicles of **38** possessing 170±30 nm diameter with a membrane thickness of 6±1 nm. DLS studies indicated almost similar particle size (190±20 nm) with a narrow distribution, which supported the TEM observation. The presence of the S–S bond in **39** enabled the incorporation of reducing agent sensitivity in the self-assembled vesicular structures.

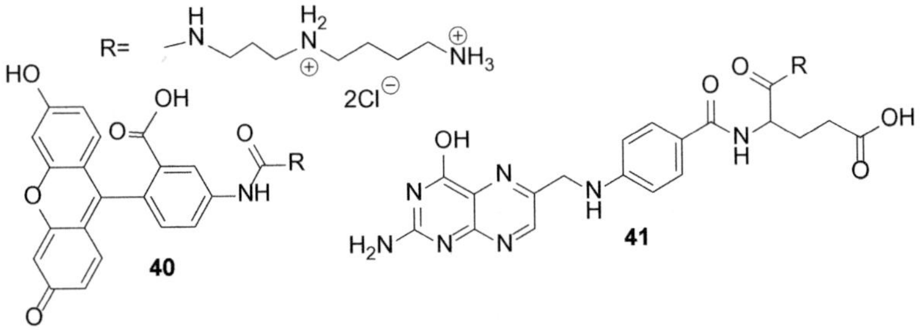

The vesicles have also been decorated with targeting ligand **40** and imaging probe **41** utilizing nondestructive non-covalent interactions along with an ability of vesicles to bind polyamines. The vesicular structures displayed excellent stability in the extracellular conditions. However, the self-assembled vesicular structures ruptured under intracellular conditions within 12 hours, which opened the potential utility of the vesicular structure in drug delivery. The potential utility of the vesicles in the targeted drug delivery was demonstrated using the anti-cancer drug doxorubicin (DOX). The vesicular structures decorated with **40** or **41** and loaded with vesicles displayed increased cytotoxicity owing to selective and efficient uptake and release of DOX through receptor-mediated endocytosis followed by reduction-induced release of a drug molecule into the cytoplasm. Besides, a variety of targeting ligands that consists of small molecules, antibodies, imaging probes, magnetic particles and fluorescent dyes, etc., have been introduced over the surface of the vesicle of the CBs. This type of vesicular structure has the potential to load a variety of drug molecules for targeted drug delivery.

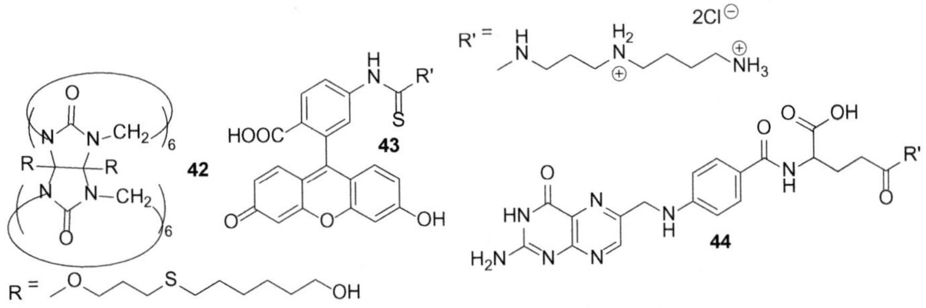

Kim et al. (Park et al. 2009) have synthesized a CB[6] derivative (**42**), which formed nanoparticles showing an average diameter of 190±50 nm. The nanoparticles were exploited for targeted drug delivery after decorating them with fluorescence probe (**43**) and targeting ligand (**44**) to trace their location after endocytosis. Significant cytotoxicity of paclitaxel was demonstrated using the nanoparticles of **42** decorated with **43** and **44**. To exploit the significance of polymeric nanostructures in drug delivery applications, Kim et al. (Kim et al. 2007) have also used the thiol-ene photopolymerization between CB[6] derivative (**45**) and dithiols (Scheme 1), which yielded polymeric nano-capsules.

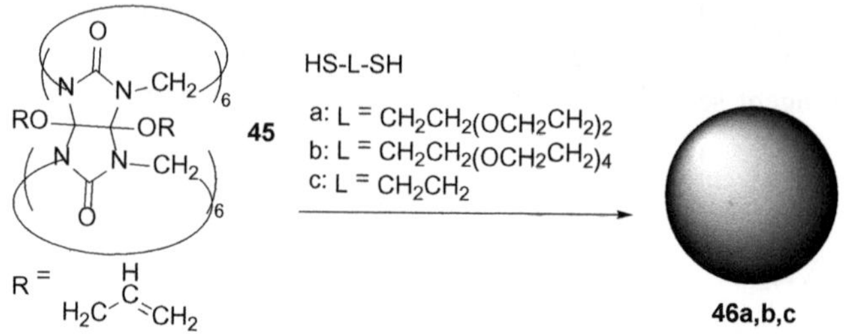

Scheme 1   Synthetic scheme for the preparation of nano-capsule of **45** (Kim et al. 2007).

The method can be used for the preparation of nano-capsules of a variety of monomers having a flat core and multiple polymerizable functional groups at the periphery. The DLS studies indicated an average diameter of 110±30 nm in the case of **45a**, which was found to agree with the results obtained with SEM and AFM studies. Bigger nanoparticles having an average diameter of 150±50 nm were observed in the case of **45b** owing to the larger dithiol group, while **45c** yielded smaller nanoparticles having an average diameter of 50±10 nm. The studies based on high resolution and cryo-TEM indicated a hollow interior within a thin shell having an average thickness of 2.1±0.3 nm. The inner shell of the polymer nanocapsules can be exploited to capture the guest molecules, while the surface of the nanoparticles exploited functional groups to capture the guest molecules through non-covalent interactions. The utility of nano-capsules in the potential drug delivery application and imaging was suggested by this study.

To further probe the applications of polymeric nano-capsules, Kim et al. (Kim et al. 2010) reported a combined theoretical and experimental study along with the mechanism of their formation. The experimental study demonstrated the effect of monomer concentration, reaction temperature and the medium. The study indicated that reactions such as amide formation can also be used to synthesize polymeric nano-capsules. The study suggested that the polymeric nano-capsules approach can not only be used to deliver drugs to the targeted cells, but also release them inside in a controlled manner. Kim et al. (Kim et al. 2010) further exploited the advantages of vesicular structures derived from CB[6]s and nanocapsules to develop a facile drug delivery vehicle for targeted drug delivery and controlled release. Kim et al. (Yun et al. 2014) recently developed a convenient method to prepare highly stable metal nanoparticles-decorated nanocapsules (M = Pd, Au, and Pt), which can be dispersed easily in water and display a narrow size distribution. The hollow interior of the nanocapsule obtained from CB[6] derivatives was exploited to introduce a variety of nanoparticles over its surface. The metal nanoparticle decorated surface prevented the self-aggregation of polymer nanocapsule and provided high stability along with water dispensability. The presence of CB[6] units along with S–S loops on the surface of the CB-PN helped in the successful conclusion of this approach. The PD@CB-PN displayed outstanding heterogeneous catalytic properties in C–C and C–N bond formation reaction with and an excelled reusability and recyclability. Apart from this, the metal nanoparticles on polymeric nano-capsules were suggested for potential use in other applications like imaging and nanomedicine.

Self-assembly of Macrocyclic Receptors based on Pillararenes

In recent years, the self-assembly of macrocycles based on pillar[*n*]arenes have drawn the attention of researchers around the world, which contains pillar[*n*]arenes electron-rich and hydrophobic sites (Ogoshi et al. 2008; Cao et al. 2009; Si et al. 2011; Xue et al. 2012; Ogoshi et al. 2016). The aromatic repeat units in pillar[*n*] arenes are connected through methylene bridges that impart a uniquely rigid architecture to pillar[*n*]arenes, which can be modified by well-known chemical reactions to produce amphiphilic derivatives for application in self-assembly, sensing, supramolecular polymers and so on (Ma et al. 2011; Zhang et al. 2011; Strutt et al. 2014; Kakuta et al. 2018). Owing to the easy availability of pillar[5] arene, this was used on most occasions to construct amphiphilic macrocycles for the construction of self-assembled structures.

Huang et al. (Yao et al. 2012) have synthesized an amphiphilic pillar[5]arene (**46**) possessing five hydrophilic amino groups and five hydrophobic alkyl groups and investigated it for self-assembling properties. The study indicated a Critical Micelle Concentration (CMC) concentration at 1.5×10^{-4} M in water for **47**, while DLS, TEM and SEM indicated the formation of self-assembled vesicles having an average radius of ~200 nm (Fig. 10.8).

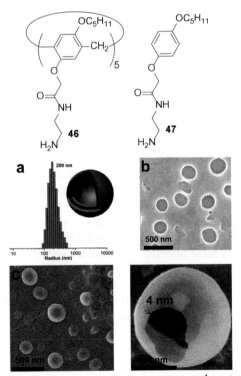

Figure 10.8 (a) The DLS study of a solution of **46** (2.0×10^{-4} M) in water at a scattering angle of 90° showing the vesicular particles formed in water. (b) SEM and (c) TEM images of self-assemblies of **46** (2.0×10^{-4} M) in water. (d) ruptured vesicles visible in the SEM image of **46**. Reproduced here with the permission of ACS (Yao et al. 2012).

The reduction in the pH resulted in the transformation of vesicular structures into micellar structures. The ruptured vesicles indicated bilayer walls with a wall thickness of ~4 nm. A decrease in the particle size from 200 nm at pH 7.0 to 10 nm at pH 3.0 was observed. To exploit the self-assembling properties of **46**, this was investigated for uptake and release of the hydrophilic guest (calcein). The calcein loaded vesicles in response to a change in pH to 4.0 exhibited a rapid and complete release of calcein accompanied by an enhancement in fluorescence emission intensity. The release of calcein was attributed to the pH-induced transformation of vesicles into micellar structures. Besides, the solution of **46** displayed floccules after 2 weeks with an increase in size and darkening with the time of incubation. No floccules were observed under similar conditions with non-cyclic structure **47**. The TEM, SEM and AFM studies with **46** indicated that the floccules were microfibers having an average diameter of ~1.2 μm (inner diameter 800 nm). The study further investigated the possible mechanism of transformation of vesicles into microtubes using TEM. The transformation of bilayer vesicles into necklace-like structures occurs, when they are in close contact with each other. The bilayer vesicles fused to form large vesicles having a sheet-like appearance that formed through the rupture of large vesicles. The necklace-like structure associates together to produce the nanotubes, while the sheet-like structural form produces microtubes through the roll-up due to H-bond interactions. The presence of strong electron-donating primary amine groups on the inner and outer walls of the microtubes led the authors to investigate the absorption of electron-deficient 2,4,6-trinitrotoluene (TNT), which is known to be an explosive substance having a hazardous effect on the environment and human beings. As expected, the TNT was absorbed easily by microtubes as indicated by a decrease in the absorbance value of the free TNT. Huang et al. (Yao et al. 2013) have used **46** to improve the stability of gold nanoparticles in water through the reduction of $HAuCl_4$ ascorbic acid in the presence of **46**. At a 0.1 [**46**]/[$HAuCl_4$] ratio, metastable nanoparticles, which precipitated in water and having a solubility in chloroform were obtained. The formation of chloroform soluble nanoparticles indicated the presence of a monolayer of **46** at the surface of the gold nanoparticles having an outward arrangement of hydrophobic alkyl group towards the chloroform solvent. However, an increase in the [**46**]/$HAuCl_4$] ratio a bilayer arrangement of molecules of **46** was observed as confirmed by thermogravimetric analysis (TGA), where the hydrophilic moieties were oriented outwards towards the water. The use of non-cyclic **47** produced metastable and chloroform soluble nanoparticles. Owing to the presence of both hydrophilic and hydrophobic moieties over the gold nanoparticles surface, they were amphiphilic. Further, the **46** stabilized nanoparticles were loaded over the exterior surface of the self-assembled microtubes of **46** at room temperature to obtained composite microtubes. The amphiphilic gold nanoparticles were also aggregated into floccules in water having a soft filament shape and wine color, 2 months after preparation of its aqueous solution that too without any external stimulus. The TEM studies established the formation of an average diameter of 600 nm and 60 nm thickness for the Self-assembled Composite Microtubes (SCMTs). The SCMTs displayed stability at a high temperature, in the presence of a strong acid, strong base and displayed excellent green catalytic activity towards the borohydride catalyzed reduction of *p*–nitroaniline to 1,4-diaminobenzene. A yield loss of less than 3% during 20 cycles was observed, which indicated excellent recyclability of the catalyst for this reaction. Diao et al. (Zhou et al. 2013) decorated the surface of Reduced Graphene Oxide (RGO) with macrocycle **46** through amide bond between the amino group of **46** and carboxylic acid group on the surface of the Graphene Oxide (GO), which was further strengthened by $\pi\cdots\pi$ interactions between the aromatic benzene rings of **46** and graphene oxide surface. The graphene oxide-**46** composite material was further transformed into water-dispersive RGO-**46** nanocomposites after reduction with hydrazine. The TGA analysis indicated the formation of RGO-**46**-AuNP nanocomposites self-assembled through the stabilization of AuNPs by self-assembly of **47** over gold nanoparticle surface due to the formation of amide bonds by showing the presence of a large amount of **46**. The RGO-**46** also displayed the electrochemical sensing properties towards guest molecules having a size similar to the cavity of the macrocycle **46**. The results further demonstrated that the ternary nanocomposite RGO-**46**-AuNPs incorporate the advantages and function in a synergetic fashion combining the properties of **46**, RGO and AuNPs. The RGO-**46**-AuNPs was demonstrated to show an excellent detection limit $(1.2 \times 10^{-8}$ M) towards dopamine in a broad concentration range $(1.5 \times 10^{-8}$ to 1.9×10^{-5} M). Xue et al. (Yao et al. 2014b) synthesized a series of amphiphilic pillar[5]arene derivatives (**46**, **48–50**) and investigated them for self-assembly in water using DLS, SEM and TEM techniques.

The DLS studies confirmed the formation of aggregates having diameters between 150–200 nm for all pillar[5]arene derivatives (**46**, **48–50**), which exceeded the corresponding pillar[5]arene derivative's molecular lengths. The results suggested the formation of vesicular entities rather than the formation of simple micellar structures. The SEM studies suggested the formation of spherical aggregates having a diameter of ~200 nm, which confirmed the DLS results, while TEM images revealed the presence of hollow spheres, which corresponds to the length of two amphiphilic pillar[5]arenes. The TEM results also indicated a thickness of 5 nm for the vesicles having bilayer walls. The DLS and TEM studies further indicated the conversion of the vesicular structure of **46** and **48** into a micellar structure having a diameter of ~15 nm on bubbling carbon dioxide gas, which reverted to the vesicular structure after bubbling nitrogen gas.

The addition of silver ions to **49** was demonstrated to produce self-assembled dendritic structures of **49**-silver complex.

Sakurai et al. (Nishimura et al. 2013) have reported the synthesis of an amphiphilic pillar[5]arene (**51**) bearing polar lysine moiety (hydrophilic part) and an alkyl chain as a hydrophobic moiety.

The macrocyclic receptor **51** displayed spontaneous self-assembly in water. The studies based on Small-Angle X-ray Scattering (SAXS), field flow fluctuation coupled with multi-angle light scattering (FFF-MALS), TEM and AFM revealed stable bimolecular micellar structures. AFM studies suggested the formation of a spherical structure having uniform heights in the range of 4.2–4.5 nm, which were also supported by TEM measurements. The DLS studies confirmed the size of the self-assembled structure by showing a mean diameter of 4.3±1.1 nm, which was in excellent agreement with the TEM and AFM studies. Huang et al. (Yu et al. 2013) have synthesized a galactose (hydrophilic part) and alkyl (hydrophobic part) substituted pillar[5]arene derivative (**52**). The macrocycle **52** self-assembled in water to produce nanotubes as confirmed by TEM, SEM and fluorescence techniques. The macrocycle display a self-assembly behavior similar to that observed in the case of pillar[5]arene derivative (**46**). Similar to **46**, the macrocycle **52** display a water solubility above CMC owing to the formation of bilayer vesicles having an average diameter of ~170 nm initially, which further aggregated into nanotubes having 100–200 nm diameter and several micrometer lengths after an incubation period of 1 week. The nanotubes obtained coated with carbohydrates on the surface by using macrocycle **52** were successfully demonstrated to act as a cell glue for agglutination of *E. coli* through carbohydrate-protein interactions (Fig. 10.9).

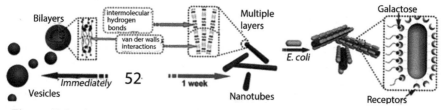

Figure 10.9 A representation of self-assembly of **52** in water and application in cell agglutination. Reproduced here with the permission of ACS (Yu et al. 2013).

Zhao et al. (Zhang et al. 2013) exploited the amphiphilic pillar[5]arene derivatives self-assembled structures as dye carriers for bio-imaging.

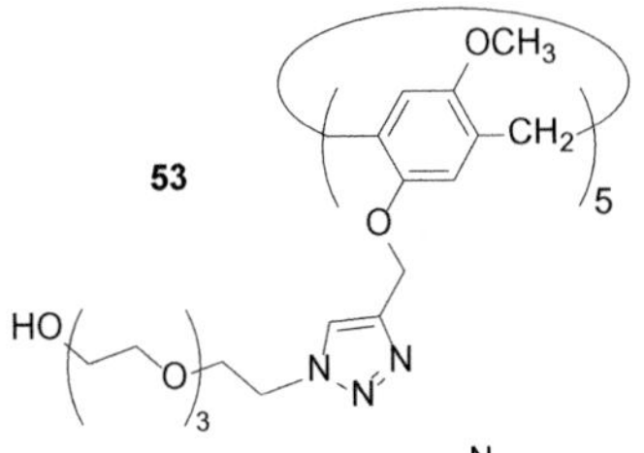

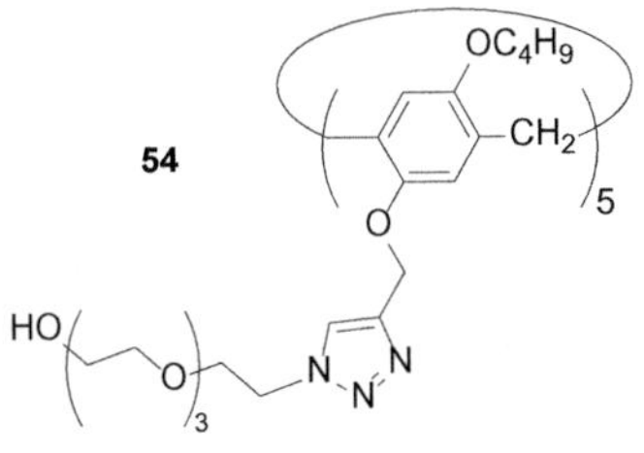

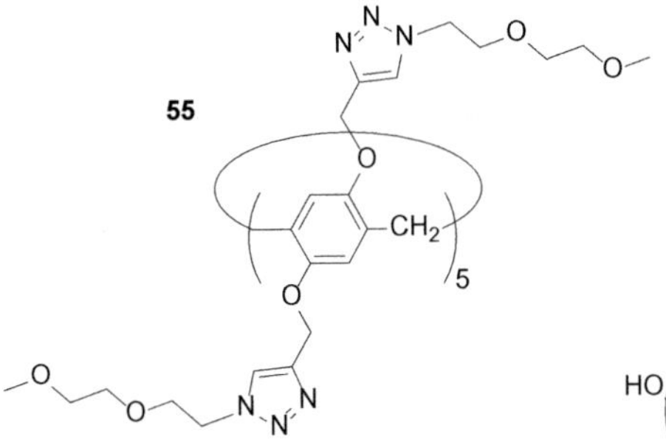

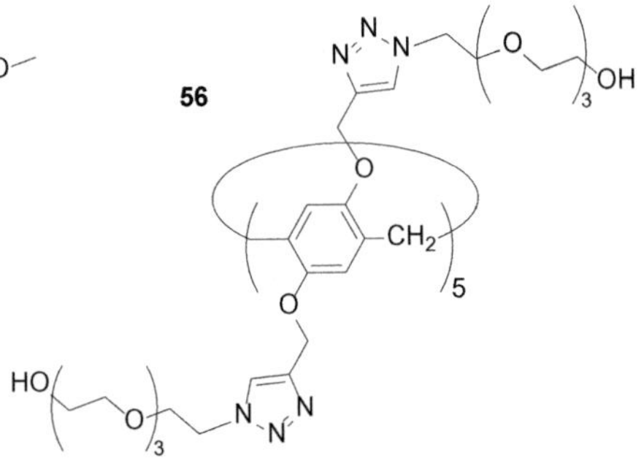

The self-assembly behavior of **53–56** was monitored using DLS and TEM measurements (Zhang et al. 2013). The tadpole-like pillar[5]arene derivatives **53** and **54** formed bilayer membranes above Critical Assembly Concentration (CAC), where alkyl groups were located inside the bilayers, while ethylene glycol moieties faced the inner and outer solution (water). The pillar[5]arene derivatives (**55, 56**) formed bola-like amphiphiles having hydrophilic ethylene glycol moieties at both ends. The amphiphilic structures of **55** and **56** aggregated above CAC into lamellar structures having water molecules in the inner part of the cores. The aggregated structures on investigation for application as delivery vehicles for hydrophilic and hydrophobic dyes revealed that performances were dependent on their structural form (Zhang et al. 2013). The delivery of **57** using pillar[5]arene (**53–56**) was monitored using ^{1}H-, 2D-NOESY (nuclear overhauser effect spectroscopy) NMR, UV-visible, MTT assay and confocal laser scanning microscopy (CLSM), which revealed no effect on the stability of assembly vesicular morphologies with **57** during bio-imaging (Fig. 10.10). The pillar[5] arene derivatives (**53–56**) delivered hydrophilic rhodamine B (RhB) to the cells. The hydrophobic fluorescein isothiocyanate (FITC) was delivered better for bio-imaging by tadpole-like amphiphilic pillar[5]arene derivatives. The studies were further conducted using mixed dyes, which revealed that the assemblies obtained from tadpole-like amphiphilic pillar[5]arenes (**53, 54**) successfully delivered both RhB and FITC to Hela Cells for bio-imaging and displayed red and green colors, respectively. The CLSM studies revealed that the assemblies constructed from pillar[5]arenes derivatives **53** and **56** were successfully delivered **57** and RhB to HeLa cells showing green and red colors, respectively. The delivery of anticancer drug DOX was also investigated using the assemblies of pillar[5]arene derivatives (**53–56**), which demonstrated the successful delivery of DOX to the cells by staining the cells with red colors in the cytoplasm and nucleus without affecting the morphologies of the assemblies. The research demonstrated the application of self-assembled structures constructed using pillar[5]arenes in the delivery of dyes *in vitro* by using biocompatible carriers such as drug carriers, which may help in advanced cancer therapy (Zhang et al. 2013).

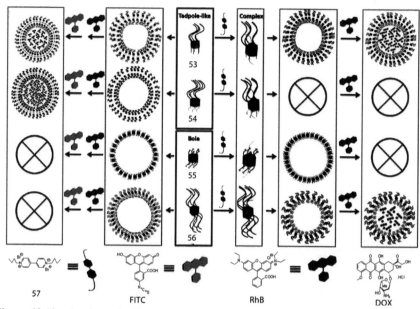

Figure 10.10 A schematic diagram showing the mechanism for dual bioimaging by using biocompatible pillar[5]arene derivatives based assemblies for delivery of **57**, FITC, and RhB. Reproduced here with the permission of ACS (Zhang et al. 2013).

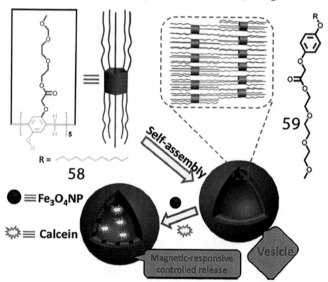

Figure 10.11 Chemical Structure of AP5-Glycol and the Schematic Representation of the Formation of Bilayer Vesicles and Further Formation of Magnetic Hybrid Vesicles in Water. Reproduced here with the permission of ACS (Zhou et al. 2014a).

Diao et al. (Zhou et al. 2014a) reported the design and synthesis of pillar[5] arene derivative **58** having glycol as a hydrophilic moiety and alkyl group as hydrophobic part.

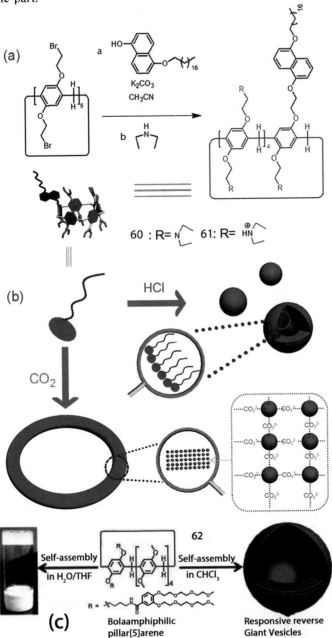

Figure 10.12 (a) Synthetic scheme for the synthesis of **60** and **61**; (b) Schematic representation of HCl and CO_2-switchable self-assembly of **60** (Jie et al. 2014) (c) The formation of assemblies of **62** in water and chloroform (Gao et al. 2013).

The studies based on DLS, TEM and SEM revealed the formation of vesicles above CAC (1.5×10^{-5} M) in water having a diameter of ~100 nm, which displayed stability even for several weeks. The TEM also indicated the formation of hollow vesicular assemblies in water having a diameter of ~120 nm. The noncyclic molecule **59** containing a structure similar to **58** displayed assemblies with a diameter of only ~5.0 nm. The vesicles were also investigated for external stimuli-responsive properties, that indicated reversible thermal and dynamic behavior. It was also observed that the oleic acid-coated magnetic iron oxide nanoparticles can be incorporated into the bilayer vesicles of **59**, which yielded hybrid magnetic responsive vesicles for potential application in the magnetic responsive release of calcein.

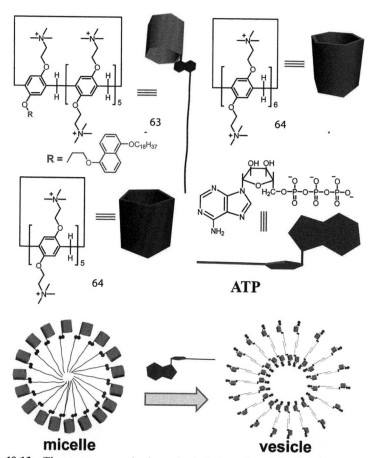

Figure 10.13 The structures and schematic depiction of amphiphilic pillar[6]arene (**63**), water-soluble amphiphilic pillar[5]arene derivative (**65**), water-soluble amphiphilic pillar[6] arene (**64**) derivative, adenosine triphosphate (ATP) and the conversion of micelles into vesicles. Reproduced here with the permission of RSC (Zhou et al. 2014b).

Huang et al. (Jie et al. 2014) also reported the synthesis of pillar[5]arene derivatives, which formed assemblies in water that were observed to be responsive towards carbon dioxide (Fig. 10.11). The DLS and TEM images revealed the presence of spherical micelles in the presence of HCl and toroid like structures on treatment with CO_2. The exposure of toroids-like structure to N_2 or heat led to the disruption of assembled structure to irregular assemblies. The assembled structure have a potential application in drug delivery application.

The glycol and alkyl group substituted pillar[5]arenes (**62**) were further investigated by Huang et al. (Gao et al. 2013). The amphiphilic pillar[5]arene displayed the formation of giant vesicles having a hydrophobic group oriented towards chloroform, while in aqueous THF solution gel was obtained.

A self-assemblies based on pillar[5]arene and pillar[5]arene derivatives (**63–65**) were reported by Diao et al. (Zhou et al. 2014b) in water. The micellar structures formed initially in water transformed into bilayer vesicles in the presence of ATP. The treatment of bilayer vesicles with alkaline phosphatase led to the collapse of micellar structures due to the hydrolysis of the ATP (Fig. 10.13).

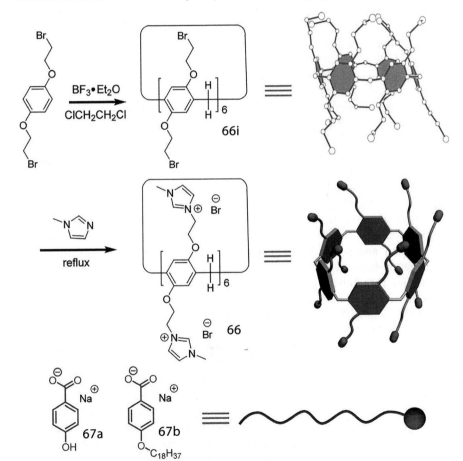

Figure 10.14 Synthesis of water-soluble ionic liquid pillar[6]arene derivative **66** along with the crystal structure of intermediate **66i**, and representation of **66** and guest molecules (**67**) used in the study. Reproduced here with the permission of RSC (Yao et al. 2014a).

Xue et al. (Yao et al. 2014a) also reported the synthesis of a cationic pillar[6] arene derivative (66) and investigated its assemblies for controlled release. Receptor 66 formed a strong complex with guest 67a in water with a high association constant ($3.21\pm0.82 \times 10^6$ M^{-1}) through non-covalent interactions. The complexation was successfully used to control the self-assembly process between 67b and 66. The vesicles formed by 66–67b were further investigated for the release of calcein. The change in pH of the solution led to the collapse of the vesicles and the release of calcein dye molecule (Fig. 10.14).

AGGREGATION INDUCED EMISSION (AIE) FOR CHEMO/BIO-SENSING

In recent years the fluorescent self-assembled materials have attracted considerable attention due to their striking performance in sensing, imaging, energy storage and optical materials. The amphiphilic macrocyclic receptors are extensively used as chemosensors to perceive ionic and molecular species (Zhao et al. 2010; Zhu et al. 2018). Chemosensors in general use an interaction between an analyte and the supramolecular macrocyclic host to produce J- or H-aggregates, which produces an optical response (Chang et al. 2013). As can be expected, the analyte that induces the formation of H-aggregated causes a hypsochromically shifted absorption band, while analytes that induces the formation of J-aggregates causes a bathochromically shifted absorption band. The formation of J-aggregates displays an increase in fluorescence intensity, while H-aggregates generally display a fluorescence quenching response (Möbius 1995; Chang et al. 2013). However, examples are available in literature, where H-aggregates display an increase in fluorescence intensity (Rösch et al. 2006; Chang et al. 2013).

Aggregation-induced emission (AIE) has been a focus of research for diverse applications for the development of solid-state fluorescent materials, sensors, bioimaging and so on (Lehn 1993; Klajn et al. 2010; Lehn 2017). The term aggregation-induced emission is generally used to describe the phenomena of molecular receptors, which in response to the presence of a specific analyte undergo a change in aggregation to produce a change in fluorescence emission (Lou and Yang 2018). The supramolecular macrocyclic ligands are ideal candidates to act as building blocks for material generation, which can respond to the external stimuli as they can be converted into amphiphilic structures (Lou and Yang 2018). This arrangement of aggregation-induced emission with macrocyclic ligand chemistry has already produced a variety of systems such as smart functional materials, host-guest complexes, supramolecular nanoparticles and supramolecular polymers. The last several decades have seen an enormous increase in the number of reports in the areas dealing exclusively with molecular recognition and self-assembly through non-covalent interactions (Kumar et al. 2019). Such interactions between a host and the guest often produced a variety of aggregates (Sahoo et al. 2017; Kumar and Kumar 2019). A variety of receptors for analytes based on J-aggregates have been reported through colorimetric and fluorometric signals, but H-aggregates have rarely been exploited for similar applications (Möbius 1995; Rösch et al. 2006; Sreejith et al. 2008; Chang et al. 2013; Kumar et al. 2017). The use of macrocycles in host-guest chemistry is an important part of supramolecular chemistry. A typical host-guest system consists of a macrocyclic host unit that coordinates a specific guest-forming inclusion complex in most cases using non-covalent interactions and a signaling unit. The functional groups attached to the macrocyclic units not only impart selectivity towards a particular analyte, but also induces the sensitivity in the molecular unit towards aggregation. The desired selectivity and sensitivity observed in macrocyclic host-guest systems makes them ideal candidates for the construction of smart materials with controllable properties through appropriate stimuli. Various macrocyclic compounds such as calix[n]arenes, crown ether, cyclodextrins, curcurbit[n]urils and pillararenes have been reported in good yield allowing tunability through appropriate modification along with desired properties, which either served as building blocks for the reversible complex formation or stimuli-responsive smart materials. For example, macrocyclic systems have been used to obtain smart functional materials with controllable emission properties. The fluorescent properties of macrocyclic systems can also be controlled through the formation of an inclusion complex with the analyte (Mudliar and Singh 2016). The self-assembly of amphiphilic macrocyclic receptors through inclusion complex often produces J- or H-type of aggregates (Gao et al. 2006; Ma et al. 2007; Gadde et al. 2008; Kiba et al. 2009; Guo and Liu 2014) (Figs. 10.15–10.16).

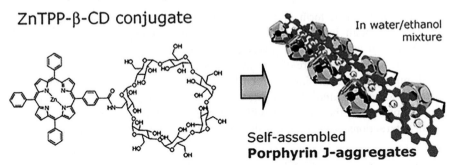

Figure 10.15 Supramolecular J-aggregates of Zn-Tetraphenyl porphyrin-β-CD conjugate. Reproduced here with the permission of ACS (Gao et al. 2006, Kiba et al. 2009).

The calix[4]arene sulfonate has also been used as a template to induce the H-aggregation of thiazole orange dye, which leads to enhancement in fluorescence intensity. It should be noted that thiazole orange does not exhibit fluorescence in aqueous solution (Lau and Heyne 2010) (Fig. 10.16).

The interaction between auramine-O with the macrocyclic receptor (β-cyclodextrin), which led to the formation of aggregates of the molecule has been reported by Awasthi and Singh (Awasthi and Singh 2017). It was observed that the formation of the inclusion complex is not responsible for J-aggregate formation. However, the negatively charged group (sulfonate) over the macrocyclic receptor (β-cyclodextrin) acted as a template for the aggregation of auramine-O dye, which produces a strong emission (Fig. 10.17).

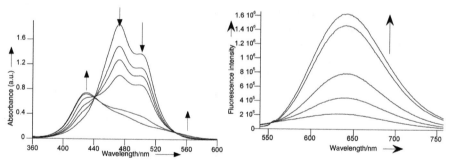

Figure 10.16 Change in the fluorescence and absorption spectra of thiazole orange on the addition of *p*-tetrasulfonylcalix[4]arene. Reproduced here with the permission of ACS (Lau and Heyne 2010).

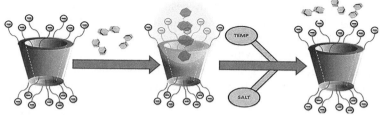

Figure 10.17 Schematic representation of J-aggregation of auromine-O induced by *β*-cyclodextrin. Reproduced here with the permission of ACS (Awasthi and Singh 2017).

The aggregated form of a guest can be easily disrupted by a macrocyclic receptor through inclusion complex formation. Gadde et al. reported through experimental and theoretical investigations that the size of the macrocyclic host molecule and the structure of dye dictated the disaggregation of pseudoisocyanine and pinacyanol. The larger cucurbit[7]uril disrupted the H-aggregates of pseudoisocyanine and J-aggregates of pinacyanol through inclusion complex formation, while the shallow cavities of cucurbit[6]uril displayed comparatively poor interactions with the guest dyes (Fig. 10.18) (Gadde et al. 2008).

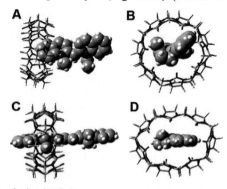

Figure 10.18 The optimized minimum energy structures of the CB[7] complexes with pseudoisocyanine (A and B) and pinacyanol (C and D). The binding energies calculated using B3LYP/6-31G(d) methods were found to be −38.3 and −89.4 kcal/mol, respectively. Reproduced here with the permission of ACS (Gadde et al. 2008).

Pillar[*n*]arene-based Supra-amphiphiles for Biological Imaging

Fluorescent systems having near-infrared emission (NIR) are an interesting class of molecular receptors due to their potential applications in biological imaging (Liu et al. 2013; Zhu et al. 2018). Owing to interesting self-assembling properties of amphiphilic pillararenes, Huang et al. (Shi et al. 2016a) reported the application of self-assembled nanoparticles of pillar[5]arene (**68**) derivative with NIR emission in the biochemical application.

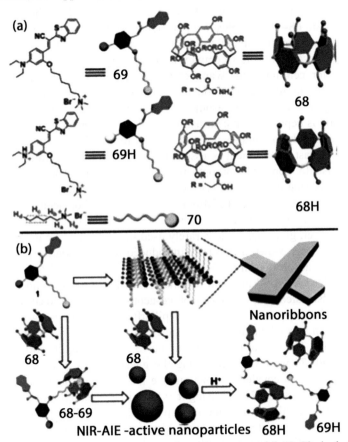

Figure 10.19 (a) A schematic representation of structures of **68–70**, including their protonated derivatives. (b) Schematic representations of self-assembled structures of **68**, **59** and **68–69** conjugates as well as pH responsiveness of nanoparticles obtained from **68–69** conjugate. Reproduced here with the permission of ACS (Shi et al. 2016a).

The –CN group substituted merocyanines are known for their absorption in the visible region and emission in the NIR region. The fluorescence emission in the NIR region enhances due to the aggregation-induced emission phenomena. Hence the study exploited a cyano-substituted merocyanine skeleton and introduced a cationic group to convert it into a suitable guest for pillar[5]arene derivative **69**. An aqueous solution of **69** in its non-aggregated form displayed negligible emission, while it aggregated to form nano-ribbons to produce a slight enhancement in fluorescence emission. However, the formation and inclusion complex between pillar[5]arene (**68**) and **69** produced a significant increase in the fluorescence band due to complexation induced emission. The host-guest complex aggregated to form nanoparticles with pH-responsive properties. The pH-responsive properties were due to the precipitation of pillar[5]arene (**68**) under acidic conditions. Besides during photoexcitation of nanoparticles, biological samples did not suffer any significant damage, which prompted the application of nanoparticles with NIR emission in cell imaging. The confocal image analysis indicated that the incubation of **68–69** complex nanoparticles produced bright red fluorescence emission in the cytoplasm of the bEnd.3 and Hela cells. It was proposed that pillararene based host-guest complex have great potential to develop fluorescent materials for cell imaging. Wang et al. (Zhang et al. 2006) and Chen et al. (Li et al. 2015) highlighted the unique advantages of the photoacoustic image in the monitoring of diseases by ultrasonic waves detection in thermo-elastic expansion. The method is dependent on contrasting agents or probes for good performance to obtain high-resolution images in greater depth for efficient monitoring of disease. A large contrast requirement is essential to achieve high performance imaging, which is usually dependent on optical/acoustic efficiency. Therefore, the efficient rate of conversion of absorbed laser energy into the thermos-acoustic wave is an important parameter (Shi et al. 2016c). To amplify the photoacoustic signal during imaging, the use of graphene oxide has been reported by Dai et al. (Robinson et al. 2011). However, graphene oxide is rarely used for photoacoustic imaging owing to structural defects, which are usually introduced by the oxidative processes along with its poor absorption in the NIR region (Li et al. 2015). Chen et al. (Yu et al. 2018a) utilized a hybrid material of pillararene and graphene oxide, to solve the problems associated with GO during photo-acoustic imaging. To construct a design in an appropriate system, a substituted pillar[6]arene (**73**) having bicarbonate as counter anion was taken and its host-guest complex with a pyrene derivative (**74**) was prepared. The size of the pillar[n]arene **73** good enough to form an inclusion complex with **74** was confirmed using the guest **76**. The pyrene derivative (**74**) was chosen to non-covalently functionalize the graphene oxide surface with the supramolecular system (**73–74**) using $\pi \cdots \pi$ stacking interactions (Yu et al. 2018a). The hybrid material was reported to display enhanced NIR absorption, favorable photo-thermal effect, etc., which were superior in comparison to the properties displayed by native graphene oxide.

The hybrid material was also used as a NIR mediated heating source to decompose the bicarbonate anion and produce carbon dioxide nano-bubbles. The carbon dioxide nanotubes were found to reflect the scattering ultrasound to increase the vibration of the medium, which render the hybrid material as a contrasting agent and acted as a nano-amplifier to boost the photoacoustic signal. The hybrid material was further applied during *in vivo* experiments on a tumor-bearing nude mouse, which led to a significant increase in the photoacoustic signal of the tumor cells treated with the hybrid materials. The GO@**74–75** produced poorer results in comparison of GO@**73–75** hybrid material. The work demonstrated the effect of supra-amphiphiic systems in improving the capacity of graphene oxide in the indispensable enhancement of photoacoustic signal (Yu et al. 2018a).

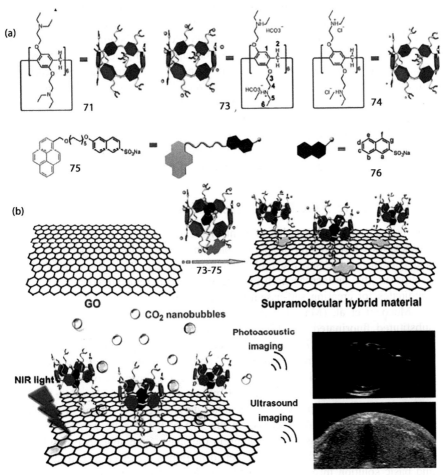

Figure 10.20 (a) Structures of the various chemical entities (**73–76**) were used in the study. (b) Scheme for the preparation of supramolecular hybrid material (GO@**73–75**) exhibiting NIR light-triggered photoacoustic signal and ultrasound imaging enhancement. Reproduced here with the permission of RSC (Yu et al. 2018a).

CONTROL OF pH OVER SELF-ASSEMBLY

The pH-controlled assembly of macrocyclic receptors represents an interesting area with potential applications in ditopic sensing, logic gates, on-off switchable systems, molecular machines, controlled release of guest from nanocontainers, switchable colorimetric/fluorometric sensing electronics and biological systems (Ashton et al. 1997; Jiang et al. 2007; Leung et al. 2009). To control the self-assembly of macromolecular systems in template-directed or 'bottom-up' approaches, a pH change provides the trigger that controls the process. The measurement of pH of the solution on addition of a readily available acid or a base can provide quantitative information (Zimmerman and Corbin 2000). The pH-responsive group located on the macrocyclic system responds to the change in the concentration of H^+ ions, which changes the non-covalent interactions such as H-bond, van der Waals and $\pi \cdots \pi$ interactions (Hoeben et al. 2005). Groups like amine, phenolate, pyridine, etc., if part of the macrocyclic structure competes for protonation leading to an equilibrium between a protonated and unprotonated state, which allows pH control of self-assembly. A variation in pH value of the solution containing macrocyclic receptors allow control of coulombic or electrostatic repulsive forces between two states. A variation in pH can also control the conformational switching within the individual molecules through variation in non-covalent and $\pi \cdots \pi$ interactions, which in turn can control the self-assembly of macrocyclic receptors.

Calix[n]arene based pH Responsive Assemblies

The use of pH as a stimulus to control the morphology of the nanostructures requires the incorporation of specific building blocks into the molecular structure that determines the specific molecular self-assembly. The presence of phenolic groups in the molecular architecture of calix[n]arenes makes them suitable candidates for creating self-assembled structures.

Mecozzi et al. (Martin and Mecozzi 2007) reported a series of upper rim substituted fluorinated calix[4]arene derivatives (**76–77**) containing hydrophilic groups at the lower rim and evaluated their pH-responsive self-assembly in different solvents. A combination of DLS and TEM investigations was used to find the influence of pH (pH = 1.2–12) on the aggregation behavior of the calix[4]arene derivatives. Overall, the change in pH influenced the octa-carboxylate derivative of calix[4]arene much more than any other derivative by a change in the charged state of the molecule. An increase in the size of the micelle was observed with increasing the pH along with a change in the shape from spherical to cylindrical. The study also revealed that the bilayer formation is preferred in extremely acidic and basic solution in comparison to the neutral pH conditions.

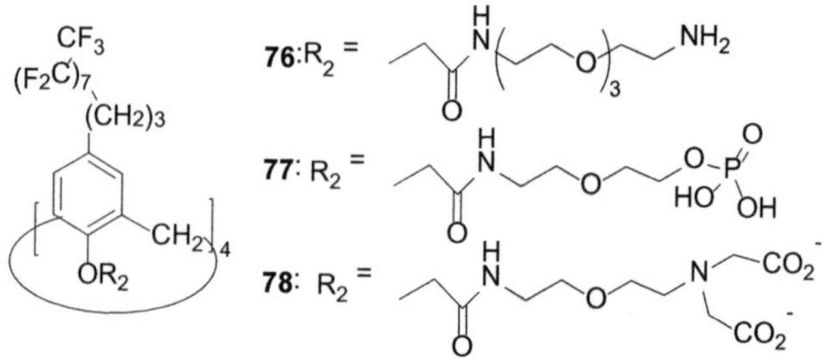

Cho et al. (Cho et al. 2008) have synthesized a calix[4]arene derivative (**79**) having four alanine moieties at the upper rim and a long alkyl chain at the lower rim. The aggregation behavior of the calix[4]arene derivative (**79**) was investigated at two different pH (3.0, 7.0, and 9.0) using field emission scanning electron microscopy and energy-filtered transmission electron microscopy. At a pH value of 3.0, the calix[4]arene displayed spherical structures of 200–250 nm diameter, which was not observed at pH values of 7.0 and 9.0 (Fig. 10.21). At pH 7.0 necklace like aggregated structures connected were obtained having a diameter of ca. 500 nm. IR spectroscopic results indicated that the carboxylic acid group formed strong intermolecular H-bonds. The metallization studies of the self-assembled structures of **79** were performed by addition of $AgNO_3$ at pH 7.0 to successfully generate silver-coated 3D dendritic nanostructures, which indicated that the calix[4]arene derivative (**79**) can be used to stabilize the inorganic materials (Cho et al. 2008).

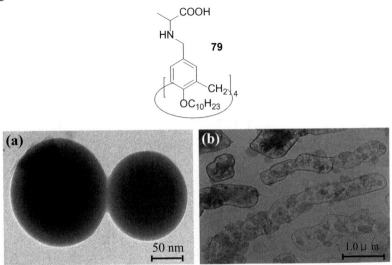

Figure 10.21 Energy Filtered Transmission Electron Microscopy (EF-TEM) images of self-assembled structures observed at (a) pH 3.0 and (b) pH 7.0 in an aqueous solution of the macrocycle (**79**). Reproduced here with the permission of ACS (Cho et al. 2008).

Lee et al. (Lee et al. 2004) have synthesized a series of amphiphilic calix[4] arene derivatives and investigated their self-assembling behavior using DLS, TEM and FE-SEM techniques. The TEM and DLS studies indicated that the calix[4]arene derivatives **80a** and **80b** aggregated to form 200 nm and 35 nm size particles in aqueous solutions, respectively. The calix[4]arene derivative **80c** displayed 5–6 nm particle size owing to the presence of longer hydrophilic ethoxyethanol group at the upper rim. Besides, the studies with **80b** indicated that a change in pH value from 7.0 to 5.0 led to the transformation of vesicles structure into micelles (Fig. 10.22). The pH-responsive transformation was attributed to the quaternization of the amines, which caused an increase in hydrophilic surface area that triggered the conversion of vesicle structure into micellar structures (Fig. 10.23). The application of such structures for the release of calcein was successfully achieved. The change in the structure of the aggregate may be used for the release of drug molecules for cancer treatment owing to the difference in the pH of the infected and normal tissues.

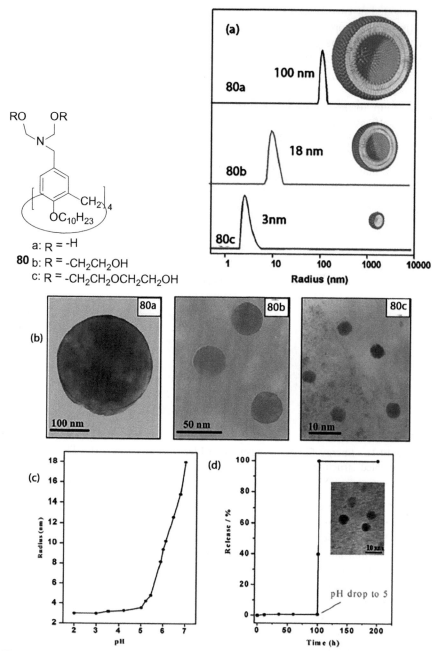

Figure 10.22 The calix[4]arene derivatives **80a-c** used in the study to obtain (a) the DLS results of the aggregated structures at a scattering angle of 90° formed in aqueous solution, (b) TEM images (5 mg/ml in aqueous solution), (c) Hydrodynamic radius of the aggregated structure of **80b** (0.5 mM) at different pH (RT). (d) The percent release of calcein from the aggregated vesicular structure of **80b** with time. Reproduced here with the permission of ACS (Lee et al. 2004).

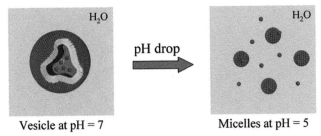

Vesicle at pH = 7   Micelles at pH = 5

Figure 10.23 A representation of a transition of vesicle-to-micelle observed in the case of **80b** with a change in pH followed by the release of encapsulated calcein. Reproduced here with the permission of ACS (Lee et al. 2004).

On similar lines, Ramita et al. (Houmadi et al. 2007) converted a calix[6]arene skeleton into an amphiphilic macrocycle (**81**) by substituting three imidazolyl moiety on the lower rim and three sulfonate groups on the upper rim. The AFM, DLS and TEM investigations of self-assembly behavior at different pH indicated switching of the architecture of the molecular aggregates with a change in the pH. At pH value of 7.8 and 10^{-4} M concentration, the vesicles of **81** displayed 40–250 nm diameter and 5–250 nm shell thickness. Sonication leads to the formation of unilamellar vesicles having less than 50 nm size along with bilayer shells. A decrease in pH value to 6.5 produced monodispersed vesicles containing an average diameter of 50 nm. At pH 8.5, TEM images indicated the formation of vesicles having 450 nm diameter. In vesicle size and morphology were attributed to the protonation state of imidazole present in the macrocyclic molecule having a pKa of 6. The protonation or deprotonation of imidazole leads to a change in amphiphilic balance within the molecule. The addition of silver ions led to the collapse of vesicles into micelles having a diameter of 2.5 nm, which was attributed to the binding of silver ions to the imidazole moiety present within the molecular architecture leading to changes in geometrical and electronic distribution with the macrocyclic molecule. The example demonstrated that the incorporation of an imidazole moiety within the hydrophobic core of amphiphilic calix[n]arenes (**81**) provides pH or metal control self-assembly of the entire molecular architecture (Fig. 10.24) (Houmadi et al. 2007).

Mecca et al. prepared amphiphilic calix[8]arene derivatives (**82**), which displayed self-assembled structures in the form of a pH-responsive gel (Mecca et al. 2013). The hydrophobic upper rim of the calix[8]arene was modified to incorporate the amino or carboxylic acid functional groups. It was observed that a balance of hydrophobic and hydrophilic parts is essential in the successful self-assembly of macrocyclic structures like calix[8]arene. The change in the functional group and length of the alkyl chain was observed to be responsible for the stability of the assembled structure that was in the form of a gel. The concentration required for the formation of aggregates was also observed to be dependent on the concentration of the macrocyclic receptor. Electrostatic forces between the charged and uncharged groups in macrocyclic structure were observed to be responsible for the formation of assembled structures, which were easily controlled through variation in the pH (Fig. 10.25) (Mecca et al. 2013).

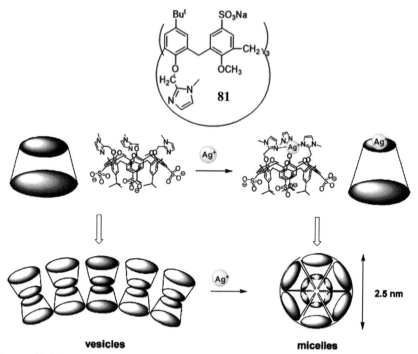

Figure 10.24 A representation of the modification of the morphologies of the self-assemblies of **81** observed the presence of silver ions. Reproduced here with the permission of ACS (Houmadi et al. 2007).

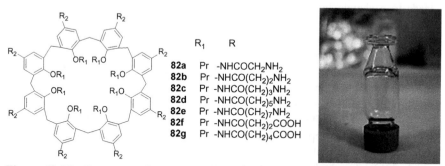

Figure 10.25 Structures of calix[8]arene derivatives **82a-g** along with a vial having 0.2% w/v of the hydrogel of **82c**. Reproduced here with the permission of ACS (Mecca et al. 2013).

Pillar[n]arenes based pH Responsive Assemblies

Macrocyclic receptors based on pillar[n]arenes are being investigated for their potential as building blocks for the stimuli-responsive supramolecular self-assembled system, which can be tuned by a change in the shape and electron density of the cavity present in their molecular architecture. Due to the chirality associated with pillar-shaped structure, pillar[n]arenes are useful for the development of stimuli-responsive supramolecular assemblies. The discussion provided here can be useful for researchers working in the area of macrocyclic receptors and other pursing host-guest supramolecular assemblies. Although the pH dependence of a variety of self-assembled systems used for the release of the bound substrate based on amphiphilic pillar[n]arene have been described earlier in this chapter, additional systems not described earlier have been discussed in the following section. Huang et al. (Yu et al. 2012) prepared a water-soluble pillar[6] arene derivative (**83**) and investigated its complexation with **86** and compared it with pillar[5]arene derivative (**84**).

The pillar[5]arene (**84**) possess a smaller cavity dimension, which was found to be smaller for the guest such as **86**, but **83** was found to have a size that matches the dimension of the guest molecule (**86**). Firstly the complex formation between **83** and **86a** was established using spectroscopic techniques like UV-Visible and 1D (^{1}H) and 2D NOESY NMR. It was followed by an investigation of pH-responsive interaction and self-assembly between amphiphilic **86b** and **83** using DLS, TEM.

It was also seen that the interaction between **83** and **86b** reduced the CAC of **86b** from $3.44 \pm 021 \times 10^{-4}$ M to $1.95 \pm 021 \times 10^{-5}$ M. The TEM images of **83–86b** indicated the formation of vesicles of an average diameter of 170 nm, where the thickness of the hollow vesicles was found to be 9 nm. The study indicated an antiparallel packing arrangement of pseudorotaxanes in **83–86b**. The change in pH from 7.4 to 6.0 led to the formation of micellar structures. The DLS studies further confirmed the pH-responsive character of the self-assembly of **83–86b**, which indicated an average diameter of 185 nm at a pH of 7.4, which decreased to 10.5 nm at pH 6.0 indicating the conversion of vesicular structures to micellar structures. A further change in the pH of the solution increased the average diameter to 203 nm, which again indicated the formation of vesicular structures. The study further investigated the controlled release of calcein from self-assembled structures of **83–86b**, which indicated an efficient release of calcein from the interior of the hollow self-assembled structure with a change in pH from 7.4 to 6.0.

Wang et al. (Duan et al. 2013) investigated the self-assembled structures of **83** and ferrocene derivative (N-1-decyl-ferrocenylme-thylamine: NDFT) for pH-responsive release of model drug molecule mitoxantrone (MTZ). The complex inclusion formation between **83** and NDFT was established through NMR and UV-visible spectroscopy. Although **83** and NDFT did not display any tendency to aggregate in their aqueous solution, the **83**-NDFT at a molar ratio of 1:20 displayed the best aggregation tendency to form vesicles in water (pH = 7.4) showing an average diameter of 167 nm. The TEM images indicated the formation of hollow spherical morphology showing an average diameter of ~130 nm and a thickness of ~7.0 nm. A reduction in pH to 6.0 led to the disappearance of the vesicular structure from TEM images. The pH-responsive change in aggregation behavior was exploited to encapsulate and release a model drug molecule mitoxantrone (MTZ). Keeping a solution of MTZ with vesicles of **83**-NDFT overnight and removal of unloaded MTZ using dialysis led to the preparation of pure **83**-NDFT vesicles loaded with MTZ drug. The UV-visible spectroscopy displayed an increase in the intensity of the absorption band of the vesicles of **83**-NDFT in the 610–660 nm region on loading the vesicles with MTZ. The color change from colorless (unloaded vesicles) to blue (loaded vesicles) was observed during the preparation of MTZ vesicles of **83**-NDFT showing an encapsulation efficiency of 11.2%. The change in pH of the solution to 6.5 and 4.0 led to a release efficiency of 71% and 95% within 24 hours, respectively. *In vitro* studies of delivery of MTZ drug to cancer cells demonstrated that the cancer cells were able to disassemble the vesicular structure to release the drug molecule successfully.

Huang et al. (Yao et al. 2014c) have further developed hybrid organic/inorganic nanostructures based by stabilizing gold nanoparticles with **83**–**86** self-assembled structures, where **86c** was substituted with varying hydrophobic chain length. The gold nanoparticles were prepared through the reduction of $HAuCl_4$ with $NaBH_4$ in the presence of a varying concentration of **83**. The formation of **83** stabilized gold nanoparticles were confirmed using Surface Plasmon Resonance (SPR) by indicating a peak at ~520 nm. At a concentration of 0.4 mM of **83**, the gold nanoparticle surface was completely covered by **83**. The TEM images assigned an average diameter of 2.84±0.32 nm to the spherical nanoparticles. The **83** stabilized gold nanoparticles were treated with a 0.04 mM solution of **86c** to yield **83**–**86c** stabilized gold nanoparticles. The DLS studies indicated an increase in the average diameter of the nanoparticles from 2.7 (AuNPs-**83**) to 3.0 nm (AuNPs-**83**–**86c**). On increasing the concentration of **86c** further, an increase in the average diameter of the nanoparticles was observed. At 0.25 mM **86c**, a typical hollow vesicular morphology of the AuNPs-**83**–**86c** was observed in the study through SEM image analysis, which displayed a diameter of 400 nm, similar to the size observed in DLS studies. The studies based on TEM, suggested bilayer vesicles having an average thickness of 10 nm. The vesicles were found to encapsulate hydrophilic calcein within the vesicular structures and released it on the reduction of pH from neutral to acidic through the addition of HCl. Recently, a similar study based on **83** also reported the inclusion of boronic acid derivatives followed by a pH and reactive singlet oxygen responsive self-assembly into vesicular structures (Hao et al. 2020). The size of the self-assembled structure was found to be dependent on the ratio of the **83** and the guest molecule. The vesicular structure displayed a good pH-responsive Nile red loading release profile, which demonstrated its potential as a drug carrier to the cancer cells.

Cyclodextrin based pH Responsive Assemblies

Owing to the presence of D-glucose units in cyclodextrins, they are often incorporated into materials used for the development of nanostructures (Zhang et al. 2020). The introduction of heterocyclic moiety containing a protonable/ deprotonable functional group into the cyclodextrin skeleton is an attractive choice to obtain a pH-responsive self-assembled structure (Zhang et al. 2020). A variety of pH and CO_2 responsive systems have been reported by incorporating the concept of protonation and pK_a substituent attached to the cyclodextrin skeleton (Peng et al. 2017). The acid dissociation constant (pK_a) has a direct influence on the protonation-deprotonation state of the macrocyclic receptor in solution, which dictates the structure-activity relationship of the chemical entity in the solution state. Precise control over the protonation and deprotonation equilibria in solution leads to control over primary electrostatic interactions, which simulate the secondary supramolecular interactions to influence physiological events in macromolecular assemblies (Zhang et al. 2020). For example, Hao et al. (Zhang et al. 2010) synthesized a γ-hydroxybutyric-β-cyclodextrin, which formed a complex with methyl orange. The cyclodextrin derivative-methyl organic complex on an investigation using TEM microscopy displayed vesicles type self-assembled structures of ~120 nm diameter in water. The DLS studies indicated an average diameter of ~150 nm at a pH of 8.49. The reduction in pH resulted in particles having an average diameter of 30 nm (TEM studies), which was attributed to the protonation of methyl orange leading to the destruction of aggregated state.

$$R = COCH_2CH_2CH_2OH$$

Zhang et al. (He et al. 2013) reported an α-cyclodextrin (**88**) based-platform to engineer pH-responsive nanoplatforms. The stability of the nanoparticles was successfully controlled through pH-controlled hydrolysis of acetal groups. The oil in water emulsion solvent evaporation techniques successfully produced nanoparticles of **88**. The study successfully demonstrated the potential use of pH-controlled nanoparticles in the targeted delivery of cancer drug paclitaxel (PTX) to cancer cells (Scheme 2). The nanoparticles were shown to release 87% drug molecules at pH 5.0 within 8 hours. The macrocycle **88** based-nanoparticles demonstrated effective drug loading and antitumor efficiency and lower side effects.

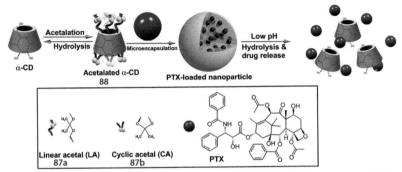

Scheme 2 Schematic representation of the formation of pH-responsive PTX nano formulation based on acetalated a-CD **88** (Ac-α-CD). Reproduced here with the permission of Elsevier (He et al. 2013).

REDOX RESPONSIVE ASSEMBLY OF MACROCYCLES

A wide distribution of charge potential and redox-active states in intra/extra-cellular environment mandatary for the development of self-assembled structures possessing groups having sensitivity toward specific sites have been focus of considerable attention (Zhang et al. 2020). The redox sensitivity can be incorporated into a macrocycle through functionalization with a disulfide bond, electroactive ferrocenyl or quinone moieties (Chawla et al. 2010; Lee et al. 2015). Redox switching is an interesting phenomenon in macrocyclic chemistry, where a variation in the membrane potential leads to the formation of assemblies.

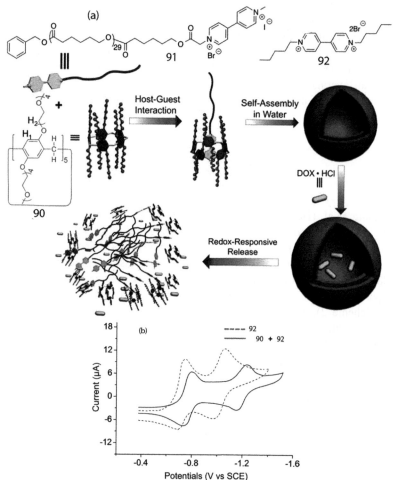

Figure 10.26   (a) Structures of **90, 92**, polymer **91** and schematic representation of polymeric vesicles preparation along with the process depicting redox-controlled DOX·HCl release (b) Cyclic voltammogram (0.1 V/s) of 1.0 mM **92** recorded in the presence of one equivalent macrocycle **90** (red solid line), while the black dashed line depicts the voltammogram of **92** in the absence of **90** (black dashed line; SCE: saturated calomel electrode. Reproduced here with the permission of ACS (Chi et al. 2015).

Huang et al. (Chi et al. 2015) reported the formation of redox-responsive supra-amphiphilic systems based on pillar[5]arene-paraquat polymer conjugate. The formation of the inclusion complex between **90** and paraquat derivative **92** was established using NMR spectroscopy and cyclic voltammetry (Figs. 26a, b). The complex formation between **90** and **92** was investigated further to develop redox responsive vesicles using the inclusion complex of macrocycle **90** and polymer **91**. The complex between macrocyclic pillaran[5]arene (**90**) and a guest (**91**) also displayed redox switching properties, which were observed through DLS and TEM studies. The size of the aggregated structure (189 nm to 300 nm) was observed to change under the reducing influence of $Na_2S_2O_4$. The aggregated structure displayed a controlled release of doxorubicin-induced by $Na_2S_2O_4$. Besides the drug-loaded vesicles displayed lower toxicity in comparison to the free drug molecules.

On similar lines, Yang et al. (Cui et al. 2019) reported an amphiphilic pillar[5] arene based pseudo[1]rotaxane having a redox responsive disulfide bond. The pseudo[1]rotaxanes displayed self-assembling behavior in an aqueous solution, which was demonstrated to be disrupted by GSH. The self-assembled nanoparticles successfully encapsulated DOX efficiently for release in a high GSH concentration environment.

The use of disulfide was also reported for cyclodextrin based-systems to control the self-assembly process, which can be developed into functional biomaterials. Huang et al. (Yu et al. 2018b) have also reported the development of cyclodextrin based assemblies containing disulfide linkage as a redox responsive nano-carrier for cancer diagnostics and therapeutics.

SULFONATOCALIX[*n*]ARENES-BASED SUPRAMOLECULAR AMPHIPHILES WITH PHOTO-RESPONSIVENESS

Light as a stimulus attracts considerable attention to influence the properties of a system as it is convenient without any need for an external additive. The use of light as a stimulus to control host-guest interaction has displayed considerable promise in the area of environmental pollution. In the context of stimuli-responsive self-assembly, light can be considered as a clean, rapid and cost alternative to chemical methods to alter the molecular aggregates through alternation in molecular structure, which in turn influences the intermolecular interactions (Zhu et al. 2018). Through the coupling of photo-reversible systems and macrocyclic receptors, photo alterable amphiphilic macrocyclic systems have been constructed. For instance, Liu et al. (Wang et al. 2015) fabricated macrocyclic amphiphilic systems by assembling a photolyzable host-guest inclusion complex between calix[4]arene derivative (**93**) and a molecular unit based on substituted anthracene (**94**). The self-assembly of the inclusion complex **93–94** was reported to produce mean particle size of 266 nm (polydispersity = 0.01). The **93** formed a 1:1 complex with **94** through $\pi\cdots\pi$ interactions, which led to an enhancement in fluorescence intensity by inhibition of photo-induced electron transfer mechanism. The authors reported calix[4]arene derivative induced critical aggregation concentration at ca. 0.06–0.1 mM in water. It was further reported that the size of the particles can be modulated through temperature variation, while UV-light exposure led to the dissipation of the assembly through the accelerated conversion of **94** into anthraquinone and alkanol. The rate of decomposition of **94** into anthraquinone and alkanol was observed to slow down in absence of calix[n]arene induced assembled structure. The photolysis reaction may be attributed to light exposure induced reaction between oxygen and the activated **94** (Bowen 1953; Barnett and Needham 1971; Powell 1987; Kohtani et al. 2005). The photolysis of the **93–94** complex was observed to be slower in the inert Argon atmosphere, which indicated the important role played by oxygen during the reaction. Besides the photolysis reaction was also accomplished using visible light in the presence of photosensitizers like Eosin Y (ESY). The photolysis reaction was attributed to the self-assembly of **93–94**, which leads to the formation of a hydrophobic core with an extended lifetime of singlet oxygen as well as an increase in the contact time of the singlet oxygen within the confined space of the nanoparticles. Moreover the inhibition of the PET mechanism also promoted photosensitization. Such systems are proposed to play an important role in the area of photo-degradation of pollutants and photodynamic therapy.

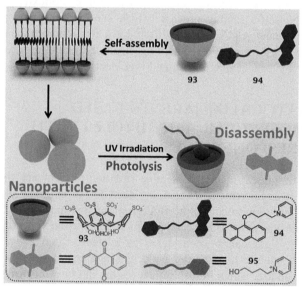

Figure 10.27 A schematic representation of an amphiphilic photolyzable supramolecular assembly. Reproduced here with the permission of ACS (Wang et al. 2015).

Similarly, Zhang et al. (Zhang et al. 2011) prepared a series of the compound by substitution of ethyl group in 1-ethyl-1′-arylmethyl-4,4′-bipyridinium compounds with 1-pyrenyl (**97**), 2-anthracenyl (**98**), 2-naphthyl (**99**) or phenyl (**100**) group. The bipyridinium derivatives substituted with **97** and **98** binds the calf thymus DNA via intercalation and groove binding modes. Besides **97** and **98** causes the photo-cleavage of plasmid (pBR 322) DNA on exposure with xenon arc lamp in air and under inert Argon atmosphere. The formation of an inclusion derivative between cucurbit[8]uril and all the aromatic donor-viologen acceptor compounds display an efficient ability to photo-cleave the DNA. A significant hypsochromic effect along with isosbestic point was observed on addition of DNA to **97** and **98** indicating strong interactions. No isosbestic point or hypsochromic shift was observed with **99** and **100** indicating weak interactions with DNA, which was attributed to the small size of their aromatic ring (Kumar et al. 2000; Ohyama et al. 2005; Hvastkovs and Buttry 2006; Modukuru et al. 2006).

The photocleavage ability was attributed to the cucurbit[8]uril induced inhibition of intramolecular backward electron transfer in aromatic donor-viologen acceptor compounds that increased the lifetime of the charge-separated excited state. The *Aryl, HO·, and ${}^1O_2$ were found to be responsible for the photo-cleavage of DNA. The light exposure caused an electron transfer from the excited state of the aromatic ring-like anthracene, pyrene or benzene to the viologen fragment to produce a charge-separated *Aryl-CH$_2$-EV$^{+·}$ species as the excited state (Sun et al. 2010). The EV$^{+·}$ reduced the oxygen dissolved in the aqueous solution to the reactive singlet oxygen that oxidizes the DNA with *Aryl radical species. The system demonstrated its utility in the development of a system to cleave DNA using light through the exploitation of host-guest macrocyclic chemistry. However, the EV$^{+·}$ species lifetime is very short due to the backward intermolecular electron transfer. The cucurbit[8]uril inhibits this intermolecular electron transfer and thereby increases the lifetime of EV$^{+·}$, which in turn improves its DNA photo-cleavage efficiency.

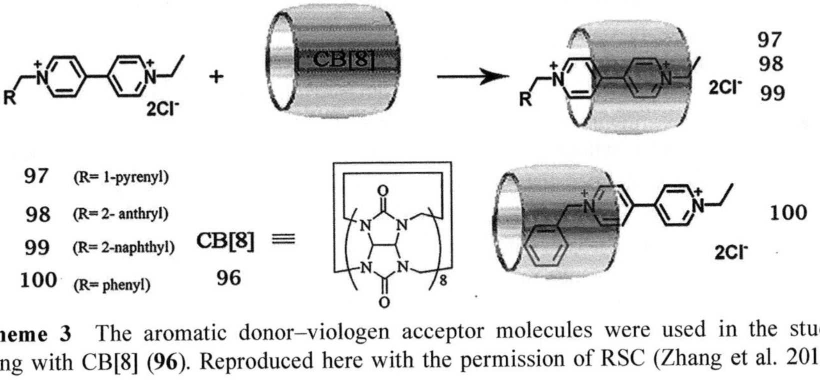

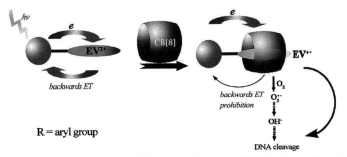

Scheme 3 The aromatic donor–viologen acceptor molecules were used in the study along with CB[8] (**96**). Reproduced here with the permission of RSC (Zhang et al. 2011).

Scheme 4 The mechanism proposed the photocleavage of DNA. Reproduced here with the permission of RSC (Zhang et al. 2011).

CONCLUSIONS

In the past few decades, increased activity in the area of aggregation phenomena in macrocyclic amphiphiles to yield well-defined architectures have been seen. A variety of self-assembled structures having interesting morphologies have been developed, which can be tuned with the help of external stimuli such as pH, light or temperature. The self-assembled architectures constructed using macrocyclic amphiphilic structures display superior properties in many aspects as compared to conventional self-assembled structures. For example, well-defined morphologies, selectivity and the recognition of analyte species, transport phenomena, drug delivery and imaging without any requirement of a tedious synthetic process have been reported. Owing to different advantages associated with macrocyclic self-assembled architectures, they have been combined with polymers, inorganic and other materials to develop multi-functional futuristic materials.

REFERENCES

Amodio, A., A.F. Adedeji, M. Castronovo, E. Franco and F. Ricci. 2016. pH-controlled assembly of DNA tiles. J. Am. Chem. Soc. 138(39): 12735–12738.

Aoyama, Y., T. Kanamori, T. Nakai, T. Sasaki, S. Horiuchi, S. Sando and T. Niidome. 2003. Artificial viruses and their application to gene delivery. size-controlled gene coating with glycocluster nanoparticles. J. Am. Chem. Soc. 125(12): 3455–3457.

Ashton, P.R., R. Ballardini, V. Balzani, M. Gómez-López, S.E. Lawrence, M.V. Martínez-Díaz, M. Montalti, A. Piersanti, L. Prodi, J.F. Stoddart and D.J. Williams. 1997. Hydrogen-bonded complexes of aromatic crown ethers with (9-anthracenyl)methylammonium derivatives. Supramolecular photochemistry and photophysics. pH-controllable supramolecular switching. J. Am. Chem. Soc. 119(44): 10641–10651.

Assaf, K.I. and W.M. Nau. 2015. Cucurbiturils: from synthesis to high-affinity binding and catalysis. Chem. Soc. Rev. 44(2): 394–418.

Auzely-Velty, R., F. Djedaini-Pilard, S. Desert, B. Perly and T. Zemb. 2000. Micellization of hydrophobically modified cyclodextrins. 1. Micellar structure. Langmuir 16(8): 3727–3734.

Avvakumova, S., P. Fezzardi, L. Pandolfi, M. Colombo, F. Sansone, A. Casnati and D. Prosperi. 2014. Gold nanoparticles decorated by clustered multivalent cone-glycocalixarenes actively improve the targeting efficiency toward cancer cells. Chem. Commun. 50(75): 11029–11032.

Awasthi, A.A. and P.K. Singh. 2017. Stimulus-responsive supramolecular aggregate assembly of auramine o templated by sulfated cyclodextrin. J. Phys. Chem. B 121(25): 6208–6219.

Barnett, W.E. and L.L. Needham. 1971. 9-Anthroxy. A protecting group removable by singlet oxygen oxidation. J. Org. Chem. 36(26): 4134–4136.

Bowen, E. 1953. Reactions in the liquid phase. Photochemistry of anthracene. Part 1—The photo-oxidation of anthracene in solution. Discuss. Faraday Soc. 14: 143–146.

Cao, D., Y. Kou, J. Liang, Z. Chen, L. Wang and H. Meier. 2009. A facile and efficient preparation of pillararenes and a pillarquinone. Angew. Chem., Int. Ed. Engl. 48(51): 9721–9723.

Chang, J., Y. Lu, S. He, C. Liu, L. Zhao and X. Zeng. 2013. Efficient fluorescent chemosensors for HSO_4^- based on a strategy of anion-induced rotation-displaced H-aggregates. Chem. Commun. 49(56): 6259–6261.

Chawla, H.M., N. Pant, S. Kumar, N. Kumar and C. Black David St. 2010. Calixarene-based materials for chemical sensors. pp. 117–200. *In*: G. Korotcenkov [ed.]. Chemical Sensors Fundamentals of Sensing Materials, Vol. 3. Momentum Press, New York: USA.

Chécot, F., S. Lecommandoux, Y. Gnanou and H.A. Klok. 2002. Water-soluble stimuli-responsive vesicles from peptide-based diblock copolymers. Angew. Chem., Int. Ed. Engl. 41(8): 1339–1343.

Chi, X., G. Yu, X. Ji, Y. Li, G. Tang and F. Huang. 2015. Redox-responsive amphiphilic macromolecular [2]pseudorotaxane constructed from a water-soluble pillar[5]arene and a paraquat-containing homopolymer. ACS Macro Lett. 4(9): 996–999.

Cho, E.J., J.K. Kang, W.S. Han and J.H. Jung. 2008. Stimuli-responsive supramolecular nanostructure from amphiphilic calix[4]arene and its three-dimensional dendritic silver nanostructure. Langmuir. 24(10): 5229–5232.

Cui, Y.-H., R. Deng, Z. Li, X.-S. Du, Q. Jia, X.-H. Wang, C.-Y. Wang, K. Meguellati and Y.-W. Yang. 2019. Pillar[5]arene pseudo[1]rotaxane-based redox-responsive supramolecular vesicles for controlled drug release. Mater. Chem. Front. 3(7): 1427–1432.

Czikklely, V., H. Forsterling and H. Kuhn. 1970. Extended dipole model for aggregates of dye molecules. Chem. Phys. Lett. 6(3): 207–210.

Donohue, R., A. Mazzaglia, B.J. Ravoo and R. Darcy. 2002. Cationic β-cyclodextrin bilayer vesicles. Chem. Commun. (23): 2864–2865.

Duan, Q., Y. Cao, Y. Li, X. Hu, T. Xiao, C. Lin, Y. Pan and L. Wang. 2013. pH-Responsive supramolecular vesicles based on water-soluble pillar[6]arene and ferrocene derivative for drug delivery. J. Am. Chem. Soc. 135(28): 10542–10549.

Falvey, P., C.W. Lim, R. Darcy, T. Revermann, U. Karst, M. Giesbers, A.T.M. Marcelis, A. Lazar, A.W. Coleman, D.N. Reinhoudt and B.J. Ravoo. 2005. Bilayer vesicles of amphiphilic cyclodextrins: host membranes that recognize guest molecules. Chem. Eur. J. 11(4): 1171–1180.

Fang, J. and K.F. Böhringer. 2008. Self-assembly. pp. 403–429. *In*: Y.B. Gianchandani, O. Tabata and H. Zappe [eds]. Comprehensive Microsystems: Fundamentals, Technology, and Applications. Oxford: Elsevier.

Gadde, S., E.K. Batchelor, J.P. Weiss, Y. Ling and A.E. Kaifer. 2008. Control of H- and J-aggregate formation via host–guest complexation using cucurbituril hosts. J. Am. Chem. Soc. 130(50): 17114–17119.

Gao, Y.A., X. Zhao, B. Dong, L. Zheng, N. Li and S. Zhang. 2006. Inclusion complexes of β-cyclodextrin with ionic liquid surfactants. J. Phys. Chem. B. 110(17): 8576–8581.

Gao, L., B. Zheng, Y. Yao and F. Huang. 2013. Responsive reverse giant vesicles and gel from self-organization of a bolaamphiphilic pillar[5]arene. Soft Matter. 9(30): 7314–7319.

Guan, B., M. Jiang, X. Yang, Q. Liang and Y. Chen. 2008. Self-assembly of amphiphilic calix[6]crowns: from vesicles to nanotubes. Soft Matter 4(7): 1393–1395.

Guan, B., Q. Liang, Y. Zhu, M. Qiao, J. Zou and M. Jiang. 2009. Functional nanohybrids self-assembled from amphiphilic calix[6]biscrowns and noble metals. J. Mater. Chem. 19(41): 7610–7613.

Gulik, A., H. Delacroix, D. Wouessidjewe and M. Skiba. 1998. Structural properties of several amphiphile cyclodextrins and some related nanospheres. An X-ray scattering and freeze-fracture electron microscopy study. Langmuir. 14(5): 1050–1057.

Guo, D.-S. and Y. Liu. 2014. Supramolecular chemistry of p-sulfonatocalix[n]arenes and its biological applications. Acc. Chem. Res. 47(7): 1925–1934.

Gutsche, C.D. 1998. Calixarenes Revisited. Cambridge, Royal Society of Chemistry.

Gutsche, C.D. 2008. Calixarenes: An Introduction. Royal Society of Chemistry.

Haider, J.M. and Z. Pikramenou. 2005. Photoactive metallocyclodextrins: sophisticated supramolecular arrays for the construction of light activated miniature devices. Chem. Soc. Rev. 34(2): 120–132.

Hao, Q., Y. Kang, J.-F. Xu and X. Zhang. 2020. pH/ROS Dual-responsive supramolecular vesicles fabricated by carboxylated pillar[6]arene-based host-guest recognition and phenylboronic acid pinacol ester derivative. Langmuir. 36(15): 4080–4087.

Hayashida, O., K. Mizuki, K. Akagi, A. Matsuo, T. Kanamori, T. Nakai, S. Sando and Y. Aoyama. 2003. Macrocyclic glycoclusters. Self-aggregation and phosphate-induced agglutination behaviors of calix[4]resorcarene-based quadruple-chain amphiphiles with a huge oligosaccharide pool. J. Am. Chem. Soc. 125(2): 594–601.

He, H., S. Chen, J. Zhou, Y. Dou, L. Song, L. Che, X. Zhou, X. Chen, Y. Jia, J. Zhang, S. Li and X. Li. 2013. Cyclodextrin-derived pH-responsive nanoparticles for delivery of paclitaxel. Biomaterials. 34(21): 5344–5358.

Helttunen, K. and P. Shahgaldian. 2010. Self-assembly of amphiphilic calixarenes and resorcinarenes in water. New. J. Chem. 34(12): 2704–2714.

Hoeben, F.J., P. Jonkheijm, E. Meijer and A.P. Schenning. 2005. About supramolecular assemblies of π-conjugated systems. Chem. Rev. 105(4): 1491–1546.

Houmadi, S., D. Coquière, L. Legrand, M.C. Fauré, M. Goldmann, O. Reinaud and S. Rémita. 2007. Architecture-controlled "SMART" calix[6]arene self-assemblies in aqueous solution. Langmuir. 23(9): 4849–4855.

Howard, J. and A.A. Hyman. 2003. Dynamics and mechanics of the microtubule plus end. Nature. 422(6933): 753–758.

Hu, Q.-D., G.-P. Tang and P.K. Chu. 2014. Cyclodextrin-based host–guest supramolecular nanoparticles for delivery: from design to applications. Acc. Chem. Res. 47(7): 2017–2025.

Hvastkovs, E.G. and D.A. Buttry. 2006. Minor groove binding of a novel tetracationic diviologen. Langmuir. 22(25): 10821–10829.

Isaacs, L., S.-K. Park, S. Liu, Y.H. Ko, N. Selvapalam, Y. Kim, H. Kim, P.Y. Zavalij, G.-H. Kim and H.-S. Lee. 2005. The inverted cucurbit[n]uril family. J. Am. Chem. Soc. 127(51): 18000–18001.

Izatt, R.M., J.D. Lamb, R.T. Hawkins, P.R. Brown, S.R. Izatt and J.J. Christensen. 1983. Selective M+-H+ coupled transport of cations through a liquid membrane by macrocyclic calixarene ligands. J. Am. Chem. Soc. 105(7): 1782–1785.

Jiang, X., Z. Ge, J. Xu, H. Liu and S. Liu. 2007. Fabrication of multiresponsive shell cross-linked micelles possessing pH-controllable core swellability and thermo-tunable corona permeability. Biomacromolecules. 8(10): 3184–3192.

Jie, K., Y. Yao, X. Chi and F. Huang. 2014. A CO_2-responsive pillar[5]arene: synthesis and self-assembly in water. Chem. Commun. 50(41): 5503–5505.

Jie, K., Y. Zhou, Y. Yao and F. Huang. 2015. Macrocyclic amphiphiles. Chem. Soc. Rev. 44(11): 3568–3587.

Jin, T., F. Fujii, H. Sakata, M. Tamura and M. Kinjo. 2005. Amphiphilic p-sulfonatocalix[4] arene-coated CdSe/ZnS quantum dots for the optical detection of the neurotransmitter acetylcholine. Chem. Commun. (34): 4300–4302.

Jon, S.Y., N. Selvapalam, D.H. Oh, J.-K. Kang, S.-Y. Kim, Y.J. Jeon, J.W. Lee and K. Kim. 2003. Facile synthesis of cucurbit[n]uril derivatives via direct functionalization: expanding utilization of cucurbit[n]uril. J. Am. Chem. Soc. 125(34): 10186–10187.

Kakuta, T., T.-a. Yamagishi and T. Ogoshi. 2018. Stimuli-responsive supramolecular assemblies constructed from pillar[n]arenes. Acc. Chem. Res. 51(7): 1656–1666.

Kasha, M., H. Rawls and M. Ashraf El-Bayoumi. 1965. The exciton model in molecular spectroscopy. Pure Appl. Chem. 11(3-4): 371–392.

Kiba, T., H. Suzuki, K. Hosokawa, H. Kobayashi, S. Baba, T. Kakuchi and S.-I. Sato. 2009. Supramolecular j-aggregate assembly of a covalently linked zinc porphyrin−β-cyclodextrin conjugate in a water/ethanol binary mixture. J. Phys. Chem. B. 113(34): 11560–11563.

Kim, K., N. Selvapalam, Y.H. Ko, K.M. Park, D. Kim and J. Kim. 2007. Functionalized cucurbiturils and their applications. Chem. Soc. Rev. 36(2): 267–279.

Kim, D., E. Kim, J. Kim, K.M. Park, K. Baek, M. Jung, Y.H. Ko, W. Sung, H.S. Kim, J.H. Suh, C.G. Park, O.S. Na, D.-k. Lee, K.E. Lee, S.S. Han and K. Kim. 2007. Direct synthesis of polymer nanocapsules with a noncovalently tailorable surface. Angew. Chem., Int. Ed. Engl. 46(19): 3471–3474.

Kim, D., E. Kim, J. Lee, S. Hong, W. Sung, N. Lim, C.G. Park and K. Kim. 2010. Direct synthesis of polymer nanocapsules: self-assembly of polymer hollow spheres through irreversible covalent bond formation. J. Am. Chem. Soc. 132(28): 9908–9919.

Kim, E., D. Kim, H. Jung, J. Lee, S. Paul, N. Selvapalam, Y. Yang, N. Lim, C.G. Park and K. Kim. 2010. Facile, template-free synthesis of stimuli-responsive polymer nanocapsules for targeted drug delivery. Angew. Chem., Int. Ed. Engl. 49(26): 4405–4408.

Klajn, R., J.F. Stoddart and B.A. Grzybowski. 2010. Nanoparticles functionalised with reversible molecular and supramolecular switches. Chem. Soc. Rev. 39(6): 2203–2237.

Kohtani, S., M. Tomohiro, K. Tokumura and R. Nakagaki. 2005. Photooxidation reactions of polycyclic aromatic hydrocarbons over pure and Ag-loaded BiVO$_4$ photocatalysts. Appl. Catal. 58(3-4): 265–272.

Kueh, H.Y. and T.J. Mitchison. 2009. Structural plasticity in actin and tubulin polymer dynamics. Science. 325(5943): 960–963.

Kumar, C.V., E.H. Punzalan and W.B. Tan. 2000. Adenine-thymine base pair recognition by an anthryl probe from the DNA minor groove. Tetrahedron. 56(36): 7027–7040.

Kumar, A., K. Prakash, P.R. Sahoo and S. Kumar. 2017. Visible light controlled aggregation of a spiropyran-HSO$_4^-$ complex as a strategy for reversible detection in water. ChemistrySelect. 2(27): 8247–8252.

Kumar, A., A. Kumar, P.R. Sahoo and S. Kumar. 2019. Colorimetric and fluorescence-based detection of mercuric ion using a benzothiazolinic spiropyran. Chemosensors. 7(3): 35.

Kumar, A. and S. Kumar. 2019. A benzothiazolinic spiropyran for highly selective, sensitive and visible light controlled detection of copper ions in aqueous solution. J. Photochem. Photobiol. A 384: 112265.

Lau, V. and B. Heyne. 2010. Calix[4]arene sulfonate as a template for forming fluorescent thiazole orange H-aggregates. Chem. Commun. 46(20): 3595–3597.

Lee, M., S.-J. Lee and L.-H. Jiang. 2004. Stimuli-Responsive Supramolecular Nanocapsules from Amphiphilic Calixarene Assembly. J. Am. Chem. Soc. 126(40): 12724–12725.

Lee, H.-K., K.M. Park, Y.J. Jeon, D. Kim, D.H. Oh, H.S. Kim, C.K. Park and K. Kim. 2005. Vesicle formed by amphiphilc cucurbit[6]uril: versatile, noncovalent modification of the vesicle surface, and multivalent binding of sugar-decorated vesicles to lectin. J. Am. Chem. Soc. 127(14): 5006–5007.

Lee, M.H., J.L. Sessler and J.S. Kim. 2015. Disulfide-based multifunctional conjugates for targeted theranostic drug delivery. Acc. Chem. Res. 48(11): 2935–2946.

Lehn, J.M. 1993. Supramolecular chemistry. Science. 260(5115): 1762.

Lehn, J.-M. 2017. Supramolecular chemistry: where from? where to? Chem. Soc. Rev. 46(9): 2378–2379.

Leung, K.C.F., C.P. Chak, C.M. Lo, W.Y. Wong, S. Xuan and C.H. Cheng. 2009. pH-controllable supramolecular systems. Chem. Asian J. 4(3): 364–381.

Li, Z., M. Wang, F. Wang, Z. Gu, G. Du, J. Wu and J. Chen. 2007. γ-Cyclodextrin: a review on enzymatic production and applications. Appl. Microbiol. Biotechnol. 77(2): 245.

Li, H., P. Zhang, L.P. Smaga, R.A. Hoffman and J. Chan. 2015. Photoacoustic probes for ratiometric imaging of copper(II). J. Am. Chem. Soc. 137(50): 15628–15631.

Liang, Q., G. Chen, B. Guan and M. Jiang. 2011. Structural factors of amphiphilic calix[6] biscrowns affecting their vesicle–nanotube transitions in self-assembly. J. Mater. Chem. 21(35): 13262–13267.

Lim, C.W., B.J. Ravoo and D.N. Reinhoudt. 2005. Dynamic multivalent recognition of cyclodextrin vesicles. Chem. Commun. (45): 5627–5629.

Liu, Y., M. Chen, T. Cao, Y. Sun, C. Li, Q. Liu, T. Yang, L. Yao, W. Feng and F. Li. 2013. A cyanine-modified nanosystem for in vivo upconversion luminescence bioimaging of methylmercury. J. Am. Chem. Soc. 135(26): 9869–9876.

Lou, X.-Y. and Y.-W. Yang. 2018. Manipulating aggregation-induced emission with supramolecular macrocycles. Adv. Opt. Mater. 6(22):1800668.

Ma, H.L., J.J. Wu, W.J. Liang and J.B. Chao. 2007. Study on the Association Phenomenon of Cyclodextrin to Porphyrin J-aggregates by NMR Spectroscopy. J. Incl. Phenom. Macrocycl. Chem. 58(3): 221–226.

Ma, Y., X. Ji, F. Xiang, X. Chi, C. Han, J. He, Z. Abliz, W. Chen and F. Huang. 2011. A cationic water-soluble pillar[5]arene: synthesis and host–guest complexation with sodium 1-octanesulfonate. Chem. Commun. 47(45): 12340–12342.

Markowitz, M.A., R. Bielski and S.L. Regen. 1989. Ultrathin monolayers and vesicular membranes from calix[6]arenes. Langmuir. 5(1): 276–278.

Martin, O.M. and S. Mecozzi. 2007. Synthesis and pH-dependent self-assembly of semifluorinated calix[4]arenes. Tetrahedron 63(25): 5539–5547.

Martin, A.D., R.A. Boulos, K.A. Stubbs and C.L. Raston. 2011. Phosphonated calix[4]arene-based amphiphiles as scaffolds for fluorescent nano-fibres. Chem. Commun. 47(26): 7329–7331.

Mecca, T., G.M.L. Messina, G. Marletta and F. Cunsolo. 2013. Novel pH responsive calix[8] arene hydrogelators: self-organization processes at a nanometric scale. Chem. Commun. 49(25): 2530–2532.

Möbius, D. 1995. Scheibe aggregates. Adv. Mater. 7(5): 437–444.

Modukuru, N.K., K.J. Snow, B.S. Perrin, A. Bhambhani, M. Duff and C.V. Kumar. 2006. Tuning the DNA binding modes of an anthracene derivative with salt. J. Photochem. Photobiol. A 177(1): 43–54.

Mon, M., T. Grancha, M. Verdaguer, C. Train, D. Armentano and E. Pardo. 2016. Solvent-dependent self-assembly of an oxalato-based three-dimensional magnet exhibiting a novel architecture. Inorg. Chem. 55(14): 6845–6847.

Mudliar, N.H. and P.K. Singh. 2016. Fluorescent H-aggregates hosted by a charged cyclodextrin cavity. Chem. Eur. J. 22(22): 7394–7398.

Mura, P. 2014. Analytical techniques for characterization of cyclodextrin complexes in aqueous solution: a review. J. Pharm. Biomed. Anal. 101: 238–250.

Nakai, T., T. Kanamori, S. Sando and Y. Aoyama. 2003. Remarkably size-regulated cell invasion by artificial viruses. saccharide-dependent self-aggregation of glycoviruses and its consequences in glycoviral gene delivery. J. Am. Chem. Soc. 125(28): 8465–8475.

Ni, X.-L., X. Xiao, H. Cong, L.-L. Liang, K. Cheng, X.-J. Cheng, N.-N. Ji, Q.-J. Zhu, S.-F. Xue and Z. Tao. 2013. Cucurbit[n]uril-based coordination chemistry: from simple coordination complexes to novel poly-dimensional coordination polymers. Chem. Soc. Rev. 42(24): 9480–9508.

Nishimura, T., Y. Sanada, T. Matsuo, T. Okobira, E. Mylonas, N. Yagi and K. Sakurai. 2013. A bimolecular micelle constructed from amphiphilic pillar[5]arene molecules. Chem. Commun. 49(29): 3052–3054.

Nogales, E. and R. Zhang. 2016. Visualizing microtubule structural transitions and interactions with associated proteins. Curr. Opin. Struct. Biol. 37: 90–96.

Ogoshi, T., S. Kanai, S. Fujinami, T.-a. Yamagishi and Y. Nakamoto. 2008. Para-bridged symmetrical pillar[5]arenes: their lewis acid catalyzed synthesis and host–guest property. J. Am. Chem. Soc. 130(15): 5022–5023.

Ogoshi, T., T.-a. Yamagishi and Y. Nakamoto. 2016. Pillar-shaped macrocyclic hosts pillar[n] arenes: new key players for supramolecular chemistry. Chem. Rev. 116(14): 7937–8002.

Ohyama, T., H. Mita and Y. Yamamoto. 2005. Binding of 5, 10, 15, 20-tetrakis (N-methylpyridinium-4-yl)-21H, 23H-porphyrin to an AT-rich region of a duplex DNA. Biophys. Chem. 113(1): 53–59.

Osaki, F., T. Kanamori, S. Sando, T. Sera and Y. Aoyama. 2004. A quantum dot conjugated sugar ball and its cellular uptake. On the size effects of endocytosis in the subviral region. J. Am. Chem. Soc. 126(21): 6520–6521.

Park, K.M., K. Suh, H. Jung, D.-W. Lee, Y. Ahn, J. Kim, K. Baek and K. Kim. 2009. Cucurbituril-based nanoparticles: a new efficient vehicle for targeted intracellular delivery of hydrophobic drugs. Chem. Commun. (1): 71–73.

Park, K.M., D.W. Lee, B. Sarkar, H. Jung, J. Kim, Y.H. Ko, K.E. Lee, H. Jeon and K. Kim. 2010. Reduction-sensitive, robust vesicles with a non-covalently modifiable surface as a multifunctional drug-delivery platform. Small. 6(13): 1430–1441.

Parrot-Lopez, H., C.C. Ling, P. Zhang, A. Baszkin, G. Albrecht, C. De Rango and A.W. Coleman. 1992. Self-assembling systems of the amphiphilic cationic per-6-amino-. beta.-cyclodextrin 2, 3-di-O-alkyl ethers. J. Am. Chem. Soc. 114(13): 5479–5480.

Pashirova, T.N., E.M. Gibadullina, A.R. Burilov, R.R. Kashapov, E.P. Zhiltsova, V.V. Syakaev, W.D. Habicher, M.H. Rümmeli, S.K. Latypov, A.I. Konovalov and L.Y. Zakharova. 2014. Amphiphilic O-functionalized calix[4]resocinarenes with tunable structural behavior. RSC Adv. 4(20): 9912–9919.

Peng, L., S. Liu, A. Feng and J. Yuan. 2017. Polymeric nanocarriers based on cyclodextrins for drug delivery: host–guest interaction as stimuli responsive linker. Mol. Pharmaceutics. 14(8): 2475–2486.

Peyratout, C., E. Donath and L. Daehne. 2001. Electrostatic interactions of cationic dyes with negatively charged polyelectrolytes in aqueous solution. J. Photochem. Photobiol. A 142(1): 51–57.

Powell, M.F. 1987. Facile aryl ether hydrolysis: kinetics and mechanism of 9-anthryl ether cleavage in aqueous solution. J. Org. Chem. 52(1): 56–61.

Ravoo, B.J. and R. Darcy. 2000. Cyclodextrin bilayer vesicles. Angew. Chem., Int. Ed. Engl. 39(23): 4324–4326.

Ravoo, B.J., J.-C. Jacquier and G. Wenz. 2003. Molecular recognition of polymers by cyclodextrin vesicles. Angew. Chem., Int. Ed. Engl. 42(18): 2066–2070.

Robinson, J.T., S.M. Tabakman, Y. Liang, H. Wang, H. Sanchez Casalongue, D. Vinh and H. Dai. 2011. Ultrasmall reduced graphene oxide with high near-infrared absorbance for photothermal therapy. J. Am. Chem. Soc. 133(17): 6825–6831.

Rodik, R.V., A.S. Klymchenko, N. Jain, S.I. Miroshnichenko, L. Richert, V.I. Kalchenko and Y. Mély. 2011. Virus-sized DNA nanoparticles for gene delivery based on micelles of cationic calixarenes. Chem. Eur. J. 17(20): 5526–5538.

Rösch, U., S. Yao, R. Wortmann and F. Würthner. 2006. Fluorescent H-aggregates of merocyanine dyes. Angew. Chem. Int. Ed. 45(42): 7026–7030.

Ruben, M., D. Payer, A. Landa, A. Comisso, C. Gattinoni, N. Lin, J.-P. Collin, J.-P. Sauvage, A. De Vita and K. Kern. 2006. 2D Supramolecular assemblies of benzene-1,3,5-triyl-tribenzoic acid: temperature-induced phase transformations and hierarchical organization with macrocyclic molecules. J. Am. Chem. Soc. 128(49): 15644–15651.

Sahoo, P.R., K. Prakash, A. Kumar and S. Kumar. 2017. Efficient reversible optical sensing of water achieved through the conversion of H-aggregates of a merocyanine salt to J-aggregates. ChemistrySelect. 2(21): 5924–5932.

Samanta, D. and R. Klajn. 2016. Aqueous light-controlled self-assembly of nanoparticles. Adv. Opt. Mater. 4(9): 1373–1377.

Sansone, F., M. Dudič, G. Donofrio, C. Rivetti, L. Baldini, A. Casnati, S. Cellai and R. Ungaro. 2006. DNA condensation and cell transfection properties of guanidinium calixarenes: dependence on macrocycle lipophilicity, size, and conformation. J. Am. Chem. Soc. 128(45): 14528–14536.

Shahgaldian, P., M.A. Sciotti and U. Pieles. 2008. Amino-substituted amphiphilic calixarenes: self-assembly and interactions with DNA. Langmuir. 24(16): 8522–8526.

Shi, B., K. Jie, Y. Zhou, J. Zhou, D. Xia and F. Huang. 2016a. Nanoparticles with near-infrared emission enhanced by pillararene-based molecular recognition in water. J. Am. Chem. Soc. 138(1): 80–83.

Shi, W., L. Bai, J. Guo, X. Wang, N. Xue, C. Zhou and Y. Zhao. 2016b. Self-assembly film of azobenzene and layered double hydroxide and its application as a light-controlled reversible sensor for the detection of Be^{2+}. Sens. Actuators B. 223: 671–678.

Shi, Y., H. Qin, S. Yang and D. Xing. 2016c. Thermally confined shell coating amplifies the photoacoustic conversion efficiency of nanoprobes. Nano Res. 9(12): 3644–3655.

Si, W., L. Chen, X.-B. Hu, G. Tang, Z. Chen, J.-L. Hou and Z.-T. Li. 2011. Selective artificial transmembrane channels for protons by formation of water wires. Angew. Chem., Int. Ed. Engl. 50(52): 12564–12568.

Song, S., J. Wang, H.-T. Feng, Z.-H. Zhu and Y.-S. Zheng. 2014. Supramolecular hydrogel based on amphiphilic calix[4]arene and its application in the synthesis of silica nanotubes. RSC Adv. 4(47): 24909–24913.

Sreejith, S., P. Carol, P. Chithra and A. Ajayaghosh. 2008. Squaraine dyes: a mine of molecular materials. J. Mater. Chem. 18(3): 264–274.

Strobel, M., K. Kita-Tokarczyk, A. Taubert, C. Vebert, P.A. Heiney, M. Chami and W. Meier. 2006. Self-assembly of amphiphilic calix[4]arenes in aqueous solution. Adv. Funct. Mater. 16(2): 252–259.

Strutt, N.L., H. Zhang, S.T. Schneebeli and J.F. Stoddart. 2014. Functionalizing pillar[n]arenes. Acc. Chem. Res. 47(8): 2631–2642.

Sukegawa, T., T. Furuike, K. Niikura, A. Yamagishi, K. Monde and S.-I. Nishimura. 2002. Erythrocyte-like liposomes prepared by means of amphiphilic cyclodextrin sulfates. Chem. Commun.(5): 430–431.

Sun, Y., C.-G. Yan, Y. Yao, Y. Han and M. Shen. 2008. Self-assembly and metallization of resorcinarene microtubes in water. Adv. Funct. Mater. 18(24): 3981–3990.

Sun, S., S. Andersson, R. Zhang and L. Sun. 2010. Unusual partner radical trimer formation in a host complex of cucurbit[8]uril, ruthenium(II) tris-bipyridine linked phenol and methyl viologen. Chem. Commun. 46(3): 463–465.

Sun, Y., Y. Yao, C.-G. Yan, Y. Han and M. Shen. 2010. Selective decoration of metal nanoparticles inside or outside of organic microstructures via self-assembly of resorcinarene. ACS Nano 4(4): 2129–2141.

Szejtli, J. 1998. Introduction and general overview of cyclodextrin chemistry. Chem. Rev. 98(5): 1743–1754.

Tanaka, Y., M. Miyachi and Y. Kobuke. 1999. Selective vesicle formation from calixarenes by self-assembly. Angew. Chem., Int. Ed. Engl. 38(4): 504–506.

Versluis, F., I. Tomatsu, S. Kehr, C. Fregonese, A.W.J.W. Tepper, M.C.A. Stuart, B.J. Ravoo, R.I. Koning and A. Kros. 2009. Shape and release control of a peptide decorated vesicle through pH sensitive orthogonal supramolecular interactions. J. Am. Chem. Soc. 131(37): 13186–13187.

Vico, R.V., J. Voskuhl and B.J. Ravoo. 2011. Multivalent interaction of cyclodextrin vesicles, carbohydrate guests, and lectins: a kinetic investigation. Langmuir 27(4): 1391–1397.

Voskuhl, J. and B.J. Ravoo. 2009. Molecular recognition of bilayer vesicles. Chem. Soc. Rev. 38(2): 495–505.

Voskuhl, J., T. Fenske, M.C.A. Stuart, B. Wibbeling, C. Schmuck and B.J. Ravoo. 2010a. Molecular recognition of vesicles: host-guest interactions combined with specific dimerization of zwitterions. Chem. Eur. J. 16(28): 8300–8306.

Voskuhl, J., M.C.A. Stuart and B.J. Ravoo. 2010b. Sugar-decorated sugar vesicles: lectin–carbohydrate recognition at the surface of cyclodextrin vesicles. Chem. Eur. J. 16(9): 2790–2796.

Wang, H., S. Wang, H. Su, K.J. Chen, A.L. Armijo, W.Y. Lin, Y. Wang, J. Sun, K.I. Kamei and J. Czernin. 2009. A supramolecular approach for preparation of size-controlled nanoparticles. Angew. Chem., Int. Ed. Engl. 121(24): 4408–4412.

Wang, Y.-X., Y.-M. Zhang and Y. Liu. 2015. Photolysis of an amphiphilic assembly by calixarene-induced aggregation. J. Am. Chem. Soc. 137(13): 4543–4549.

Ward, M.D. and P.R. Raithby. 2013. Functional behaviour from controlled self-assembly: challenges and prospects. Chem. Soc. Rev. 42(4): 1619–1636.

Xiao, T., L. Qi, W. Zhong, C. Lin, R. Wang and L. Wang. 2019. Stimuli-responsive nanocarriers constructed from pillar[n]arene-based supra-amphiphiles. Mater. Chem. Front. 3(10): 1973–1993.

Xue, M., Y. Yang, X. Chi, Z. Zhang and F. Huang. 2012. Pillararenes, a new class of macrocycles for supramolecular chemistry. Acc. Chem. Res. 45(8): 1294–1308.

Yang, H., S. Zhang, K. Liu and Y. Fang. 2016a. Calix[4]arene-based low molecular mass gelators to form gels in organoalkoxysilanes. RSC Adv. 6(111): 109969–109977.

Yang, Y., K. Achazi, Y. Jia, Q. Wei, R. Haag and J. Li. 2016b. Complex assembly of polymer conjugated mesoporous silica nanoparticles for intracellular pH-responsive drug delivery. Langmuir 32(47): 12453–12460.

Yao, Y., M. Xue, J. Chen, M. Zhang and F. Huang. 2012. An amphiphilic pillar[5]arene: synthesis, controllable self-assembly in water, and application in calcein release and TNT adsorption. J. Am. Chem. Soc. 134(38): 15712–15715.

Yao, Y., M. Xue, Z. Zhang, M. Zhang, Y. Wang and F. Huang. 2013. Gold nanoparticles stabilized by an amphiphilic pillar[5]arene: preparation, self-assembly into composite microtubes in water and application in green catalysis. Chem. Sci. 4(9): 3667–3672.

Yao, Y., J. Li, J. Dai, X. Chi and M. Xue. 2014a. A water-soluble pillar[6]arene: synthesis, host–guest chemistry, controllable self-assembly, and application in controlled release. RSC Adv. 4(18): 9039–9043.

Yao, Y., P. Wei, S. Yue, J. Li and M. Xue. 2014b. Amphiphilic pillar[5]arenes: influence of chemical structure on self-assembly morphology and application in gas response and λ-DNA condensation. RSC Adv. 4(12): 6042–6047.

Yao, Y., Y. Wang and F. Huang. 2014c. Synthesis of various supramolecular hybrid nanostructures based on pillar[6]arene modified gold nanoparticles/nanorods and their application in pH- and NIR-triggered controlled release. Chem. Sci. 5(11): 4312–4316.

Yoshino, N., A. Satake and Y. Kobuke. 2001. An artificial ion channel formed by a macrocyclic resorcin[4]arene with amphiphilic cholic acid ether groups. Angew. Chem., Int. Ed. Engl. 40(2): 457–459.

Yu, G., X. Zhou, Z. Zhang, C. Han, Z. Mao, C. Gao and F. Huang. 2012. Pillar[6]arene/ paraquat molecular recognition in water: high binding strength, ph-responsiveness, and application in controllable self-assembly, controlled release, and treatment of paraquat poisoning. J. Am. Chem. Soc. 134(47): 19489–19497.

Yu, G., Y. Ma, C. Han, Y. Yao, G. Tang, Z. Mao, C. Gao and F. Huang. 2013. A sugar-functionalized amphiphilic pillar[5]arene: synthesis, self-assembly in water, and application in bacterial cell agglutination. J. Am. Chem. Soc. 135(28): 10310–10313.

Yu, G., J. Yang, X. Fu, Z. Wang, L. Shao, Z. Mao, Y. Liu, Z. Yang, F. Zhang, W. Fan, J. Song, Z. Zhou, C. Gao, F. Huang and X. Chen. 2018a. A supramolecular hybrid material constructed from graphene oxide and a pillar[6]arene-based host–guest complex as an ultrasound and photoacoustic signal nanoamplifier. Mater. Horiz. 5(3): 429–435.

Yu, G., Z. Yang, X. Fu, B.C. Yung, J. Yang, Z. Mao, L. Shao, B. Hua, Y. Liu, F. Zhang, Q. Fan, S. Wang, O. Jacobson, A. Jin, C. Gao, X. Tang, F. Huang and X. Chen. 2018b. Polyrotaxane-based supramolecular theranostics. Nat. Commun. 9(1): 766.

Yun, G., Z. Hassan, J. Lee, J. Kim, N.-S. Lee, N.H. Kim, K. Baek, I. Hwang, C.G. Park and K. Kim. 2014. Highly stable, water-dispersible metal-nanoparticle-decorated polymer nanocapsules and their catalytic applications. Angew. Chem., Int. Ed. Engl. 53(25): 6414–6418.

Zakharova, L.Y., V.V. Syakaev, M.A. Voronin, F.V. Valeeva, A.R. Ibragimova, Y.R. Ablakova, E.K. Kazakova, S.K. Latypov and A.I. Konovalov. 2009. NMR and spectrophotometry study of the supramolecular catalytic system based on polyethyleneimine and amphiphilic sulfonatomethylated calix[4]resorcinarene. J. Phys. Chem. C. 113(15): 6182–6190.

Zhang, H.F., K. Maslov, G. Stoica and L.V. Wang. 2006. Functional photoacoustic microscopy for high-resolution and noninvasive *in vivo* imaging. Nat. Biotechnol. 24(7): 848–851.

Zhang, H., L. Sun, Z. Liu, W. An, A. Hao, F. Xin and J. Shen. 2010. pH-responsive vesicle-like particles based on inclusion complexes between cyclodextrins and methyl orange. Colloids Surf. A Physicochem. Eng. Asp. 358(1): 115–121.

Zhang, Z., Y. Luo, J. Chen, S. Dong, Y. Yu, Z. Ma and F. Huang. 2011. Formation of linear supramolecular polymers that is driven by C–H···π interactions in solution and in the solid state. Angew. Chem., Int. Ed. Engl. 50(6): 1397–1401.

Zhang, T., S. Sun, F. Liu, Y. Pang, J. Fan and X. Peng. 2011. Interaction of DNA and a series of aromatic donor-viologen acceptor molecules with and without the presence of CB[8]. Phys. Chem. Chem. Phys. 13(20): 9789–9795.

Zhang, H., X. Ma, K.T. Nguyen and Y. Zhao. 2013. Biocompatible pillararene-assembly-based carriers for dual bioimaging. ACS Nano. 7(9): 7853–7863.

Zhang, Y.-M., Y.-H. Liu and Y. Liu. 2020. Cyclodextrin-based multistimuli-responsive supramolecular assemblies and their biological functions. Adv. Mater. 32(3): 1806158.

Zhao, Y.S., H. Fu, A. Peng, Y. Ma, Q. Liao and J. Yao. 2010. Construction and optoelectronic properties of organic one-dimensional nanostructures. Acc. Chem. Res. 43(3): 409–418.

Zhou, J., M. Chen, J. Xie and G. Diao. 2013. Synergistically enhanced electrochemical response of host-guest recognition based on ternary nanocomposites: Reduced graphene oxide-amphiphilic pillar[5]arene-gold nanoparticles. ACS Appl. Mater. Interfaces 5(21): 11218–11224.

Zhou, J., M. Chen and G. Diao. 2014a. Magnetic-responsive supramolecular vesicles from self-organization of amphiphilic pillar[5]arene and application in controlled release. ACS Appl. Mater. Interfaces 6(21): 18538–18542.

Zhou, J., M. Chen and G. Diao. 2014b. Synthesis of the first amphiphilic pillar[6]arene and its enzyme-responsive self-assembly in water. Chem. Commun. 50(80): 11954–11956.

Zhu, H., L. Shangguan, B. Shi, G. Yu and F. Huang. 2018. Recent progress in macrocyclic amphiphiles and macrocyclic host-based supra-amphiphiles. Mater. Chem. Front. 2(12): 2152–2174.

Zimmerman, S.C. and P.S. Corbin. 2000. Heteroaromatic modules for self-assembly using multiple hydrogen bonds. pp. 63–94. *In*: M. Fuiita (ed.). Molecular Self-Assembly Organic Versus Inorganic Approaches. Berlin, Heidelberg: Springer.

Chapter 11

Stimuli-responsive Macrocyclic Receptors

INTRODUCTION

Macrocyclic receptors are a family of cyclic cage-like hosts that consist of repeating units to create specific cavities, which include calixarenes, cyclodextrins, cucurbiturils, cyclotriveratrylenes, etc., (Gutsche 1998; Gutsche 2008; Chawla et al. 2010; Busschaert et al. 2015). Owing to the specific size of the cavity such host molecules display enhanced selectivity towards an analyte species. An analyte species is held within the cavity of the macrocyclic host through the non-covalent interactions like Van der Waals, hydrophobic, hydrogen bond, and electrostatic interactions, that work synergistically to provide a basis of molecular recognition. The pre organized cavity of such a macrocyclic host molecule yields a highly selective sensing system for ionic and molecular species with an optical signal.

The extent of molecular recognition is contingent on the size, shape and orientation of the molecules involved. Artificial receptors are not only useful for increasing selectivity and sensitivity in extractions and other selective separations methods, but molecular recognition is also a great tool for creating novel sensors (Lehn 2011). A great deal of understanding and quantifying these molecular recognition interactions is required to create better, more accurate and faster analysis and develop novel sensing materials with endless applications. Recent work in this emerging multidisciplinary field was acknowledged with a 1987 Nobel Prize to C.J. Pederson, (Pedersen 1967, 1988) Donald J. Cram (Cram and Cram 1974) and J.-M. Lehn (Dietrich et al. 1969, 1973) for their work on crown ethers and cryptands.

Since that time, continued efforts have been undertaken to synthesize and investigate the coordination chemistry of related host compounds for ionic recognition based on the principles of complementarity and pre organization. For instance, cyclodextrins, (Chen and Jiang 2011; Guo et al. 2011; Jana and Bandyopadhyay

2011; Schönbeck et al. 2011) glycoluril, (Hof et al. 2000; Pinjari and Gejji 2009; She et al. 2009) calixarenes, (Chawla et al. 2010) cyclotriveratrylenes (Steed et al. 1996; Brotin et al. 2005; Szumna 2010) and other cyclophanes (Atwood 2017) follow fundamental principles of supramolecular chemistry.

The macrocyclic receptors binding to an analyte may be compared to placing an analyte in a container of nano-size dimensions. If this situation is compared to a daily life situation, where an object is placed in a container having a cap or lid. The object can be held inside the closed container while opening the lid allowing one to place the object inside the container and take it out when needed. The addition of this feature to molecular receptors leads to the development of stimuli-controlled macrocyclic receptors for practical application in the storage and release of chemical species. The rate of capture or release of an analyte (chemical species) can be controlled by using an appropriate stimulus.

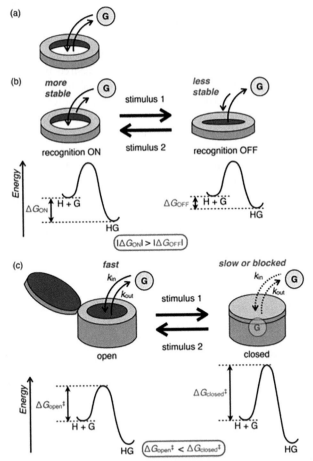

Figure 11.1 Schematic drawings of controlled host-guest binding. (a) Guest uptake/ release of a conventional host, (b) thermodynamic control of host-guest binding, (c) kinetic control of host-guest binding. Reproduced here with the permission of the Chemical Society Japan (Akine and Sakata 2020).

The control of binding and release of an analyte by a macrocyclic host is a function of increasing or decreasing of $\Delta G^{\ddagger}$ by external stimuli. The area of the stimuli-responsive system in the case of stimuli controlled macrocyclic systems is limited to few molecules, while the thermodynamic alteration of the binding constant to control it towards a guest species is well established through photoreaction, redox reaction or pH changes, etc. Kinetic and thermodynamic factors are responsible for the turn on and off tendencies of the macrocyclic receptors. In the absence of thermodynamic factors, the release of strongly bound guests is usually not achieved even when the structure of the receptor's lid is open. Normally a difference in thermodynamic factors leads to a change in binding affinity, which allows uptake and release of bound guest molecules or ions. The thermodynamic control is a well-established control for on and off binding of the guest by the macrocyclic receptors. However, only a few examples of kinetic control on the guest binding by macrocyclic receptors are available that deal with uptake and release of guest molecules or ions.

Besides the fundamental understanding of molecular interactions and structural transformations in response to external stimuli is another aspect of the rational design of macrocyclic receptors with desired properties for targeted applications. A large number of publications in recent decades have appeared on various aspects of macrocyclic receptors that involve functionalization to achieve the desired attributes (Chawla et al. 2010). A common feature of these architectures is their ability to encapsulate only those guests whose size, shape and chemical exterior complement the hosts. Their interior can be regarded as a new phase of matter which is capable of controlling the flow of reactant, transient species, products and catalyzing chemical or biochemical reactions.

A variety of stimuli like light, pH, temperature and pressure are used to transform a receptor structurally from one form to another. In the following sections, the receptors classification based on different stimuli are described.

LIGHT RESPONSIVE RECEPTORS

Light responsive receptors respond to light by switching between structurally different states having significantly different absorption spectra. Usually, such structurally different forms display variable affinity towards a guest species. In this context, photochromic materials attracted much attention in the last decades due to their potential applications as optical fibers, switches, optical memories, calcium oscillation mimics and other molecular devices (Crano and Guglielmetti 1999; Bouas-Laurent and Dürr 2001; Dürr and Bouas-Laurent 2003). Attachment of even a single photochromic unit to a macrocyclic receptor creates receptors that respond to the presence of light of an appropriate wavelength. Photochromism is defined as the reversible transformation of a chemical species between two configurations, with different absorption spectra, shifted in one or both directions by absorption of electromagnetic radiation of an appropriate wavelength (Bertelson 1971; Crano and Guglielmetti 1999). A variety of photochromic systems such as spiropyrans and spirooxazines, oxazines, naphthopyran azobenzenes, diarylethenes and fulgides have been reported (Bertelson 1971; Crano and Guglielmetti 1999). The photochromic molecules usually display light, pH, temperature, ion or molecule responsive behavior. The photochromic compounds normally undergo an alteration in their molecular structure on exposure to an appropriate stimulus. The photochromic molecule such as spiropyran, spirooxazine and naphthopyran display two typical forms, which are known as colorless Spiro (SP) and a Colored Merocyanine (MC) form. The structural forms of photochromic oxazines are called OX (colorless) and IN (colored) forms. The polar MC form exists as a zwitterion in polar solution and has more complicated isomers based on the *cis–trans* isomerization of the conjugated bonds such as TTC (*trans*-transoid-*cis*), CTC, CTT and TTT. The colored IN form of photochromic oxazines display less complicated chemistry during structural alteration.

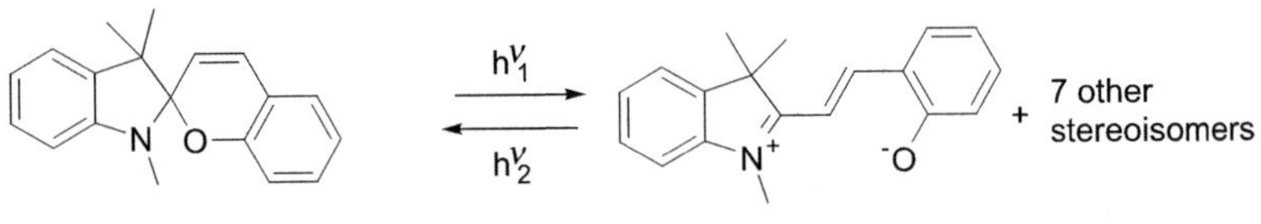

1 Closed form, Colourless   **1** Open form, Coloured

Figure 11.2 A photo-induced interconversion of spiropyran.

Photochromic molecules can be modified through chemical reactions to develop them into useful and selective receptors for an analyte like metal ions, anions or molecules. The binding can be turned off and on by the light of a particular wavelength caused by a change in the configuration of the photochromic scaffold. The covalent fusion of a macrocyclic molecule like a crown ether or calixarene with a photochromic skeleton-like spiropyran offers a great variety of properties, including the control over binding of a chemical species through photoisomerization of the spiropyran (or another photochromic unit) ring (Kawai 1998; Takeshita and Irie 1998; Takeshita et al. 1998; Tanaka et al. 2000, 2001). A variety of macrocyclic receptors possessing photochromic units have appeared in literature (Kawai 1998; Takeshita and Irie 1998; Takeshita et al. 1998; Tanaka et al. 2000, 2001). Next different stimuli-responsive macrocyclic receptors for a different type of chemical species are described.

While developing receptors, the following features of photochromic molecules must be kept in mind: (1) only the MC isomers strongly absorb visible light, (2) the MC isomers-analyte complex are thermodynamically more stable than the SP isomers under dark conditions, and (3) the energy barrier for the isomerization of the MC to SP isomers is relatively low. Thus, dark-adapted merocyanine in aqueous media will readily isomerize to closed-form on exposure to safe, visible light. The type of interconversion under mild conditions is advantageous when considering the applications mentioned above. Additionally, the other requirements of a photoreversible receptor for an analyte are:

(4)   The compound must be switched on or off with light (Alward 1999).

(5)   The compound must be able to bind in the dark with a large binding constant with an observable colorimetric or fluorometric signal. Light can be used to dissociate the metal ion.

(6) The compound should bind to a metal ion of interest.

(7) The complex should be photosensitive to allow possible pathways for analyte release. If only the ligand is photosensitive, then light can influence the analyte-ligand binding indirectly (Alward 1999).

Classification of Photoactive Receptors

The structural alteration in the photoactive receptors proceeds through a single step electrocyclic transformation during light irradiation (Fig. 11.2). Several examples of photoactive molecules are available in literature (Moorthy et al. 2013). Out of many photochromic species, naphthopyrans, spiropyrans and spirooxazines display enhanced activation in response to metal chelation (Shao et al. 2009; Yagi et al. 2009; Huang et al. 2015; Baldrighi et al. 2016). Another class of photoswitches has also been reported in literature such as acyl hydrazones developed by Heccht et al. (van Dijken et al. 2015). Such small molecule switches exhibit excellent fatigue resistance with more than 100 nm band separation between E and Z isomers. Hecht et al. have also revamped the fatigue resistance of diarylethenes by introducing electron-withdrawing groups, such as trifluoromethyl substituents into the core system (Herder et al. 2015). Visible light-triggered 1,2-dicyanoethene substituted blinkers have been developed, which can undergo reversible transformation between *cis* and *trans* geometries (Guo et al. 2015; Zhou et al. 2015). The excellent electronic nature of reversible dicyanoethenes bears the potential to act as an electron carrier in novel sunlight responsive semiconductor materials.

Advantages of Photoactive Receptors

Unlike other sensors, photoactive receptors provide a viable structural design to induce metal ion binding and consequent release with visible light (Kumar et al. 2008). Owing to their swift switching response under the influence of light, photoactive molecules operated directly in many practical fields with a large turnover worldwide, mainly memory storage, nonlinear optics, rewritable inkjet appliances, sensitive optical fibers, etc. (Thies et al. 2012; Doddi et al. 2015; Genovese et al. 2015; Wang et al. 2015; Barachevsky 2016; Han et al. 2016; Lee et al. 2016; Lin et al. 2016; Mohan Raj et al. 2016; Vlasceanu et al. 2016; Mutoh et al. 2017).

The detection of the metal ion is an intensively investigated research area (Satapathy et al. 2012; Hessels et al. 2015; Jiménez-Sánchez et al. 2015; Liu et al. 2016; Yan et al. 2016; Yang et al. 2016; Chen et al. 2017). A variety of receptors have been developed for the detection of metal ions, which are relevant both for biology and the environment. Metal ions are ubiquitous in one's environment. Some metal ions are essential to drive biological processes at low concentration, while others are highly toxic (Galloway et al. 1982; De Silva et al. 1997; Quang and Kim 2010; Aragay et al. 2011; Saleem and Lee 2015). Besides, the biologically relevant metal ions display toxicity when present in excess, and affect the health of living organisms. Excessive use of industrial products in daily life universally have

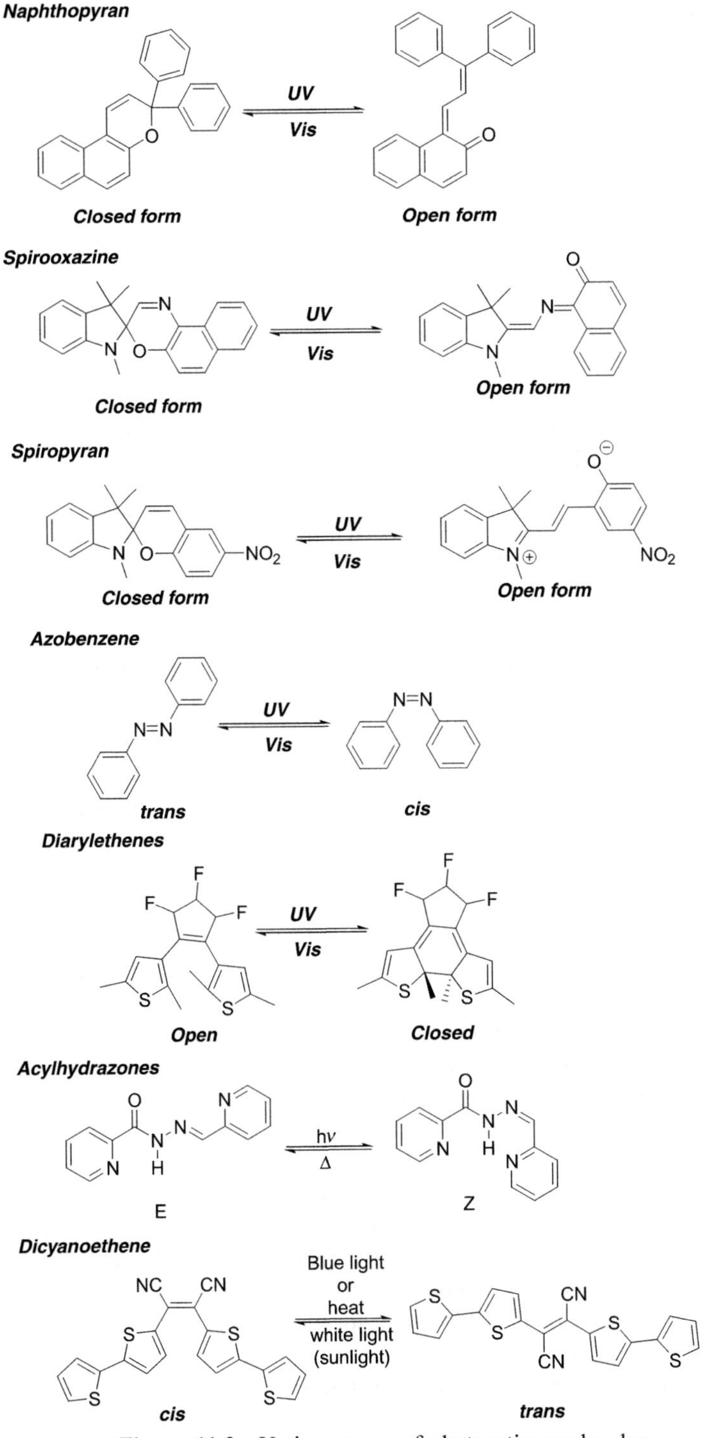
Naphthopyran
UV
Vis
O
O
Closed form
Open form
Spirooxazine
N
N
O
UV
Vis
O
N
N
Closed form
Open form
Spiropyran
N
O
NO2
UV
Vis
O
N
NO2
Closed form
Open form
Azobenzene
N=N
UV
Vis
N=N
trans
cis
Diarylethenes
F
F
F
UV
Vis
F
F
F
S
S
S
S
Open
Closed
Acylhydrazones
O
N
N
N
H
N
hν
Δ
O
N
N
N
H
N
E
Z
Dicyanoethene
NC
CN
S
S
S
S
Blue light
or
heat
white light
(sunlight)
CN
CN
S
S
S
S
cis
trans

Figure 11.3 Various types of photoactive molecules.

created new sources of metal pollution, which require the scientific community to broaden the detection techniques to deal with this growing environmental problem (Dean et al. 1972). Although several receptors have been reported for the detection of metal ions, among them the macrocyclic receptors are unique as they provide a pre organized cavity, which leads to selectivity and sensitivity during the detection process. The receptor is one of the most important and expensive components during any detection process. Therefore, the re use of such a receptor can improve the cost of the detection process. However, there are many advanced forms of photoreversible studies that reflect the heavy metal ions sensing consciousness of the past (Brown 1971; Dürr and Bouas-Laurent 2003). This material described in this chapter is far from static, bringing freshness in highlighting the elaborative binding mechanism and the fundamental principle behind tracking. It includes metal ions from diverse arenas starting from transition metal ions like Fe^{3+} to lanthanide metal ions notably Eu^{3+}, Gd^{3+}, and Tb^{3+}, etc. The photo-reversibility feature as an advanced tool for sensing heavy metal ions will also be addressed here. The focus is on macrocyclic receptor design based on well-known macrocyclic molecules (crown ethers, calixarene, cyclodextrins etc.) conjugated to photochromic molecules such as oxazines, spirooxazines, spiropyrans and naphthopyrans, which enable stimuli responsive behavior.

Light Responsive Metal Ions Macrocyclic Receptors

The covalently attached photochromic moiety to the pre organized macrocyclic site, in general, allows the development of a light-responsive reversible macrocyclic receptor. The photochromic unit attached to the calixarene skeleton could undergo a change in polarity when closed-form gets converted to open the form in response to light. This remarkable change in polarity and structure (especially planarity) between SP and MC forms of spiropyran could be utilized in sensing metal ions. Thus a photo-switchable macrocyclic sensor has great potential to act as a sensor of organic/inorganic compounds. These photoreversible sensors would be able to detect/bind or release (preconcentration) heavy metal ions. Combination of calixarene and photochromic materials could provide the required selectivity and specificity in metal ion detection and removal from various samples. The technique exploits interactions and cages typical of supramolecular chemistry for sensing metal ions, sensitively and selectively to achieve better limits of detection along with improved accuracy. Such a sensing device can reduce time to minute scale and is economically favorable with larger coverage. Macrocyclic photoreversible receptors have increasingly found it feasible to tackle the adverse atmospheric remedies that bear such features on a single platform with added flexibility. The highlights of macrocyclic photoreversible receptors involve alternation in structure between two different forms of a photochromic moiety as a part of the macrocyclic structure, which results in different absorption peaks on the screen of a UV recorder in a reversible manner; in a sense the latter one (MC) can easily come back to the initial one (SP) within a short timescale (Crano and Guglielmetti 1999). The pivotal function of the light-mediated sensor is an authentic selection of metal species. The so-called lacunae in theory and practice make it difficult to launch an all-rounder inbuilt encapsulator. By changing the appropriate functionality around the base structure, the target species may exhibit a new twist to the interesting phenomena of detection (Brown 1971). That is why probably today's scientists and technologists are gearing up for more advanced sensors. The mandate is to conceptualize sensors surveillance capacity that can be enhanced by manipulating crown ethers (Kimura et al. 2000), calix[*n*]arene (Grün et al. 2008), fullerene activated functionality, etc., (Straight et al. 2005). More often, light sensitivity makes such receptors conspicuous, while binding to the metal ions.

Lead selective receptors

Owing to high toxicity associated with lead ions, Stauffer et al. developed a naphthopyran derivative (**2**) covalently linked to a crown ether (Stauffer et al. 1997). The photoreversible macrocycle displayed a selective binding affinity toward lead ions in aqueous methanol under dark conditions. Benzo 15-crown-5 ether moiety has been observed to bind the lead ions assisted by the naphthopyran unit to produce a voltammetric signal. In response to an increase in the concentration of receptor (**2**), the Pb^{2+} ions reduction band decreased in intensity. A broad signal due to the formation of a **2**-Pb^{2+} complex has been observed at ~600 mV (cathodic reduction). The binding was again monitored in the presence of UV-light, which displayed a reduction peak potential for free Pb^{2+} ions at ~420 mV due to the release of lead ions by the naphthopyran-crown ether (**2**). The results indicated that the association-dissociation cycle can be easily controlled using UV light (Scheme 1). The electrochemical results have been supported by ^{1}H and ^{13}C-NMR spectroscopies, which indicated a chemical shift in the methylene protons of the crown ether moiety on the addition of lead ions. The corresponding experiments with the dimethoxy analogue of **2** (without crown ether moiety) displayed no electrochemical reduction peak potential for lead ions.

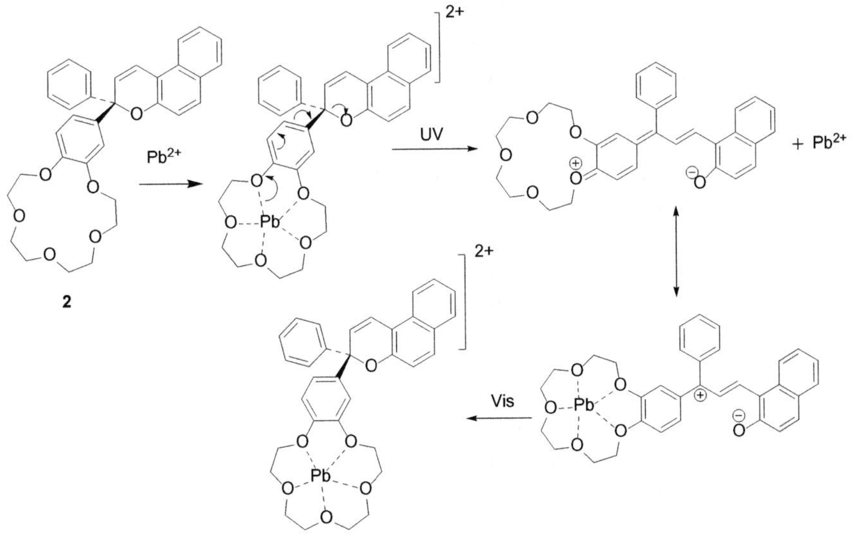

Scheme 1 Light-induced isomerization of naphthopyran receptor **2**.

Light responsive receptors for lanthanides and actinides

The complex of gadolinium ions with synthetic molecular receptors find applications as contrasting agents (Weinmann et al. 1984; Aime and Caravan 2009). Keeping this fact in view, Louie et al. (Tu et al. 2009) developed an aza crown modified spiropyran **3**, which was observed to coordinate the Gd^{3+} ions using crown ether moiety assisted by three carboxylate ions and the phenolic oxygen atom present in the merocyanine form of the spiropyran moiety. The 3-Gd^{3+} complex has been observed to be sensitive to the presence of visible light and reverted to the *spiro* form of the spiropyran after visible light absorption to produce a decrease in relaxivity to 18%, which can be easily controlled by light through the spiro-merocyanine from equilibrium. The 3-Gd^{3+} complex also produced a relaxivity decrease to 26% in the presence of NADH. However, the process has been observed to be irreversible in the presence of NADH. The redox sensitivity of the 3-Gd^{3+} complex along with contrasting ability opened a new avenue towards its utility in a variety of areas along with the potential application as an MRI (Magnetic Resonance Imaging) contrast agent.

Lanthanum complexes have been detected to play a diverse role in different cellular and biological functions like lipid peroxidation, ATPase activity, antitumor agents, phosphate ester hydrolysis and diagnosis of hyperphosphatemia along with a variety of other applications (Haiduc and Silvestru 1990; Zheng et al. 2000; Ganjali et al. 2003; Wang et al. 2014; Areti et al. 2015; Gálico et al. 2016). The size equivalence of the Gd^{3+} ions and Ca^{2+} ions, allow the interaction of Gd^{3+} ions with the binding sites of Ca^{2+} in protein for efficient mimicking (Corbalan-Garcia et al. 1992; Areti et al. 2015). The lanthanum complexes also find diverse applications in industries, which made determination of La^{3+}ion a necessity.

In view of the statement given above, Kimura et al. developed a crown ether modified spiropyran probe (**4**) and examined its affinity towards transition, alkali, alkaline and lanthanide metal ions (Kimura et al. 2000). The spiropyran based macrocycle (**4**) detected the presence of La^{3+} ions in the presence of metal ions, such as Cd^{2+}, Pb^{2+}, Eu^{3+}, Li^+, Na^+, K^+, Mg^{2+}, Ca^{2+} in acetonitrile and produced an intense absorption band at 500–600 nm in the dark. The appearance of an absorption band in the 500–600 nm region indicated the presence of the merocyanine form of the spiropyran, which formed a complex with the La^{3+} ion. The formation of the complex may be attributed to the coordination of lanthanum ions with the phenolic oxygen atom, which induced the ring-opening of the *spiro* form producing a merocyanine form along with a shift in an absorption band in the visible region. The formation of the complex has been observed to be dependent on the size of the crown ether moiety. The smaller size crown ether cavity connected to spiropyran (**4a**) prefers to bind La^{3+} ions. During the studies of complex formation between lanthanum ions and **4a** in the presence of other metal ions, a competition with smaller size alkali and alkaline earth metal ions has also been observed. However, the macrocyclic probe **4b** having two spiropyran moieties displayed better selectivity and formed a complex with the La^{3+} ion without interference from other metal ions. The complex formation between **4b** and La^{3+} ions was confirmed using mass spectroscopy, which displayed an intense peak at m/z: 565 indicating the La^{3+}ion complexation. The competitive mass spectrum studies also indicated the selective formation of a complex between macrocyclic probe (**4b**) and La^{3+} ions in the presence of different metal ions, thereby confirming the selectivity. The ^{139}La NMR spectrum was also used to confirm the interaction between the probe and the lanthanum ion, which indicated the interaction of the phenolate ion of the merocyanine unit of the spiropyran with La^{3+} ions. The mass spectrometry studies further revealed the difference in the affinity of the macrocyclic probes **4a** and **4b** toward metal ions through the peak intensity analysis of the receptor-metal complex.

The formation of a complex between **4b** and lanthanum ion has been observed to be controlled using visible light in acetonitrile (Scheme 2). The **4b**-La^{3+} complex in the presence of 100 fold potassium ion on exposure to visible light (>500 nm) produced a decrease in the intensity of the absorption band at 500 nm, which has been attributed to the conversion of the merocyanine form into the spiro form. The mass spectrometry also indicated a reduction in the intensity of the peak due to **4b**-La^{3+} with a simultaneous increase in the intensity of the **4b**-K^+ complex ion peak complex on exposure to visible light, when studies were conducted in the presence of 100-fold potassium ion concentration. The **4b**-La^{3+} ion peak in the mass spectrum was observed to regenerate in the dark after light exposure. Both mass spectrometry and UV-visible spectroscopy indicated the release of La^{3+} ion from the probe on exposure to visible light along with the binding of the potassium ion (Scheme 2).

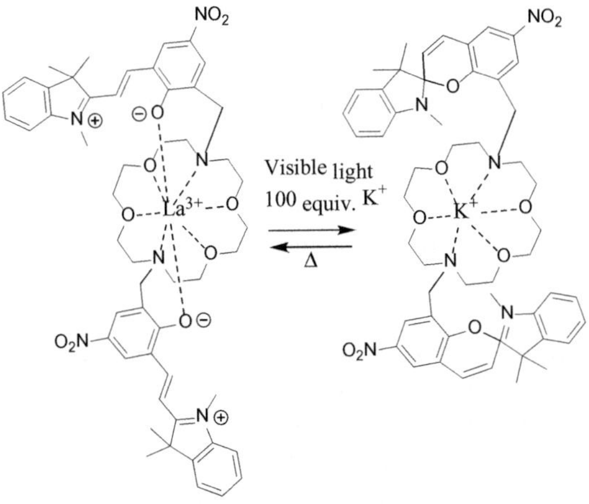

Scheme 2 The light-controlled release of La^{3+} ion from the receptor **4b**.

Light controlled complex with multiple ions

Chebun'kova et al. synthesized morpholine and aza-crown substituted with naphthopyran to produce photoreversible probes (**5a-c**) for the recognition of metal ions. The probes were found to produce a hypsochromic shift in their absorption spectra on the addition of Mg^{2+}, Ba^{2+}, and Pb^{2+} ions in acetonitrile (Chebun'kova et al. 2005). However, the non-macrocyclic probe **5a** produced hypsochromic shift in its absorption spectra on addition of very high metal ions concentration (0.1 M). The result suggested the formation of complex between the crown ether moiety and the metal ions. The solution of metal ions and probes on exposure to UV-light produce a shift in the absorption band to the visible region due to the formation of the merocyanine form of the photochromic unit associated with the crown ether. The formation of 1:1 complex between probe **5b, c** led to a large bathochromic shift along with a reduction in the rate constant for the conversion of the merocyanine colored from to the colorless spiro form. The experiment indicated the interaction between the phenolic oxygen atom of the merocyanine form and the metal ions. The better binding between the metal ions and the probe **5b, c** highlighted the role played by the macrocyclic moieties during the binding process as compared to the probe **5a**.

Paramonov et al. (Paramonov et al. 2010) also reported the synthesis of a chromene-crown ether-based probe **6**, which was found to coordinate metal ions like Mg^{2+}, Ba^{2+}, and Pb^{2+} in acetonitrile (Scheme 3). The complex formation was observed to strictly follow the host-guest size fit model, where a smaller size ion was observed to be recognized by a probe having a small cavity dimension that matched the ion and formed 1:1 complex. The larger ions were observed to form sandwiched type complex concerning the crown ether moiety of probes to form 2:1 (H:G) complex. The metal ions produced a small hypsochromic shift in the absorption spectra of the probes indicating the coordination between the crown ether moiety and the metal ion. The metal ion complexes of the probe, when exposed to UV-light produced an absorption in the visible region, which was attributed to the formation of a colored form of the chromene moiety. The complex formation between metal and the probe **6** decreased the rate of thermal conversion of the colored form to the colorless *spiro* form. The ^{1}H-NMR spectroscopy further confirmed the interaction between the probe and the metal ion. The change in chemical shift on light exposure to form *merocyanine* from-metal ion complex was also confirmed using ^{1}H-NMR spectroscopy at the molecular level.

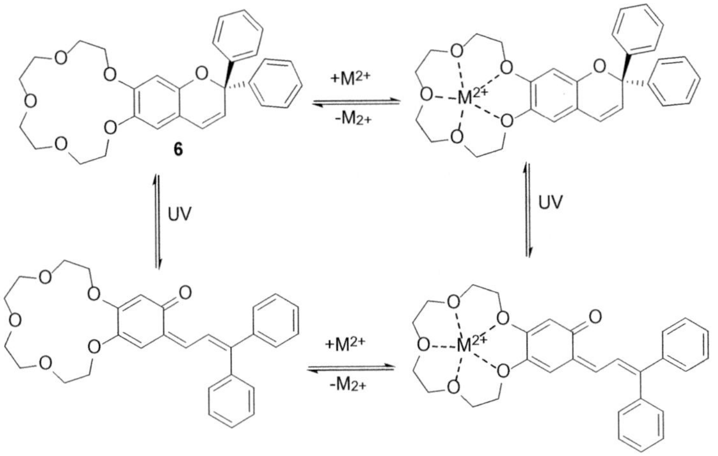

Scheme 3 Isomerization of naphthopyran receptor **6**.

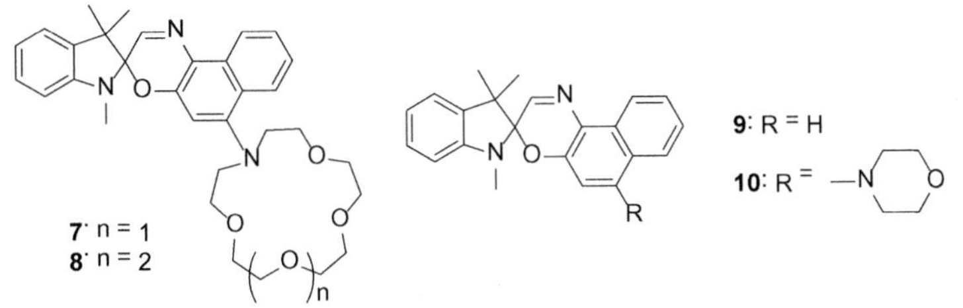

7: n = 1
8: n = 2

9: R = H
10: R = —N(morpholine)

Fedorova et al. (Fedorova et al. 2005) synthesized aza-crown substituted spirooxazine based probes **7–8** along with non-macrocyclic analogues (**9, 10**) and compared them for affinity toward different metal ions. During the studies, affinities of the probes toward lead were compared with either lanthanide metal ions or alkaline earth metal ions (Mg^{2+}, Ca^{2+}, Ba^{2+}) in acetonitrile. The substituted spirooxazine based probes (**7–10**) were observed to form an open colored merocyanine that was induced by Pb^{2+} ion or other lanthanide metal ions (La^{3+}, Eu^{3+}, Tb^{3+}). The alkaline earth metal ions were observed to form complex only with the crown ether moiety with no interaction with the spirooxazine unit of the probes. (Scheme 4). The merocyanine form or probes were observed to produce two types of complex geometries with the metal ions (lead and lanthanide ions). At an equimolar concentration of metal ions and the probe, the crown ether moiety of the probe were observed to form a complex with the metal ion, while on addition of excessive concentration of metal ions to a solution of the probe led to a complex that displayed an interaction between the phenolic oxygen atom of the merocyanine unit and crown ether. The complex formation between the probes **7–8** and lead or lanthanide ions led to an increase in the stability of the colored merocyanine form along with a bathochormically shifted absorption band. The solution of the complexes produced absorption spectra of closed *spiro* form on irradiation with the visible light (Fedorova et al. 2005). The ^{1}H-NMR (Fig. 11.4) and SERRS (Fig. 11.5) spectroscopic techniques have also been used to evaluate the complex formation between the metal ions and the probes. The ^{1}H-NMR spectroscopic studies for probe **9** and its complex with lead ions were performed in CD_3CN, which revealed the appearance of a merocyanine form with a simultaneous disappearance of the closed *spiro* form on addition of lead ions. Storing a solution of probe **9** and lead ions led to the formation of a stable complex (Fig. 11.5). A similar pattern was also observed for other probes during the studies by the authors. The association constants have been calculated using UV-visible spectroscopy for probes **7, 9** with lead and magnesium ions. Probe **7** produced a 1:1 complex with a metal ion [(For **8**; $\log K_{11}$ = 0.6 (Mg^{2+}); 2.8 (Pb^{2+})], while probe **9** displayed both 1:1 and 1:2 complex geometries with lead ions [(For **9**; $\log K_{11}$ = 2.5 (Mg^{2+}), 4.0 (Pb^{2+}) along with $\log K_{12}$ = 4.4 (Pb^{2+}) where K_{12} refer to $9:2Pb^{2+}$ association stoichiometry] (Fedorova et al. 2005).

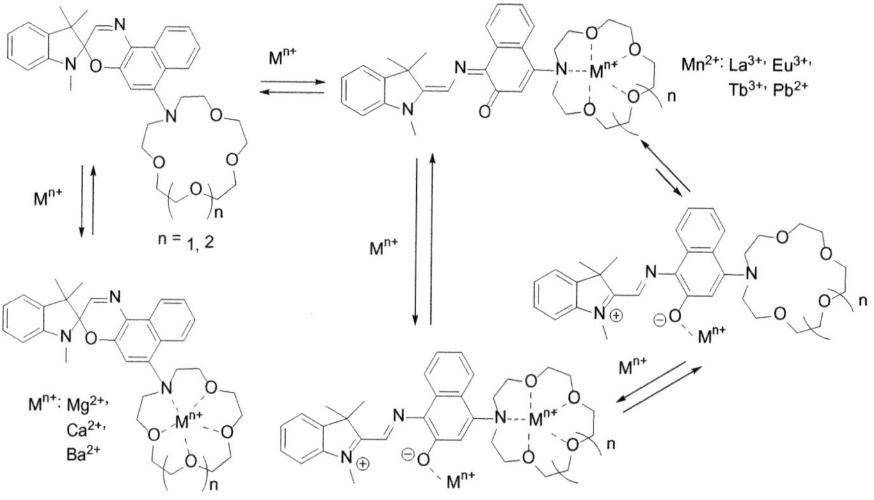

Scheme 4   The formation of a complex between the probes **7–8** and different metal ions (Fedorova et al. 2005).

The complex formation between Eu^{3+} ion and probe **7** has also been investigated using the SERRS technique (Fig. 11.5). The Raman bands in the SERRS spectra were observed to be dominated by the presence of an open merocyanine form, which consists of indoline and naphthalene moieties (Fedorova et al. 2005). As could be expected owing to the absence of any resonance to produce electronic transitions in the aza crown ether moieties, no vibrations in the spectra were observed due to the presence of the aza crown ether moieties. The transoid stereoisomers *TTC* (*trans-trans-cis*) and *CTC* (*cis-trans-cis*) were observed in the SERRS spectra, which were confirmed due to the presence of vibration bands at 1540 cm^{-1} and 1558 cm^{-1} (Fedorova et al. 2005). The addition of Eu^{3+} ions in the M to L ratio from 1–10 to the probe **7** led to an alteration in the relative intensities and positions of vibrations bands observed in the SERRS spectra. However, no change in the intensities of the vibration band and positions

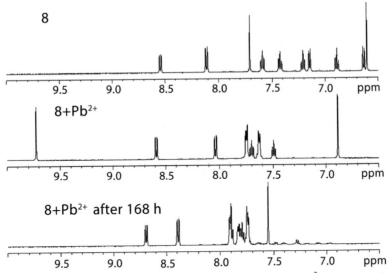

Figure 11.4 Partial ^{1}H-NMR spectra of the probe **7** ([**7**] = 5.0 × 10^{-3} M) on the addition of one equivalent lead ion in acetonitrile-d_3. Reproduced with the permission of the publisher (John Wiley and Sons) (Fedorova et al. 2005).

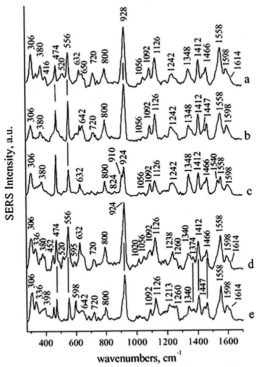

Figure 11.5 The SERRS spectra of **7** with with Eu^{3+} (a) [Eu^{3+}]/[**7**] = 1; (b) [Eu^{3+}]/[**7**] = 10; (c) free **7**; (d) [Eu^{3+}]/[**9**] = 10; (e) free **9**. Reproduced with the permission of the publisher (John Wiley and Sons) (Fedorova et al. 2005).

in the SERRS spectra was observed when the metal/ligand ratio were taken beyond 10. Besides, the increased metal ion concentration stabilized the *CTC* stereoisomer of probe **7** (Fedorova et al. 2005). A comparison in the SERRS spectra of **7 and 9** and their Eu^{3+} complexes revealed remarkable similarity, which indicated the participation of merocyanine oxygen atom in the complex formation as this is the only atom common in both the probes that show interaction with the metal ion. Therefore it was concluded that the major changes in the probes originated due to the interaction between the phenolic oxygen atom and the metal ions.

Calixarenes are also an interesting class of receptors, a modification of a calixarene skeleton with a photoresponsive unit, which can produce an excellent photo-reversible macrocyclic receptor for metal ions. The basic calixarene molecular framework at the upper or lower rim can be easily functionalized by simple chemical reactions to provide highly efficient molecular receptors with tunable selectivity and specificity (Chawla et al. 2010). A few examples of calixarene possessing photochromic units have appeared in literature, however, no report has appeared where photo-responsive calixarenes are used for detection and extraction of heavy metal ions.

To exploit the advantages of the calixarene skeleton and spiropyran unit, Liu et al. (Liu et al. 2006) reported calixarene derivatives coupled to two spiropyran units at the lower rim to create a new photoreversible probe (**12**). The synthesized probe (**12**) was screened for affinity towards important lanthanides (La^{3+}, Pr^{3+}, Eu^{3+}, Gd^{3+}, and Er^{3+}), transition (Fe^{3+}, Cu^{2+}, and Zn^{2+}), alkali (Na^+ and K^+) and alkaline earth (Mg^{2+} and Ca^{2+}) metal ions. Probe (**12**) was observed to form a complex with lanthanide metal ions like La^{3+}, Pr^{3+}, Eu^{3+}, Gd^{3+} and Er^{3+} that led to the development of a visible color change from purple to yellow, which can be easily seen by naked eyes (Fig. 11.6). The color change was attributed to the formation of the merocyanine form, which binds the metal ions in acetonitrile (Scheme 5) (Liu et al. 2006).

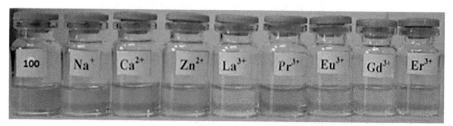

Figure 11.6   Change in the color of the solutions of the probe **12** (5.0×10^{-5} M) on the addition of nitrate salt of different metal ions in acetonitrile. Reproduced here with the permission of Elsevier (Liu et al. 2006).

The UV-visible spectroscopy has been further used to study the formation of a complex between the probe and the metal ions, which indicated the formation of merocyanine form of probe **12** as inferred due to the presence of an absorption band at 550 nm that shifted hypsochromically in the presence of metal ions used in the study. In the presence of lanthanide metal ions, the probe displayed a significant UV-visible spectral shift, while minimal or no spectral shift was observed on the addition of alkali, alkaline earth or transitions metal ions. The complex formation between lanthanide metal ions and probe **12** produced a 68–84 nm shift in the UV-visible absorption band. The magnitude of spectral change was determined to follow the order $Yb^{3+} > Er^{3+} > Gd^{3+} > Dy^{3+} > Eu^{3+} > Pr^{3+} > La^{3+}$, which indicated that the complex formation was largely dependent on the size-fit effect.

The ^{1}H-NMR spectroscopy was also used to investigate the complex between the probe (**12**) and the lanthanide metal ions (Fig. 11.7).

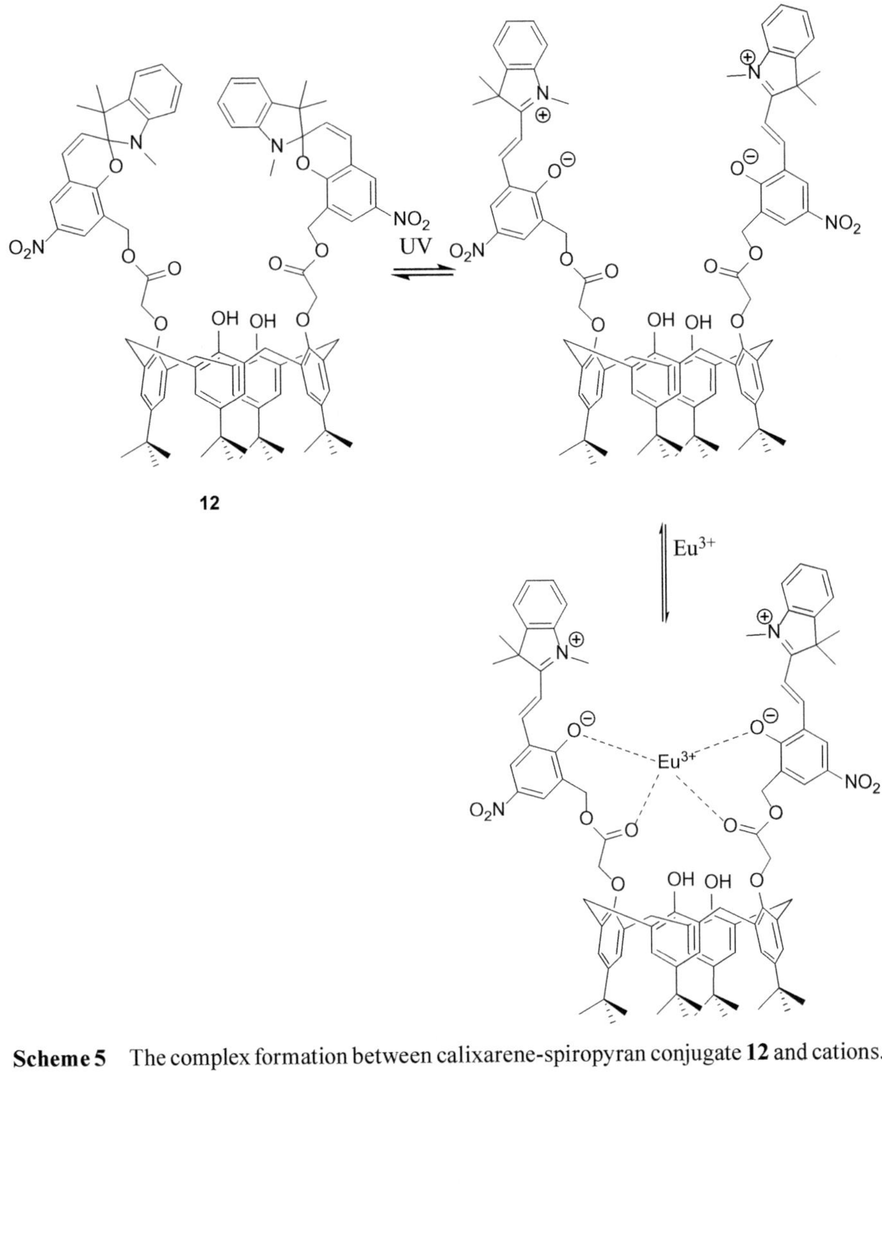

Scheme 5 The complex formation between calixarene-spiropyran conjugate **12** and cations.

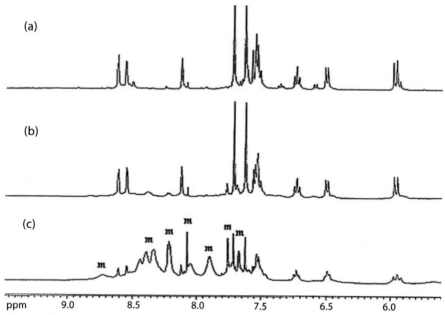

Figure 11.7 Partial ^{1}H-NMR spectra of probe (**12**) recorded in CD$_3$CN at 400 MHz (a) Free probe **12** (b) 2 hours after the addition of Eu^{3+} ions under dark conditions (c) 18 hours after the addition of the Eu^{3+} ions under dark conditions. The peak marked with the letter 'm' indicates the peaks due to the merocyanine form.

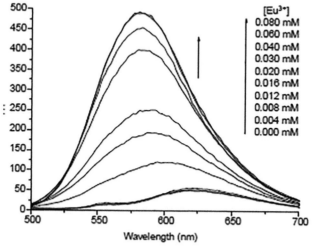

Figure 11.8 The change in fluorescence intensity of the receptor **12** (20 mM) on incremental addition of Eu (NO$_3$)$_3$ in an acetonitrile solution; [Eu (NO$_3$)$_3$] = 0-80 mM.

The open merocyanine form of probe **12** responded to the exposure of visible light (475 nm) to produce an intense fluorescence band with maxima at 555 nm (Fig. 11.8). The stereoisomers of the open form of probe **12** formed a complex with the lanthanide ions in the dark, which resulted in a hypsochromic shift of

68–84 nm with prominent absorption in the visible region of the spectrum. The complex formation also produced a 42 nm shift in the fluorescence spectra on addition of Er^{3+} in a solution of the probe **12** in MeCN. Specifically, complexes of probe **12** with lanthanide ions produced a strong fluorescence response. An enhancement in the intensity of the emission band on the incremental addition of lanthanide ions was also observed (Fig. 11.8). Probe **12** on addition of alkali metal ions like Na^+, K^+, alkaline earth metal ions like Mg^{2+}, Ca^{2+}, and transition metal ions like Fe^{3+}, Cu^{2+}, Zn^{2+} did not show any indication of the formation of complex even in the presence of UV or visible light.

Machitani et al. also reported the synthesis of a photochromic calix[4] arene ligand bearing a spirobenzopyran and three ethyl-ester moieties (**13**). The photocontrol of alkali metal ion extractability of **13** was investigated and compared with a non-photo controlled analogue (**14**) (Koji et al. 2010). Macrocycle **13** displayed a gradual shift in the absorption band to the visible region on the incremental addition of sodium ions in the dark. The ^{1}H-NMR spectroscopy indicated the inclusion of sodium ion by the probe **13**. The carbonyl and ether oxygen atoms present in probe **13** were observed to hold the sodium ion in the cavity using non-covalent interactions. The extraction studies with alkali metal ions were performed in the presence and absence of light, which indicated an enhancement in extraction ability under UV-light as compared to visible light exposure. The enhanced extractability of the alkali metal ions under UV-light was attributed to the conversion of the *spiro* form to the merocyanine form, which led to the interaction between the phenolate part of the spiropyran. The formation of the merocyanine form improved the binding ability of the probe towards metal ions. A comparison of the extraction ability of probe **13** with probe **14** indicated an improvement in the extractability of alkali metal ion due to the formation of the merocyanine form, which used its phenolate part to largely bind the metal ions. As can be expected probe **14** displayed no effect of light on the extractability of the metal ion.

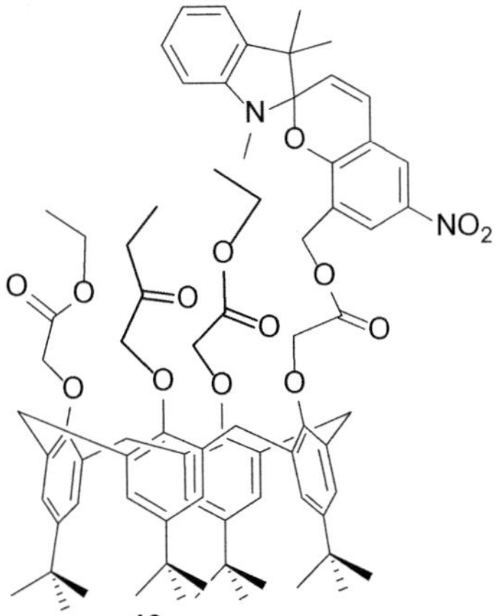

13 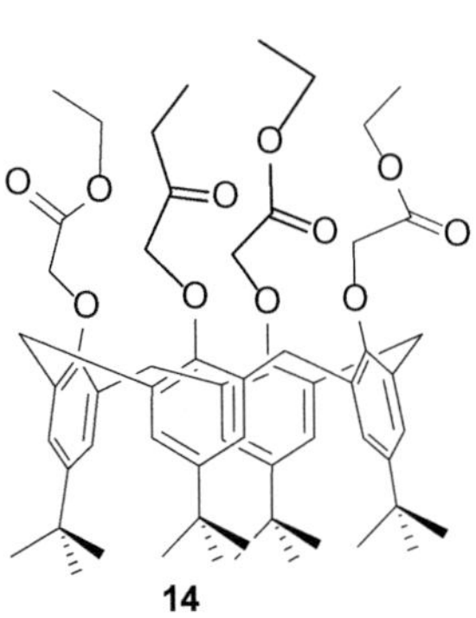						**14**

A reaction between the polyether chain and the calix[*n*]arene produces an improved ionophore for metal ions, which in turn can be converted into photoreversible receptors after covalently connecting a photochromic unit. The new type of receptor incorporates the advantages of crown ether, calixarene and photochromic units. Hence to exploit the advantages of all three molecular fragments, Grofssik et al. (Grofcsik et al. 2002) developed a calix[4]crown-5 ether derivative conjugated covalently to a spiropyran moiety using an adjacent ester group. The spiropyran–merocyanine equilibrium was not affected by the complexation of alkali-and alkaline earth metal cations, probably due to the large intramolecular distance between the crown ether binding site and the signaling unit as indicated by the calculation of rate constant for the rate of thermal bleaching.

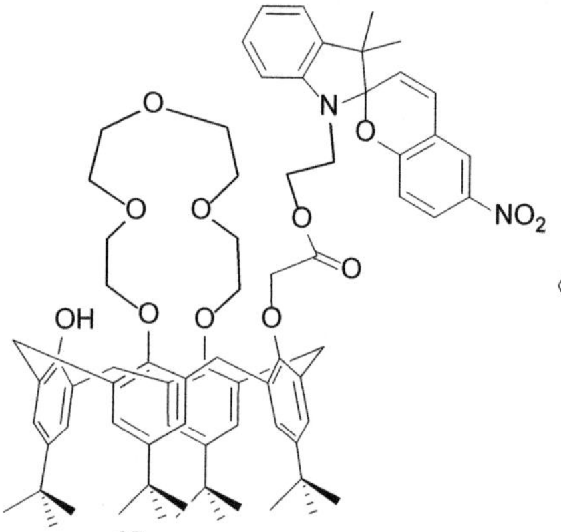

15

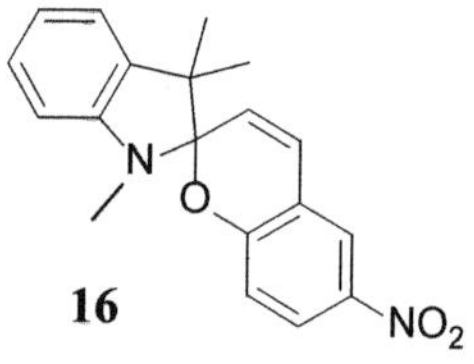

16

Grun et al. have reported the synthesis of Calix[4](aza)crown-5, -6, and -7 ionophores in a cone and 1,3-alt conformation carrying photochromic indolospirobenzopyran. It was reported that in the dark the SP–MC equilibrium shifted remarkably to the colored merocyanine form by cations of high positive charge density (alkaline earth metal ions) (Grün et al. 2008). The cone form (**17a-c**) has been observed to be more effective in coordinating metal ions (Mg^{2+}, Ca^{2+}, Ba^{2+}, Zn^{2+}, Cu^{2+}, Pb^{2+}) as compared to 1,3-alternate conformer (**18a-c**).

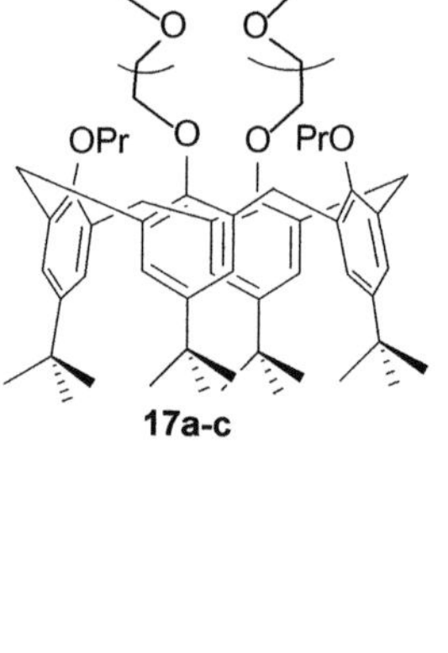

a: (n,m=1)
b: (n=1, m=2)
c: (n,m=2)

17a-c

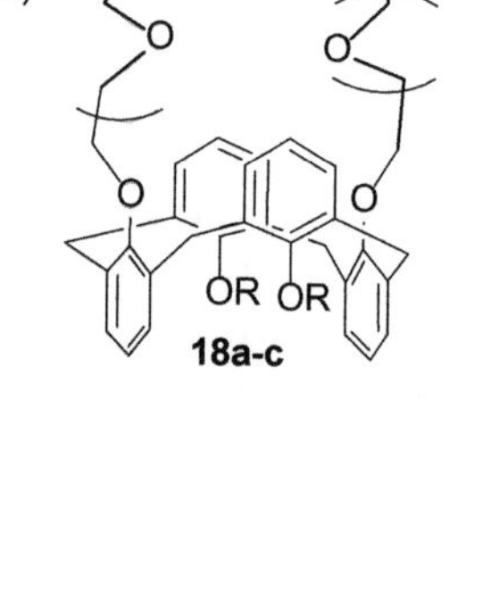

18a-c

THE pH-RESPONSIVE RECEPTORS

Crown ethers are among the first in a family of pH-responsive macrocyclic receptors. The change in pH is the most common way of changing the receptor structurally. The change in pH generally requires the addition of an acid or a base to alter the binding preference of the macrocyclic receptor toward the analyte to regenerate the original receptor and release the analyte. Generally harsher conditions are required for their regeneration (Tewari and Singh 2000; Kumar et al. 2001a, b) that severely affect the composition of the sample. It is also important to note here that the change in pH or using very high acid concentration requires the use of an additive that further increases the pollution. In a pH-controlled guest binding to a host macrocycle, the pH-responsive group can be present either on the host or the guest species. However, in the case of metal ion binding, the pH-responsive group is substituted over the macrocyclic ligand. A change in the protonation state of the responsive group provides the necessary stimulus to turn-on or off the host-guest association through perturbation in the non-covalent factors (Zimmerman and Corbin 2000; Aakeröy and Beatty 2001; Cooke and Rotello 2002; Cantrill et al. 2005; Hoeben et al. 2005; Ye et al. 2005; Lv et al. 2008). A change in the pH value leads to a change in the overall charge on the macrocyclic host or the analyte in the solution through an equilibrium between protonated and deprotonated state. The equilibrium between two states controlled through variation in the pH leads to a change in the coulombic or electrostatic repulsive/attractive forces between the host and guest molecule, which can be used to control the guest binding or release. The competing association of the macrocyclic host can be defined using Scheme 6, where K_f denotes the formation constant between the macrocyclic host and the guest molecule, K_{dH} denotes the dissociation of acid, while K_{fH} denotes the protonation of the macrocyclic host.

$$MH + G \xrightleftharpoons{K_f} MH\text{-}G \quad \text{------} \qquad K_f = \frac{[MH\text{-}G]}{[MH][G]}$$

$$A\text{-}H \, (acid) \xrightleftharpoons{K_{dH}} A^- + H^+ \quad \text{------} \qquad K_{dH} = \frac{[A^-][H^+]}{[A\text{-}H]}$$

$$MH + H^+ \xrightleftharpoons{K_{fH}} MH\text{-}H^+ \quad \text{------} \qquad K_{fH} = \frac{[MH\text{-}H^+]}{[MH][H^+]}$$

Scheme 6   Association and dissociation constants are defined in scheme 6, where K_{asson} is the association.

For example, Lehn et al. (Hriciga and Lehn 1983) demonstrated the pH-dependent preferential transport of K^+ or Ca^{2+} ions from the aqueous phase through a membrane using crown ethers. A change in pH from 2 to 9 increases the preference for Ca^{2+} over K^+ ions.

The pH-responsive host-guest systems based on pH-responsive receptors are attractive candidates for the development of drug-delivery systems (Zhou et al. 2016). The therapeutic materials with a pH-responsive characteristic is a prevailing approach due to acidification of the local environment of tissue by the disease or endosomal processing of internalized carriers (Schmaljohann 2006). For example, the acidic environment of tumor cells promotes the release of bound drug molecule encapsulated by a pH-responsive drug delivery vehicle (Santha Moorthy et al. 2017). The pH response can be imparted to a macrocyclic receptor by covalent linking positive and negatively charged moieties, which allow the macrocyclic receptor to aggregate and align itself to create an extended cavity to accommodate small molecular guests. The bound guest can be released through an adjustment in the pH to destroy the complementarity of the charge, which disrupt the aggregation (Zhou et al. 2016; Braegelman and Webber 2019). Different types of macrocyclic receptors possess certain features and properties in their molecular architectures, which may further contribute towards their utility in drug delivery applications. For instance, calix[*n*]arenes have an extendable cavity dimension with hydrophobic and hydrophilic cavities. The hydrophobic cavities have been modified through substitution with a sulfonate group, which not only improved the binding affinity but also improved the water solubility and biocompatibility along with a reduction in the cytotoxicity (Makha and Raston 2001; Baldini et al. 2007; Xue et al. 2013; Braegelman and Webber 2019). Like calix[*n*]arenes, pillar[*n*]arenes can also be modified to produce drug delivery vehicles (Ogoshi et al. 2016). For instance, a carboxylate group substituted pillar[*n*]arenes responded to a change in pH of the medium. Huang et al. (Xiao-fan et al. 2017) reported a pillar[5]arene substituted with carboxylate group as a pH responsive macrocycle. The complex formation between metal ions like Zn^{2+} caused reduction in the solubility of substituted pillar[5]arene in water. Du et al. (Jiang et al. 2017) exploited the properties of carboxylate substituted pillar[5]arene to develop a pH, GSH, zinc ion, CO_2 and hexanediamine responsive controlled release system. The cyclodextrins-based receptors due to their hydrophobic electron-rich interior and hydrophilic exterior provide a unique environment to develop pH-responsive drug delivery systems particularly for hydrophobic drugs (Crini 2014). The macrocyclic receptors based on cucurbiturils display a remarkable stability in a range of chemical environments. However, the cucurbiturils consisting of odd glycoluril monomers display water solubility and biocompatibility required for biological applications (Wheate and Limantoro 2016; Liu 2017).

THERMO-RESPONSIVE RECEPTORS

Temperature controlled release of an analyte appears from the small and negative values of $\Delta H°$ and $\Delta S°$ during the complex formation process between the host and guest, which makes $\Delta G°$ dependent on temperature based on $\Delta G° = \Delta H° - T\Delta S°$. The opposing effect of $\Delta H°$ and $\Delta S°$ can be exploited to generate the thermos-responsive receptor. The opposing effect was found to be significant for polymer-metal ion interaction in a heterogeneous medium. Based on this fact, Warshawsky and Kahana (Warshawsky and Kahana 1982) demonstrated the temperature-controlled release of alkali metal salts from polymeric crown ethers. Polymeric benzo-18-crown-6 (**19**) was used to pack a column followed by saturating the column with KCl at room temperature. An increase in temperature by 40°C resulted in a threefold increase in the release of potassium ions as an eluent.

The phenomenon has a potential use in the desalination processes, temperature-controlled phase-transfer catalysis and in the thermos-responsive regulation of polymeric delivery vehicle of sodium or potassium ions.

REDOX DRIVEN RECEPTORS

The redox-active macrocyclic probes are designed to generate the electrochemical signal with the binding of an analyte to a macrocyclic host possessing a redox responsive moiety within the molecular architecture (Fig. 11.9) (Beer 1989). The coupling of electrochemical signals with the selectivity associated with the macrocyclic receptor led to the development of a variety of receptors. However, the majority of studies exploited the crown ether framework for the development of selective receptors for environmentally and biologically important analytes. The oxidation-reduction cycles in redox responsive receptors can be represented as shown in Scheme 7. Modeling electron transfer processes in macrocyclic receptors generate considerable interest as it could lead to understanding similar processes in biological systems (Beer 1989, McConnell et al. 2015).

For successful recognition of an analyte using a redox responsive receptor, the binding should produce a large or significantly observable shift in redox potential for the macrocyclic host to avoid the effect of an experimental error ($\sim\pm5$ mV) on the results. The redox potential shift is generally obtained using the K_{ox}/K_{red} ratio, which is dependent on the influence of redox reaction during the complex formation as shown in Fig. 11.10. The different pathways or combinations involve (a) the electrostatic interaction between the redox center and the guest molecule through space (b), communication through a bond (like a conjugated bond) that link the binding to the redox center (c) the bond formation between a guest and the redox center (d) perturbation of the redox center due to a guest molecule binding (e) disruption of the interaction between redox centers by the guest molecule.

The endogenous reducing agents like glutathione (GSH) and oxidizing agents like H_2O_2 are often targeted for interaction with redox responsive groups within the architecture of the macrocyclic receptors.

The redox responsive macrocyclic receptors can be subdivided into different groups depending on the macrocyclic receptor and redox responsive group that is reduced or oxidized electrochemically during the complex formation.

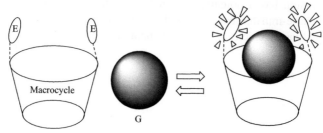

Figure 11.9   A response is generated by redox responsive macrocyclic receptor on binding a guest.

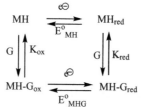

Scheme 7   The graphical representation of the oxidation-reduction cycle in redox responsive macrocycles.

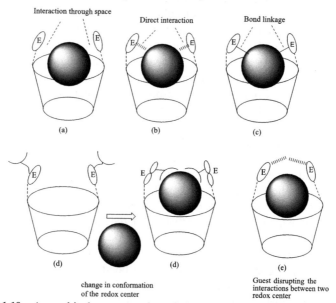

Figure 11.10   A graphical representation of electrochemical and complex formation reactions that occur in macrocyclic receptors.

Disulfide-Bond Dependent Redox Responsive Macrocyclic Receptor

The disulfide linkage is a moiety abundantly found in natural and synthetic organic materials. The disulfide linkage when incorporated into a macrocyclic receptor, impart it redox sensitivity. Redox responsive macrocycles have been reported based on disulfide bonds (Dandekar et al. 2012; Chen et al. 2016; Zhu et al. 2018). For instance, Yu et al. (Tamura and Yui 2017) constructed reduction sensitive, cleavable polyrotaxanes consisting of cyclodextrins threaded on a linear polymer chain with bulky stopper molecular units at the chain terminal. The stimuli (reduction) cleavable polyrotaxane dissociate into constituent molecular fragments in response to reducing intracellular or acidic lysosomal environments. Such a design finds an application in the delivery of drugs of likely organs of the body for treating Niemann-Pick type C (NPC) diseases. Recently, Chen et al. (Yu et al. 2018a) also demonstrated the application of nanocarriers based on polyrotaxanes constructed with cyclodextrin derivative and disulfide linkages in diagnostics and therapeutics.

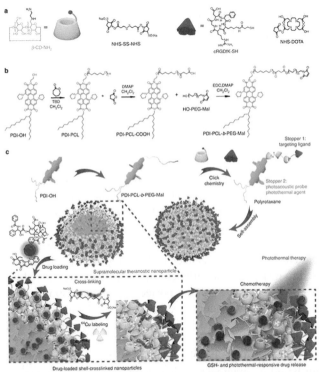

Figure 11.11 Schematic representation of synthesis and fabrication shell cross-linked NPs for application as supramolecular theranostics (**a**) The structures and representations of the different building blocks (β-CD-NH$_2$, NHS-SS-NHS, cRGDfK-SH and NHS-DOTA) (**b**) Synthetic route to the polyrotaxane (PDI-PCL-*b*-PEG-RGD⊃β-CD-NH$_2$) (**c**) Schematic illustrations of the preparation of drug-loaded SCNPs and dual-responsive drug release. Reproduced here with the permission of Nature under Creative Commons license 4.0 (Yu et al. 2018a).

The multicomponent nanoparticles (NPs) shown in Fig. 11.11 were fabricated through self-assembly of CD based polyrotaxane using perylene diimide and Arg-Gly-Asp peptide at the terminal end. A stimuli responsive (temperature and redox) release of drug molecules (paclitaxel and camptothecin) was successfully demonstrated in case of breast cancer treatment. The experiment used a cooperative effect of target localization of photocoustic imaging. Benefiting from the study Chen et al. (Yu et al. 2018b) also developed a theranostic nanosystem based on polyrotaxanes constructed using complex of substituted β-CDs (containing disulfide linkages) and anticancer drug camptothecin (CPT). The nanosystem incorporated the advantages of both the targeting ligand and imaging agent. It successfully delivered the drug molecules to the cancerous cell based on reductive cleavage of disulfide bond due to the presence of GSH-overexpressing in cancerous cells. The research demonstrated the dynamic nature of nano assemblies based on macrocyclic ligands in overcoming the physiological disadvantages of the drugs and reducing the adverse side effect of chemotherapeutics.

Ferrocene Based Macrocyclic Receptor

Ferrocene has been used in several reports as a substitute to modify macrocycles like crown ether, calixarenes, cucurbiturils, pillararenes and other macrocycles due to its shape/size. Ferrocene moiety in combination with Hydrogen Peroxide (H_2O_2) has been found to be useful in stimuli responsive anti-cancer drug delivery systems due to higher than normal concentration of H_2O_2 in the cancerous cells (Zhang et al. 2020). Based on this, Ge et al. (Wang et al. 2018) developed micellar NPs based on self-assembly of β-CD incorporated into a block copolymer PEG-b-P(PLG-g-CD), Ascorbyl Palmitate (PA) and ferrocenecarboxylic acid hexadecyl ester (DFc). The block copolymer PEG-b-P(PLG-g-CD) was cross-assembled with a block copolymer of DFc and PA to produce core-shell nanoparticles by an inclusion complex between β-CD and ferrocene moiety within DFc polymer. The multifunctional micellar nanoparticles displayed excellent stability in serum media. The PA present in the nanoparticles assisted in H_2O_2 generation in tumor tissues, while ferrocene moiety helped in the transformation of H_2O_2 into hydroxyl radical to achieve a cancer cell-killing ability. The nanoparticles were demonstrated to display highly efficient tumor cell suppression ability. On similar lines, Li et al. (Kang et al. 2017) reported the redox responsive supramolecular assemblies based inclusion complex between PEGylated-β-CD and ferrocene modified with CPT (linked through a S–S bond). The nano-assemblies displayed an excellent inhibition effect on tumor cells. The *in-vivo* results demonstrated the excellent efficacy of the developed nano-system without any side effect. The nano-system based on macrocyclic ligands in a nano-system may serve as a new type of promising drug delivery vehicles.

Redox Responsive Systems Based on Electroactive Guests

An alternative strategy for detection based on electrochemical output can be designed if the guest analyte has a redox active moiety within its molecular framework. Based on this strategy, Kim et al. (Jeon et al. 2002) reported inclusion of redox responsive methylviologen dication in curbit[8]uril (Fig. 11.12). The redox chemistry was exploited to alter the complex stoichiometry reversibly between 1:1 and 1:2 (H:G) by reduction or oxidation of the guest.

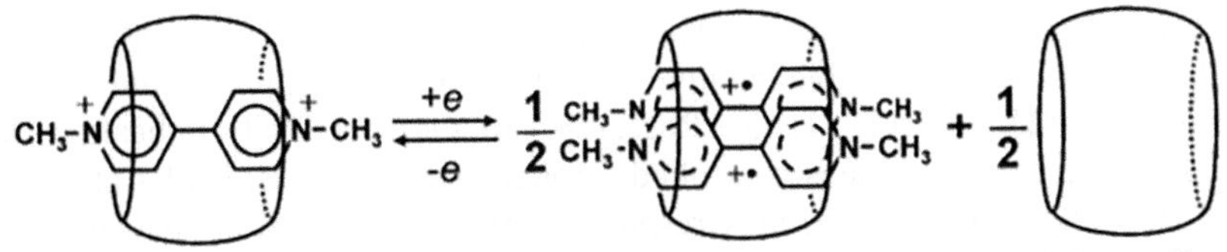

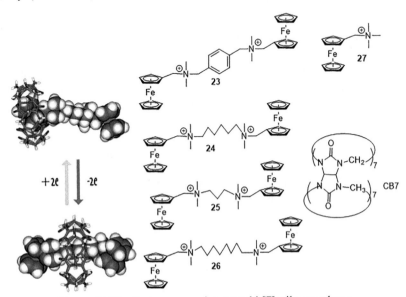

Figure 11.12　The schematic representation of redox responsive methylviologen dication inclusion by cucurbit[8]uril. Reproduced here with permission of RSC (Jeon et al. 2002).

Based on this work, Kaifer et al. (Ling et al. 2007) reported a study demonstrating the redox responsive complex formation between cucurbiturils (**20**) and guests (**21, 22**). Significantly different binding responses towards both guest molecules was observed to a difference in planarity, charge distribution and size. Out of the two guests, **22** was slightly bigger and non-planar, which led to a reduction in the binding affinity between **20** and **22** in comparison to binding affinity between **20** and **21**.

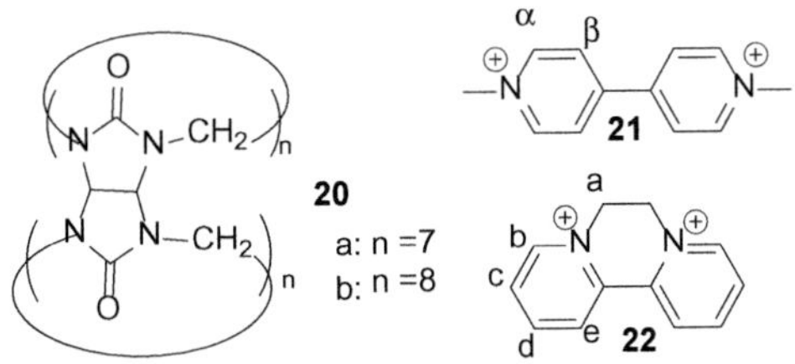

One electron reduction improves the stability of CB7 (**20a**) complex with **22** in comparison to **21**. The study demonstrated that **20b** tolerated the charge distribution on **22**$^{2+}$ and its one electron reduction product showing a good binding affinity (K>10^4 M^{-1}).

Figure 11.13　Redox responsive cucurbit[7]urils complexes.

Kaifer et al. (Sobransingh and Kaifer 2006) took the work further by developing redox controlled reversible complex between cucurbit[7]urils and ferrocene derivatives (**23–26**) (Fig. 11.13). Research can find potential application in the area of molecular machines controlled through redox process. Kaifer et al. (Yi et al. 2011) also reported the redox responsive recognition of ferrocenylammonium guests (**28–29**) using CB7 (**20a**) and CB8 (**20b**) (Fig. 11.14). The study indicated that **28** forms a strong and stable complex with macrocycle **20b** (CB8) with tempo moiety of the guest occupying the inner cavity of CB8, while ferrocene moiety of **30** occupies the cavity of the **20b**. Investigation of complex formation between **30** and CB8 revealed competition between TEMPO and ferrocene moieties for occupation of CB8 cavity. The complex formation between CB7 (**20a**) and guest (**29–30**) revealed inclusion of ferrocene moiety inside the host molecule.

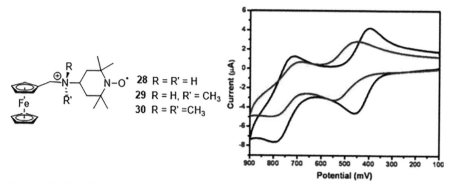

Figure 11.14   The chemical structure of the guests (**28–29**) used with cucurbiturils (**20**) and cyclic voltammetric response of CB7 (**20a**) on addition of one equivalent guest 29^+ (1.0 mM) in 50 mM NaCl (red) and free guest 29^+ (black). Reproduced here with the permission of RSC (Yi et al. 2011).

The inclusion of the guests (**31–32**) by CB7 was also investigated by Kaifer et al. (Sindelar et al. 2007). A moderately stable complex was observed between the guest and the host (CB7) due to the steric effect owing to the size of the guest, which does not allow them to fit within the cavity of CB7. The addition of CB7 to a solution of the guest inhibit their redox process without affecting the redox potential (Fig. 11.15). The change in redox properties of the guest during inclusion complex formation was exploited by Kaifer et al. (Sindelar et al. 2005). The study demonstrated the formation of a highly complex between $Me2DAP^{2+}$ and CB8 (**21b**). Moreover a stable charge transfer complex also formed between the $Me2DAP^{2+}$ and catechol inside the cavity of CB8 (Fig. 11.16). The study has a potential application in sensitive detection of catecholamine neurotrasmitters.

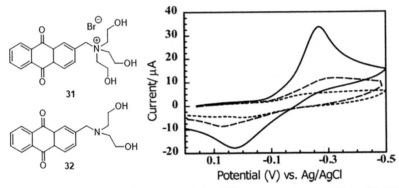

Figure 11.15 Structure of the guest and the CV response of guest **32** in 0.1 M HCl solution (continuous line) and on addition of 1.0 equivalent (discontinuous line) and 2.0 equivalent (dotted lined) of CB7 (**20a**). Scan rate was kept at 0.1 V/s. Reproduced here with the permission of RSC (Sindelar et al. 2007).

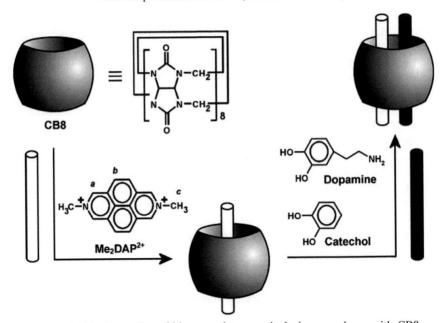

Figure 11.16 Formation of binary and ternary inclusion complexes with CB8. Reproduced here with permission of John Wiley and Sons (Sindelar et al. 2005).

CONCLUSIONS

Various modifications to macrocyclic systems have been reported in the context of their potential applications in wider areas of diagnostics, medicine and the environment. Appropriate modification of the macrocyclic receptor opens numerous opportunities for engineering smart stimuli-responsive functional supramolecular systems to achieve desired properties in the developed material.

Stimuli-responsive macrocycles not only serve to achieve the desired function like a normal receptor, but also respond to appropriate stimuli. This chapter highlighted the different types of stimuli-responsive receptors and their host-guest chemistry. Besides the application of macrocyclic receptors as a drug delivery vehicle and sensitizer in photodynamic therapy were also discussed. Stimuli-responsive systems have the potential to be developed into a new type of smart material for advanced applications in medicine and environment.

REFERENCES

Aakeröy, C.B. and A.M. Beatty. 2001. Crystal engineering of hydrogen-bonded assemblies-a progress report. Aust. J. Chem. 54(7): 409–421.

Aime, S. and P. Caravan. 2009. Biodistribution of gadolinium-based contrast agents, including gadolinium deposition. J. Magn. Reson. 30(6): 1259–1267.

Akine, S. and Y. Sakata. 2020. Control of guest binding kinetics in macrocycles and molecular cages. Chem. Lett. 49(4): 428–441.

Alward, M.R. 1999. The synthesis, analysis and applications of photoreversible metal-ion chelators. PhD. University of Pittsburgh.

Aragay, G., J. Pons and A. Merkoçi. 2011. Recent trends in macro-, micro-, and nanomaterial-based tools and strategies for heavy-metal detection. Chem. Rev. 111(5): 3433–3458.

Areti, S., S. Bandaru, R. Teotia and C.P. Rao. 2015. Water-soluble 8-hydroxyquinoline conjugate of amino-glucose as receptor for La^{3+} in HEPES buffer, on whatman cellulose paper and in living cells. Anal. Chem. 87(24): 12348–12354.

Atwood, J.L. 2017. Comprehensive Supramolecular Chemistry II. Elsevier, Oxford, UK.

Baldini, L., A. Casnati, F. Sansone and R. Ungaro. 2007. Calixarene-based multivalent ligands. Chem. Soc. Rev. 36(2): 254–266.

Baldrighi, M., G. Locatelli, J. Desper, C.B. Aakeröy and S. Giordani. 2016. Probing metal ion complexation of ligands with multiple metal binding sites: The case of spiropyrans. Chem. Eur. J. 22(39): 13976–13984.

Barachevsky, V. 2016. Electrical properties of photochromic organic systems (review). High Energy Chem. 50(5): 371–388.

Beer, P.D. 1989. Meldola medal lecture. Redox responsive macrocyclic receptor molecules containing transition metal redox centres. Chem. Soc. Rev. 18(0): 409–450.

Bertelson, R. (ed.). 1971. Photochromism, Techniques in Chemistry. Wiley-Interscience, New York.

Bouas-Laurent, H. and H. Dürr. 2001. Organic photochromism (IUPAC technical report). Pure Appl. Chem. 73(4): 639–665.

Braegelman, A.S. and M.J. Webber. 2019. Integrating stimuli-responsive properties in host-guest supramolecular drug delivery systems. Theranostics 9(11): 3017–3040.

Brotin, T., V. Roy and J.-P. Dutasta. 2005. Improved synthesis of functional CTVs and cryptophanes using Sc (OTf) 3 as catalyst. J. Org. Chem. 70(16): 6187–6195.

Brown, G.H. 1971. Photochromism. Wiley-Interscience, New York.

Busschaert, N., C. Caltagirone, W. van Rossom and P.A. Gale. 2015. Applications of supramolecular anion recognition. Chem. Rev. 115(15): 8038–8155.

Cantrill, S.J., K.S. Chichak, A.J. Peters and J.F. Stoddart. 2005. Nanoscale borromean rings. Acc. Chem. Res. 38(1): 1–9.

Chawla, H.M., N. Pant, S. Kumar, N. Kumar and C. Black David St. 2010. Calixarene-based materials for chemical sensors. pp. 117–200. *In*: G. Korotcenkov [ed.]. Chemical Sensors Fundamentals of Sensing Materials, Vol. 3. Momentum Press, New York: USA.

Chebun'kova, A.V., S.P. Gromov, T.M. Valova, Y.P. Strokach, M.V. Alfimov, O.A. Fedorova, V. Lokshin and A. Samat. 2005. Investigation of the azacrown-ether substituted naphtopyranes. Mol. Cryst. Liq. Cryst. 430(1): 67–73.

Chen, G. and M. Jiang. 2011. Cyclodextrin-based inclusion complexation bridging supramolecular chemistry and macromolecular self-assembly. Chem. Soc. Rev. 40(5): 2254–2266.

Chen, L.-X., Y.-M. Zhang, Y. Cao, H.-Y. Zhang and Y. Liu. 2016. Bridged bis(β-cyclodextrin)s-based polysaccharide nanoparticles for controlled paclitaxel delivery. RSC Adv. 6(34): 28593–28598.

Chen, Y., W. Zhang, Y. Cai, R.T. Kwok, Y. Hu, J.W. Lam, X. Gu, Z. He, Z. Zhao and X. Zheng. 2017. AIEgens for dark through-bond energy transfer: design, synthesis, theoretical study and application in ratiometric Hg^{2+} sensing. Chem. Sci. 8: 2047–2055.

Cooke, G. and V.M. Rotello. 2002. Methods of modulating hydrogen bonded interactions in synthetic host–guest systems. Chem. Soc. Rev. 31(5): 275–286.

Corbalan-Garcia, S., J. Teruel and J. Gomez-Fernandez. 1992. Characterization of ruthenium red-binding sites of the Ca^{2+}-ATPase from sarcoplasmic reticulum and their interaction with Ca^{2+}-binding sites. Biochem. J. 287(3): 767–774.

Cram, D.J. and J.M. Cram. 1974. Host-guest chemistry. Science. 183(4127): 803–809.

Crano, J.C. and R.J. Guglielmetti. 1999. Organic Photochromic and Thermochromic Compounds. Springer, US.

Crini, G. 2014. Review: A history of cyclodextrins. Chem. Rev. 114(21): 10940–10975.

Dandekar, P., R. Jain, M. Keil, B. Loretz, L. Muijs, M. Schneider, D. Auerbach, G. Jung, C.-M. Lehr and G. Wenz. 2012. Cellular delivery of polynucleotides by cationic cyclodextrin polyrotaxanes. J. Control. Release. 164(3): 387–393.

De Silva, A.P., H.N. Gunaratne, T. Gunnlaugsson, A.J. Huxley, C.P. McCoy, J.T. Rademacher and T.E. Rice. 1997. Signaling recognition events with fluorescent sensors and switches. Chem. Rev. 97(5): 1515–1566.

Dean, J.G., F.L. Bosqui and K.H. Lanouette. 1972. Removing heavy metals from waste water. Environ. Sci. Technol. 6(6): 518–522.

Dietrich, B., J.M. Lehn and J.P. Sauvage. 1969. Les cryptates. Tetrahedron Lett. 10(34): 2889–2892.

Dietrich, B., J.M. Lehn and J.P. Sauvage. 1973. Cryptates—XI: Complexes macrobicycliques, formation, structure, proprietes. Tetrahedron. 29(11): 1647–1658.

Doddi, S., B. Ramakrishna, Y. Venkatesh and P.R. Bangal. 2015. Photo-driven near-IR fluorescence switch: synthesis and spectroscopic investigation of squarine-spiropyran dyad. RSC Adv. 5(118): 97681–97689.

Dürr, H. and H. Bouas-Laurent. 2003. Photochromism: Molecules and Systems. Gulf Professional Publishing.

Fedorova, O.A., A.V. Koshkin, S.P. Gromov, Y.P. Strokach, T.M. Valova, M.V. Alfimov, A.V. Feofanov, I.S. Alaverdian, V.A. Lokshin and A. Samat. 2005. Transformation of 6'-aminosubstituted spironaphthoxazines induced by Pb(II) and Eu(III) cations. J. Phys. Org. Chem. 18(6): 504–512.

Gálico, D., T. Fraga-Silva, J. Venturini and G. Bannach. 2016. Thermal, spectroscopic and in vitro biological studies of the lanthanum complex of naproxen. Thermochim. Acta. 644: 43–49.

Galloway, J.N., J.D. Thornton, S.A. Norton, H.L. Volchok and R.A. McLean. 1982. Trace metals in atmospheric deposition: A review and assessment. Atmos. Environ. 16(7): 1677–1700.

Ganjali, M.R., A. Daftari, M. Rezapour, T. Puorsaberi and S. Haghgoo. 2003. Gliclazide as novel carrier in construction of PVC-based La(III)-selective membrane sensor. Talanta. 59(3): 613–619.

Genovese, M.E., A. Athanassiou and D. Fragouli. 2015. Photoactivated acidochromic elastomeric films for on demand acidic vapor sensing. J. Mater. Chem. A 3(44): 22441–22447.

Grofcsik, A., P. Baranyai, I. Bitter, A. Grün, É. Köszegi, M. Kubinyi, K. Pál and T. Vidóczy. 2002. Photochromism of a spiropyran derivative of 1,3-calix[4]crown-5. J. Mol. Struct. 614(1):69–73.

Grün, A., P. Kerekes and I. Bitter. 2008. Synthesis, characterization and cation-induced isomerization of photochromic calix[4](aza) crown-indolospiropyran conjugates. Supramol. Chem. 20(3): 255–263.

Guo, P., Y. Su, Q. Cheng, Q. Pan and H. Li. 2011. Crystal structure determination of the β-cyclodextrin–p-aminobenzoic acid inclusion complex from powder X-ray diffraction data. Carbohydr. Res. 346(7): 986–990.

Guo, X., J. Zhou, M.A. Siegler, A.E. Bragg and H.E. Katz. 2015. Visible-light-triggered molecular photoswitch based on reversible E/Z isomerization of a 1, 2-dicyanoethene derivative. Angew. Chem. Int. Ed. 54(16): 4782–4786.

Gutsche, C.D. 1998. Calixarenes Revisited. Royal Society of Chemistry, Cambridge, UK.

Gutsche, C.D. 2008. Calixarenes: An Introduction. Royal Society of Chemistry, Cambridge, UK.

Haiduc, I. and C. Silvestru. 1990. Metal compounds in cancer chemotherapy. Coord. Chem. Rev. 99: 253–296.

Han, G.D., S.S. Park, Y. Liu, D. Zhitomirsky, E. Cho, M. Dincă and J.C. Grossman. 2016. Photon energy storage materials with high energy densities based on diacetylene-azobenzene derivatives. J. Mater. Chem. A. 4(41): 16157–16165.

Herder, M., B.M. Schmidt, L. Grubert, M. Pätzel, J. Schwarz and S. Hecht. 2015. Improving the fatigue resistance of diarylethene switches. J. Am. Chem. Soc. 137(7): 2738–2747.

Hessels, A.M., P. Chabosseau, M.H. Bakker, W. Engelen, G.A. Rutter, K.M. Taylor and M. Merkx. 2015. eZinCh-2: A versatile, genetically encoded FRET sensor for cytosolic and intraorganelle Zn^{2+} imaging. ACS Chem. Biol. 10(9): 2126–2134.

Hoeben, F.J., P. Jonkheijm, E. Meijer and A.P. Schenning. 2005. About supramolecular assemblies of π-conjugated systems. Chem. Rev. 105(4): 1491–1546.

Hof, F., C. Nuckolls and J. Rebek. 2000. Diversity and selection in self-assembled tetrameric capsules. J. Am. Chem. Soc. 122(17): 4251–4252.

Hriciga, A. and J.-M. Lehn. 1983. pH regulation of divalent/monovalent Ca/K cation transport selectivity by a macrocyclic carrier molecule. Proc. Natl. Acad. Sci. U.S.A. 80(20): 6426–6428.

Huang, Y., F. Li, C. Ye, M. Qin, W. Ran and Y. Song. 2015. A photochromic sensor microchip for high-performance multiplex metal ions detection. Sci. Rep. 5: 9724.

Jana, M. and S. Bandyopadhyay. 2011. Hydration properties of α-, β-, and γ-cyclodextrins from molecular dynamics simulations. J. Phys. Chem. B. 115(19): 6347–6357.

Jeon, W.S., H.-J. Kim, C. Lee and K. Kim. 2002. Control of the stoichiometry in host–guest complexation by redox chemistry of guests: Inclusion of methylviologen in cucurbit[8] uril. Chem. Commun. (17): 1828–1829.

Jiang, L., X. Huang, D. Chen, H. Yan, X. Li and X. Du. 2017. Supramolecular vesicles coassembled from disulfide-linked benzimidazolium amphiphiles and carboxylate-substituted pillar[6]arenes that are responsive to five stimuli. Angew. Chem., Int. Ed. Engl. 129(10): 2699–2703.

Jiménez-Sánchez, A., B. Ortíz, V.O. Navarrete, N. Farfán and R. Santillan. 2015. Two fluorescent schiff base sensors for Zn^{2+}: the Zn^{2+}/Cu^{2+} ion interference. Analyst. 140(17): 6031–6039.

Kang, Y., X. Ju, L.-S. Ding, S. Zhang and B.-J. Li. 2017. Reactive oxygen species and glutathione dual redox-responsive supramolecular assemblies with controllable release capability. ACS Appl. Mater. Interfaces. 9(5): 4475–4484.

Kawai, S.H. 1998. Photochromic bis(monoaza-crown ether)s. Alkali-metal cation complexing properties of novel diarylethenes. Tetrahedron Lett. 39(25): 4445–4448.

Kimura, K., H. Sakamoto, S. Kado, R. Arakawa and M. Yokoyama. 2000. Studies on metal-ion complex formation of crown ether derivatives incorporating a photoionizable spirobenzopyran moiety by electrospray ionization mass spectrometry. Analyst. 125(6): 1091–1095.

Koji, M., S. Megumi, N. Yoshio and K. Keiichi. 2010. Synthesis of novel photochromic calix[4]arene ligand bearing a spirobenzopyran and three ethyl-ester moieties and photocontrol of its alkali metal ion extractability. Bull. Korean Chem. Soc. 83(9): 1107–1112.

Kumar, M., D. Rathore and A.K. Singh. 2001a. Pyrogallol immobilized amberlite XAD-2: A newly designed collector for enrichment of metal ions prior to their determination by flame atomic absorption spectrometry. Microchim. Acta 137(3–4): 127–134.

Kumar, M., D.P. Rathore and A.K. Singh. 2001b. Quinalizarin anchored on amberlite XAD-2: A new matrix for solid-phase extraction of metal ions for flame atomic absorption spectrometric determination. Fresenius J. Anal. Chem. 370(4): 377–382.

Kumar, S., C. Chau, G. Chau and A. McCurdy. 2008. Synthesis and metal complexation properties of bisbenzospiropyran chelators in water. Tetrahedron. 64(30–31): 7097–7105.

Lee, K.E., J.U. Lee, D.G. Seong, M.-K. Um and W. Lee. 2016. Highly sensitive ultraviolet light sensor based on photoactive organic gate dielectrics with an azobenzene derivative. J. Phys. Chem. C. 120(40): 23172–23179.

Lehn, J.-M. 2011. Supramolecular Chemistry: Concepts and Perspectives. John Wiley & Sons, Germany.

Lin, L., Z. Zhang, Z. Lu, Y. Guo and M. Liu. 2016. Two-photon-induced isomerization of spiropyran/merocyanine at the air/water interface probed by second harmonic generation. J. Phys. Chem. A. 120(40): 7859–7864.

Ling, Y., J.T. Mague and A.E. Kaifer. 2007. Inclusion Complexation of diquat and paraquat by the hosts cucurbit[7]uril and cucurbit[8]uril. Chem. Eur. J. 13(28): 7908–7914.

Liu, Z., L. Jiang, Z. Liang and Y. Gao. 2006. A selective colorimetric chemosensor for lanthanide ions. Tetrahedron. 62(14): 3214–3220.

Liu, F., J. Du, M. Xu and G. Sun. 2016. A highly sensitive fluorescent sensor for palladium and direct imaging of its ecotoxicity in living model organisms. Chem. Asian J. 11(1): 43–48.

Liu, L. 2017. Controlled release from cucurbituril. J. Incl. Phenom. Macrocycl. Chem. 87(1): 1–12.

Lv, J., H. Liu and Y. Li. 2008. Self-assembly and properties of low-dimensional nanomaterials based on π-conjugated organic molecules. Pure Appl. Chem. 80(3): 639–658.

Makha, M. and C.L. Raston. 2001. Direct synthesis of calixarenes with extended arms: p-phenylcalix[4,5,6,8]arenes and their water-soluble sulfonated derivatives. Tetrahedron Lett. 42(35): 6215–6217.

McConnell, A.J., C.S. Wood, P.P. Neelakandan and J.R. Nitschke. 2015. Stimuli-responsive metal-ligand assemblies. Chem. Rev. 115(15): 7729–7793.

Mohan Raj, A., F.M. Raymo and V. Ramamurthy. 2016. Reversible disassembly-assembly of octa acid-guest capsule in water triggered by a photochromic process. Org. Lett. 18(7): 1566–1569.

Moorthy, J.N., S. Mandal, A. Mukhopadhyay and S. Samanta. 2013. Helicity as a steric force: stabilization and helicity-dependent reversion of colored o-quinonoid intermediates of helical chromenes. J. Am. Chem. Soc. 135(18): 6872–6884.

Mutoh, K., Y. Kobayashi and J. Abe. 2017. Efficient coloration and decoloration reactions of fast photochromic 3H–naphthopyrans in PMMA-b-PBA block copolymer. Dyes Pigm. 137: 307–311.

Ogoshi, T., T.-a. Yamagishi and Y. Nakamoto. 2016. Pillar-shaped macrocyclic hosts pillar[n]arenes: new key players for supramolecular chemistry. Chem. Rev. 116(14): 7937–8002.

Paramonov, S., S. Delbaere, O. Fedorova, Y. Fedorov, V. Lokshin, A. Samat and G. Vermeersch. 2010. Structural and photochemical aspect of metal-ion-binding to a photochromic chromene annulated by crown-ether moiety. J. Photochem. Photobiol. A 209(2): 111–120.

Pedersen, C.J. 1967. Cyclic polyethers and their complexes with metal salts. J. Am. Chem. Soc. 89(10): 2495–2496.

Pedersen, C.J. 1988. The discovery of crown ethers (Noble Lecture). Angew. Chem., Int. Ed. Engl. 27(8): 1021–1027.

Pinjari, R.V. and S.P. Gejji. 2009. Inverted cucurbit[n]urils: density functional investigations on the electronic structure, electrostatic potential, and NMR chemical shifts. J. Phys. Chem. A. 113(7): 1368–1376.

Quang, D.T. and J.S. Kim. 2010. Fluoro- and chromogenic chemodosimeters for heavy metal ion detection in solution and biospecimens. Chem. Rev. 110(10): 6280–6301.

Saleem, M. and K.H. Lee. 2015. Optical sensor: a promising strategy for environmental and biomedical monitoring of ionic species. RSC Adv. 5(88): 72150–72287.

Santha Moorthy, M., S. Bharathiraja, P. Manivasagan, K.D. Lee and J. Oh. 2017. Crown ether triad modified core-shell magnetic mesoporous silica nanocarrier for pH-responsive drug delivery and magnetic hyperthermia applications. New J. Chem. 41(19): 10935–10947.

Satapathy, R., Y.-H. Wu and H.-C. Lin. 2012. Novel dithieno-benzo-imidazole-based Pb^{2+} sensors: Substituent effects on sensitivity and reversibility. Chem. Commun. 48(45): 5668–5670.

Schmaljohann, D. 2006. Thermo- and pH-responsive polymers in drug delivery. Adv. Drug Deliv. Rev. 58(15): 1655–1670.

Schönbeck, C., P. Westh, J.C. Madsen, K.L. Larsen, L.W. Städe and R. Holm. 2011. Methylated β-cyclodextrins: influence of degree and pattern of substitution on the thermodynamics of complexation with tauro- and glyco-conjugated bile salts. Langmuir. 27(10): 5832–5841.

Shao, N., X. Gao, H. Wang, R. Yang and W. Chan. 2009. Spiropyran-based optical approaches for mercury ion sensing: Improving sensitivity and selectivity via cooperative ligation interactions using cysteine. Anal. Chim. Acta. 655(1): 1–7.

She, N.-F., M. Gao, X.-G. Meng, G.-F. Yang, J.A.A.W. Elemans, A.-X. Wu and L. Isaacs. 2009. Supramolecular rhombic grids formed from bimolecular building blocks. J. Am. Chem. Soc. 131(33): 11695–11697.

Sindelar, V., M.A. Cejas, F.M. Raymo, W. Chen, S.E. Parker and A.E. Kaifer. 2005. Supramolecular assembly of 2,7-dimethyldiazapyrenium and cucurbit[8]uril: A new fluorescent host for detection of catechol and dopamine. Chem. Eur. J. 11(23): 7054–7059.

Sindelar, V., S.E. Parker and A.E. Kaifer. 2007. Inclusion of anthraquinone derivatives by the cucurbit[7]uril host. New J. Chem. 31(5): 725–728.

Sobransingh, D. and A.E. Kaifer. 2006. Electrochemically switchable cucurbit[7]uril-based pseudorotaxanes. Org. Lett. 8(15): 3247–3250.

Stauffer, M., D. Knowles and S. Weber. 1997. Optical control over Pb^{2+} binding to a crown ether-containing chromene. Chem. Commun. (3): 287–288.

Steed, J.W., H. Zhang and J.L. Atwood. 1996. Inclusion chemistry of cyclotriveratrylene and cyclotricatechylene. Supramol. Chem. 7(1): 37–45.

Straight, S.D., J. Andréasson, G. Kodis, A.L. Moore, T.A. Moore and D. Gust. 2005. Photochromic control of photoinduced electron transfer. Molecular double-throw switch. J. Am. Chem. Soc. 127(8): 2717–2724.

Szumna, A. 2010. Inherently chiral concave molecules—from synthesis to applications. Chem. Soc. Rev. 39(11): 4274–4285.

Takeshita, M. and M. Irie. 1998. Photoresponsive tweezers for alkali metal ions. Photochromic diarylethenes having two crown ether moieties. J. Org. Chem. 63(19): 6643–6649.

Takeshita, M., C.F. Soong and M. Irie. 1998. Alkali metal ion effect on the photochromism of 1,2-bis(2,4-dimethylthien-3-yl)-perfluorocyclopentene having benzo-15-crown-5 moieties. Tetrahedron Lett. 39(42): 7717–7720.

Tamura, A. and N. Yui. 2017. Rational design of stimuli-cleavable polyrotaxanes for therapeutic applications. Polym. J. 49(7): 527–534.

Tanaka, M., K. Kamada, H. Ando, T. Kitagaki, Y. Shibutani and K. Kimura. 2000. Synthesis and photochromism of crowned spirobenzothiapyran: Facilitated photoisomerization by cooperative complexation of crown ether and thiophenolate moieties with metal ions. J. Org. Chem. 65(14): 4342–4347.

Tanaka, M., M. Nakamura, M.A.A. Salhin, T. Ikeda, K. Kamada, H. Ando, Y. Shibutani and K. Kimura. 2001. Synthesis and photochromism of spirobenzopyran derivatives bearing an oxymethylcrown ether moiety: Metal ion-induced switching between positive and negative photochromisms. J. Org. Chem. 66(5): 1533–1537.

Tewari, P.K. and A.K. Singh. 2000. Amberlite XAD-7 impregnated with xylenol orange: a chelating collector for preconcentration of Cd(II), Co(II), Cu(II), Ni(II), Zn(II) and Fe(III) ions prior to their determination by flame AAS. Fresenius J. Anal. Chem. 367(6): 562–567.

Thies, S., H. Sell, C. Bornholdt, C. Schütt, F. Köhler, F. Tuczek and R. Herges. 2012. Light-driven coordination-induced spin-state switching: Rational design of photodissociable ligands. Chem. Eur. J. 18(51): 16358–16368.

Tu, C., E.A. Osborne and A.Y. Louie. 2009. Synthesis and characterization of a redox-and light-sensitive MRI contrast agent. Tetrahedron. 65(7): 1241–1246.

van Dijken, D.J., P. Kovaříček, S.P. Ihrig and S. Hecht. 2015. Acylhydrazones as widely tunable photoswitches. J. Am. Chem. Soc. 137(47): 14982–14991.

Vlasceanu, A., S.L. Broman, A.S. Hansen, A.B. Skov, M. Cacciarini, A. Kadziola, H.G. Kjaergaard, K.V. Mikkelsen and M.B. Nielsen. 2016. Solar-thermal energy storage in a photochromic macrocycle. Chem. Eur. J. 22(31): 10796–10800.

Wang, S., L. Ding, J. Fan, Z. Wang and Y. Fang. 2014. Bispyrene/surfactant-assembly-based fluorescent sensor array for discriminating lanthanide ions in aqueous solution. ACS Appl. Mater. Interfaces. 6(18): 16156–16165.

Wang, X., J. Hu, G. Liu, J. Tian, H. Wang, M. Gong and S. Liu. 2015. Reversibly switching bilayer permeability and release modules of photochromic polymersomes stabilized by cooperative noncovalent interactions. J. Am. Chem. Soc. 137(48): 15262–15275.

Wang, Y., W. Yin, W. Ke, W. Chen, C. He and Z. Ge. 2018. Multifunctional polymeric micelles with amplified fenton reaction for tumor ablation. Biomacromolecules. 19(6): 1990–1998.

Warshawsky, A. and N. Kahana. 1982. Temperature regulated release of alkali metal salts from novel polymeric crown ether complexes. J. Am. Chem. Soc. 104(9): 2663–2664.

Weinmann, H.-J., R.C. Brasch, W.-R. Press and G.E. Wesbey. 1984. Characteristics of gadolinium-DTPA complex: a potential NMR contrast agent. Am. J. Roentgenol. 142(3): 619–624.

Wheate, N.J. and C. Limantoro. 2016. Cucurbit[n]urils as excipients in pharmaceutical dosage forms. Supramol. Chem. 28(9-10): 849–856.

Xiao-fan, J., X. Dan-yu, Y. Xu-zhou, W. Hu and H. Fei-he. 2017. Supramolecular polymer materials based on crown ether and pillararene host-guest recognition motifs. Acta Polym. Sin. 1: 9–18.

Xue, Y., Y. Guan, A. Zheng and H. Xiao. 2013. Amphoteric calix[8]arene-based complex for pH-triggered drug delivery. Colloids Surf. B 101: 55–60.

Yagi, S., S. Nakamura, D. Watanabe and H. Nakazumi. 2009. Colorimetric sensing of metal ions by bis(spiropyran) podands: Towards naked-eye detection of alkaline earth metal ions. Dyes Pigm. 80(1): 98–105.

Yan, L., T. Qing, R. Li, Z. Wang and Z. Qi. 2016. Synthesis and optical properties of aggregation-induced emission (AIE) molecules based on the ESIPT mechanism as pH- and Zn^{2+}-responsive fluorescent sensors. RSC Adv. 6(68): 63874–63879.

Yang, J., H. Rong, P. Shao, Y. Tao, J. Dang, P. Wang, Y. Ge, J. Wu and D. Liu. 2016. Highly selective ratiometric peptide-based chemosensors for zinc ions and applications in living cell imaging: A study for reasonable structure design. J. Mater. Chem. B. 4(36): 6065–6073.

Ye, B.-H., M.-L. Tong and X.-M. Chen. 2005. Metal-organic molecular architectures with 2,2'-bipyridyl-like and carboxylate ligands. Coord. Chem. Rev. 249(5-6): 545–565.

Yi, S., B. Captain and A.E. Kaifer. 2011. The importance of methylation in the binding of (ferrocenylmethyl)tempammonium guests by cucurbit[n]uril (n = 7, 8) hosts. Chem. Commun. 47(19): 5500–5502.

Yu, G., Z. Yang, X. Fu, B.C. Yung, J. Yang, Z. Mao, L. Shao, B. Hua, Y. Liu, F. Zhang, Q. Fan, S. Wang, O. Jacobson, A. Jin, C. Gao, X. Tang, F. Huang and X. Chen. 2018a. Polyrotaxane-based supramolecular theranostics. Nat. Commun. 9(1): 766.

Yu, G., X. Zhao, J. Zhou, Z. Mao, X. Huang, Z. Wang, B. Hua, Y. Liu, F. Zhang, Z. He, O. Jacobson, C. Gao, W. Wang, C. Yu, X. Zhu, F. Huang and X. Chen. 2018b. Supramolecular polymer-based nanomedicine: High therapeutic performance and negligible long-term immunotoxicity. J. Am. Chem. Soc. 140(25): 8005–8019.

Zhang, Y.-M., Y.-H. Liu and Y. Liu. 2020. Cyclodextrin-based multistimuli-responsive supramolecular assemblies and their biological functions. Adv. Mater. 32(3): 1806158.

Zheng, H.-L., Z.-Q. Zhao, C.-G. Zhang, J.-Z. Feng, Z.-L. Ke and M.-J. Su. 2000. Changes in lipid peroxidation, the redox system and ATPase activities in plasma membranes of rice seedling roots caused by lanthanum chloride. Biometals. 13(2): 157–163.

Zhou, J., X. Guo, H.E. Katz and A.E. Bragg. 2015. Molecular switching via multiplicity-exclusive E/Z photoisomerization pathways. J. Am. Chem. Soc. 137(33): 10841–10850.

Zhou, T., N. Song, S.-H. Xu, B. Dong and Y.-W. Yang. 2016. Dual-responsive mechanized mesoporous silica nanoparticles based on sulfonatocalixarene supramolecular switches. ChemPhysChem. 17(12): 1840–1845.

Zhu, H., L. Shangguan, B. Shi, G. Yu and F. Huang. 2018. Recent progress in macrocyclic amphiphiles and macrocyclic host-based supra-amphiphiles. Mater. Chem. Front. 2(12): 2152–2174.

Zimmerman, S.C. and P.S. Corbin. 2000. Heteroaromatic modules for self-assembly using multiple hydrogen bonds. pp. 63–94. *In*: M. Fuiita (ed.). Molecular Self-Assembly Organic Versus Inorganic Approaches. Springer, Berlin, Heidelberg.

Index

About the Authors

Satish Kumar is an Assistant Professor in the Department of Chemistry, St. Stephen's College, University of Delhi. During the last 18 years, Dr. Satish Kumar has worked in interdisciplinary areas covering theoretical chemistry, nanochemistry, and the application of principles of molecular recognition to design molecular receptors. Dr. Satish Kumar has published several research papers related to the development of receptors for neutral molecules, anions, and cations. He has almost 16 years of teaching experience gained at CSU-Los Angeles, University of Memphis, and St. Stephen's College.

Priya Ranjan Sahoo received a B.Sc. degree in Chemistry from Utkal University (2011) and an M.Sc. degree from the University of Delhi (2013). He obtained his Ph.D. from the University of Delhi under the supervision of Dr. Satish Kumar. He has undertaken contract research in Jubilant Generics Limited (India). He studied magnesium dynamics at Tohoku University (Japan) as Tokyo biochemical research foundation fellow. He is currently a postdoctoral fellow at Ecole Normale Supérieure, Sorbonne Université (Paris).

Violet Rajeshwari Macwan is an Assistant Professor of Chemistry at St. Stephen's College, University of Delhi, and has taught undergraduate students for 16 years. She pursued her graduation, postgraduation, MPhil in Chemistry, and Ph.D. in Education from the University of Delhi. She did her doctoral research in the field of science education at the tertiary level. Her area of specialization is inorganic chemistry and her research interests include corrosion science, environmental chemistry, and chemistry education at the tertiary level.

Jaspreet Kaur is B.Sc and M.Sc (H.S) in Chemistry from Punjab University; M.Tech and Ph.D. from IIT Delhi. She has teaching and research experience of more than 11 years. She is currently teaching at St. Stephen's College and previously taught at LPU, IP University, and ARSD College (University of Delhi). She is actively involved in research and has received research grants from DU and DST. She has published many papers in national and international Journals.

Mukesh has done M.Sc., M.Tech., and Ph.D. in Chemistry from IIT Delhi and has teaching and research experience of more than 15 years including that of working in the API Industry. He is presently working as Assistant Professor at Keshav Mahavidyalaya (University of Delhi). He is actively involved in research, has got few research grants, and published few papers in national/international journals. He has attended various national/international conferences and won a couple of prizes for paper presentations.

Rachana Sahney is currently working as an Associate Professor at Amity University Uttar Pradesh, Noida, India. She received her Ph.D. degree in Chemistry from B.H.U., India under the supervision of Prof. R.L. Gupta in 2000. She then worked as a postdoctoral research fellow at Brandies University, USA with Prof. Irving R. Epstein. Her current research interest is focused on the design and synthesis of nanostructured materials for sensing applications.